INTRODUCTION TO GAS LASERS: POPULATION INVERSION MECHANISMS

With Emphasis on Selective Excitation Processes

BY

Colin S. Willett, B.Sc., Ph.D., M.Inst.P.

Harry Diamond Laboratories,
Department of the Army,
Washington, DC 20438

PERGAMON PRESS

OXFORD · NEW YORK · TORONTO · SYDNEY

Pergamon Press Ltd., Headington Hill Hall, Oxford

Pergamon Press Inc., Maxwell House, Fairview Park, Elmsford, New York 10523

Pergamon of Canada Ltd., 207 Queen's Quay West, Toronto 1

Pergamon Press (Aust.) Pty. Ltd., 19a Boundary Street, Rushcutters Bay, N.S.W. 2011, Australia

First edition 1974

Library of Congress Cataloging in Publication Data

Willett, Colin S
   Gas lasers: population inversion mechanisms.

   (International series of monographs in natural philosophy, v. 67)
   Includes bibliographies.
   1. Gas lasers. I. Title.
TA1695.W54 1974   621.36'63   73-21813
ISBN 0-08-017803-0

Printed in Hungary

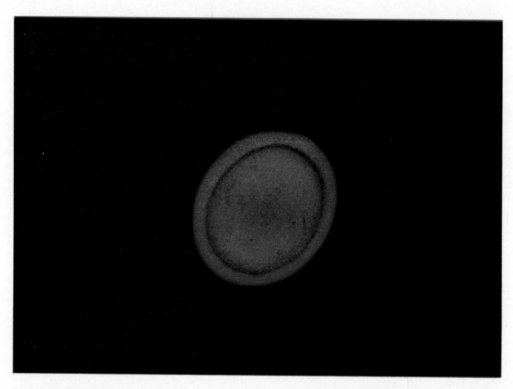

Frontispiece. Two-wavelength pulsed output characteristic of the first (He - Hg) ion laser. Oscillation in the green center at 567.7 nm is surrounded by multimode annular oscillation in the red at 615.0 nm. Original kindly supplied by A. L. Bloom of Coherent Radiation, Palo Alto, California.

Contents

CONTENTS

Preface

IN THIS book the intention is to make it clear what the important processes are in gas discharge lasers, and to bring together for the first time into one source information on basic atomic collision processes that operate in a gas laser and their relation to gas discharge processes and parameters.

Numerous figures with full captions are used to illustrate the text. This means that the reader does not have to go to the original literature for experimental detail that he might require, and further, by examination of the figures, can come to his own conclusions about a system or process that is being discussed. While most of the figures have been chosen to illustrate the text, some have been deliberately chosen to stimulate interest and speculation in areas where the author considers that questions need to be asked and satisfactory answers sought.

The references also are numerous. Though they have been selected because of personal preferences, they nevertheless provide an invaluable lead into the literature for a newcomer to the field, and help the reader who might disagree with the interpretation of the author and who might wish to probe more deeply into the subject. Throughout, operating characteristics and discharge conditions are given of the lasers discussed in illustrating various selective excitation processes. The book can be used, therefore, as a reference to the behavior of gas-discharge lasers by anyone already in the field. In addition, the listing in the appendix of atomic and molecular laser transitions reported up to the end of 1972 constitutes a reference source in its own right.

The book is directed at newcomers starting research in gas lasers, basic atomic collision processes, or gas discharges who wish to obtain a background to the subject, and also at specialists who might have specialized (as is usual) in one area of gas lasers to the exclusion of others. The mathematics content is minimal as it is considered that it is not needed here where the emphasis is on physical concepts. It is believed that any reader with an elementary knowledge of atomic physics and spectroscopy could benefit from reading part or all of the text and be brought up to the current boundary of knowledge in atomic-collision and gas-laser processes.

This book did not develop in the usual way from material prepared for university courses, though much of the content has been presented at the graduate level to university and government atomic-, plasma-physics, and laser research groups. Started over 5 years ago, it was based on my Ph.D. thesis submitted to the University of London in 1967. At the commencement of my graduate research work in 1963, I was concerned with one's inability to predict with any reasonable success possible gas-laser systems, or even the behavior of actual working systems from the extensive literature available on gas discharge or atomic processes. It seemed clear, and slightly disquieting to a newcomer to research, that it was just not known with any certainty what were dominant collision processes in even a straightforward gas discharge, let alone in a laser gas discharge. From the beginning of my interest in gas lasers in 1963 it was not known to my satisfaction what processes the addition of helium as an often-called "buffer gas" affected in, or introduced to a gas discharge, or what factors determined the electron temperature in a discharge and what processes the electron temperature itself affected. The reason for the almost universal $1/D$ gain relationships observed in gas lasers was lost on me, and further, I was not content to accept that helium was not playing a major role in a number of gas-laser systems when it was clear that its presence was essential in producing population inversion. Whereas it seemed clear that resonant-energy excitation transfer from excited species of helium was responsible for selective excitation in the well-known red 0.6328-μm He–Ne laser, it was not clear what were the mechanism or mechanisms that were introduced by the addition of helium to the relatively high gain 0.6150-μm mercury, the visible iodine, and the 0.4416-μm cadmium ion lasers, all of which required helium in their operation. Throughout the development of gas lasers, and even gas discharges, much has been attributed to electron-temperature effects when other processes could just as well be invoked to explain certain observations.

Now, with only a few exceptions, it is known what the principal processes are that are responsible for producing population inversion in all the reported gas laser systems. Rather surprisingly to most, in view of the complexity of processes occurring in gas discharges, the situation invariably is that there is one process, or at the most only a few processes that can be pointed out as being responsible for producing population inversion in a gas discharge laser or for producing certain characteristics of a gas discharge. As most of the gross questions about the effect of electron temperature, presence of helium, the reason for the $1/D$ gain relationship and other controversial issues about gas discharge lasers have now been resolved, it seemed an opportune time to collate this material and publish it in a book form that should be of some lasting value.

Acknowledgements

I WISH to acknowledge my indebtedness to the Harry Diamond Laboratories, Department of the Army, for their cooperation and help during the writing and preparation of this book.

A number of persons have been involved directly or indirectly in various ways in the preparation of the manuscript, be it in supplying prepublication or unpublished material; critically reviewing; suggesting improvements of content, clarity, or style in the manuscript; or simply in encouraging me over the years through discussions. Those to whom I express my gratitude include: Drs. A. L. Bloom, L. Brown, W. B. Bridges, and A. von Engel; Mr. H. F. Gibson; Drs. T. J. Gleason and E. I. Gordon; Mr. R. D. Hatcher; Professor O. S. Heavens; Mr. J. S. Kruger; Drs. E. F. Labuda and C. A. Morrison; Mr. H. Ogata; Dr. S. Rayment; Professor W. W. Robertson; Mr. A. R. Saunders; Professor A. E. Siegman; Dr. W. T. Silfvast; Professor V. Smiley; Mr. J. Tompkins; and Drs. A. L. Ward and D. E. Wortman.

Especial thanks are due to Dr. R. T. Young for his encouragement and guidance as my supervisor from 1967 to 1969 at the Harry Diamond Laboratories, for permission to use unpublished material included in Chapters 3 and 4, material presented as Appendix Figs. A1 to A9, and for reviewing an earlier but vital version of the manuscript.

Lastly, I cannot adequately express my appreciation to Mrs. Thelma Hulsey for the effort that she has put into the typing of the final and various versions of the manuscript, without her full cooperation and willingness to add additional work to her normal duties this book would not have been possible.

Keedysville, Maryland COLIN S. WILLETT

Notes Regarding Sections, Equations, References, and Notation

Sections

Each chapter is divided into sections and subsections referred to as Section 2.2, Section 3.3.1, respectively. Sections are further subdivided to cover individual items where the subject matter covers a wide area of interest.

Equations

These are numbered serially within each chapter and are quoted in parentheses thus, eqn. (2.1), eqn. (4.16), etc.

References

These are numbered serially in each chapter as superscripts and are collected and tabulated at the end of each chapter. In the case of references to the Appendix tables, these are collected and tabulated as references to "Atomic Laser Transitions" and "Molecular Laser Transitions".

Collisional and Spectroscopic Notation

Throughout the text most of the symbols used are defined in the locality of their use. There is overlap in the use of certain symbols from one section to another, but it is believed that this should be obvious to the reader and should not lead to any confusion.

In order to describe inelastic collisions as concisely as possible and to make it clear just what species are involved, the following symbols have been adapted (following the system used by J. B. Hasted, *Physics of Atomic Collisions*, Butterworth, 1964):

Electron	e
Neutral (unexcited) atom	A or O
Electronically excited atom	A' or O'
Metastable atom	A^*
Ionized atom (ground state)	A^+ or 1
Excited ionized atom	$A^{+\prime}$ or $1'$
Molecule or atom (unexcited)	M
Difference in potential energies (or energy discrepancy) at infinite nuclear separation of isolated atoms or molecules	ΔE_∞

Other symbols are defined in the text.

In common with what is fairly well-established laser practice, the upper state of a transition is given before a lower state. Roman numerals I, II, III, etc., are used as in standard spectroscopic notation to refer to neutral, singly ionized, doubly ionized, trebly ionized lines or transitions, respectively. Thus a neutral helium line is referred to as a He-I line. Normal chemical abbreviations are used for the elements with the exception that Ar (instead of A) is used as an abbreviation for argon. This usage is introduced to make the text easier to read by referring to an argon atom as an Ar atom rather than an A atom. The metric system is used throughout with wavelengths given in nanometers (nm) or micrometers (μm), and dimensions given in mm or cm. Ångström units (Å) are used for convenience of reference in figures for internuclear separations instead of subunits of meters.

Mixed spectroscopic notation is used in the text. In referring to noble-gas laser transitions, both Paschen and Racah notation is used whereas Racah notation only is used for other transitions. Unless otherwise stated the energy levels are as given in *Atomic Energy Levels*, Vols. I, II, and III, Natl. Bur. Stds. Circular 467 (1949, 1952, and 1958) by Charlotte E. Moore, U.S. Government Printing Office, Washington, D.C. 20402. For information on coupling schemes and selection rules in atomic spectra the reader is referred to *Atomic Spectra* (Longmans, 1962) by H. G. Kuhn; and "Forbidden Transitions", by R. H. Garstang, in *Atomic and Molecular Processes* (ed. D. R. Bates, Academic Press, New York, 1962), pp. 1–46.

CHAPTER 1

Historical Development and Basic Principles of Gas Lasers

1.1. Introduction

One would hardly have dared to predict with the development of the first gaseous optical laser in 1961 by Javan, Bennett, and Herriott[1] that within 11 years, gas lasers would be developed that operate at the gigawatt-power level. But this indeed has occurred, the milliwatt-power level of the near-infrared He–Ne laser of Javan et al.[1] has been increased to the gigawatt-power level of the 10.6-μm CO_2 laser operated under pulsed conditions and to 60 kW under CW conditions.[2a, 2b]

Presently, laser oscillation or spatially coherent stimulated emission has been observed at wavelengths that range from the vacuum ultraviolet at around 0.1 μm to beyond 1000 μm in the submillimeter region of the spectrum. In all, stimulated emission has been reported to occur on approximately 1000 electronic transitions in atomic species of over forty elements, and on innumerable vibrational/rotational/electronic transitions of molecular species of about forty compounds.

Figure 1.1 shows those elements in which oscillation has been achieved in atomic species. It covers all the laser transitions reported in both neutral and ionized atomic species that are given in Appendix Tables 1 to 46. Molecular compounds in which oscillation has been

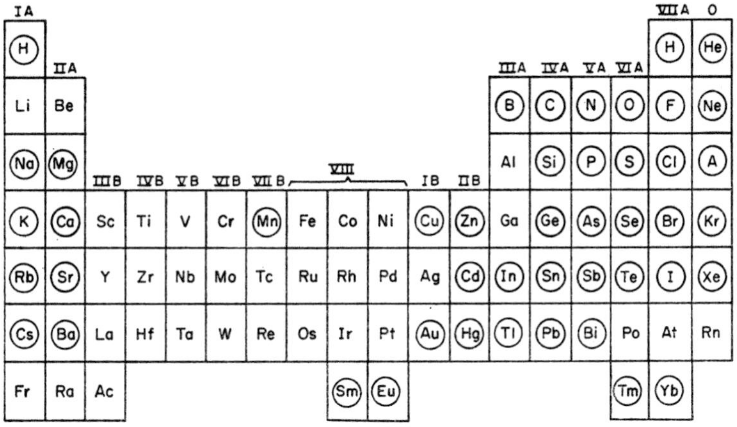

FIG. 1.1. Partial periodic table showing elements (circled) in which laser oscillation in atomic species has been reported.

reported are not conveniently displayed in the form of a figure. Their laser transitions are given in Appendix Tables 47 to 85.

Although much of the work on gas lasers has been technological, the borderline between technology and pure physics has been such that an increased understanding of basic atomic physics has followed each technological advance. Particularly important, one has been made painfully aware of those areas in which there is little basic understanding, but in which areas the physicist, chemist, engineer, spectroscopist can converse and all contribute.

The discussion of gas lasers in this book is restricted to those that are excited by an electrical discharge, with the one exception that gas lasers whose upper laser levels are selectively excited by optical line-absorption are also covered. Although the processes that occur in a number of molecular gas lasers that rely on chemical reactions to produce population inversion are often intimately related to collision processes discussed here, because of the limitation of space of a book of this type, the emphasis is on the interaction of electrical properties of gas laser discharge media and the atomic-collision mechanisms operative in them that produce population inversion.

In this chapter, prior to going into the physics of gas lasers, a brief chronology is given of major advances that have been made in gas-laser development up to 1973. Included also in the chronology are details of important basic selective excitation processes that have been determined to be operative in gas lasers and which the author considers to be milestones in gas laser development just as much as the technological developments.

The chronology is followed in Section 1.3 by Fundamental Processes: covering mean free paths, collision cross-sections, collision frequency, velocity and energy distributions, and basic collision processes and principles that are of importance in establishing population inversion and which are referred to frequently in the chapters that follow.

Brief treatments are then given in Section 1.4 of The Interaction of Radiation with Matter, and in Section 1.5, Oscillation Conditions with an emphasis on the dependence of gain on wavelength in gas lasers.

It is the author's intention to familiarize the reader with fundamentals only, and to provide a framework for the appreciation of atomic collision processes that are involved in producing selective excitation and population inversion. For more detailed information on what is considered here, and for information on basic properties of lasers common to gas lasers, and applications of lasers the reader is referred to other texts such as:

Optical Masers, by G. Birnbaum (Academic Press, New York, 1964);
Optical Physics, by M. Garbuny (Academic Press, New York, 1965);
The Laser, by W. V. Smith and P. P. Sorokin (McGraw-Hill, New York, 1966);
An Introduction to Lasers and Masers by A. E. Siegman (McGraw-Hill, New York, 1971);
Quantum Electronics, by A. Yariv (Wiley, New York, 1967);
Introduction to Optical Electronics, by A. Yariv (Holt, Rinehart and Winston, 1971); and
Gas Lasers, by A. L. Bloom (Wiley, New York, 1968).

1.2. A Chronology of Major Advances and Knowledge Gained of Important Basic Selective Excitation Processes

1954

Gordon, Zeiger, and Townes[3] construct microwave ammonia maser.

1958

Schawlow and Townes[4] extend maser techniques to the infrared and optical region, lay down ground rules for obtaining laser action, and propose a practical optically pumped laser system.

1959

Javan[5] proposes the use of electron-impact collisions and atom–atom excitation transfer in a gas discharge to produce population inversion.

1960

Javan, Bennett, and Herriott[6] obtain oscillation in the near-infrared in the afterglow of a discharge in a He–Ne mixture, and establish that atom–atom excitation transfer from He* 2^3S_1 metastables can be used as a gas-laser selective excitation process.

1961

Javan, Bennett, and Herriott[7] obtain CW oscillation on five transitions in neutral neon in the near infrared using rf-discharge excitation.

Fox and Li[8] develop the mode theory of the Fabry–Perot interferometer as an optical resonator.

1962

Boyd and Kogelnik[9] calculate high- and low-loss regions for laser cavity resonators.

Brangaccio[10] and Rigrod et al.[11] introduce the use of Brewster's angled end-windows thus enabling cavity mirrors to be used external to the gas discharge tube. This allows wavelength selectivity of laser resonators to be made easily without modifications to the whole laser structure as was required with internal mirror resonators.

Rigden and White[12] obtain CW laser oscillation in the visible at 632.8 nm in an rf-excited He–Ne mixture using excitation transfer between He* 2^1S_0 metastables and neon atoms as a strong upper laser level selective excitation mechanism.

Bennett,[13] and with Faust and McFarlane,[14] obtained practical amounts of gain in an rf-excited discharge in pure neon and showed that electron-impact excitation alone could be used as a selective excitation mechanism to produce oscillation in pure gases in the absence of atom–atom resonant-excitation energy-transfer.

1963

Boot and Clunie[15] initiate the utilization of high-current, high-voltage pulsed (radar) techniques to produce laser oscillation (in atomic carbon and neon) and later achieve oscillation in a high-pressure discharge.[16] By depositing large amounts of electrical energy into a gas in a short time-interval they were able to investigate a wide range of discharge and afterglow conditions that might be suitable for producing population inversion.

Mathias and Parker produce the first oscillation in a molecular species (in N_2) in the near-infrared[17] and later (in CO) in the visible[18] using short rise-time, pulsed excitation. In so doing, thus introduce the first self-terminating lasers in which the lifetime of the lower laser levels are substantially longer than those of the upper laser levels.

Bloom, Bell, and Rempel[19] report oscillation at 3.39 μm on a transition in Ne-I in a He–Ne mixture with an extremely high gain of 20 dB/m.

White and Gordon[20, 21] establish that excitation of the upper laser level of the 632.8-nm (He–Ne) laser line is mainly by excitation transfer from $He^* \, 2^1S_0$ metastables.

1964

Bell,[22] early in 1964, obtains oscillation in an ion specie for the first time. Oscillation was observed in the visible at 615.0 nm in the singly ionized spectrum of mercury in a high current pulsed discharge in a He–Hg mixture.

Bridges,[23] Convert, Armand, and Martinot-Lagarde[24, 25] observe and identify pulsed visible oscillations in mercury–noble gas mixtures as transitions in single ionized argon.

Bennett et al.[26] obtain quasi-CW oscillation in the visible in singly ionized argon.

Gordon, Labuda, and Bridges[27] obtain true CW oscillation in singly ionized argon, and also in krypton and xenon.

Patel, Faust, and McFarlane[28] obtain low-power CW oscillation in CO_2 at approximately 10 μm.

Patel;[29, 30] Legay-Sommaire, Henry, and Legay,[31] and Barchewitz et al.[32, 33] analyse oscillation in the CW CO_2 laser.

Legay and Legay-Sommaire[34] and Patel[35] use molecule–molecule vibrational energy transfer from excited N_2 molecules to selectively excite the upper laser level of the 10.6-μm CO_2 laser line.

Fowles and Jensen[36, 37] and Jensen and Fowles[38] observe visible laser oscillation in singly ionized iodine in a pulsed helium–iodine mixture and attribute the selective excitation, together with that responsible for selective excitation in the 615.0-nm He–Hg laser discovered by Bell,[22] to asymmetric thermal-energy charge transfer from ground state He^+ ions.

1965

Patel[39] raises CW-output of 10.6-μm CO_2 lasers from the milliwatt-power level to 10 W.

Moeller and Rigden[40] add helium to CO_2, and CO_2–N_2 in a CO_2 laser and considerably increase the output power.

Patel, Tien and McFee[41] increase CW-output power of 10.6-μm CO_2 lasers to 100 W by using a flowing CO_2–N_2–He gas mixture.

Leonard[42] uses transverse electrical excitation to give multikilowatt pulses in the u.v. from molecular nitrogen.

Fowles and Silfvast[43] report pulsed oscillation in the ionic spectra of zinc and cadmium, and Silfvast, Fowles, and Hopkins[44] obtain pulsed oscillation at 441.6 nm in singly ionized cadmium in the presence of helium.

Dyson[45] shows that thermal-energy charge-transfer from He^+ ions can be the only mechanism responsible for selective excitation of the upper level of the singly ionized 615.0-nm mercury laser line in a He–Hg mixture.

1966

Walter et al.[46] calculate optimum characteristics of transient lasers.

1967

Sharma and Brau[47] show that vibrational energy transfer between N_2^* and CO_2 in the 10.6-μm CO_2-N_2 laser is dominated by a long-range dipole–quadrupole interaction and not by a short-range repulsive force.

Sobolev and Sokovikov[48] relate the optimum electron temperature of approximately 2 eV[49] in 10.6-μm N_2-CO_2-He lasers to the processes of resonant electron-impact excitation of N_2^* and subsequent vibrational energy transfer to CO_2.

Fowles and Hopkins[50] report quasi-CW laser oscillation at 441.6 nm in Cd-II in a He–Cd discharge.

Treanor, Rich, and Rehm[51] develop theory of anharmonic CO–CO pumping and relaxation.

1968

Silfvast reports efficient CW-oscillation at 441.6 nm in Cd-II in a He–Cd discharge.[52]

1969

Sosnowski[53] and Goldsborough[54] use cataphoretic pumping of cadmium vapor from anode to cathode in a simple glow discharge in 441.6-nm He–Cd ion lasers.

Goldsborough[55] reports CW oscillation at 325.0 nm in Cd-II in a He–Cd ion laser.

Nighan and Bennett[56] calculate the efficiency of 10.6-μm CO_2 lasers from a consideration of the electron–energy-distribution functions and vibrational excitation rates.

Jensen, Collins, and Bennett[57] establish that charge transfer is a strong selective excitation process in the He–Zn ion laser.

Tiffany, Targ, and Foster[58] obtain 1 kW/m of CW power at 10.6 μm in crossed-flow, transversely excited CO_2 laser.

1970

Beaulieu,[59] and Dumanchin and Rocca-Serra[60] develop transversely excited atmospheric pressure CO_2 (TEA) lasers using resistor-loaded and preionization techniques, respectively.

Silfvast and Klein[61] develop visible and near-infrared multi-wavelength He–Se ion laser in which asymmetric thermal-energy charge transfer from He^+ ions is the upper laser level selective excitation.

Hodgson,[62] Waynant et al.[63] report laser oscillation in the vacuum-ultraviolet in the Lyman band (0.1400–0.1650 μm).

Basov et al.[64] produce stimulated emission in the ultraviolet (0.1600–0.1700 μm) in liquid xenon by means of a pulsed high-energy electron beam.

Osgood, Eppers, and Nichols[65] attribute population inversion in the 5-μm CO laser to the anharmonic pumping process of Treanor, Rich, and Rehm.[51]

1971

P. W. Smith[66] develops waveguide gas lasers.

Schuebel[67] reports CW oscillation of the 615.0-nm Hg-II line in a hollow cathode discharge.

Silfvast[68] establishes that the Penning reaction is responsible for excitation of the 441.6-nm Cd-II laser line in a He–Cd discharge.

Riseberg and Schearer[69] establish that both Penning reactions and charge transfer are responsible for selective excitation in the He–Zn ion laser.

1972

Hodgson and Dreyfus[70] and Waynant[71] obtain superradiation emission in the Werner band of H_2 at 0.1000–0.1200 μm using electron-beam and travelling-wave excitation, respectively.

Fenstermacher et al.[72] publish detailed results on electron-beam-controlled discharges for pumping large volumes of CO_2 laser media at high pressure.

1.3. Fundamental Processes

In a gas-discharge laser we are concerned with two basic collective collision processes. The two processes are intimately related to the production or maintenance of population inversion.

The first process is ionization. This is necessary for the maintenance of an electrical discharge under CW conditions. When population inversion occurs in the afterglow of a pulsed discharge, ionization has to precede it in order to provide ion-pairs and metastable species for the recombination and metastable-to-atom/molecule energy-transfer processes that produce the population inversion.

The second basic collective collision process is excitation of the upper and lower laser levels by processes that are (ideally) strong selective excitation processes for the upper level alone. Such processes include excitation energy-transfer (atom–atom or molecule–molecule), charge transfer, Penning reactions, and those processes described in Chapter 2. Included also in the second basic collective collision process are processes that (ideally) de-excite only the lower laser level and do not adversely affect the steady-state population of the upper laser level.

These two collective processes of ionization and excitation invariably have conflicting requirements for their optimization. The energy of electrons, which are the principal energy carriers in a gas discharge and whose presence is necessary to produce ionization, is higher than that required to produce excitation of atoms or molecules so that a compromise has to be sought between the conditions required for ionization and the excitation processes that determine the laser output power. Preionization of pulsed gas lasers[72] is one way in which the differing conditions for ionization can be divorced from those required for the optimization of excitation and population inversion. Basically, ionization is produced by fast electrons[72, 60] and then excitation alone is produced by the application of an electric field that accelerates the slow electrons produced in the ion-pair ionization process to electron energies that are ideal for excitation but which are too low to produce further ionization.

The average energy of electrons in a discharge is determined by the average energy that is acquired from the applied electric field between inelastic collisions. The actual distribution of electron energies is determined by the interaction of the electrons among themselves and with atoms or molecules. The average electron energy is a function of the E/N ratio where E is the electric field and N is the atom or molecule density. Since N is directly proportional to the gas pressure p at constant temperature and as the units of N (per unit volume) are more cumbersome to use than p (in torr), E/N is often replaced as a ratio by E/p. The relationship between average electron energy or electron temperature of a particular electron energy distribution and the E/p ratio in a weakly ionized plasma is considered in Section 3.2.2 and is given by eqns. (3.12) to (3.14). The average electron energy in a weakly ionized plasma bounded by a cylindrical dielectric wall can also be related to the product of the gas pressure p and the tube diameter D (see eqn. (3.15), Figs. 3.5 to 3.7 and Appendix Figs. 1 to 9).

Factors of importance in maintaining the supply of electrons and the upper laser level excitation are: the ground state atom or molecule density N, the electron concentration n_e, the actual shape of the electron-energy distribution, and the cross-sections for ionization and excitation. Specifically the production rate R (in cm^{-3} sec^{-1}) for a process with cross-section σ is given by

$$R = N \int_0^\infty \sigma(E)f(E)\,dE \qquad (1.1a)$$

where N is the ground state particle density, $\sigma(E)$ is the cross-section for a particular reaction, which is a function of the electron energy, and $f(E)$ is the energy distribution of the particles.

For a process involving electron impact excitation on a collection of N particles, the production rate of a particular excited state becomes

$$R_e = N \int_0^\infty \sigma_e(E)f_e(E)\,dE \qquad (1.1b)$$

where $\sigma_e(E)$ is the electron-impact excitation cross-section for the particular excited state and $f_e(E)$ is the electron-energy distribution, and $\int f_e(E)\,dE = n_e$.

The exact electron energy dependence of a cross-section is often not known, and a velocity-averaged cross-section $\bar{\sigma}_e$ for a particular reaction is assumed. The production rate of an excited state is given then by the relationship

$$R = Nn_e\bar{\sigma}_e\bar{v}_e \qquad (1.2a)$$

where n_e is the electron concentration, N is the atom or molecule density, and $\bar{v}_e$ is the average velocity of the electrons relative to the atoms or molecules.

For a general process involving energy transfer between one specie and another, the production rate is simply

$$R = N_1N_2\bar{\sigma}\bar{v} \qquad (1.2b)$$

where N_1 and N_2 are the density of particles to which energy is transferred and density of energy carriers, respectively, and $\bar{v}$ is their average relative velocity.

Throughout the following chapters, cross-sections for particular reactions that occur in gas lasers are often referred to. Without some idea of the physical size that atoms or molecules present to each other in the absence of interactions, the quoted cross-section tells us little physically about the reaction. With a knowledge of the hard-sphere cross-section

7

considering the reaction as a collision between elastic spheres, the magnitude of an experimentally determined cross-section considered relative to the hard-sphere cross-section enables one to ascertain whether or not the reaction involves a long-range interaction, and can indicate the nature of the energy-transfer interaction. The magnitude of the cross-section of course determines the importance of the reaction as one of a multitude of collision processes operating in a gaseous plasma. The velocity averaged cross-section $\bar{\sigma}$ can be related to the mean free path λ between collisions. The relationship is discussed in Section 1.3.1.

In the chapters that follow, we often refer to electron energy distributions, average electron temperature, and loosely just electron temperature. While it is the actual shape of the electron energy distribution that is of prime importance in an electrical discharge in the establishment of excited state populations, it is often convenient to refer simply to the average electron temperature. The assumption is that the distribution has a particular mathematical form that is determined by the interactions of the electrons with each other and it is then meaningful to talk about an electron temperature. Particular types of electron-energy distributions are described in Section 1.3.2.

In practice, the shape of the distribution is a result of interactions of the electrons with atoms and molecules in the discharge as well as the result of interactions of the electrons amongst themselves. In weakly ionized plasmas this is particularly true. For example, a distribution can be depleted of electrons of a certain energy where the gas atoms or molecules in the discharge have large inelastic cross-sections for excitation from the ground state, or the distribution can have a small excess of electrons of particular energies where "superelastic collisions" of electrons with excited atoms or molecules result in the electrons acquiring the potential energy of the excited atoms or molecules.

Depletion of electrons of certain energies in a distribution is most pronounced in discharges in molecular gases. This arises through the existence of inelastic resonant collision processes in molecular gases that have large cross-sections for excitation by electrons of low energy of a few eV[73] (Figs. 6.4 and 6.5). It is as a result of these resonant inelastic collision processes, which are dominant processes at low electron energies in certain molecular gases, that the average electron temperature, which is optimum for producing upper state excitation in at least two middle-infrared molecular gas lasers, is closely related to, and approximately numerically equal to the electron energy at the maxima of their upper level excitation functions. (See Section 6.1.1 "Nitrogen–Carbon Dioxide Laser" and Section 6.1.2 "Pure Carbon Dioxide Laser".)

An excess of electrons over that expected in a normal distribution at particular energies due to superelastic collisions are not as common as a depletion of electrons due to inelastic collisions. The reason is that excited state populations in weakly ionized plasmas are typically 10^5 times smaller than ground state densities of about 10^{16} cm^{-3} (at 1 torr of pressure). This means that the cross-sections have to be very large before a detectable change can be made in the electron-energy distribution. Such excesses of electrons in an electron-energy distribution have been observed in hollow cathode discharges where metastable state populations can be high and superelastic collisions are more numerous.[74, 97] An electron-energy distribution in a Ne–H$_2$ laser in which both a depletion and an excess of electrons at different energies occurs is illustrated in Fig. 4.5.

1.3.1. MEAN FREE PATHS AND COLLISION CROSS-SECTIONS

Since the outer electronic shells of atoms and molecules have radii of the order of 10^{-8} cm, the hard-sphere elastic interaction distances we talk about are of this dimension. In kinetic theory it is of more convenience to use the diameter d than the radius since on impact the atomic or molecular centers approach each other to within twice the radii of the atoms, assuming that the atoms or molecules are alike. In considering the hard sphere interaction of an electron and an atom, since the electron radius is about 10^{-5} smaller than that of the atom or molecule, the closest approach of the electron and the atom is closely that of the radius of the atom.

Atomic cross-sections can be related to the mean free path between collisions of one type of atom with another, like or unlike, where the mean free path λ is the average distance travelled between collisions. In a mixture of particles in thermal equilibrium where $\lambda > d$, the value of λ depends somewhat on the relative velocity of the colliding particles. Correcting for the relative motion of the particles, the mean free path of a particle, type 1 (radius r_1) and mean energy $E_1 = mv_1^2/2$ when it is present in a small concentration in a gas consisting of N particles cm^{-3} of type II (radius r_2) and mean energy $E_2 = mv_2^2/2$, where v_1 and v_2 are the random velocities, is given by[75, 76]

$$\lambda = \frac{1}{N\pi(r_1+r_2)^2(1+v_2^2/v_1^2)^{1/2}}. \tag{1.3}$$

For collision of atoms or molecules among themselves, $E_1 = E_2$, $r_1 = r_2$ and eqn. (1.3) becomes

$$\lambda = 1/N\pi d^2\sqrt{2} = 1/4\sqrt{2}\cdot N\sigma \tag{1.4}$$

where d is the diameter of the atom or molecule and σ is its collision cross-section. For electrons moving in gases $r_1 = r_e \ll r_2$ and $v_1 = v_e \gg v_2$, and (1.3) reduces to

$$\lambda_e = 1/N\pi r_2^2 = 1/N\sigma_e \tag{1.5}$$

where σ_e is the cross-section of the atom presented to the electron (i.e. the electron-impact cross-section).

It can be seen from eqn. (1.5) that λ_e is inversely proportional to the cross-section σ_e and the atom or molecule density N and it follows from eqns. (1.4) and (1.5) that

$$\lambda_e = 4\sqrt{2}\cdot\lambda. \tag{1.6}$$

The electron mean free path is therefore about 5.6 times larger than the mean free path of atoms moving among themselves. This can only be used as a guide; λ_e can actually be larger or smaller than λ, since λ_e is found to depend on the energy of the electron (the Ramsauer–Townsend effect).

The total cross-section Q for an elastic collision between like particles in unit volume of gas is related to the collision cross-section of a single particle σ by

$$Q = N\sigma \tag{1.7}$$

where N is again the particle density. Q is usually referred to 1 torr and 273 K, and N is then equal to 3.54×10^{16} molecules cm^{-3}. The unit of σ is often taken as $\pi r_0^2 = 0.88 \times 10^{-16}$ cm^2, where r_0 is the first Bohr radius.

Mean free paths of a number of atoms and molecules, and their collision cross-sections, are given in Table 1.1.

TABLE 1.1. *Mean free paths λ of atoms and molecules in their own gas, and cross-sections σ from eqn. (1.2), at 1 torr and 273 K*[a]

Gas:	He	Ne	Ar	Kr	Xe	N$_2$	O$_2$	Hg
λ (10^{-3} cm)	17.6	12	8.1	6.6	5.6	6.7	7	≈ 3
σ (10^{-16} cm^2)	2.9	4.3	6.3	7.7	9.5	7.6	7.3	≈ 17

[a] The values of λ are taken from von Engel,[77] and were obtained from viscosity measurements. The cross-sections have been calculated from the values of λ using eqn. (1.4).

Table 1.1 shows that the collision cross-sections of the noble gases at 1 torr and 273 K are in the approximate range 3×10^{-16} cm^2 to 10×10^{-16} cm^2, and are about 7×10^{-16} cm^2 for the two molecular gases N$_2$ and O$_2$. These values are somewhat lower than more recently reported values of gas-kinetic cross-sections,[78] but are approximately the same as some of the corresponding cross-sections given in Hasted.[76]

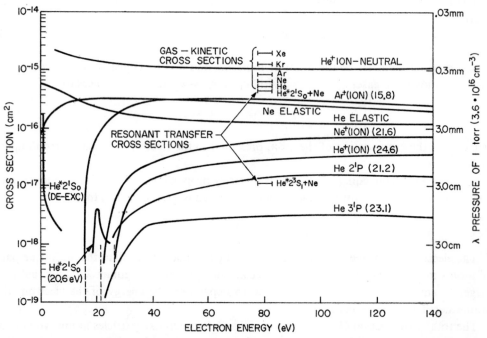

FIG. 1.2. Various inelastic cross-sections, excitation and ionization functions together with gas kinetic cross-sections enabling comparisons between them to be easily made. (After J. Y. Wada and H. Heil,[78] by courtesy of the IEEE.)

10

Figure 1.2 presents cross-sections of a number of inelastic and elastic collision processes, a number of which are mentioned throughout the text. Gas kinetic cross-sections for He, Ne, Ar, Kr, and Xe are included in the figure to facilitate comparisons between them and experimentally determined cross-sections. Values of mean free paths appropriate to the magnitude of the cross-sections are also given in the figure. One might note that the two $He^* + Ne$ resonant transfer cross-sections are smaller than gas kinetic cross-sections for both He and Ne. The He^* $2^3S_1 + Ne$ cross-section is more than an order of magnitude smaller than the He or Ne gas kinetic cross-sections while the He^* $2^1S_0 + Ne$ cross-section is only slightly smaller. The difference in magnitude of the two $He^* + Ne$ cross-sections, and the fact that both are smaller than the gas kinetic cross-section between unexcited He atoms, probably arises because of the existence of potential barriers of differing height in the molecular potentials of each pair of colliding atoms that give rise to the dependence of their cross-sections on gas temperature as shown in Figs. 2.7 and 2.8.

Collision frequency. The number of collisions each particle makes per second is given by

$$\nu_c = v_r/\lambda \tag{1.8}$$

where v_r is the random velocity of the particle.

For collisions of an atom or molecule in its own gas, the number of collisions it makes per second can be obtained from eqns. (1.8) and (1.4) as

$$\nu_c = 4\sqrt{2} \cdot N\sigma v_r. \tag{1.9}$$

At 1 torr, ν_c for an O_2 molecule is approximately 6×10^6 sec^{-1}.

For collisions of an electron in a gas, from eqns. (1.8) and (1.5),

$$\nu_e = N\sigma_e v_e \tag{1.10}$$

where N is the atom or molecule concentration and v_e is the electron random velocity. The total number of collision $\Sigma \nu_e$ made per second by all electrons is then

$$\Sigma \nu_e = n_e N\sigma_e v_e. \tag{1.11}$$

The collision frequency of electrons is not as readily calculable as the collision frequency of atoms or molecules since σ_e depends on the electron velocity.

Ionization efficiency. The ionization efficiency η_e of electrons is defined as the number of ion pairs produced by one electron per cm of path reduced to 1 torr and 273 K. It is related to the probability f_e of ionization of an atom or molecule by

$$\eta_e = f_e/\lambda_e. \tag{1.12}$$

Since $\lambda_e = 1/N\sigma_e$, and f_e is the ratio of the effective ionization cross-section q_e and the cross-section of the atom σ_e presented to the electron (i.e. $f_e = q_e/\sigma_e$), from eqn. (1.12)

$$\eta_e = Nq_e. \tag{1.13}$$

The ionization efficiency, therefore, is numerically equal to the total cross-section Q for ionization of the atoms or molecules by electrons since $Q = Nq_e$. Curves of ionization efficiencies, which are obtained directly from experiment, are given in Fig. 3.3.

1.3.2. VELOCITY AND ELECTRON ENERGY DISTRIBUTIONS

Particles in thermal equilibrium moving randomly and interacting strongly among themselves, be they atoms, molecules, or electrons, have a velocity distribution which is characteristic only of the average energy and masses of the particles. Since the energy is constant under equilibrium conditions and is independent of the mass of the particles and varies only with temperature, an energy distribution is produced that is dependent only on the absolute temperature. Boltzmann has shown that under equilibrium conditions in the

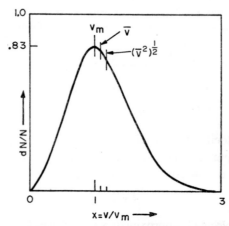

FIG. 1.3. Maxwellian velocity distribution. The ordinate dN/N is the fraction of molecules whose velocity v relative to the most probable velocity lies in an interval dv about the value of v/v_m; v_m is the most probable velocity, $\bar{v}$ the average velocity, and $(\bar{v}^2)^{1/2}$ the r.m.s. velocity. (Adopted from a figure given by L. Loeb.[79])

presence of numerous hard-sphere elastic collisions of the particles, only one characteristic distribution can exist. This distribution is the same about the average or most probable value, and only the average value is displaced with temperature. Maxwell considered a statistical ensemble of particles such as molecules, and arrived at the same result. The result that only one characteristic distribution exists is known as the Maxwell–Boltzmann distribution law, and the distribution itself is called a Maxwellian distribution (Fig. 1.3).[79]

The Maxwellian velocity distribution $Nf(v)\,dv$ of molecules in a gas at a temperature T is given by

$$dN/N = (4/\pi^{1/2})x^2 e^{-x^2}\,dx \qquad (1.14)$$

with $x = v/v_m$; where dN/N is the fraction of molecules with velocities in the range $v+dv$. The most probable velocity v_m is the maximum of the distribution where $mv_m^2/2 = kT$, hence

$$v_m = (2kT/m)^{1/2}. \qquad (1.15)$$

The r.m.s. velocity is given by

$$(\overline{V^2})^{1/2} = (3kT/m)^{1/2} = 1.22 v_m. \qquad (1.16)$$

The average velocity $\bar{V}$ is given by

$$\bar{V} = V_m(2/\pi^{1/2}) = 1.13 v_m. \qquad (1.17)$$

12

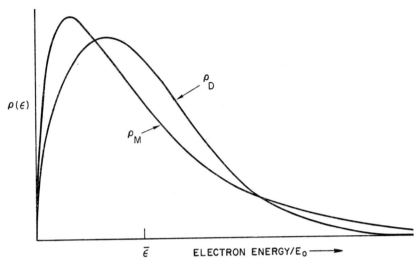

Fig. 1.4. Comparison of Maxwellian ϱ_M and Druyvesteyn ϱ_D electron energy distributions. Normalized to the same mean energy $\bar{\varepsilon}$ and the same number of electrons.

A Maxwellian distribution with the positions of the three velocity values indicated is presented in Fig. 1.4.

The number of molecules whose *energy* lies in the interval $E+dE$ is equal to the energy distribution $Nf(E) \cdot dE$; since $Nf(E) \cdot dE = Nf(v)m \cdot v \cdot dv$, the velocity distribution must be modified to convert it to an energy distribution, giving

$$dN/N = (4/\pi^{1/2})x^{1/2}e^{-x}\,dx \tag{1.18}$$

where $x = E/kT$.

If we equate some mean energy $\bar{E}$ to $m/2$ multiplied by the (r.m.s. velocity)^2 from eqn. (1.16) we obtain

$$\bar{E} = \frac{mv^2}{2} = \frac{3kT}{2}. \tag{1.19a}$$

Considering a Maxwellian distribution of electrons interacting with themselves at an equivalents electron temperature, eqn. (1.19a) becomes

$$\bar{E} = \frac{3kT_e}{2} \tag{1.19b}$$

where k is the Boltzmann constant. The conversion between electron volts and T_e in degrees K, for a Maxwellian distribution is

$$1 \text{ eV (mean energy)} = 7733 \text{ K.} \tag{1.20}$$

The Maxwellian distribution applies when the particles comprising the ensemble are numerous and interact strongly with each other and readily transfer kinetic energy one to another in collisions. Under such conditions Maxwellian distributions of electrons would only be expected to occur in dense plasmase where electron collision frequencies are high.

In a weakly ionized plasma ($n_e < 10^{10}$ cm^{-3}) where electron collision frequencies are low, because of the disparities between the mass (m_e) of an electron and the mass (m) of an atom, energy exchange and rearrangement does not readily occur, and the electron distribution takes on another form that differs from a Maxwellian distribution. The form it takes is the Druyvesteyn distribution:

$$dn_e/n_e = 1.63x^{1/2}e^{(-x)^2}\, dx \qquad (1.21)$$

with $x = E/E_0$, where n_e is the electron density and

$$E_0 = (4m_e/3m)^{1/2}\, e\lambda_e X/p,$$

where λ_e is the electron mean free path at 1 torr, and X/p is the electric field per unit pressure. The Druyvesteyn distribution (1.21) differs from the Maxwellian distribution (1.18) in that it depends on $\exp(-E/E_0)^2$, while the Maxwellian distribution depends on $\exp-(E/E_0)$. The Druyvesteyn distribution terminates more sharply at high energy values than does the Maxwellian distribution. For the same mean energy, therefore, the Maxwellian distribution contains a larger fraction of electrons at higher energies than the Druyvesteyn (Fig. 1.4).[79, 80] The actual shape of electron energy distributions can be experimentally determined by means of measurements of the current drawn by metal probes at different potentials inserted into a plasma.[81] Whereas simple direct measurements of the current drawn by a probe versus probe potential do enable some idea to be gained of the gross features of an electron

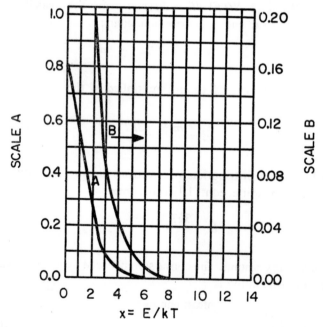

FIG. 1.5. Integrals of the Maxwell–Boltzmann energy distribution law, i.e. $\int_x^\infty x^{1/2}e^{-x}\, dx$ from x to ∞ plotted against values of $x = E/kT$. The ordinates must be multiplied by 1.13 to give the correct M–B integrals.[85] The curves marked A and B correspond to the similarly lettered scale. (From L. Loeb[79] by courtesy of Wiley, Inc.)

energy distribution, more refined techniques such as those used by Boyd and Twiddy[82, 83] and Twiddy[84] are needed to detect the important fine detail of a distribution.

Given a Maxwellian energy distribution in an ensemble of particles such as atoms or electrons at a particular mean energy or electron temperature, one often needs to know the number of particles there are that have an energy exceeding a particular value. This information is required for calculating the amount of excitation and ionization that is being produced given a knowledge of appropriate excitation and ionization functions, and is required for calculating excitation rates of reactions when activation energies or potential barriers for a reaction are involved.

The fraction of particles whose value exceeds a definite value E given by E/kT relative to kT, where T is any mean temperature of a Maxwellian distribution, can be obtained from integrals of the Maxwell–Boltzmann energy distribution curve $(4/\pi)^{1/2} \int_x^\infty x^{1/2} e^{-x} \, dx$. The integrals of $\int_x^\infty x^{1/2} e^{-x} \, dx$ from x to ∞ for various values of $x = E/kT$ have been evaluated by Loeb.[79] They are given in Figs. 1.5 and 1.6. Figure 1.6 is an expanded logarithmic scale version of Fig. 1.5 that enables more accurate evaluations to be made for larger values of E/kT between 2 and 14. The scale values in the figures must be multiplied by $(4/\pi)^{1/2} = 1.13$ to give the full values of the integrals $(4/\pi)^{1/2} \int_x^\infty x^{1/2} e^{-x} \, dx$ from x to ∞.[85]

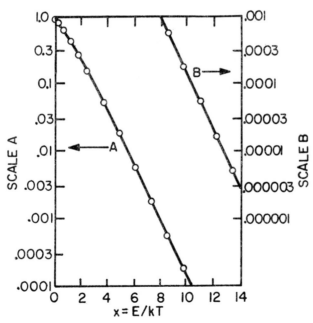

FIG. 1.6. Integrals of the Maxwell–Boltzmann energy distribution law, i.e. $\int_x^\infty x^{1/2} e^{-x} dx$ from x to ∞ plotted logarithmically against values of $x = E/kT$. The ordinates must be multiplied by 1.13 to give the correct M–B integrals.[85] These enable one to evaluate the number of particles at any average energy or temperature whose value exceeds a definite value E given by E/kT relative to kT. Such data are of use in estimating ionization and excitation from excitation functions. (After L. Loeb,[79] by courtesy of Wiley, Inc.)

1.3.3. TYPES OF COLLISION

Electron–atom, electron–molecule; and atom–atom collisions for our purposes can be separated into

1. Collisions of the first kind; and
2. Collisions of the second kind.

Collisions of the first kind are collisions in which kinetic energy of one specie is converted into potential energy of another specie. In collisions of the second kind, potential energy is converted into some other form of energy (other than radiation) such as kinetic energy, or is transferred as potential energy in the form of electronic, vibrational, or rotational energy to another unlike or like specie.

Collisions of the first kind

A collision of the first kind between an electron and an atom, by definition, leads to excitation or ionization of the atom. It can only occur when the energy of the electrons exceeds the excitation or ionization potential of the atom. In addition to linear momentum having to be conserved in the collision, angular momentum about the common center of mass must be conserved. The change of angular momentum Δp during the collision must balance the difference in angular momentum between the atom in its final and initial states, i.e.

$$\Delta p = (h/2\pi)\,\Delta F \tag{1.22}$$

where ΔF is the change in the inner quantum number. Because of this requirement for conservation of angular momentum, the probability of excitation of an atom occurring at the threshold of excitation is very small. However, as the kinetic energy of the electron increases above the threshold energy it is easier to satisfy the requirement for conservation of angular momentum as the electron can move away from the atom, and the probability of excitation increases.

The probability of excitation or cross-section for excitation of a particular state is a strong function of the electron energy. Optically allowed transitions have a broad maximum in their excitation functions at a few to ten times the threshold energy for excitation, whereas forbidden transitions have excitation functions that rise sharply to a maximum close to threshold and then fall rapidly. This general behavior is illustrated in Fig. 1.2 by the excitation function of the 2^1P state of neutral helium at 21.2 eV above the ground state for an optically allowed transition, and by the sharp excitation function of the optically forbidden He* 2^1S_0 to the ground state transition ($\Delta L = 0$, and $J = 0 \to J = 0$) close to 20 eV.

The sharpness of excitation functions of forbidden transitions is attributed to the narrow range of electron energies between which electron exchange can occur. For a triplet transition in helium the total spin S changes from 0 to 1. This necessitates a reversal of spin of one of the electrons and is achieved by electron exchange.

Close to various thresholds of excitation of both atoms and molecules, numerous resonances are increasingly being detected in studies with low energy electrons.[86–89] Two such sets of resonances occur in N_2 between 1.8 and 3.0 eV (Fig. 6.5) and in CO between 1 and 3 eV (Fig. 6.32). These resonances are associated with the formation of short-lived negative-ion states that decay rapidly (in times of the order of 10^{-13} sec) to excited states of the atom or molecule. The reactions can have large cross-sections.[73] A negative-ion state is produced by the incident electron attaching itself to the atom or molecule during the collision.

Ionization functions of atoms and molecules are similar to excitation functions of optically allowed transitions. Ionization occurs when the electron energy reaches the ionization energy and the probability of ionization increases rapidly to a broad maximum at about 100 eV as the electron energy increases, and afterwards decreases. The probability of ionization with electron energy is usually expressed as curves of ionization efficiency η against electron energy. The ionization efficiency is the number of ion pairs produced by one electron in moving 1 cm, and, as we have seen from eqn. (1.13), is numerically equal to the ionization cross-section Q. A set of ionization efficiency curves is given in Fig. 3.3.

Collisions of the second kind

Collisions of the second kind occur in two-body collisions between excited atoms and between excited atoms and electrons.

Atom–atom excitation-transfer (Section 2.2)

$$A' + B \rightleftarrows A + B' \pm \Delta E_\infty; \tag{1.23}$$

and Penning ionization (Section 2.4)

$$A^* + B \underbrace{\hspace{2cm}}_{} \begin{array}{l} \nearrow \; A + B^{+\prime} + e^- \;(+\text{kinetic energy}), \tag{1.24a} \\ \\ \searrow \; A + B^+ + e^- \;(+\text{kinetic energy}), \tag{1.24b} \end{array}$$

are two examples of collisions of the second kind between atoms.

In the atom–atom excitation-transfer reaction (1.23) it is required that ΔE_∞, the difference in potential energy between the excited states A' and B', must be small for the reaction to have a cross-section that approaches gas kinetic. Another requirement for atom–atom excitation transfer to occur is that the Wigner spin rule be obeyed.[90, 91] This rule is known as the conservation of total spin.

Provided that electron spin reversal is negligible in the collision, that is to say the electrons are strongly coupled, the Wigner spin rule states that the total electron spin must be conserved in the collision for the reaction to have any significant probability. The rule is illustrated diagrammatically in Fig. 1.7 for the collision between an atom in a triplet state

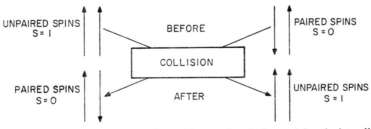

Fig. 1.7. Diagrammatic illustration of the Wigner spin rule for a triplet–singlet collision.

and one in a singlet state. The only direction of reasonable probability of the reaction is in the direction shown.

Though there has been considerable discussion as to its validity since it did not appear to be obeyed in one important case involving excitation transfer in helium (where electron spins depending on the states involved are strongly coupled and spin reversal would not be expected and the rule would be anticipated to be a good rule), the rule has been shown experimentally to be valid and to be a good rule. In the particular case in point of excitation transfer in helium where it did appear that spin was not conserved,[92] Abrams and Wolga[93] have now shown in a direct demonstration that the spin rule is indeed obeyed.

As yet, we do not understand in detail how excitation transfer proceeds, though clearly the answer lies in determining the interaction potentials involved in the collision process.

In the Penning reaction (1.24), given as the second example of a collision of the second kind, there is no requirement for the reaction to be resonant as an electron is available as a third body in the exit channel to remove any excess energy. Experimentally it has been found that Penning reactions have cross-sections that are as large as 10^{-15} cm^2, which are closely gas-kinetic (Table 2.2). In well documented cases,[94, 95] it appears that the Wigner spin rule is also obeyed.

The so-called "superelastic" reactions

$$e^- \text{ (slow)} + A' \quad \begin{cases} \longrightarrow e^- \text{ (fast)} + A, & (1.25a) \\ \longrightarrow e^- \text{ (fast)} + A'', & (1.25b) \end{cases}$$

represent collisions of the second kind between an electron and an excited atom. The potential energy of the excited atom A' is converted into kinetic energy of the electron e^- (fast) and in the process A' becomes de-excited to A or A''. Since spin can always be conserved in the reaction, the reaction should have a high probability.

Though undoubtedly the reaction does occur, observations of its occurrence are extremely limited. It has been observed to occur in mercury vapor,[96] and in the plasma of a hollow cathode discharge,[97] where high metastable atom and slow electron concentrations increase the likelihood of the reaction occurring. In one important superelastic reaction

$$e^- \text{ (slow)} + \text{He}^* \, 2^1S_0 \rightarrow e^- \text{ (slow} + 0.79 \text{ eV)} + \text{He}^* \, 2^3S_1, \qquad (1.26)$$

which is believed to be responsible for gain saturation in the the 632.8-nm and 3.39-μm He–Ne lasers (referred to extensively in Section 4.1.1.6 "Operating mechanisms and characteristics" of the He–Ne 632.8 nm and 3.39-μm laser transitions), spin exchange is involved in "converting" the singlet $\text{He}^* \, 2^1S_0$ metastable to a triplet $\text{He}^* \, 2^3S_1$ metastable. This makes it a highly resonant reaction for slow, thermal-energy electrons having small angular momenta. The cross-section for (1.26) for 300 K electrons has been found[98, 99] to be the extremely large one of 3.0×10^{-14} cm^2. The magnitude of this cross-section and resonant nature of the reaction indicates that it probably proceeds via the formation of a short-lived negative ion state of helium in the reaction

$$e^- + \text{He}^* \, 2^1S_0 \rightarrow (\text{He})^- \rightarrow e^- + \text{He}^* \, 2^3S_1. \qquad (1.27)$$

The Franck–Condon principle

In both first and second kind collisions that involve transfer of potential energy to a molecule, the Franck–Condon principle applies. This principle states that during an electronic transition in a molecule induced by the influence of an external perturbation as in electron impact, or in absorption or emission of radiation, the internuclear separation of the molecule and velocity of relative nuclear motion alter to a negligible extent and the transition takes place so quickly that the nuclei do not have time to move any appreciable distance. This follows because of the large ratio of nuclear mass of a molecule to the mass of an electron or mass of an atom (if it is a light atom). The transition, therefore, will be represented as a vertical line on a molecular potential energy diagram, as illustrated in Fig. 1.8. The initial state is taken to be the ground vibrational state for which the inter-

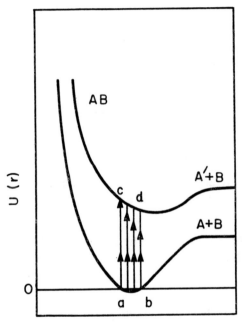

FIG. 1.8. Illustrating vertical electronic transitions in a molecule from an initial state to a final state in accordance with the Franck–Condon principle.

nuclear separation lies between the limits *a* and *b*. As the molecule spends most of its time at the classical turning points *a* and *b* of its nuclear motion, electronic transitions are most likely to occur from these points. According to the Franck–Condon principle the transitions are represented as two vertical lines *ac* and *bd* that intersect the upper and final electronic state at *c* and *d*. The positions of *c* and *d* determine the extent to which the upper state is vibrationally excited.

While the Franck–Condon principle gives a qualitative picture of the transitions, and the extent of the vibrational excitation that occurs in the upper state, it does not tell us anything quantitative about the probability of a transition between *ab* to any of the vibrational states of the upper electronic level. The probability, however, can be calculated quantum mechanically. It is given by the so-called "overlap integral" which is concerned with overlap of the wave functions of the ground state and the vibrational states of the upper electronic state. Calculated values of the overlap integrals are known as Franck–Condon factors. These give directly the transition probability for a vertical transition, or the converse process of emission of radiation.

Tables of such Franck–Condon factors for transitions in H_2 and N_2 appear in the text as Tables 6.6 and 6.7, and 6.13 to 6.15.

1.4. The Interaction of Radiation with Matter

There are three types of transition processes by which radiation interacts with matter in the establishment of thermal equilibrium amongst energy states. For our purposes it is

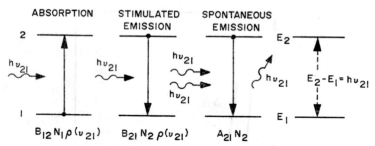

FIG. 1.9. Transition processes operative in a two-level atomic system.

sufficient to discuss the interaction of the radiation with two levels of a quantized system; an upper level 2 with a population N_2, and a lower level 1 with a population N_1. The three transition processes in Fig. 1.9 are:

(a) Absorption of a photon of energy $h\nu$ by an atom in the lower level 1 raising it to level 2. The rate at which this occurs is given by $B_{12}N_1\varrho(\nu)$, where $\varrho(\nu)$ is the energy density of the radiation field at the frequency ν, and B_{12} is the Einstein B coefficient for absorption.

(b) Stimulated emission from excited atoms, in the level 2, which then decay to level 1, is induced by the radiation $h\nu$. The rate at which this occurs is given by $B_{21}N_2\varrho(\nu)$ where B_{21} is the transition probability for induced emissions. The stimulated emission is coherent with radiation that induces it.

(c) Spontaneous emission from excited atoms in level 2, which decay to level 1 in a statistically random way that is independent of the radiation density. The rate at which spontaneous emission occurs is $A_{21}N_2$, where A_{21} is the Einstein A coefficient for spontaneous emission. The Einstein coefficient will be defined later.

At thermal equilibrium the population N_2 in level 2 at a potential energy E_2 is given by the Boltzmann distribution

$$N_2 = Ng_2 e^{-E_2/kT} / \Sigma_j g_j e^{-E_j/kT} \tag{1.28}$$

where g_2 is the statistical weight of energy level 2 (equal to $2J+1$).

20

Similarly,

$$N_1 = Ng_1e^{-E_1/kT}/\Sigma_j g_j e^{-E_j/kT}. \tag{1.29}$$

Thus we obtain the relationship between the population in levels 2 and 1,

$$N_2g_1/N_1g_2 = \exp\left[-(E_2-E_1)kT\right] = \exp\left(-h\nu/kT\right). \tag{1.30}$$

When $E_2 > E_1$, T is either positive or negative depending on the ratio of N_2g_1/N_1g_2. Taking the statistical weights of the levels to be equal, for all positive values of T we have $N_2 < N_1$. Under conditions of thermodynamic equilibrium, this is the normal situation whatever the statistics we use to describe the population.

When levels 1 and 2 are not in thermodynamic equilibrium, a situation can exist in which $N_2 > N_1$. If we still use eqn. (1.30) to describe the situation between levels 2 and 1, we ascribe a negative value to the temperature T, and thus introduce the concept of negative temperature. It would appear that the situation is more descriptively described as one of population inversion.

An effective temperature, however, may be defined in terms of the population N_2 and N_1 when levels 1 and 2 are not in thermodynamic equilibrium with each other. In a gas discharge a number of temperatures can exist: T_e (electron temperature); T_g (gas temperature); T_{vib} (vibrational temperature); and T_{rot} (rotational temperature).

Invariably $T_e \neq T_g \neq T_{vib} \neq T_{rot}$. Each temperature is determined by the component or components with which each is coupled and with which each is in equilibrium. For example, a situation can exist where the excitation of one type of species is mainly by electron impact, and the excitation will then be in equilibrium with the kinetic energy of the electrons rather than with atoms. If a second specie is closely coupled to a state or states of the first specie (for example by atom–atom excitation transfer) but is only weakly coupled to the energy of the electrons, the second specie will be in equilibrium with the atom temperature. In this way, in the same environment, energy levels of different species with the same potential energy can be populated according to different temperatures.

Returning to our ideal case of a two-level system in thermodynamic equilibrium with the radiation field, equating the rate of upward (absorption) transitions to the rate of downward (spontaneous and stimulated emissions) transitions, we have

$$B_{12}N_1\varrho(\nu_{21}) = A_{21}N_2 + B_{21}N_2\varrho(\nu_{21}). \tag{1.31}$$

Combining this with the Boltzmann relation between N_1 and N_2 from eqn. (1.30) we obtain the expression for the radiation density

$$\varrho(\nu_{2k}) = (A_{21}/B_{21})\left[(B_{12}g_1/B_{21}g_2)e^{h\nu/kT} - 1\right]^{-1}. \tag{1.32}$$

This must be identical with the Planck radiation formula

$$\varrho(\nu_{21}) = (8\pi h\nu^3/c^3)\left[e^{h\nu/kT} - 1\right]^{-1}. \tag{1.33}$$

In which case

$$B_{12}g_1 = B_{21}g_2, \tag{1.34}$$

and when the statistical weights g_1 and g_2 are equal

$$B_{12} = B_{21} \tag{1.35}$$

and

$$A_{21}/B_{21} = 8\pi h v_{21}^3/c^3, \tag{1.36}$$

from which

$$A_{21} = (8\pi v_{21}^2/c^3)hvB_{21}. \tag{1.37}$$

Relationships (1.35) and (1.36) between B_{12} and B_{21} and A_{21} and B_{21} and the radiation frequency are basic to the interaction of radiation and matter. Relationship (1.35) asserts that the probability of absorption of radiation by an atom in a lower level is as likely as the probability for stimulated emission by an atom in a higher level when both levels are radiatively coupled by hv.

Relationship (1.36) asserts that the probability for spontaneous emission is directly proportional to the cube of the transition frequency.

In eqn. (1.37) the factor in parenthesis is the number of radiation modes p per unit frequency range per unit volume where $p(v)\,dv$ is the number of modes between v and $v+dv$ and is given by

$$p(v_{21})\,dv = 8\pi v_{21}^2 V\,dv/c^3 \tag{1.38}$$

where V is the volume occupied by the modes.

It follows from eqn. (1.37) that the spontaneous transition probability equals the product of the number of radiation modes and the stimulated emission rate (or absorption rate) for a single quantum hv per mode. The dependence of the spontaneous emission probability on v^3 and its relation to the stimulated transition probability arises because an excited atom can radiate spontaneously into a number of radiation modes while stimulated emission can only be induced into only the applied mode at the resonant frequency v.

For electric dipole transitions the probability that an atom in an upper level 2 will radiate and decay to a lower level 1 is given by the relationship determined from quantum mechanics,

$$P_{21} = (64\pi^4 v_{21}/3hc^3)\mu_{21}^2 + (8\pi^3/3h^2)\,\varrho(v_{21})\mu_{21}^2 \tag{1.39}$$

where μ_{21} is the dipole matrix element for the transition, defined by

$$\mu_{21}^2 = e^2 \left| \int \psi^* r \psi_1\,d\tau \right|, \tag{1.40}$$

where ψ_2, and ψ_1 are wave functions for the states 2 and 1 respectively and r is a radius vector. μ_{21}^2 is related to the line strength of the transition.[100-1]

The first term in eqn. (1.39) is the spontaneous emission term, and is equal to the transition probability for spontaneous emission. The second term involves the radiation density $\varrho(v_{21})$ and is equal to the transition probability for an absorptive transition. The two terms are directly related to the Einstein coefficients for spontaneous emission, absorption and stimulated emission, introduced at the beginning of Section 1.4.

Following Heavens,[100] if there are N_2 atoms in the upper state, the spontaneous emission rate is $A_{21}N_2$, and the probability of emission per atom is A_{21}, this is directly related to the first term in (1.39), hence

$$A_{21} = (64\pi^4 v_{21}^3/3hc^3)\mu_{21}^2. \tag{1.41}$$

If there are N_1 atoms in the lower state 1, then the number of absorptions per unit time is $B_{12}N_1\varrho(\nu_{21})$. The transition probability per atom is thus $B_{12}\varrho(\nu_{21})$, which is equal to the absorptive term in the transition probability expression (1.39). Hence

$$B_{12} = B_{21} = (8\pi^3/3h^2)\mu_{21}^2. \tag{1.42}$$

Reintroducing the statistical weights g_2 and g_1, eqns. (1.41) and (1.42) become

$$A_{21} = (64\pi^4\nu_{21}^3/3hg_2c^3)\mu_{21}^2; \tag{1.43}$$
$$B_{21} = (8\pi^3/3h^2g_2)\mu_{21}^2; \tag{1.44}$$

and from eqn. (1.35)

$$B_{12} = (8\pi^3/3h^2g_1)\mu_{21}^2. \tag{1.45}$$

From eqns. (1.43) and (1.44) we can obtain the relationship of the ratio of the Einstein A and B coefficients for spontaneous and stimulated emission

$$A_{21}/B_{21} = 8\pi h\nu_{21}^3/c^3 \tag{1.46}$$

in agreement with eqn. (1.36), which was determined by appealing to the Planck radiation law.

A consideration is given in Chapter 2 to the theoretical calculation of pure radiative transition probabilities and their inverse, radiative lifetimes, and their agreement with experimentally determined values.

1.5. Oscillation Conditions

In an electrically excited gas laser the discharge medium supplies the stimulated emission that builds up coherent oscillations from the usual dominant spontaneous emission on a particular transition or transitions. If the gain on one pass between the resonator mirrors is sufficient to compensate for losses in the system, the intensity in the resonator mode increases until stimulated emission builds up and competes with spontaneous emission for atoms whose center frequencies are within a natural linewidth of the mode.[102]

The requirements for oscillation were first put forward in a paper by Schawlow and Townes,[4] in which they equated the power produced by stimulated emission in the cavity mode to the rate at which energy decayed. The same result is obtained by equating the gain and loss per unit length. The gain by stimulated emission and its dependence on line shape and loss to absorption is calculable using classical methods. In most gas discharges operating at low gas pressures the primary source of line broadening is the thermal motion of the atoms.

Following the treatment of Mitchell and Zemansky,[103] when the natural linewidth and pressure broadening ≪ the Doppler width, the expression for the fractional gain/unit distance α at the center of a Doppler-broadened line is given by

$$\alpha = (\ln 2/\pi)^{1/2}(g_2A_{21}/4\pi)(N_2/g_2-N_1/g_1)(\lambda_0^2/\Delta\nu_D) \tag{1.47}$$

where λ_0 is the center frequency of the transition and $\Delta\nu_D$ is the linewidth, and where α determines the intensity I of a plane wave at λ_0 according to $I = I_0e^{\alpha l}$, where l is the length

23

over which gain occurs. It is evident from eqn. (1.47) that to obtain gain we must have the inequality

$$N_2/g_2 - N_1/g_1 > 0. \tag{1.48}$$

This inequality imposes limitations on the lifetime requirements of levels that are suitable operating levels for CW laser operation.

The most favorable case occurs when only the upper level is excited and the lower level is populated only radiatively by the laser transition. The density of atoms N_1 in the lower state satisfies the equality[106]

$$\dot{N}_1 = -R_1 N_1 + A_{21} N_2 \tag{1.49}$$

where R_1 represents the *total* decay rate of the lower level. A steady-state solution to (1.49) exists such that $N_2/g_2 > N_1/g_1$ if the total decay rate R_1 obeys the inequality

$$R_1 > A_{21}(g_2/g_1). \tag{1.50}$$

For obtaining laser oscillation this condition is not sufficient; enough inversion has to exist so that the optical gain in a single pass exceeds the total losses per single pass.

For small losses, where the gain does not have to be high and $e^{\alpha l}$ in $I = I_0 e^{\alpha l}$ approximates to $1 + \alpha l$ so that the gain $= \alpha l$, the oscillation condition becomes

$$((g_1/g_2)N_2 - N_1) > \frac{\text{Loss}}{l} \cdot \frac{\Delta \nu_D}{\lambda_0^2} \cdot \frac{4\pi}{g_2 A_{21}(\ln 2/\pi)^{1/2}} \cdot \tag{1.51}$$

The steady-state population densities N_2 and N_1 are given by $\dot{N}_2 = n_2 - N_2/\tau_2 = 0$ and $\dot{N}_1 = n_1 - N_1/\tau_1 = 0$, where τ_2 and τ_1 are the lifetimes of the levels and n_2 and n_1 are the rates of excitation per unit volume of the two levels respectively that we are considering.

Hence $N_2 = n_2\tau_2$ and $N_1 = n_1\tau_1$ and we obtain from eqn. (1.51) the oscillation condition

$$((g_1/g_2)n_2\tau_2 - n_1\tau_1) > \frac{\text{Loss}}{l} \cdot \frac{\Delta \nu_D}{\lambda_0^2} \cdot \frac{4\pi}{g_2 A_{21}(\ln 2/\pi)^{1/2}} \cdot \tag{1.52}$$

It can be seen from (1.52) that even if the lifetimes τ_2 and τ_1 and the statistical weights are favorable oscillation cannot be achieved unless the excitation rate n_2 to the upper level is large enough. The ways in which this is achieved, through selective excitation mechanisms that have high probabilities of occurrence in a discharge environment, are considered in detail in Chapter 2.

Clearly of comparable importance to the excitation rates of levels 2 and 1 in establishing the oscillation condition are the lifetimes τ_2 and τ_1. They too are considered in Chapter 2.

1.5.1. DEPENDENCE OF GAIN ON WAVELENGTH

It is of considerable interest and importance in the operation of laser systems to know if there is any dependence of the gain on the wavelength of operation λ_0. From the gain expression of eqn. (1.47),

$$\alpha = (\ln 2/\pi)^{1/2} (g_2 A_{21}/4\pi) (N_2/g_2 - N_1/g_1) (\lambda_0^2/\Delta \nu_D),$$

it would appear that since α is directly proportional to λ_0^2 and inversely proportional to the linewidth $\Delta\nu_D$, the gain realizable at long wavelengths with the attendant small Doppler linewidths should be considerably more than at short wavelengths. That is, we would expect the gain to exhibit a $\lambda_0^2/\Delta\nu_D$ dependence. This "expectation" has been used consistently in the literature to explain the considerably higher gains that have been observed on long-wavelength transitions such as the 3.5-μm Xe-I and 3.39-μm Ne-I laser transitions, compared with shorter-wavelength transitions such as the 1.15-μm and 0.6328-μm Ne-I laser transitions. We need, however, to consider the gain expression more closely.

Faust and McFarlane[104] have considered in detail the possible dependence of gain on the wavelength of a transition, and we follow their treatment closely.

The gain expression given above and as (1.47) can be rewritten as

$$\alpha[(N_2/g_2-N_1/g_1)g_2]^{-1} = 3.75\times10^{-10}\lambda_0^2(A_{21}/\Delta\nu_D) \tag{1.53}$$

where the left-hand side of the equation is the ratio of the gain constant α to the volume density of inversion, divided by g_2, and in which the units of α are cm^{-1}, N_2 and N_1 are in cm^{-3}; the g's are the degeneracies, $2J+1$ for each state; and A_{21} and $\Delta\nu_D$ are in sec^{-1}, and λ_0 is in μm. At a gas temperature of 400 K, we take

$$\Delta\nu_D = [\lambda_0 \text{ (mass number)}^{1/2}]^{-1}\times4.3 \text{ GHz} \tag{1.54}$$

where λ_0 is in μm.

The Einstein A coefficient can be expressed in terms of the line strength S as[105]

$$A_{21} = (64\pi^4/3h)(1/g_2\lambda_0^3)S \tag{1.55a}$$

which simplifies numerically to

$$A_{21} = (2.02\times10^6/g_2\lambda_0^3)S. \tag{1.55b}$$

In eqn. (1.55b), λ_0 is in microns and S is in atomic units, $a_0^2e^2$. Substituting for A_{21} and $\Delta\nu_D$ from eqns. (1.55b) and (1.54) in (1.53) we obtain

$$\alpha(N_2/g_2-N_1/g_1)^{-1} = 1.76\times10^{-13} \text{ (mass number)}^{1/2}S. \tag{1.56}$$

It can be seen from (1.56) that the gain/inversion density ratio is directly proportional to the line strength S and that there is no explicit dependence of the ratio of gain to inversion density upon wavelength. It happens also that any systematic implicit dependence upon λ_0 buried in S disappears for long wavelengths.[104]

For long wavelengths (say, beyond 2 μm) the inversion is not expected to depend systematically upon wavelength as processes other than spontaneous emission on the laser transition dominate in determining N_2 and N_1. At long wavelengths the linewidth is not a Doppler linewidth (which varies as $1/\lambda_0$), and the appropriate linewidth can remain constant, whereby $\alpha/[(N_2/g_2-N_1/g_1)]$ in the limit would vary as $1/\lambda_0$. As pointed out by Faust and McFarlane,[104] while the output of spontaneous emission sources at progressively greater wavelengths decreases as $1/\lambda_0$, there is no comparable adverse wavelength dependence for laser operation.

Measured values of gain, gain/inversion density ratio and calculated values of line strength of Ne-I and Xe-I laser transitions are given in Table 1.2.

TABLE 1.2. *Line strengths and gain inversion ratios of laser lines for which α the gain constant has been measured*[a]

Laser line	Measured gain α (cm^{-1})	Line strength S	Gain/inversion density $\alpha/[(N_2/g_2 - N_1/g_1)]$
1.153-μm, Ne-I	1.2×10^{-3}	10	7.9×10^{-12}
3.392-μm, Ne-I	$\geqslant 4 \times 10^{-2}$	56	4.4×10^{-11}
0.6328-μm, Ne-I	5.3×10^{-4}	0.52	4.1×10^{-13}
2.027-μm, Xe-I	7.6×10^{-3}	15	2.9×10^{-11}
3.508-μm, Xe-I	0.16	73	1.5×10^{-10}

[a] Taken from ref. 104.

The table can be used to show that there is extremely good correlation between the gain/inversion-density ratio and the line strengths. For example, the ratio of the values of $\alpha(N_2/g_2 - N_1/g_1)$ for the 3.39-μm and 0.6328-μm Ne-I laser lines is $4.4 \times 10^{-11}/4.1 \times 10^{-13} =$ 107, and the ratio of their line strengths S is 56/0.52 = 108. Similarly there is remarkable agreement between the same ratio for the 1.153-μm and 0.6328-μm Ne-I lines of 19.3 compared to 19.2 for the ratio of their line strengths. There is also close agreement between the ratios of gain/inversion-density ratios and the ratio of the line strengths for the high-gain Xe-I laser lines at 3.508 μm and 2.027 μm of 5.17 compared to 4.85.

It can be concluded that the high gains observed on a number of infrared laser lines are associated with large line strengths of the transitions involved, and not through any λ^2 dependence of the gain.

References

1. JAVAN, A., BENNETT, W. R. Jr., and HERRIOTT, D. R., *Phys. Rev. Lett.* **6**, 106 (1961).
2a. RHEAULT, F., LACHAMBRE, J. L., GILBERT, J., FORTIN, R., and BLANCHARD, M., Paper Q. 2, Q.E.C. Montreal (1972), reported 2-GW power generation under pulsed conditions.
2b. GERRY, E. T., *Laser Focus* (Dec. 1970), p. 27, reports 600,000 W of CW power.
3. GORDON, J. P., ZEIGER, H. J., and TOWNES, C. H., *Phys. Rev.* **99**, 1264 (1955).
4. SCHAWLOW, A. L. and TOWNES, C. H., *Phys. Rev.* **112**, 1940 (1958).
5. JAVAN, A., *Phys. Rev. Lett.* **3**, 86 (1959).
6. JAVAN, A., BENNETT, W. R., Jr., and HERRIOTT, D. R. *Advances in Quantum Electronics*, pp. 18–49, Columbia Univ. Press, N.Y., 1960.
7. JAVAN, A., BENNETT, W. R. Jr., and HERRIOTT, D. R., *Phys. Rev. Lett.* **6**, 106 (1961).
8. FOX, A. G. and LI, T., *Bell Systems Tech. J.* **40**, 453 (1961).
9. BOYD, G. D. and KOGELNIK, H., *Bell System Tech. J.* **41**, 1347 (1962).
10. BRANGACCIO, D. J., *Rev. Sci. Instr.* **33**, 921 (1962).
11. RIGROD, W. W., KOGELNIK, H., BRANGACCIO, D. J., and HERRIOTT, D. R., *J. Appl. Phys.* **33**, 743 (1962).
12. RIGDEN, J. D., and WHITE, A. D., *Proc. IRE* **50**, 7 (1962).
13. BENNETT, W. R., Jr., *Bull. Am. Phys. Soc.* **7**, 15 (1962).
14. BENNETT, W. R., Jr., FAUST, W. L., and McFARLANE, R. A. (unpublished work, discussed in "Gaseous optical masers" by W. R. BENNETT, Jr., in *Appl. Optics*: Supplement I on *Optical Masers* (ed. O. S. Heavens), pp. 24–62 (1962).

15. Boot, H. A. H. and Clunie, D. M., *Nature* **197**, 173 (1963).
16. Boot, H. A. H. and Clunie, D. M., *Nature* **4882**, 773 (1963).
17. Mathias, L. E. S. and Parker, J. T., *Appl. Phys. Lett.* **3**, 16 (1963).
18. Mathias, L. E. S. and Parker, J. T., *Phys. Lett.* **7**, 194 (1963).
19. Bloom, A. L., Bell, W. E. and Rempel, R. C., *Appl. Optics* **2**, 317 (1963).
20. White, A. D. and Gordon, E. I., *Appl. Phys. Lett.* **3**, 197 (1963).
21. Gordon, E. I. and White, A. D., *Appl. Phys. Lett.* **3**, 199 (1963).
22. Bell, W. E., *Appl. Phys. Lett.* **4**, 34 (1964).
23. Bridges, W. B., *Appl. Phys. Lett.* **4**, 128 (1964).
24. Convert, G., Armand, M. and Martinot-Lagarde, P., *Compt. Rend.* **258**, 3259 (1964).
25. Convert, G., Armand, M. and Martinot-Lagarde, P., *Compt. Rend.* **258**, 4467 (1964).
26. Bennett, W. R., Jr., Knutson, J. W., Mercer, G. N. and Detch, J. L., *Appl. Phys. Lett.* **4**, 180 (1964).
27. Gordon, E. I., Labuda, E. F. and Bridges, W. B., *Appl. Phys. Lett.* **4**, 178 (1964).
28. Patel, C. K. N., Faust, W. L. and McFarlane, R. A., *Bull. Am. Phys. Soc.* **9**, 500 (1964).
29. Patel, C. K. N., *Phys. Rev.* **136A**, 1187 (1964).
30. Patel, C. K. N., *Phys. Rev. Lett.* **12**, 588 (1964).
31. Legay-Sommaire, N., Henry, L. and Legay, F., *Compt. Rend.* **260**, 339 (1964).
32. Barchewitz, P., Dorbec, L., Farrenq, R., Truffert, A. and Vautier, P., *Compt. Rend.* **260**, 3581, (1965).
33. Barchewitz, P., Dorbec, L., Truffert, A. and Vautier, P., *Compt. Rend.* **260**, 5491 (1965).
34. Legay, F. and Legay-Sommaire, N., *Compt. Rend.* **260**, 3339 (1964).
35. Patel, C. K. N., *Phys. Rev. Lett.* **13**, 617 (1964).
36. Fowles, G. R. and Jensen, R. C., *Proc. IEEE* **52**, 851 (1964).
37. Fowles, G. R. and Jensen, R. C., *Appl. Optics* **3**, 1191 (1964).
38. Jensen, R. C. and Fowles, G. R., *Proc. IEEE* **52**, 1350 (1964).
39. Patel, C. K. N., *Appl. Phys. Lett.* **7**, 15 (1965).
40. Moeller, G. and Rigden, J. D., *Appl. Phys. Lett.* **7**, 274 (1965).
41. Patel, C. K. N., Tien, P. K. and McFee, J. H., *Appl. Phys. Lett.* **7**, 290 (1965).
42. Leonard, D. A., *Appl. Phys. Lett.* **7**, 4 (1965).
43. Fowles, G. R. and Silfvast, W. T., *IEEE J. Quantum Electr.* **QE-1**, 131 (1965).
44. Silfvast, W. T., Fowles, G. R. and Hopkins, R. D., *Appl. Phys. Lett.* **8**, 318 (1965).
45. Dyson, D. J., *Nature* **207**, 361 (1965).
46. Walter, W. T., Solimene, N., Piltch, M. and Gould, G., *IEEE J. Quant. Electr.* **QE-2**, 474 (1966).
47. Sharma, R. D. and Brau, C. A., *Phys. Rev. Lett.* **19**, 1273 (1967).
48. Sobolev, N. N. and Sokovikov, V. V., *Sov. Phys. Usp.* **10**, 153 (1967).
49. Clark, P. O. and Smith, M. R., *Appl. Phys. Lett.* **9**, 367 (1966).
50. Fowles, G. R. and Hopkins, B. D., *IEEE J. Quant. Electr.* **QE-3**, 419 (1967).
51. Treonor, C. E., Rich, J. W. and Rehm, R. G., *J. Chem. Phys.* **48**, 1798 (1968).
52. Silfvast, W. T., *Appl. Phys. Lett.* **13**, 169 (1968).
53. Sosnowski, T. P., *J. Appl. Phys.* **40**, 5138 (1969).
54. Goldsborough, J. P., *Appl. Phys. Lett.* **15**, 159 (1969).
55. Goldsborough, J. P., *IEEE J. Quant. Electr.* **QE-5**, 133 (1969).
56. Nighan, W. L. and Bennett, J. H., *Appl. Phys. Lett.* **14**, 240 (1969).
57. Jensen, R. C., Collins, G. J. and Bennett, W. R., Jr., *Phys. Rev. Lett.* **23**, 363 (1969).
58. Tiffany, W. B., Targ, R. and Foster, J. D., *Appl. Phys. Lett.* **15**, 91 (1969).
59. Beaulieu, A. J., *Appl. Phys. Lett.* **16**, 504 (1970).
60. Dumanchin, R. and Rocca-Serra, J., Paper 18-6 presented at the 6th Q.E.C., Kyoto, Japan, Sept. 1970.
61. Silfvast, W. T. and Klein, M., *Appl. Phys. Lett.* **17**, 400 (1970).
62. Hodgson, R. T., *Phys. Rev. Lett.* **25**, 494 (1970).
63. Waynant, R. W., Shipman, J. D., Elton, R. C. and Ali, A. W., *Appl. Phys. Lett.* **17**, 383 (1970).
64. Basov, N. G., Danilychev, V. A., Popov, Yu. M. and Khodevich, D. D., *Sov. Phys. JETP Lett.* **12**. 329 (1970).
65. Osgood, R. M., Jr., Eppers, W. C., Jr. and Nichols, E. R., *IEEE J. Quant. Electr.* **QE-6**, 145 (1970).
66. Smith, P. W., *Appl. Phys. Lett.* **19**, 132 (1971).
67. Schuebel, W. K., *IEEE J. Quant. Electr.* **QE-7**, 39 (1971).
68. Silfvast, W. T., *Phys. Rev. Lett.* **27**, 1489 (1971).
69. Riseberg, L. A. and Shearer, L. D., *IEEE. J. Quant Electr.* **QE-7**, 40 (1971).

70. HODGSON, R. T. and DREYFUS, R. W., *Phys. Rev. Lett.* **28**, 536 (1972).
71. WAYNANT, R. W., *Phys. Rev. Lett.* **28**, 533 (1972).
72. FENSTERMACHER, C. A., NUTTER, M. J., LELAND, W. T. and BOYER, K., *Appl. Phys. Lett.* **20**, 56 (1972). See also BASOV, N. G., *Laser Focus* (Sept. 1972), pp. 45–47.
73. SPENCE, D., MAUER, J. L. and SCHULZ, G. J., *J. Chem. Phys.* **57**, 5516 (1972) and references therein.
74. AFANASEVA, V. L., LUKIN, A. V. and MUSTAFIN, K. S., *Sov. Phys. Tech. Phys.* **12**, 233 (1967).
75. BUSH, V. and CALDWELL, S. H., *Phys. Rev.* **38**, 1898 (1931).
76. HASTED, J. B., *Physics of Atomic Collisions*, p. 7, Butterworth, Washington (1964).
77. VON ENGEL, A., *Ionized Gases*, 2nd ed., p. 31., Clarendon Press, Oxford, 1955.
78. WADA, J. Y. and HEIL, H., *IEEE J. Quant. Electr.* **QE-1**, 327 (1965) and references therein.
79. LOEB, L. B., *Atomic Structure*, pp. 344–51, Wiley Inc., New York, 1938.
80. DRUYVESTEYN, M. J. and PENNING, F. M., *Revs. Mod. Phys.* **12**, 87 (1940).
81. DRUYVESTEYN, M. J., *Z. Phys.* **64**, 793 (1930).
82. BOYD, R. L. F. and TWIDDY, N. D., *Proc. Roy. Soc.* A **250**, 53 (1959).
83. BOYD, R. L. F. and TWIDDY, N. D., *Proc. Roy. Soc.* A **259**, 145 (1960).
84. TWIDDY, N. D., *Proc. Roy. Soc.* A **275**, 338 (1963).
85. I wish to thank Dr. R. T. Young for bringing this to my attention. The ordinates of Figs. 97B and 97C of ref. 79 have been mistakenly given as $(4/\pi)^{1/2}f(x)$ instead of just $f(x)$.
86. SCHULZ, G. J. and PHILBRICK, N. W., *Phys. Rev. Lett.* **13**, 477 (1964).
87. CHAMBERLAIN, G. E., *Phys. Rev.* **155**, 46 (1967).
88. PICHANICK, F. M. J. and SIMPSON, J. A., *Phys. Rev.* **168**, 64 (1968).
89. SCHULZ, G. J., *Phys. Rev.* **135**, 988 (1964).
90. WIGNER, E., *Gott. Nachr.* 375 (1927).
91. MASSEY, H. S. W. and BURHOP, E. H. S., *Electronic and Ionic Impact Phenomena*, pp. 427–30, Clarendon Press, Oxford, England, 1955.
92. LEES, J. H. and SKINNER, H. W. B., *Proc. Roy. Soc.* A **137**, 186 (1932).
93. ABRAMS, R. L. and WOLGA, G. J., *Phys. Rev. Lett.* **19**, 1411 (1967).
94. SCHEARER, L. D., *Phys. Rev. Lett.* **22**, 629 (1969).
95. SCHEARER, L. D. and RISEBERG, L. A., *Phys. Rev. Lett.* **26**, 599 (1971).
96. LATYSCHEW, G. D. and LEIPUNSKI, A. S., *Z. Phys.* **65**, 111 (1930).
97. SOLDATOV, A. N., *Optics Spectrosc.* **31**, 97 (1971).
98. PHELPS, A. V., *Phys. Rev.* **99**, 1307 (1955).
99. PHELPS, A. V. and MOLNAR, J. P., *Phys. Rev.* **89**, 1203 (1053).
100. CONDON, E. U. and SHORTLEY, G. H., *Theory of Atomic Spectra*, Cambridge University Press, 1965.
101. HEAVENS, O. S., *Optical Masers*, p. 17, Methuen, London, 1964.
102. BENNETT, W. R., Jr. in *Applied Optics: Supplement on Chemical Lasers* (1965), pp. 3–33.
103. MITCHELL, A. and ZEMANSKY, M. W., *Resonance Radiation and Excited Atoms*, Cambridge Univ. Press, 1961.
104. FAUST, W. L. and MCFARLANE, R. A., *J. Appl. Phys.* **35**, 2010 (1964).
105. BATES, D. R. and DAMGAARD, A., *Phil. Trans. Proc. Roy. Soc.* (London), A **242**, 101 (1950).
106. BENNETT, W. R., Jr., in *Applied Optics*: Supplement on Optical Masers (1962), pp. 24–62.

CHAPTER 2

Selective Excitation Processes in Gas Discharges

2.1. Theoretical Considerations

A number of analytical studies of population inversion in gaseous discharges have been made.[1-4] In all cases the studies start with assumed cross-sections for electronic or collisional and de-excitation processes and an assumed Maxwellian electron-energy distribution. In most cases, the degree of agreement of experimental results with the predictions based on these studies, in which one plays off one nonthermal equilibrium process against another in order to achieve population inversion, has not instilled confidence in this approach to the complex processes of the gaseous laser system. These analytic uncertainties, stemming from assumed cross-sections, reinforce what has been clear from the beginning of the development of the gas laser in 1961;[5] more data on collision cross-sections of all types is required, providing that one first examines those reactions that clearly have large cross-sections. Because of the recognized importance of resonant processes in numerous atomic collision phenomena, the assumed existence of Maxwellian electron energy distributions in gas discharges also needs to be verified before any theoretical treatment of collective discharge processes can realistically commence. Indeed, as some experimental results show, even small deviations in otherwise smoothly varying electron energy distributions appear to be responsible for the strong laser oscillation observed in particular species.

In spite of the many gaps in our knowledge of even apparently simple collisional processes or systems, it is possible in a qualitative way to understand the operation of gaseous lasers by considering the dominant processes that are operative, or appear to be operative, in producing population inversion. It is these dominant processes that have become apparent in retrospect as gas laser research and development has progressed.

As shown in Chapter 1 the steady state populations of particular excited energy levels are determined by

(1) the effective lifetime (τ) of these states in the presence of radiative decay and collisional deactivation and

(2) the rate of excitation (n) to such states by inelastic collisions or other excitation mechanisms such as optical pumping.

Both lifetimes and inelastic collisions in some cases can be treated theoretically to yield valid results.

2.1.1. TRANSITION PROBABILITIES

Theoretical calculations of pure radiative transition probabilities, which are inversely proportional to radiative lifetimes, can be made using the central field approximation of Bates and Damgaard.[6] The calculation of transition probabilities between levels of an atomic system entails the evaluation of radial transition integrals for the levels concerned. Bates and Damgaard have shown that almost the whole contribution to the transition integrals involved arises from the region of the atom for which a Coulomb approximation for the potential is valid. Basically, in this approximation the charge of the nucleus and other electrons is considered as lumped within a small radius, and for a radius or less than a critical radius r_c the contribution to the integral is considered negligible. Bates and Damgaard take r_c as the value of r for which the true potential differs by no more than one percent from the Coulomb potential as the limit of $r_c \rightarrow r_\infty$ (Fig. 2.1).

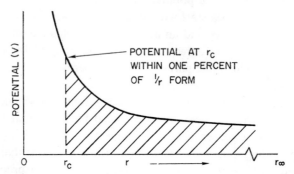

FIG. 2.1. Illustrating the Coulomb potential for an atom. The shaded area $r_e \rightarrow r_\infty$ indicates the part that contributes principally to the transition integral.

In spite of the simplifications inherent in the central field approximation, it is found that the method yields results in extremely good agreement with experiment, and in close agreement with results calculated from the more complex Hartree–Fock differential equations as shown in Table 2.1.[6]

The method results in the following agreement being obtained between calculated and experimentally determined pure radiative transition probabilities:

(a) *simple, light atomic systems*—accurate on all transitions; agreement within 10 percent for p–s transitions with elements up to atomic number 19, and for p–d and d–f transitions with elements having higher atomic numbers (a simple system is taken as being an atom having one electron outside a closed shell);

(b) *simple, heavy atomic systems*—not subject to interference between levels; for atomic numbers higher than 19 (e.g. for Cu-I or Ca-I), the agreement is to within about 50 percent;

(c) *complex atomic systems*—fair agreement is shown; even Tl-I (atomic number 81) is not expected to differ by a factor greater than 2.

The probability per unit time of an excited atom making a spontaneous emission from state A to state B, $A(A, B)$, depends on the interaction of the atom with the radiation field.

It is given by Condon and Shortley[7] as

$$A(A, B) = (64\pi^4 \nu^3/3h)\, S(A, B)/(2J_a+1) \qquad (2.1)$$

where $S(A, B)$ is the line strength of a particular transition, J_a is the total angular momentum of state A (the upper state) and ν the wave number of the light emitted in cm^{-1}, h is Planck's constant and the term $2J_a+1$ is the statistical weight of the upper state. The line strength is given by

$$S(A, B) = S(M), S(L), \sigma^2 \qquad (2.2)$$

where $S(M)$ and $S(L)$ derived from the integral over the angular part of the wave functions represent the relative strengths of the multiplet and the line within a multiplet respectively.

TABLE 2.1. *Values of σ^2 for a number of transitions*[a]

Transition		Theory		Experiment
		Coulomb approx.	Hartree method	
Li-I	$2s$–$2p$	5.5	5.5	5.5
	$2s$–$3p$	0.018	0.020	0.020
Na-I	$3s$–$3p$	6.1	6.3	6.5, 6.8
	$3s$–$5p$	0.0067	0.0069	0.0066
K-I	$4s$–$4p$	8.3	9.0	8.4
	$4p$–$4d$	0.0001	0.0004	0.0003
Cu-I	$4s$–$4p$	2.9	—	3.4
Cs-I	$6s$–$6p$	9.9	—	9.3
Zn-I	$4s\,^1S$–$4p\,^1P$	1.8	—	1.1, 1.4
Cd-I	$5s\,^1S$–$5p\,^1P$	1.9	—	1.5
Hg-I	$6s\,^1S$–$6p\,^1P$	1.5	4.8	1.2
Tl-I	$6p\,^2P_{1/2}$–$7s\,^2S_{1/2}$	0.57	—	0.53
	$6p\,^2P_{1/2}$–$10s\,^2S_{1/2}$	0.0065	—	0.024, 0.0080

[a] Where σ^2 is directly proportional to the transition probability (eqn. (2.5)).

The σ^2 depends on the radial portions of the wave functions in terms of the two azimuthal quantum numbers and energy level values and is expressed in the form

$$\sigma^2 = (4l^2-1)^{-1}\left[\int_0^\infty R_i R_f r\, dr\right]^2. \qquad (2.3)$$

Here l is the larger of the two azimuthal quantum numbers involved in the transition. The functions R_i and R_f, which refer to the initial and final state of the atom, respectively, satisfy a differential equation of the form

$$\frac{d^2R}{dr^2} + \left(2V - \frac{l(l+1)}{r^2} - \varepsilon\right) R = 0, \qquad (2.4)$$

31

V being the potential, l the azimuthal quantum number, and ε an energy parameter that is adjusted to be an eigenvalue when the equation is solved by numerical integration outwards from the origin. Values of σ^2 are calculable directly from the tables of Bates and Damgaard.

With ν expressed in Rydbergs, the transition probability can be obtained from the determination of σ^2 via

$$A(A, B) = 2.662 \times 10^9 \, k\nu^3\sigma^2 \, \sec^{-1}, \tag{2.5}$$

where k depends on the line of the multiplet considered.[7, 8]

This theoretical treatment yields only the best value, pure radiative lifetimes, not effective lifetimes. In a gaseous-discharge environment it is the effective lifetimes that determine the degree of selective excitation, population inversion, oscillation, and gain conditions. These effective lifetimes can differ appreciably from purely radiative ones due to the combined processes of spontaneous decay, radiation trapping, and inelastic collisions. For a consideration of the determination of effective lifetimes the reader is referred to "Measurement of excited state relaxation rates" by Bennett, Kindlmann and Mercer.[9] It should be noted here that laser oscillation has been reported on a number of transitions for which the theoretical transition probabilities are very small or zero[10, 11] so that in determining the feasibility of a possible laser system recourse must needs be made to the experimental determination of these effective lifetimes.

2.1.2. SELECTIVE EXCITATION MECHANISMS

Theoretical treatments of inelastic as well as elastic collisions have been made by Massey and Burhop,[12] Townsend,[13] Mott and Massey,[14] and Bates.[15] Full reviews of these theoretical treatments together with experimental analyses are given in Hasted,[16] McDaniel,[17] Massey,[18] Mott and Massey,[14] and Bates and McCarrol.[19] These provide a good background to a number of inelastic collision mechanisms that result in selective excitation in gas laser systems.

Particular processes that have been used to give selective excitation of desired energy states include:

(1) resonant excitation-energy transfer involving atom–atom, atom–ion, and molecule–molecule reactions;
(2) charge transfer;
(3) Penning reactions;
(4) dissociative excitation transfer;
(5) electron impact;
(6) charge neutralization.

Additional selective excitation mechanisms that, although not inelastic collision mechanisms, are conveniently considered here, are:

(7) line absorption, and molecular photodissociation absorption; and
(8) radiative cascade-pumping.

2.2. Resonant Excitation-energy Transfer

This type of energy transfer is applicable to the transfer of energy in both atom–atom and molecule–molecule collisions and results in electronic and vibrational energy transfer. It also has some applicability, as will be made evident in Chapter 5, to energy transfer in ion–atom collisions.

2.2.1. EXCITATION TRANSFER (atom–atom)

When atoms A and B are irradiated by resonance radiation of A, emission of radiation by the atoms of B often can be observed subsequently. This is the phenomenon of sensitized fluorescence.[20] The classic example of this is the transfer of electronic energy that occurs between excited atomic mercury and ground-state sodium atoms in the reaction

$$\text{Hg } 6p\ ^3P_1 + \text{Na } 3s\ ^2S_{1/2} \rightarrow \text{Hg } 6s^2\ ^1S_0 + \text{Na } 9s\ ^2S_{1/2} - \Delta E_\infty. \tag{2.6}$$

Since the potential energy of an excited Hg $6p\ ^3P_1$ atom is less than that of a sodium atom in the Na $9s\ ^2S_{1/2}$ state, energy must be supplied to enable the reaction to proceed. This energy can be obtained from the kinetic energy of the colliding atoms.

Beutler and Josephy[21] found that when a mixture of mercury and sodium was excited

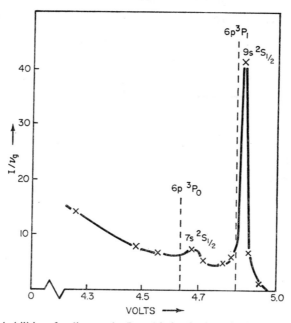

Fig. 2.2. Excitation probabilities of sodium to the $7s$ and $9s$ levels via excitation-transfer collisions with mercury atoms in the $6p\ ^3P_0$ and $6p\ ^3P_1$ states. The excitation probability is indicated by the intensity (I) of lines from these levels divided by the appropriate frequency (ν) and the statistical weight (g). The abscissa shows the energy of the levels and the broken lines the energy of the Hg $6p\ ^3P_0$ and Hg $6p\ ^3P_1$ states. The lines on which observations were made occur in the wavelength range 253.7–579.0 nm. (After H. Beutler and B. Josephy,[21] by courtesy of J. Springer.)

with mercury resonance radiation at 253.7 nm, a number of sodium lines were emitted. Sodium has a number of energy levels in the narrow range 4.26–4.94 eV above the ground state close to the $6p\ {}^3P$ state of mercury at 4.860 eV. With such a small energy range, no differences in intensity in the absence of selective excitation processes would be expected from lines with upper levels in this range. Figure 2.2 shows that the $9s\ {}^2S_{1/2}$ state of sodium (within 0.020 eV of the Hg $6p\ {}^3P_1$ state) is preferentially excited. A small maximum in emission also occurs at the sodium $7s\ {}^2S_{1/2}$ state (within 0.045 eV of the metastable Hg* $6p$ ${}^3P_0$ state) showing further the resonant quality of the reaction. In spite of the "tidiness" of this explanation, however, it is possible that the stated atomic energy coincidences are fortuitous and that an intermediate, long-lived molecular state of mercury is in resonance with the Na states. This appears to be the case in sensitized fluorescence of Hg–Cd and Hg–Zn mixtures that had been attributed originally to atom–atom excitation transfer.[22]

The overall resonant-excitation energy-transfer process of the type

$$A' + B \rightleftarrows A + B' \pm \varDelta E_\infty \tag{2.7a}$$

where $\varDelta E_\infty$ is the difference in potential energy at infinite separation between the excited species A' and B', and A and B are ground-state species in which potential energy is transferred from one atom to another, has been studied theoretically by Massey and Burhop.[12] It is known as a "collision of the second kind"[23,24] (a collision of the first kind being one in which kinetic energy is converted into potential energy). It will be noted from detailed balancing that the reaction is reversible. When A' is a metastable specie (A^*), and B' can decay radiatively, the reaction proceeds principally in the forward direction

$$A^* + B \rightarrow A + B' \pm \varDelta E_\infty. \tag{2.7b}$$

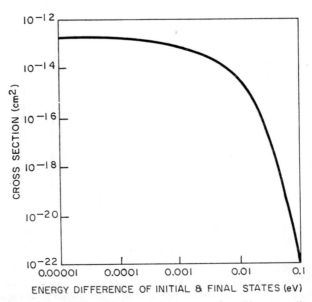

FIG. 2.3. Theoretical variation of cross-section for excitation transfer with energy discrepancy between the initial and final states of the colliding atoms. (After N. F. Mott and H. S. W. Massey,[14] by courtesy of Oxford University Press.)

Theory shows that this type of collision has a cross-section (σ) that is

(1) Highly resonant for near exact energy coincidence and larger than 10^{-14} cm². With ΔE_∞ of the order of a few kT cross-sections of 10^{-16} cm² (of gas-kinetic dimensions) are realized; whereas the cross-sections are negligible for $\Delta E_\infty > 0.1$ eV (as illustrated in Fig. 2.3).
(2) Dependent on the energy of relative motion of the colliding atoms or molecules (Fig. 2.4).
(3) Larger for collisions of a similar kind when the Wigner spin rule[25] is obeyed than when it is disobeyed. Given the reactions

$$A'(\uparrow\uparrow)+B(\uparrow\downarrow) \left\{ \begin{array}{ll} \longrightarrow A(\uparrow\downarrow)+B'(\uparrow\uparrow)\pm\Delta E_\infty, & (2.8a) \\ \longrightarrow A(\uparrow\downarrow)+B'(\uparrow\downarrow)\pm\Delta E_\infty, & (2.8b) \end{array} \right.$$

eqn. (2.8a), in which the total spin of 1 is conserved, is the more likely reaction channel.

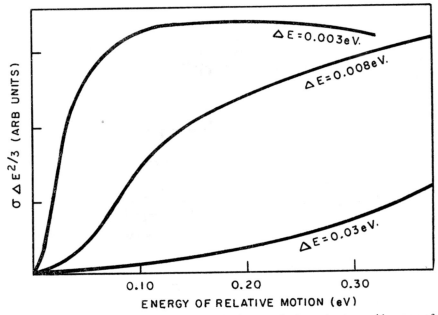

FIG. 2.4. Theoretical variation of cross-section for excitation transfer between atoms with energy of relative motion of the colliding atoms for different resonant defects ΔE. (After N. F. Mott and H. S. W. Massey,[14] by courtesy of Oxford University Press.)

Experimentally it has been found that the probability of excitation transfer occurring in a collision appears to be dependent on at least four factors:

(1) the magnitude of ΔE_∞;
(2) spin conservation (the Wigner spin rule);
(3) the size of the colliding atoms;
(4) the relative velocity of the colliding atoms A' and B (determined by the gas temperature (T_g) in a gas discharge and the individual masses of the colliding atoms).

The relative importance of these four factors in a particular reaction is not known. Excitation transfer that occurs in He–Ne laser discharges between at least three excited states of helium and neon provides most of the information on the importance of the factors.

It is clear, however, that the magnitude of ΔE_∞ (but surprisingly not the sign of ΔE_∞) and spin conservation determine the probability of excitation transfer to a large extent. A factor of prime importance, which does not lend itself to a simple rule-of-thumb interpretation, is the actual shape of the molecular interaction potentials during a collision between two atoms undergoing excitation transfer, and the interaction of the incoming, A', B, and outgoing, A, B', channels of eqns. (2.8a) and (2.8b). The basis for this is covered in the following section which considers in some detail excitation transfer between the He* 2^1S_0 and He* 2^3S_1 metastable states and neon.

Figure 2.5 illustrates the energy-level coincidence for the excitation-transfer reaction between the metastable He* 2^1S_0 state and the four Ne-3s states (Paschen notation), and

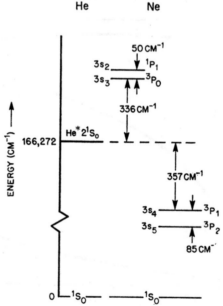

FIG. 2.5. Partial energy levels for helium and neon showing the energy-level coincidence between the metastable He* 2^1S_0 state and the four Ne-3s states.

shows that the He* 2^1S_0 level is situated approximately in the middle of the Ne-3s states. Here selective excitation occurs preferentially to the Ne-$3s_2$ state[26, 27] (the uppermost 3s state), which is higher in potential energy than the He* 2^1S_0 state, in the endothermic reaction

$$\mathrm{He^*}\,2\,^1S_0(\uparrow\downarrow)+\mathrm{Ne}\,^1S_0(\uparrow\downarrow) \rightarrow \mathrm{He}\,^1S_0(\uparrow\downarrow)+\mathrm{Ne}\,3s_2(\uparrow\downarrow)-\Delta E_\infty \quad (386 \text{ cm}^{-1}), \qquad (2.9)$$

in which the total spin S of 0 is conserved, and a ΔE_∞ of 386 cm^{-1} (about 2 kT) has to be provided in the form of kinetic energy by the discharge.

Figure 2.6 illustrates the similar coincidence between the He* 2^3S_1 state and the 2s states of neon for the reaction

$$\text{He}^* \, 2\,^3S_1(\uparrow\uparrow) + \text{Ne}\,^1S_0(\uparrow\downarrow) \;\rightarrow\; \text{He}\,^1S_0(\uparrow\downarrow) + \text{Ne}\,2s_{2,\,3,\,4,\,5} + \Delta E_\infty \quad (313 - 1247\ \text{cm}^{-1}). \quad (2.10)$$

Unlike the energy coincidence between the Ne-3s states and the He* 2^1S_0 metastable state, all the Ne-2s levels lie below the He* 2^3S_1 state in energy. The $2s_5$ state is lower by as much as 1247 cm^{-1} (or about 0.14 eV). In spite of the large magnitude and range of energy discrepancy of $+313$ to $+1247$ cm^{-1} between the Ne-2s states and the He* 2^3S_1 state where

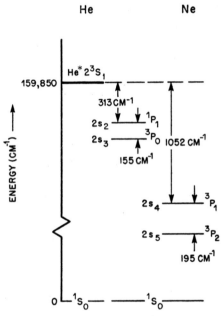

FIG. 2.6. Partial energy levels for helium and neon showing the energy-level coincidence between the metastable He* 2^3S_1 state and the four Ne-2s states.

excitation transfer would appear to be small or negligible according to theory,[12] significant energy transfer is observed.[28] In the latter case, no strong preference for selective excitation of particular states is observed, and all the states are selectively excited via excitation transfer.[26] In the case of excitation transfer to the $2s_{3,\,4,\,5}$ states, spin is, and has been observed experimentally, to be conserved.[28] With the $2s_2$ state, spin is not conserved. Presumably here, on a simple basis, the smaller energy discrepancy compensates for the spin-rule violation. In the excitation transfer reaction between He* 2^1S_0 metastables and the Ne-3s states, spin is conserved only for the upper-lying Ne-$3s_2$ state. Presumably here spin conservation is more important than the energy discrepancy ΔE_∞.

   In the light of present-day knowledge, the significance of ΔE_∞ in determining excitation transfer cross-sections is not known, though spin conservation does appear to be important. Jones and Robertson,[29] from measurements in the afterglow of a He–Ne discharge at various temperatures, have shown that the cross-section (σ) for the excitation transfer reactions between He* 2^1S_0 and He* 2^3S_1 metastable atoms and ground-state neon atoms both follow the relationship

$$\sigma = \sigma_0 \exp\left(-E_a/kT\right) \tag{2.11}$$

where σ_0 is the cross-section at infinite temperature. Plots of the experimental results that lead to this relationship are given in Figs. 2.7 and 2.8, where σ is plotted against reciprocal temperature T. In (2.11), E_a has the form of an activation energy (or dissociation energy) in which E_a is not equal to the appropriate ΔE_∞ for the reaction of concern. E_a for the endothermic reaction (2.9) is equal to 0.034 eV, whereas ΔE_∞ is 0.048 eV; and for the collective excitation-transfer reaction (2.10) to all the four Ne-$2s$ states from He* 2^3S_1 state, E_a is 0.044 eV, whereas the mean ΔE_∞ for the four levels is approximately 0.09 eV. Distribution of energy between the four Ne-$2s$ levels in atom–atom and electron–atom collisions has precluded a determination of the individual cross-sections.

The presence of an activation energy or a potential barrier in the molecular-interaction potentials that separates the free atoms from a potential well is consistent with a long-range repulsive interaction between interacting He* 2^1S_0 and He* 2^3S_1 atoms and ground-state Ne $^1S_0$ atoms in the formation of a molecule or quasi-molecule. Such an interaction would give rise to an ingoing molecular potential $A'B$ similar to that illustrated in Fig. 4.19 for the $A^1\Sigma_u^+$ state of molecular helium formed by two interacting He* 2^1S_0 and He $^1S_0$ atoms.[30]

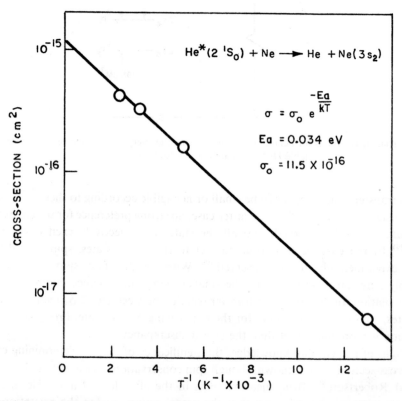

FIG. 2.7. Variation of cross-section with gas temperature (K^{-1}) for excitation transfer between He* 2^1S_0 metastables and Ne to the Ne-$3s_2$ state. (By courtesy of Jones and Robertson.[29])

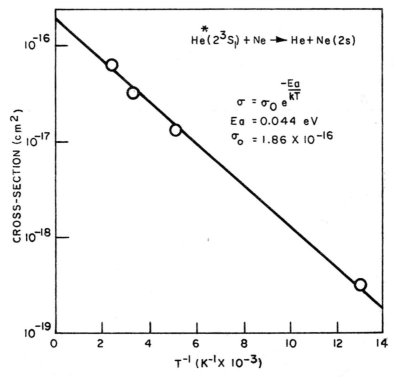

FIG. 2.8. Variation of cross-section with gas temperature (K^{-1}) for excitation transfer between He* 2^3S_1 metastables and Ne to the Ne-$2s_{2-5}$ states. (By courtesy of Jones and Robertson.[29])

The effective cross-section for molecular formation by ground state and metastable helium atoms has been given by Buckingham and Dalgarno[30] as

$$\sigma = \tfrac{1}{2}(S_1 + S_2)\pi r_0^2 \exp(-\Delta E/kT) \tag{2.12}$$

where ΔE in their notation is the height of the potential barrier in the formation of the molecule, and S_1 and S_2 are associated with the mean free paths for collisions of ground state and metastable helium atoms in helium. Equation (2.12) clearly has the same general form as (2.11) implying that there is considerable similarity between He*–He collisions and He*–Ne collisions. On a simple-minded interpretation this might have been expected because of the closed-shell electronic configuration of ground state He-$^1S_0$ and Ne-$^1S_0$ atoms, with He* 2^1S_0 atoms common to both reactions.

To account for the activation energy of both reactions (2.9) and (2.10), and the disposition of the energy levels A' and B' at infinite nuclear separation, it is proposed that the interaction potentials of each have the form presented in Figs. 2.9b and 2.9a where A', B and A, B' are the incoming (initial) and outgoing (final) states respectively. These correspond to cases of pseudo-crossing and curve-crossing respectively. If this is the case, and if other excitation transfer reactions behave in the same way, it is clear why the magnitude of ΔE_∞ tells us little about the probability of excitation transfer in a particular reaction since it is the shape of the interaction potentials close to, and at, the collision that will determine to a large extent the probability of excitation transfer. Due to the complexity of calculating interaction potentials of even simple interactions such as the $He^* 2^1S_0 + He\ {}^1S_0$ systems,[30] little hope is held for calculating the interaction potentials of more complicated systems, such as the He^*–Ne system. Finally, if excitation transfer reactions do proceed via the formation of quasi-molecules and the general form of potential barriers shown in Figs. 2.9a and 2.9b, it is not

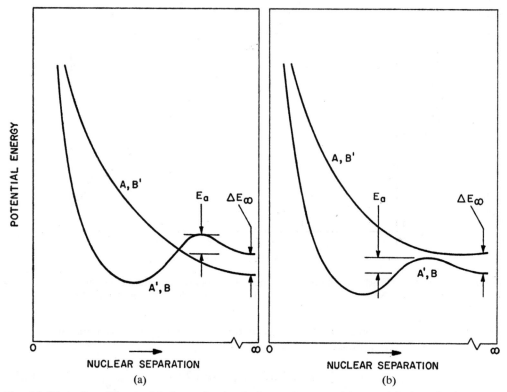

FIG. 2.9. Illustration of the possible shape of molecular interaction potentials to explain the activation-energy behavior of excitation transfer between helium metastables and neon for (a) $He^* 2^3S_1 + Ne$, (b) $He^* 2^1S_0 + Ne$.

clear that detailed balancing should apply between the initial and final states of the reaction in the presence of what could be stabilizing collisions inside the potential barriers at smaller nuclear separations.

2.2.2. VIBRATIONAL ENERGY TRANSFER (molecule–molecule)

The classical theory of vibrational energy exchange between molecular systems based on the work of Landau and Teller[31] has been described by Herzfeld and Litovitz.[32] The treatment of Landau and Teller leads to the result that the probability of energy transfer depends on and increases with the ratio of the period of vibration of the composite particle to the time of interaction of the colliding molecules in the following way:

$$1/Z = (1/Z_0) \exp{-2\pi va/v}. \tag{2.13}$$

Here Z is the collision number specifying the number of collisions required per excitation transfer, Z_0 is a pure number of the order unity, ν is the vibration frequency, a is the distance over which the interaction occurs (of the order of molecular dimensions) and v is the relative velocity of the colliding molecules. Since only the Fourier components of a disturbing impulse force $F(t)$ near the oscillator frequency can force a disturbance, the vibration frequency ν can be taken as $\Delta E/h$, where ΔE is the energy discrepancy between the vibrational quanta of the initial state E_2 and final state E_1.

Basically, the probability for vibrational transfer depends on the extent to which the collision is nonadiabatic as discussed in Massey's near-adiabatic hypothesis. This hypothesis, an extension of the Ehrenfest thermodynamic adiabatic principle,[33] applies to the response of quantized systems to gradual, nonperiodic changes of parameters such as external fields. If the changes are slow, the quantized system remains the same; if rapid, transitions to other quantum states can occur. The ratio va/v in (2.13) specifies the degree of adiabacy of the collision.

Figure 2.10, after Hancock and Smith,[34] shows the experimental probabilities of V–V energy transfer versus energy discrepancy $\Delta\nu$. P is divided by ν to allow for the increase in rate which is predicted for energy exchange between identical collision partners when the frequency increases. For added CO, N_2, and OCS, rates have been measured for endothermic processes and converted using the appropriate Boltzmann factor. With NO, de-excitation of CO ($\nu \leqslant 11$) is exothermic (denoted by ●) and of CO ($\nu = 12$ and 13) endothermic (denoted

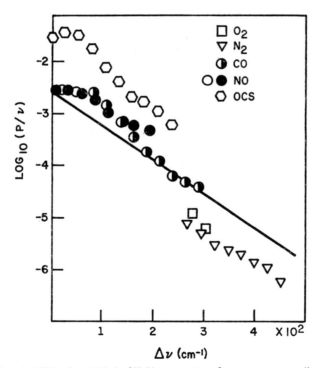

Fig. 2.10. Reduced probabilities $\log_{10}(P/\nu_0)$ of V–V energy transfer versus energy discrepancy $\Delta\nu$ between CO and CO, NO, N_2, O_2, and OCS. All the values are for V–V exchange in the exothermic direction. (After G. Hancock and I. W. M. Smith,[34] by courtesy of the Optical Society of America.)

by $\bigcirc$). The straight line represents an empirical relationship suggested by Callear[35] to correlate early results on V–V exchange with vibrational frequencies close to 200 cm^{-1}. Several interesting features of Fig. 2.10 are apparent. For processes with Δv of more than 60 cm^{-1}, each set of points show an approximately linear decrease of log (P/v) with a slope similar to the straight-line relationship deduced by Callear. For $\Delta v \leqslant 60$ cm^{-1}, the reduced probabilities are constant. This is believed to arise because the molecules are able to exchange rotational energy, and the pure vibrational-energy resonance is smeared out by rotational energy-level coincidences. Good agreement occurs between the experimental results plotted in Fig. 2.10 and theoretical results given by the treatment of Herzfeld and Litovitz[32] (Callear[34, 35]).

Vibrational energy transfer is found to be rapid provided

(1) the difference in vibrational frequency of the initial and final states is not greater than 500 cm^{-1} (or a few kT);
(2) the mass of the colliding molecules is low; this is required to ensure that the collision is of as short a duration as possible. Experiment shows also that the most probable transitions are those having unit change of vibrational quantum number, $\Delta v = 1$.

2.3. Charge Transfer

Asymmetric charge transfer is represented in its simplest form by a collision between unlike atoms of the type

$$A^+ + B \rightarrow A + B^+ \pm \Delta E_\infty \qquad (2.14)$$

in which a single electron is transferred from atom B in the neutral ground to the ion A^+ which is in the ion ground state. As in eqn. (2.7), ΔE_∞ is the energy difference between the initial and final potential-energy states of the collision system or simply the difference in ionization energies of A and B when ionization is to the ground state. The energy dependence of the cross-section for a particular charge-transfer reaction depends markedly on the magnitude of ΔE_∞ as in Massey's near-adiabatic hypothesis mentioned earlier.

As in the atom–atom excitation transfer process, the cross-section for a collision in which transfer of charge occurs is small at gas kinetic collision velocities unless ΔE is small. However, such cross-sections, in agreement with theoretical prediction, are not large compared to gas-kinetic cross-sections (Massey[18]) except in certain circumstances mentioned later.

From the Correspondence Principle for small energy differences (ΔE) between two quantized systems $\Delta E \approx h\nu$, or $|\Delta E| \cdot \Delta t \approx h$, where Δt is the oscillation period of the composite interacting ion and atom.

If a is the range over which an interaction occurs, of the order of the dimensions of an atom (10^{-8} cm), and v is the relative velocity of the interacting ion and atom, the time of interaction is equal to a/v. In the "adiabatic region" the time of interaction is longer than Δt, when Δt from above is equal to $h/\Delta t$. That is when

$$a/v \gg h/|\Delta E|. \qquad (2.15)$$

As the relative velocity of the interacting particles rises, the cross-section for the reaction increases and reaches a maximum when the time of interaction is comparable to the oscillation period Δt. This is when

$$a/v_m \approx h/|\Delta E|, \tag{2.16}$$

where v_m is the velocity at the cross-section maximum. When $|\Delta E|$ is small or zero, the maximum cross-section occurs at low relative velocities. The energy discrepancy $|\Delta E|$, however,

(a) is a function of the separation of the colliding particles and ΔE_∞, the value of ΔE at infinite separation, is normally used in the expressions above; and

(b) can be dominated by Coulomb or polarization forces and corrections must be made to give a mean energy defect averaged over the interaction path.

In addition to the uncertainty of ΔE, it is not clear what value of a should be used in (2.16). According to Drukarev,[36] the value of a for small energy defects ($\Delta E \ll I$) is proportional to $(2I)^{-1/2}$ where I is the ionization potential of the incident atom. For large energy defects ($\Delta E \approx I$), Drukarev predicted, and Perel and Daley[37] have shown, that a/v_m is not inversely proportional to $|\Delta E|$ as in (2.16), but that

$$a/v_m \propto (|\Delta E|)^{1/2}. \tag{2.17}$$

In the simple charge transfer (10/01) reaction considered, where polarization interaction is applicable, the mean energy defect $|\Delta E|$ is given by

$$|\Delta E| = \Delta E_\infty + \overline{Ep} \tag{2.18}$$

where $\overline{Ep}$ is the polarization energy.

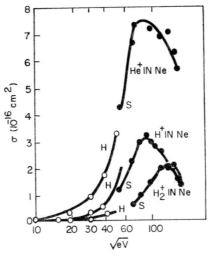

FIG. 2.11. Variation of cross-section for asymmetric charge transfer with incident ion energy (eV)$^{1/2}$, in the range where maxima in the cross-section are observed. The H and S refer to results obtained by Hasted and Stedeford[62] respectively. (After J. B. H. Stedeford and J. B. Hasted,[62] by courtesy of the Royal Society.)

A considerable number of experiments have been carried out to check the Massey near-adiabatic hypothesis, notably by Hasted and co-workers.[38]

From eqn. (2.16), and following the treatment of Hasted and Lee,[39] for an impact energy V_m (in eV), a projectile of mass m and charge e (atomic units) will have a maximum cross-section when

$$a\,|\Delta E|\,m^{1/2} = (2eV_m/300)^{1/2}, \tag{2.19}$$

resulting in

$$a = 0.057V_m^{1/2}/(m^{1/2}\Delta E), \tag{2.20}$$

where a is in nm. In the analysis of a large amount of experimental data, Hasted[40] has shown that for a number of different reactions involving asymmetric charge transfer

$$an = 0.7 \text{ nm} \tag{2.21}$$

where n is the number of electrons transferred.

Analysis of other data, and experimental work by Perel and Daley,[37] shows that the linear relationship

$$v_m = 22\times 10^7(|\Delta E|)/I^{1/2} \tag{2.22}$$

applies at low values of $|\Delta E|/I^{1/2}$, conforming with the adiabatic criterion of eqn. (2.16). For values of $|\Delta E|/I^{1/2}$ greater than 0.15 $(eV)^{1/2}$

$$v_m = 8\times 10^7(|\Delta E|/I^{1/2})^{1/2} \tag{2.23}$$

confirming the parabolic dependence of eqn. (2.17).

For asymmetric charge transfer the cross-section decreases as v decreases from the v_m corresponding to the $(eV)^{1/2}$ for the maximum cross-section σ as shown in Fig. 2.11. In the "near-adiabatic region", it appears to take the exponential form

$$10\sigma_{01} = A\exp(-Ba|\Delta E|/hv). \tag{2.24}$$

This is illustrated in Fig. 2.12 after Hasted[41] where A and B are constants. Note that $(eV)^{1/2}$ is directly proportional to the velocity v in eqn. (2.24). For low impacting energies, the probability of charge transfer is small unless ΔE is small.

Until recently the energy defect was generally assumed to be the difference in ionization potentials between A and B in eqn. (2.14), with the charge transfer occurring principally between the ground states of the ions. It is now realized that a number of charge-transfer reactions involve simultaneous excitation and ionization, and proceed via the reaction

$$A^+ + B \rightarrow A + B^{+\prime} \pm \Delta E_\infty \tag{2.25}$$

where $B^{+\prime}$ is an excited state of the ion, and ΔE_∞ is the difference in potential energy between the ionization energy of A, and the potential energy of the excited ion $B^{+\prime}$ above the neutral-atom ground state.

As $B^{+\prime}$ is an excited ion, radiative decay to lower-lying energy states can occur with the emission of radiation. This can lead to the particular increase in intensity enhancement of certain spectral lines, in the "so-called" spark-line-enhancement process.[16] The magnitude of this (10/01′) type of charge transfer reaction depends on the configuration of the level excited and ionized, and not simply on the energy defect. Duffendack and Thomson[42]

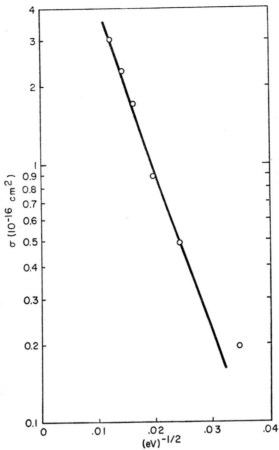

FIG. 2.12. Variation of cross-section with incident ion energy $(eV)^{-1/2}$ for asymmetric charge transfer H^+ in Ne in the near-adiabatic region, illustrating the exponential variation of the cross-section with relative velocity (v). (After J. B. Hasted,[41] by courtesy of the American Institute of Physics.)

have observed that lines from transitions from triplet levels in singly ionized copper and aluminum were enhanced in a discharge of copper and neon or aluminum and neon respectively more than transitions from singlet levels. In this case, some singlet levels are closer to the neon ion ground state than triplet levels; yet the preferred selective excitation process in the case of Cu and Ne^+, which involves simultaneous charge transfer and excitation, is

$$Ne^+ \, ^2P_{3/2}(\uparrow) + Cu \, ^2S_{1/2}(\uparrow) \rightarrow Ne \, ^1S_0(\uparrow\downarrow) + Cu^{+\prime} \, ^3P(\uparrow\uparrow) + \Delta E_\infty \qquad (2.26)$$

in which the total spin of 1 is conserved. Figure 2.13 illustrates the dependence of the cross-section (as evidenced by enhancement) on the energy discrepancy between Ne^+ and excited states of $Al^{+\prime}$ and $Cu^{+\prime}$ reported by Duffendack and Thomson.[42]

In similar processes, with gold and silver in a discharge in helium,

$$He^+ \, ^2S_{1/2}(\uparrow) + Au \, ^2S_{1/2}(\uparrow) \rightarrow He \, ^1S_0(\uparrow\downarrow) + Au^{+\prime} \, ^3D(\uparrow\uparrow) + \Delta E_\infty, \qquad (2.27)$$

$$He^+ \, ^2S_{1/2}(\uparrow) + Ag \, ^2S_{1/2}(\uparrow) \rightarrow He \, ^2S_0(\uparrow\downarrow) + Ag^{+\prime} \, ^3D(\uparrow\uparrow) + \Delta E_\infty, \qquad (2.28)$$

45

in which the total spin of 1 is again conserved. In a low-voltage helium arc, Duffendack and Thomson observed a predominance of enhancements of emission from triplet states in accordance with the Wigner spin rule as indicated by the spin vectors. In the reported case of silver enhancement, although lines are enhanced from triplet levels within 0.1 eV of the

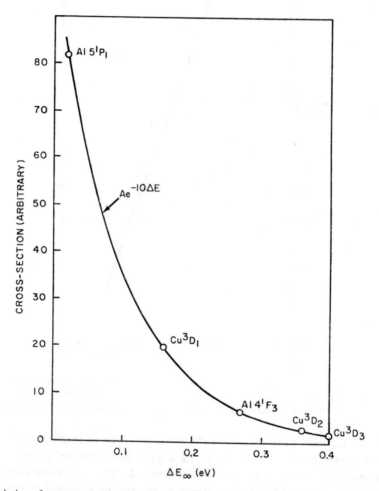

FIG. 2.13. Variation of cross-section for asymmetric charge transfer with energy discrepancy (ΔE_∞) between ground state Ne^+ ions and Al and Cu to excited singly ionized states. (After O. S. Duffendack and K. Thomson,[42] by courtesy of the American Institute of Physics.)

He^+ $^2S_{1/2}$ ion ground state, other transitions from levels approximately 2 eV away are also enhanced.[42] The latter enhancements are possibly due to selective excitation by the Penning reaction discussed in the following section and so could provide an explanation for the large energy discrepancies involved.

Similar enhancements have been observed from the s, p, d and f levels of singly ionized lead in coincidence with the singly ionized Ne^+ ground state[43] as illustrated in Fig. 2.14A. Results given in Fig. 2.14A are shown plotted semi-logarithmically in Fig. 2.14B. An ex-

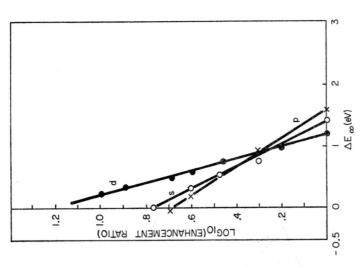

FIG. 2.14B. Variation of $\log_{10}$ (enhancement) of lines from excited s, p and d states of singly ionized Pb with energy discrepancy from the Ne^+ ion ground state. The values of the enhancements are taken from results given in Fig. 2.14A.

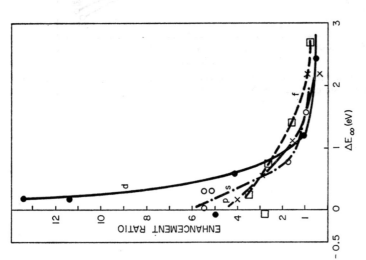

FIG. 2.14A. Variation of enhancement of lines from excited states of Pb^+ close to the Ne^+ ground state with energy discrepancy. (After W. H. Gran and O. S. Duffendack,[43] by courtesy of the American Institute of Physics.)

ponential dependence of the enhancement (or apparent cross-section) on energy discrepancy is exhibited. This behavior is similar to that observed in the near-adiabatic region of an asymmetric charge transfer reaction, in that it follows the exponential behavior of eqn. (2.24).

For an endothermic reaction, ΔE negative and larger than the ion energies expected in an arc discharge, the probability of charge transfer would be expected to be small from energy considerations. Manley and Duffendack,[44] however, find that this is not the case. In the

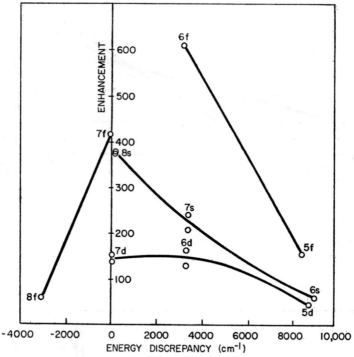

FIG. 2.15. Variation of enhancement with energy discrepancy of lines from excited states of Mg^+ in coincidence with the Ne^+ ground state. (After J. S. Manley and O. S. Duffendack,[44] by courtesy of the American Institute of Physics.)

reaction

$$Ne^+ \, {}^2P_{3/2} + Mg \, {}^1S_0 \rightarrow Ne \, {}^1S_0 + Mg^{+\prime} \, ({}^2F, \, {}^2S, \, {}^2D) \pm \Delta E_\infty \qquad (2.29)$$

at a ΔE_∞ of -0.4 eV, the cross-section for the reaction is still appreciable; although, as shown in Fig. 2.15:

(a) the apparent cross-section, evidenced by line enhancement, increases with decrease in ΔE_∞ (and spin is conserved); and

(b) the cross-section decreases more rapidly for ΔE negative than for ΔE positive. A further dependence of the enhancement on the l-value of the selectively excited states is indicated by the difference in enhancement of lines from levels differing only in configuration at the same ΔE_∞.

There is a possibility that selective excitation in these resonant or near-resonant thermal-energy charge-transfer reactions involves crossing or pseudo-crossing of energy levels in which it is not clear that the adiabatic hypothesis has any relevance.

In spite of the agreement between theory and earlier experimental results, recent work shows that there is a failure in the adiabatic criterion in ion–atom collisions. Cross-sections are observed to depend in a strongly oscillatory manner on the energy of the incident ion for symmetrical as well as asymmetrical charge-transfer reactions. This indicates that curve crossing or level-crossing can indeed be an important mechanism in charge-transfer reactions. Pending the publication of results of further work, the interested reader is referred to recent publications by researchers at the Columbia Radiation Laboratory.[45–47] It is possible that the reactions (2.26) to (2.29) involve curve or level-crossing, just as do a number of laser systems described in Chapter 5 that involve asymmetric charge-transfer in the selective excitation of their upper laser levels. The cross-sections observed in these systems are larger than those realized in atom–atom excitation transfer. For the He^+–Hg, He^+–Zn reactions, they are typically larger than 10^{-15} cm^{2} [48,49] and are comparable to atom–molecule dissociative excitation-transfer reaction and inelastic electron–molecule collision cross-sections.[50] It has only been in the last few years that charge-transfer reactions have received the attention they warranted as selective excitation mechanisms because of their large cross-sections at thermal energies.

2.4. Penning Ionization

The process of ionization of gas atoms or molecules by collisions with metastable atoms is known as the Penning effect.[51] It is represented by

$$A* + B \rightarrow A + B^{+\prime} + e^- \tag{2.30}$$

where $B^{+\prime}$ can be one of a number of excited as well as ionized states. It is energetically possible when the ionization energy of B is lower than the excitation energy of the metastable atom A^* and so can occur at thermal energies since it does not require any input of kinetic energy. Since an electron is one of the products of the reaction, it is available to carry off excess potential energy in the form of kinetic energy and make the reaction essentially a resonant one over a range of energy discrepancies between A^* and $B^{+\prime}$ of up to at least 2 eV. In an elegant set of experiments involving optically pumped He^* $2\,^3S_1$ metastables and ground state Zn, Cd and Sr atoms, Schearer[52] and Riseberg and Schearer[53] have shown that Penning collisions of the type given by eqn. (2.30) apparently obey the Wigner spin rule with total angular spin momentum conserved in the collision. In their work, A^* is a triplet helium metastable He^* $2\,^3S_1$ with the He^*–Cd Penning reaction

$$He^* \, 2\,^3S_1(\uparrow\uparrow) + Cd \, 5s^2 \,^1S_0(\uparrow\downarrow) \rightarrow He \,^1S_0(\uparrow\downarrow) + Cd^{+\prime} \, 5s^2 \,^2D_{5/2}(\uparrow) + e^-(\uparrow), \tag{2.31}$$

the excited product Cd ion in the $5s^2 \,^2D_{5/2}$ state was shown to be spin polarized with (presumably) the electron e^- polarized.[52] In the similar reaction,[53]

$$He^* \, 2\,^3S_1(\uparrow\uparrow) + Sr \, 5s^2 \,^1S_0(\uparrow\downarrow) \rightarrow He \,^1S_0(\uparrow\downarrow) + Sr^{+\prime} \, 5p \,^2P_{3/2}(\uparrow) + e^-(\uparrow), \tag{2.32}$$

spin is also conserved.

Analysis of the spectrum of electrons emitted in another Penning reaction[54]

$$He^* + Hg \rightarrow He\ {}^1S_0 + Hg^{+\prime} + e^-, \tag{2.33}$$

in which the He^* atom is a $He^*\ 2^3S_1$ metastable atom and in which the $Hg^{+\prime}$ ion is produced in the excited $6d\ ^2D_{5/2}$ and $6d\ ^2D_{3/2}$ states as well as in the ion ground state $5s\ ^2S_{1/2}$, shows that the electron carries away all the excess potential energy and that the potential energy

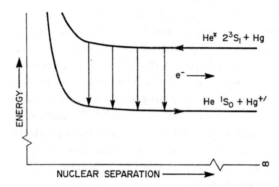

FIG. 2.16. Penning ionization of Hg by He* 2^3S_1: schematic potential energy curves. (Taken from V. Cermak and Z. Herman.[54])

curves during the collision for the reactants and the products must parallel each other and have the form of curves postulated in Fig. 2.16. On the other hand, when the He^* atom of reaction (2.33) is a metastable in the singlet $He^*\ 2^1S_0$ state, which has a higher excitation energy than the $He^*\ 2^3S_1$ state, the ionization reaction yields electrons of *lower* energy in ionization to the $^2D_{5/2}$ and $^2D_{3/2}$ states. This is interpreted as meaning that in the

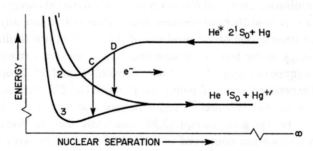

FIG. 2.17. Penning ionization of Hg by He* 2^1S_0: schematic potential energy curves. (Taken from V. Cermak and Z. Herman.[54])

region of nuclear separation where ionization occurs and an electron is ejected, the slope of the potential curve of the reactants must be larger than the slope of the products. The Penning ionization would then correspond to a vertical transition such as C in Fig. 2.17 or a transition D to a possible repulsive curve of the products $He\ ^1S_0 + Hg^{+\prime}$. It follows from a consideration of the curves in Fig. 2.17, that if the slope of the reactant potential curve 2 was *less* than that of the product curve 3, and if the potential well of reactants potential was less than that of the products $He + Hg^{+\prime}$; then after a vertical ionizing transition (at say C) the products would be stuck in the potential well and be unable to dissociate to $He + Hg^{+\prime}$, and the reaction would then be one of associative ionization

$$He^* + Hg \rightarrow HeHg^+ + e^-, \qquad (2.34)$$

the cross-sections of which are comparable to Penning reaction cross-sections.[55] Table 2.2 lists experimental cross-sections for a number of Penning reactions.

TABLE 2.2. *Penning ionization cross-sections*

Reactants	Cross-section (10^{-16} cm^2)	References
He* $2^3S_1 -$ Ar	$7.7 \pm 9\%$	a
$-$ Kr	$9 \pm 20\%$	b
$-$ Xe	$12 \pm 20\%$	b
$-$ Hg	14 ± 3	c
$-$ Cd	$45 \pm 5\%$	d
	65	e
$-$ Zn	29.1 ± 1.5	f
$-$ N$_2$	$7 \pm 15\%$	b
$-$ O$_2$	$14 \pm 7\%$	b
He* $2^1S_0 -$ Hg	14 ± 3	c

(a) C. R. Jones and W. W. Robertson, *J. Chem. Phys.* **49**, 4240 (1968).
(b) W. P. Sholette and E. E. Muschlitz, Jr., *J. Chem. Phys.* **36**, 3368 (1963).
(c) E. E. Ferguson, *Phys. Rev.* **128**, 210 (1962).
(d) L. D. Schearer and F. A. Padovani, *J. Chem. Phys.* **52**, 1618 (1970).
(e) G. J. Collins, R. C. Jensen, and W. R. Bennett, Jr., *Appl. Phys. Letters* **19**, 125 (1971).
(f) L. A. Riseberg and L. D. Schearer, *IEEE J. Quant. Electr.* **QE-8**, 40 (1971).

2.5. Dissociative Excitation Transfer (atom–molecule)

This type of inelastic collision takes the overall form

$$M' + AB \begin{cases} \rightarrow M + A' + B + \Delta E_\infty, & (2.35a) \\ \rightarrow M + A + B' + \Delta E_\infty, & (2.35b) \end{cases}$$

where M' is an excited state of an atom (usually a long-lived metastable state M^*) and AB is a diatomic (or a polyatomic) molecule. The energy discrepancy ΔE_∞, which is equal to the difference in potential energy between the excited state M' and A' (or B') and the dissociation energy of AB, appears as kinetic energy of the dissociated atoms A' or B'. A and B can be like, or unlike atoms, and A' and B' can be similar excited states. The fact that a minimum of three atoms are involved in the collision means that the necessary multidimensional potential-energy representation is more complicated than the two-dimensional representation of the molecular potential of an atom–atom interaction. Considering the fact that the molecular potential for the far simpler case of the He^*–Ne atom–atom interaction has not been calculated, it is not surprising that a quantitative theory does not exist for treating the dissociative excitation-transfer reaction. The collision, however, can be treated qualitatively.

It is assumed that relatively long-range interactions occur between M' and the molecule AB and, according to the Franck–Condon principle, cause vertical transitions between the

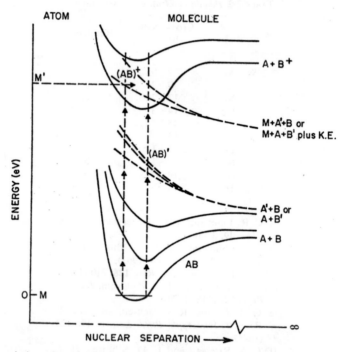

FIG. 2.18. Dissociation of a diatomic molecule into repulsive states in dissociative excitation transfer.

ground state of the molecule and excited repulsive states of the neutral molecule $(AB)'$. Having no equilibrium separation between its constituent atoms, $(AB)'$ then dissociates to give, in general, one atom in an excited state and one in the ground state (Fig. 2.18). The dashed curves represent the repulsive state of the excited molecule $(AB)'$; the dashed vertical lines indicate the region in which transitions occur from the ground state. Since there can be several repulsive $(AB)'$ curves, and since at least three atoms are involved in the collision, there is no strong requirement for close energy coincidence as in an atom–atom excitation transfer as long as total energy is conserved.

The following dissociative excitation-transfer reaction in which a close energy coincidence appears to be applicable has been reported by Leiga and McInally,[56]

$$Kr\ 1s_{2-5}+I_2 \rightarrow Kr\ ^1S_0+I'(6p\ ^4P^\circ_{3/2})+I+\Delta E_\infty\ (487\ cm^{-1}). \qquad (2.36)$$

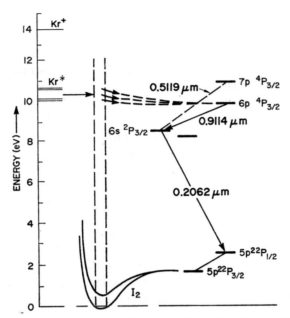

FIG. 2.19. Partial energy-level diagram and pertinent transitions for iodine and krypton illustrating the dissociative excitation-transfer reaction $Kr^* + I_2 \rightarrow Kr + I' + I$.

An extremely large enhancement of emission lines was observed at 0.9114 μm and 0.2062 μm in a pulsed discharge in a high pressure $Kr-I_2$ mixture compared with that in pure I_2. The enhancement factor (ratio of intensities in $Kr-I_2$ mixture and in pure I_2) amounted to as much as 700. A simplified energy-level diagram for iodine and krypton is presented in Fig. 2.19. It illustrates the pertinent energy levels and transitions in the $Kr-I_2$ system.

The $6p$ $^4P^{\circ}_{3/2}$ atomic iodine state decays radiatively to the $6s$ $^2P_{3/2}$ state with emission at 0.9114 μm. The $6s$ $^2P_{3/2}$ state then decays to the $5p^5$ $^2P^{\circ}_{1/2}$ state with emission at 0.2062 μm. The $6p$ $^4P^{\circ}_{3/2}$ upper state of the 0.9114-μm line is within $+487$ cm^{-1} of the metastable Kr^* $1s_5$ level, and within $+6363$ cm^{-1} of the uppermost (quasi-metastable) $1s_2$ state of the four Kr $1s_{2-5}$ states.

A line at 0.5119 μm from the $7p$ $^4P^{\circ}_{3/2}-6s$ $^2P_{3/2}$ transition was not enhanced in a $Kr-I_2$ mixture indicating that the $7p$ $^4P^{\circ}_{3/2}$ level 1196 cm^{-1} above the uppermost $Kr-1s_2$ level was not being selectively excited by the dissociative excitation-transfer process. Enhancement ratios for the 0.9114-μm line, less than those for the 0.2062-μm line under different excitation conditions, are attributed to differing opacities of the pulsed discharge at the two wavelengths. It must be pointed out that it is possible that the $6s$ $^2P_{3/2}$ upper state of the 0.2062-μm line could be excited directly by dissociative excitation transfer and not solely by cascade from the $6p$ $^4P^{\circ}_{3/2}$ state, and that this could account for the differing enhancement ratios. The observed, very broad Doppler width of the 0.9114-μm line compared with that of the 0.2062-μm line would, however, tend to preclude this. This follows since the larger energy discrepancy between the $6s$ $^2P_{3/2}$ and the Kr-1s levels would be likely to contribute significantly more translational energy to atoms in the $6s$ $^2P_{3/2}$ state than to those in the $6p$ $^4P^{\circ}_{3/2}$ state if direct dissociative excitation transfer to it was occurring, and so increase its normal linewidth. The observed broad linewidth of the 0.9114-μm line would further indicate that excess energy of reaction (2.31) is appearing as translational energy of the excited ($6p$ $^4P^\circ_{3/2}$) component of the dissociation reaction. This would be in agreement with work on another dissociative excitation-transfer reaction involved in the Ne–O_2 laser (discussed in Chapter 4).

Cross-sections of approximately 10^{-15} cm² with energy discrepancies of 1–2 eV have been reported for reactions of the dissociative excitation-transfer type involving diatomic molecules.[50]

If the products of the dissociative excitation transfer are diatomic or polyatomic radicals, they can have rotational and vibrational as well as electronic excitation. This can be a selective excitation process for vibrational/rotational states.[57]

2.6. Electron Impact

In a gaseous discharge, numerous collision mechanisms are present. Electron impact collisions upon atoms are by far the most common. Not all of the collisions, however, lead to changes of state. Some are elastic collisions. It is the inelastic collisions that are of primary interest as a means of obtaining selective excitation. They take the simple form

$$A + e^- + K.E. \rightarrow A' + e^- \tag{2.37}$$

where A can be a molecule as well as an atom. In some cases the reaction proceeds via the intermediate stage of a negatively charged compound state A^- that has a lifetime of the order of 10^{-14} sec.

The cross-section Q for direct excitation of an excited state 2 from the ground state 1 by electron impact is given by the Born approximation[58] as

$$Q_{12} \propto \int_{K_{min}}^{K_{max}} |\psi_1 e^{iKr} \psi_2^* \, d\tau|^2 \, K \, dk, \tag{2.38}$$

where K is the propagation vector for the incident electron, and ψ_2 and ψ_1 are the wave functions for the excited state 2 and the ground state 1 respectively. When the energy of the incident electron is much greater than the excitation energy threshold E_n,

$$Q_{12} \propto \left| \int \psi_1 \psi_2^* r \, d\tau \right|^2. \tag{2.39}$$

The term on the right-hand side of the proportionality is proportional to the electric dipole matrix element for a radiative transition between an excited state and the ground state. To a first approximation at electron energies where electron exchange can be neglected, the cross-section

$$Q_{12} \propto A_{21}, \tag{2.40}$$

where A_{21} is the Einstein radiative transition-probability. It follows from this that the largest cross-sections are obtained for optically allowed transitions where A_{21} is large. In the presence of radiation trapping, sizeable excited-state populations can build up in the upper level since decay to the ground state is effectively blocked. Given complete radiation trapping

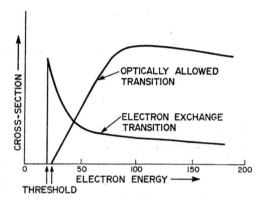

FIG. 2.20. Qualitative illustration of the energy dependence of inelastic electron excitation cross-sections of optically allowed transitions and excitation between levels of the same parity.

between the excited state 2 and the ground state 1, radiative decay can still occur between state 2 and other states that are not highly populated, even though the transitions can have a high transition probability. A considerable advantage accrues from the nature of the electron-impact excitation process. Beyond the onset of the excitation process at the threshold energy, the process is either nonresonant or broadly resonant up to approximately 100 to 150 eV or about 10 times the excitation energy (Fig. 2.20). This means that essentially all electrons with an energy above the threshold energy are available for provid-

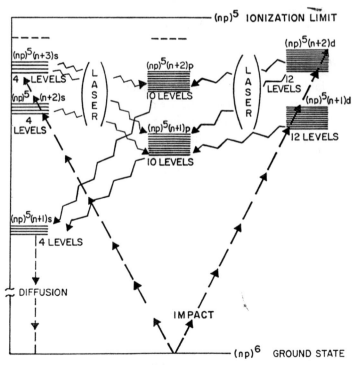

FIG. 2.21. Schematic indication of electron configurations relevant to neon, argon, krypton and xenon. (After W. R. Bennett, Jr.,[59] by courtesy of the Optical Society of America.)

ing excitation with reasonable cross-sections. After passing the maximum, the probability curves fall away slowly. In the case of the noble gases the main selective excitation of neutral states by electron impact from the ground-state occurs through the reactions:[59]

$$(np)^6 + e^- + \text{K.E.} \begin{cases} \longrightarrow (np)^5\, ms + e^-, & (2.41a) \\ \longrightarrow (np)^5\, md + e^-, & (2.41b) \end{cases}$$

where $m = n+1, n+2$, etc., and $n = 2, 3, 4$, and 5 for neon, argon, krypton, and xenon in that order. Under pressure and discharge-tube-diameter conditions sufficient to radiation trap the resonance state, and low enough to prevent excitation to other electronic configurations in collisions, radiative decay can occur only through the transitions labelled LASER in Fig. 2.21. The dominant electron-impact paths are indicated by arrows. This type of selective excitation in a four-level system where radiative decay of the lower laser level is allowed and where lifetimes are favorable can lead to the realization of continuous oscillation.

Electron exchange collisions

Electron exchange collisions can result in excitation in which spin multiplicity is changed. Except in heavy atoms, electron-exchange collisions dominate transitions involving a change of multiplicity with total electron spin differing by one-half from that of the ground state. Such collisions can have large cross-sections close to threshold in the case of transitions involving no change in azimuthal quantum number and have excitation functions with much sharper maxima than optically allowed transitions (Fig. 2.20).

Ionization by electron impact

Calculations for ionization by electron impact using the Born approximation quantum theory are similar to those for excitation and yield similar ionization functions.[60] Above threshold the ionization function of an atom increases linearly with energy to a maximum at several times the ionization energy, and then falls off gradually above it. The review article of Kieffer and Dunn[60] shows that, apart from work on helium and nitrogen, little has been published on ionization into excited states. It is of some relevance to ion lasers to know whether or not ionization functions of excited ions exhibit resonances just as do ionization functions of a number of gases close to threshold.

2.7. Charge Neutralization (ion–ion recombination)

2.7.1. DISSOCIATIVE RECOMBINATION

By "dissociative recombination" is meant the process

$$(AB)^+ + e^- \rightarrow (AB)' \rightarrow A + B \qquad (2.42)$$

in which a molecular ion $(AB)^+$ recombines with an electron and then dissociates into two neutral atoms A and B, either or both of which may be in an excited state A' or B'. The reaction occurs through an intermediate resonant molecular state $(AB)'$ formed by the capture of a slow electron by the ion $(AB)^+$. The intermediate molecular state $(AB)'$ is unstable, and the constituent atoms move apart and gain kinetic energy under the action of their mutual repulsion and remain permanently dissociated. Because the electron mass is small compared with the nuclear mass of $(AB)^+$, the nuclear separation remains unchanged during the electron capture, and a vertical transition can occur to a host of repulsive states

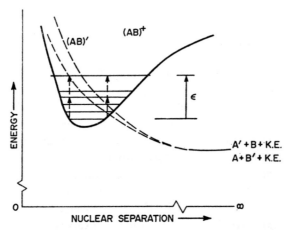

FIG. 2.22. Hypothetical potential-energy curves involved in the dissociative-recombination process, showing repulsive potentials of unstable molecules $(AB)'$ reached by vertical transitions by electron capture of energy from the stable ion AB.$^+$

$(AB)'$ that cross the initial potential curve representing the $(AB)^+ + e^-$ collision. This is illustrated in Fig. 2.22, where ε represents the kinetic energy of e^- and the dashed vertical lines indicate the region within which vertical transitions occur to dashed curves of repulsive $(AB)'$ states. The size of the cross-section for the reaction depends markedly on the distribution of the various states of $(AB)^+$ and $(AB)'$ relative to each other and the kinetic energy of the captured electron.

2.7.2. MUTUAL NEUTRALIZATION

This process is represented in a general form by

$$A^+ + B^- \begin{cases} \longrightarrow A' + B + \text{kinetic energy}, & (2.43a) \\ \longrightarrow A + B' + \text{kinetic energy}. & (2.43b) \end{cases}$$

It is essentially a charge exchange collision in which B^- donates an electron to A^+, by means of which both A^+ and B^- become neutralized to $A' + B$, or $A + B'$, or $A' + B'$.

Bates and Massey[61] have treated the process of mutual neutralization of O^+ and O^- ions as a curve-crossing process as a function of energy discrepancy (ΔE) over a range of most probable crossing points. They arrive at the interesting result that the maximum cross-section does not occur at exact resonance but at an energy discrepancy of about 1 eV.

The energy discrepancy (ΔE) is given by

$$\Delta E = A' + B - B^+ + (A) \tag{2.44a}$$

or
$$A + B' - B^+ + (A) \tag{2.44b}$$

where B^+ is the ionization energy of B, and (A) the electron affinity of A, with A' and B' excitation energies. The cross-section for the process is believed to be very large, between 10^{-12} to 10^{-13} cm^2.

2.8. Line Absorption and Molecular Photodissociation

If the center of a strong emission line falls within the Doppler half-width of a resonance transition, selective excitation of the resonance state by line absorption is possible. The transition, being a resonant one, has available to it for excitation the whole ground-state population that has a velocity distribution within that consistent with the emission line. Line absorption from states other than the ground state can also be an effective selective excitation mechanism if those states are sufficiently populated. The requirement for a close coincidence between the wavelength of the incident radiation and that of the resonance transition, up to the advent of tunable lasers, has led to limited application of this selective excitation mechanism. Line absorption itself is well covered in Mitchell and Zemansky, *Resonance Radiation and Excited Atoms* (Cambridge University Press, 1934, reprinted 1961).

Broad absorption bands of a molecule enable large amounts of optical power to be coupled into a gas. As in dissociative excitation transfer, photodissociation of diatomic molecules gives, in general, one atom in an excited state and one in the ground state, the general process being

$$AB + h\nu \rightarrow A' + B + \text{kinetic energy,}$$

or
$$\rightarrow A + B' + \text{kinetic energy.} \tag{2.45}$$

Photon energies in excess of 3 eV are required to supply the dissociation energy of the molecule AB and the excitation and kinetic energies of the dissociated atoms.

2.9. Radiative Cascade-pumping

Cascading through strong transitions, in particular laser transitions, can be used selectively to excite states that do not favor excitation by direct electron impact from the ground state. Given a strong laser or superradiant transition, together with favorable disposition of energy levels and lifetimes, a cascading series of population inversions can be induced. The technique is useful for inverting those populations that normally are difficult to invert by the processes discussed in the preceding material.

References

1. Basov, N. C. and Krokin, C. N., *Appl. Opt.* **1**, 213 (1962).
2. Javan, A. in *Quantum Electronics* (ed. C. H. Townes), pp. 564–71, N.Y., Columbia University Press, 1960.

3. WARD, R. C., *Aerospace Rept.*, TDR-930 (2250-20), TN-1 (1961).

4. FABRIKANT, V. A., *All Union Electr. Inst.* **41**, 236 (1940).

5. JAVAN, A., BENNETT, W. R., Jr. and HERRIOTT, D. R., *Phys. Rev. Lett.* **6**, 106 (1961).

6. BATES, D. R. and DAMGAARD, A., *Phil. Trans.* A **242**, 10 (1950).

7. CONDON, E. U. and SHORTLEY, G. H., *Theory of Atomic Spectra*, Cambridge Univ. Press, 1965.

8. HEAVENS, O. S., *J. Opt. Soc. Am.* **51**, 1058 (1961).

9. BENNETT, W. R. Jr., KINDLMANN, P. J. and MERCER, G. N., *Appl. Opt.* Supplement II: *Chemical Lasers* (1965), pp. 34–57.

10. KOSTER, G. F. and STATZ, H., *J. Appl. Phys.* **32,** 2054 (1961).

11. FAUST, W. L. and McFARLANE, R. A., *J. Appl. Phys.* **35**, 2010 (1964).

12. MASSEY, H. S. W. and BURHOP, E. N. S., *Electronic and Ionic Impact Phenomena*, Oxford Univ. Press, 1952.

13. TOWNSEND, J. S., *Electrons in Gases*, Hutchinsons, 1948.

14. MOTT, N. F. and MASSEY, H. S. W., *The Theory of Atomic Collisions*, 2nd ed., Oxford Univ. Press, 1952.

15. BATES, D. R., *Atomic and Molecular Processes*, p. 550, N.Y. Academic Press, 1962.

16. HASTED, J. B., *Physics of Atomic Collisions*, Butterworths, 1964.

17. McDANIEL, E. W., *Collision Phenomena in Ionized Gases*, Wiley, 1964.

18. MASSEY, H. S. W., *Repts. Progr. in Physics* **12**, 39 (1962).

19. BATES, D. R. and McCARROL, R., *Phil. Mag. Supplement* **11**, 39 (1962).

20. MITCHELL, A. C. G. and ZEMANSKY, M. W., *Resonance Radiation and Excited Atoms*, pp. 59–65, Cambridge Univ. Press, 1961.

21. BEUTLER, H. and JOSEPHY, B., *Z. Phys.* **53**, 747 (1929).

22. KRAULINYA, E. K., ARMAN, M. G., LIEPA, S. YA., SILIN, YU. A. and YANSON, U. V., *Optics Spectrosc.* **28**, 658 (1970).

23. KLEIN, O. and ROSSELAND, S., *Z. Physik* **4**, 46 (1921).

24. WILEY, E. J. B., *Collisions of the Second Kind*, Arnold, London, 1937.

25. WIGNER, E., *Nachr. Akad. Wiss. Göttingen, II. Math.-Physik Kl.* 375 (1927).

26. YOUNG, R. T., Jr., WILLETT, C. S. and MAUPIN, R. T., *J. Appl. Phys.* **41**, 2936 (1970).

27. MASSEY, J. T., SCHULZ, A. G., HICHHEIMER, B. F. and CANNON, S. M., *J. Appl. Phys.* **36, 658** (1965).

28. SCHEARER, L. D., *Phys. Letters* **27A**, 544 (1968).

29. JONES, R. C. and ROBERTSON, W. W., *Bull. Am. Phys. Soc.* **13**, 198 (1968). (At the 20th Gaseous Electronics Conference, San Francisco, Oct. 1967, material was presented on the potential-barrier behavior of the reaction He* $2^1S_0 + Ne \rightarrow He + Ne\ 3s_2 - \Delta E_\infty (386\ cm^{-1})$, as well as the reaction involving He* 2^3S_1 metastables stated in the abstract in this reference.)

30. BUCKINGHAM, R. A. and DALGARNO, A , *Proc. Roy. Soc.* (London) A **213**, 327 (1952).

31. LANDAU, L. and TELLER, E., *Physik. Z. Sowjetunion* **10**, 34 (1936).

32. HERZFELD, K. F. and LITOVITZ, T. A., *Absorption and Dispersion of Ultrasonic Waves*, p. **260**, Academic Press, 1959.

33. EHRENFEST, P., *Proc. Koninkl. Ned. Akad. Wetenschap.* **16**, 591 (1914).

34. HANCOCK, G. and SMITH, I. W. M., *Appl. Opt.* **10**, 1827 (1971).

35. CALLEAR, A. B., *Appl. Opt.* Supplement II: *Chemical Lasers* (1965), pp. 145–70.

36. DRUKAREV, G. F., in *Fifth International Conference Physics of Electronic and Atom Collisions, Leningrad, 1967*, p. 10, Nauka, Leningrad, 1967.

37. PEREL, J. and DALEY, H. L., *Phys. Rev.* A**4**, 162 (1971).

38. Reference 16 (Hasted, J. B.), pp. 420–3.

39. HASTED, J. B. and LEE, A. R., *Proc. Phys. Soc.* (London), **79**, 702 (1962).

40. HASTED, J. B., *Adv. Electronics and Electron Phys.* **13**, 1 (1960).

41. HASTED, J. B., *J. Appl. Phys.* **30**, 25 (1959).

42. DUFFENDACK, O. S. and THOMSON, K., *Phys. Rev.* **40**, 106 (1933).

43. GRAN, W. H. and DUFFENDACK, O. S., *Phys. Rev.* **51**, 804 (1937).

44. MANLEY, J. H. and DUFFENDACK, O. S., *Phys. Rev.* **47**, 56 (1935).

45. DWORETSKY, S., NOVICK, R., SMITH, W. W. and TALK, N., *Phys. Rev. Lett.* **18**, 939 (1967).

46. LIPELES, M., NOVICK, R. and TALK, N., *Phys. Rev. Lett.* **15**, 815 (1965).

47. LIPELES, M., NOVICK, R. and TALK, N., *Phys. Rev. Lett.* **15**, 690 (1965).

48. DYSON, D. J., *Nature* **207**, 361 (1965).

49. RISEBERG, L. A. and SCHEARER, L. D. *IEEE J. Quantum Electronics* **QE-7**, 40 (1971).

50. BENNETT, W. R., Jr., *Appl. Optics*, Supplement I: *Optical Masers* (1962), pp. 24–62.

51. KRUITHOF, A. A. and PENNING, F. M., *Physica* **4**, 430 (1937).

52. SCHEARER, L. D., *Phys. Rev. Lett.* **22**, 629 (1969).

53. SCHEARER, L. D. and RISEBERG, L. A., *Phys. Rev. Lett.* **26,** 599 (1971).
54. CERMAK, V. and HERMAN, *J. Chem. Phys. Lett.* **2,** 359 (1968).
55. HERMAN, Z. and CERMAK, V., *Coll. Czech. Chem. Commun.* **31,** 649 (1966). As mentioned in ref. 54, p. 359.
56. LEIGA, A. G. and McINALLY, J. A., *J. Opt. Soc. Am.* **57,** 317 (1967).
57. SHULER, K. E., CARRINGTON, T. and LIGHT, J. C., *Appl. Optics,* Supplement II: *Chemical Lasers* **8,** 81–104 (1965).
58. BORN, M., *Z. Physik* **38,** 803 (1926).
59. BENNETT, W. R., Jr., *Appl. Optics,* Supplement II: *Chemical Lasers* **8,** 3–33 (1965).
60. KIEFFER, L. J. and DUNN, G. H., *Revs. Mod. Phys.* **38,** 1 (1966).
61. BATES, D. R. and MASSEY, H. S. W., *Phil. Trans. Roy. Soc.* (London) A **239,** 269 (1943).
62. STEDEFORD, J. B. H. and HASTED, J. B., *Proc. Roy. Soc.* A **277,** 466 (1955).

CHAPTER 3

Gas Discharge Processes

3.1. Introduction

The possibility that a gas discharge could be used to give a nonequilibrium population distribution necessary to support laser oscillation appears to have been first recognized by Fabrikant[1,2] as early as 1939. It was not until 1959 and 1960, however, that specific proposals were made for laser systems utilizing the mechanisms operating in a gas discharge. These proposals were made by Javan,[3] Sanders,[4,5] Butayeva and Fabrikant,[6] and Basov and Krokhin.[7] Although Butayeva and Fabrikant[6] and Fabrikant[8] did report achievement of a "negative absorption coefficient" it was left to Javan, Bennett and Herriott[9] to be the first to achieve actual oscillation in a gas-discharge medium. In this first practical laser, rf-excitation of a mixture of helium and neon was used to produce population inversion and oscillation in the near infrared in neutral species of neon.

Since the initial proposals, all the following methods of excitation have been used in successful gas-laser systems:

(1) the glow discharge, including the
  (a) negative glow and
  (b) positive-column regions;
(2) rf-excitation;
(3) hollow-cathode discharges;
(4) pulsed discharges;
(5) miscellaneous methods, including thermal excitation and high-velocity expansion-type excitation. Though not in the form of a gas discharge the latter is mentioned because of its importance in gas laser development.

This chapter covers excitation methods (1) through (4), which are strictly gas-discharge methods, and can be related to the selective excitation mechanisms covered in Chapter 2. Miscellaneous methods of excitation will be discussed in Chapter 5 where specific gas laser systems of the molecular type are described.

3.2 The Glow Discharge

There have been a number of excellent analyses of the glow discharge, e.g. those of von Engel,[10] Francis,[11] Parker,[12] and Brown.[13] Since an analysis of the glow discharge covers most aspects of laser discharges, including those of hollow-cathode, transversely excited, and pulsed afterglow discharges, an abbreviated qualitative description of its salient features at low pressure will be given here.

Figure 3.1, after von Engel,[10] shows the spatial distribution of dark and luminous zones,

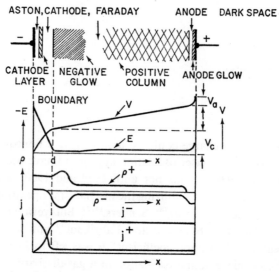

FIG. 3.1. Spatial distribution of dark and luminous zones, electric field, positive and negative space-charge densities and current densities in a glow discharge. (After A. von Engel,[10] by courtesy of the Clarendon Press.)

the electric field (E), potential (V), space-charge densities (ϱ^+ and ϱ^-), and current densities (j^- and j^+) in an abnormal glow discharge at a pressure between about 0.1 to 1 torr. Let us consider this spatial distribution produced in a glow discharge in which the cathode is unheated.

3.2.1. THE NEGATIVE GLOW

The negative glow is a beam-maintained, recombination-dominated plasma region. Whereas it is maintained by high-energy electrons that have been emitted by the cathode and accelerated across the high-field region of the cathode dark space, it is dominated by recombination mechanisms involving secondary, low-energy electrons and thermal-energy positive ions that have been produced by the high-energy electrons in ionizing collisions in ion-pair production. Going from the cathode to the anode, the negative glow is the first, large, bright, easily seen, luminous region that occurs in the glow discharge. Figure 3.1 shows that the electric field is higher in the cathode dark space that precedes the luminous-glow region of the negative glow than anywhere else in the glow discharge. The field is a minimum in the negative glow which is almost field-free due to the build-up of space charge.

Von Engel[10] has noted that if a plane cathode is mounted in a large spherical bulb and rotated with respect to a fixed anode, the negative glow swings round as if it were fixed to the cathode surface and maintains its longitudinal extent normal to the cathode surface, which shows that the negative glow is beam-maintained. The luminous region of the positive column merely fills the remaining space between the Faraday dark space and the anode. Whereas the positive column is affected by the dimensions of the discharge vessel the negative glow is not. The negative glow is, however, affected by the area of the cathode which is contributing electrons to the discharge, and by the current density and the potential across the discharge. The distinction between the so-called abnormal and normal glow discharge is due to difference in potential and current density that arises for a given cathode area. At low current densities, depending on the gas, the lateral extent of the negative glow can be less than the area of the cathode surface facing the anode. As the current is increased the lateral extent of the negative glow extends until it is as large as the whole area of the cathode, but the potential across it remains constant. All discharges to this point are referred to as normal glow discharges. Once the lateral extent of the glow is as large as the cathode area an increase in applied voltage is necessary to produce a further increase in discharge current, and a positive resistance characteristic is observed. Further increases in applied voltage and current cause the cathode dark space to contract and also cause the negative glow to extend in the direction of the anode, to increase in luminosity, and to be very well defined visually. The discharge then corresponds to what is known as an abnormal glow discharge. As shown in Fig. 3.1, the lateral extent of the negative glow corresponds to that of the cathode surface. As a subsequent discussion will show, the important positive-column region of the glow discharge is diffusion-maintained and is affected by the dimensions of the containment vessel, and is distinctly different from the negative-glow region which is electron beam-maintained. It is this distinction that determines the distinct physical differences between the two plasma regions of the negative glow and the positive column.

Let us consider an electron with low energy emitted from the cathode, perhaps thermally, or by a positive ion impinging on the cathode after having been accelerated across all or part of the field of the cathode dark space. Initially the electron is in the region of high field strength and is accelerated away from the cathode. In the Aston dark space, the electron has insufficient energy to excite or ionize; however, on reaching the cathode-layer region it has sufficient energy to produce excitation, and a small luminous region is produced. According to von Engel,[10] in the luminous cathode-layer region, the electron acquires an energy corresponding to the maxima of excitation functions. Beyond the cathode layer, the energy of the electron increases rapidly beyond values maximizing excitation, and little visible light is emitted. This is the region of the cathode dark space. The energy of the electrons in this region, depending on the potential across the cathode dark space, is in the range of 20 to 150 eV at which values their ionizing efficiency is rapidly increasing or is optimized. From Fig. 3.1 it is seen that the electric field (E) falls linearly with distance from a region close to the cathode, though of course the potential between the cathode and any point in the cathode dark space continues to rise until it reaches V_c at the beginning of the negative-glow region in an abnormal glow discharge (or V_n in a normal glow discharge). In the cathode dark space in a normal glow discharge, it appears that the average electric field per unit pressure (E/p), which is equal to the cathode fall V_n divided by the thickness of the cathode dark space and the pressure, sets itself to a value that maximizes the ionization. In an abnormal glow discharge, the values of E/p are about 3 to 10 times larger than those in a normal glow discharge. This can be inferred from tables of cathode fall in potential (V_n), reduced thickness of cathode dark space (d_n); curves of reduced thickness of dark space (d_c) as a function of the cathode fall in potential (V_c); and electron ionization coefficient (α/p) as a function of electric field (E/p) for various gases given by von Engel.[10]

Near the boundary between the cathode dark space and the negative glow, the electric field is weak, and only the fast electrons that have not lost energy in inelastic collisions will be able to ionize. Although a relatively small group of fast primary electrons reaches the edge of the negative glow, the majority of the electrons have energies only very slightly above that required to produce ionization, as shown in Table 3.1. This relatively small group of fast electrons loses its energy by inelastic ionizing collisions with gas atoms as it enters the weak-field region of the negative glow. It is these fast (beam) electrons that are responsible for the longitudinal extent of the negative glow, and for the primary ionization in the nega-tive-glow region, and are a prerequisite for its existence.

TABLE 3.1. *Electron energy distribution in the negative glow*[a]

Types of electrons	Ultimate	Secondary	Primary
Energy (eV)	0.6	7.3	25
Electron concentration (cm^{-3})	4×10^9	5×10^7	5×10^6

[a] Helium pressure, 1 torr; $j = 2 \times 10^{-5}$ A-cm^{-2} (after Pringle and Farvis[14]).

As the fast electrons lose their energy in ionizing collisions in their progress through the negative glow, they pass below energies maximizing ionization and into energies maximizing excitation and so produce extensive excitation in inelastic collisions. As the energy and density of fast or primary electrons decreases as the negative glow is traversed so the density of slow or ultimate electrons increases. Figure 3.1 shows that in the center of the negative glow the concentration of both negative and positive charges is considerably greater than in the positive-column region. The electron concentration can be 20 times larger than in the positive column.[15] Since the plasma of the negative glow is neutral, the positive-ion con-centration will also be about 20 times higher than in the positive-column region.

Further on through the negative glow, electron–ion recombination begins occurring because of the favorable concentrations and energies of both electrons and positive ions. The concentrations decrease, and the Faraday dark space starts to develop. Electrons here again are accelerated by the field, which is much less than in the cathode dark space, and acquire sufficient energy to excite again and ionize. The uniform positive column as illus-trated in Fig. 3.1 then commences, or the first striation occurs.

The actual longitudinal extent of the negative glow in the direction of the anode has been shown by Brewer and Westhaver[16] to be equal to the range of beam-electrons calculated

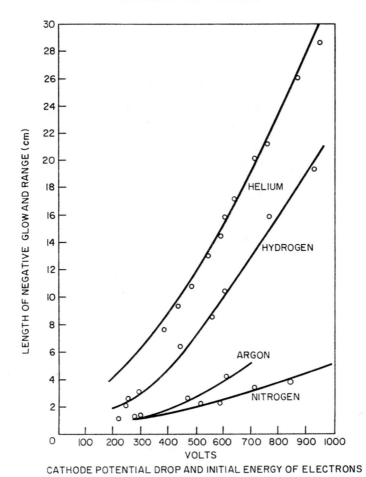

FIG. 3.2. Calculated range of electrons as a function of electron energy (lines) and observed length of nega-tive glow as a function of the full cathode-potential fall (points). (After A. K. Brewer and J. W. Westhaver,[16] by courtesy of the American Institute of Physics.)

by assuming they enter the negative glow having the full cathode-fall potential (as shown in Fig. 3.2). The range L of the primary electrons that enter the negative glow, hence the extent of the negative glow, can be calculated from the relationship due to Persson:[17]

$$L = (pV_i)^{-1} \int_0^V \frac{dV}{\beta\phi(V)} \tag{3.1}$$

where $\phi(V)$ is the inverse ionization mean free path in cm^{-1} at 1 torr, V_i is the ionization potential of the gas, p is the gas pressure in torr, V is the electron energy in electron volts at the end of the negative glow, and β is a constant greater than one. The inverse ionization mean free path (ionization efficiency) can be obtained from curves given by von Engel[10] which are shown in Fig. 3.3. In the high-energy range beyond the maxima in the curve, the inverse ionization mean free path can be approximated by the empirical relationship

$$\phi(V) = C/V. \tag{3.2}$$

65

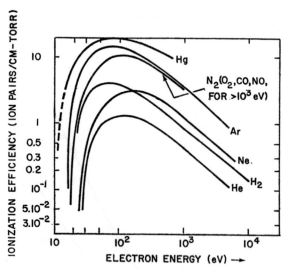

FIG. 3.3. Inverse ionization mean free path $\phi(V)$ as a function of the electron energy. $\phi(V)$ corresponds to the number of electron–ion pairs that are produced per cm per electron at a pressure of 1 torr. (After A. von Engel,[10] by courtesy of the Clarendon Press.)

Substituting for $\phi(V)$ from (3.2) in (3.1) we obtain

$$L = (V^2/2\beta V_i Cp).\tag{3.3}$$

The constant C for helium obtained from Fig. 3.3 is approximately 430 at 1 torr. From eqn. (3.3) and from the fact that β, V_i, and C are constants for a particular gas (C depends on the electron energy; $\beta \approx 1.43$ for He, ≈ 1.6 for H₂, and 1 for N₂), it can be deduced that the flux f of primary electrons in a glow discharge (which should be a function of d/L where d is the distance from the edge of the cathode dark space) will be a function of pd/V^2 for all pressures, distances, and currents. Over the pressure range in helium of 0.3 to 1 torr, a discharge current range of 0.1 to 1.0 A, and distances (d) of 3.75, 7.5 and 11.25 cm from the edge of the cathode dark space into the negative glow, Caron and Russo[18] clearly showed experimentally that the flux of primary electrons is indeed a function only of pd/V^2. Figure 3.4 is a semilogarithmic composite of all their data.[18] It clearly demonstrates that the flux is a function only of pd/V^2, and that the function is well described by the exponential relationship

$$f = A \exp\left[-\alpha(pd/V^2)\right]\tag{3.4}$$

where $\alpha = 2\beta V_i Cp$. The solid line shown in Fig. 3.4 is a best-fit to the data at the largest value of d.

The emission of radiation in the negative glow is in accord with the description of the negative glow given here. Emission from atomic species occurs most strongly on the cathode edge of the negative-glow region where primary ionization dominates and the ion concentration is highest[19] whilst emission occurs from molecular species further into the glow (Hurt[20]).

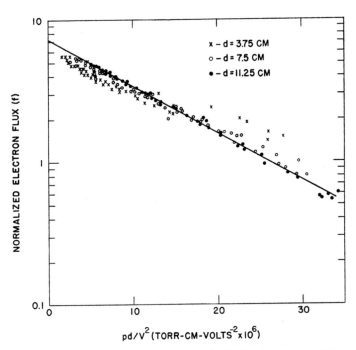

FIG. 3.4. Normalized primary electron flux (f) as a function of pd/V^2. (After Caron and Russo,[18] by courtesy of the American Institute of Physics.)

3.2.2. THE POSITIVE COLUMN

The positive column is the second luminous region of any size in the glow discharge as one goes from cathode to anode. In a glow discharge of any extent, it is usually the largest luminous region, and may present a uniform appearance or may be broken up or striated. Unlike the negative-glow region, it is a plasma region whose properties are determined by diffusion processes, and it is therefore affected by the presence of its containment vessel. That is to say, it exhibits a "wall effect".

As shown in Fig. 3.1, the axial component of the electric field in the positive column is constant and the positive ion and electron concentrations are equal throughout. Because of the smaller mobility of the positive ions, most of the current is carried by electrons. The uniform appearance of the glow is due to the random velocity (v) of the electrons that they have as a result of numerous elastic collisions in the electric field and not to their drift velocity (v_d) in the field direction. The ratio of the electron drift velocity to the random velocity is usually less than 0.1 for values of electric field per unit pressure of less than 20 V-cm^{-1} torr^{-1}.

In the steady, uniform, positive column the electric field has a value such that the production rate of electrons and ions is equal to the diffusion loss of charged particles to the walls of the discharge tube containing the plasma.

The elementary theory of the positive column gives a good quantitative description of the relation between the axial and radial electric field, the discharge-tube radius, and the nature of the gas.[10] The approach is to derive the radial distribution of charge concentration by equating the ionization rate and the rate of loss of charge by diffusion. The average electron temperature (T_e) necessary to maintain the required rate of ionization is then calculated assuming that the energy distribution of the electrons is Maxwellian. The axial field strength needed to produce this electron temperature and balance the energy losses can then be calculated.

The theory requires that the following conditions hold in the positive column:

(1) charge neutrality: $n_e = n_i$, where n_e and n_i are the electron and positive ion concentrations, respectively;
(2) the electron and ion mean free paths λ_e and λ_i are much less than the discharge tube radius R;
(3) positive ions and electrons diffuse together to the walls of the tube (ambipolar diffusion) with the electron concentration large enough so that the Debye length, $kT_e/4\pi n_e e^2$, is less than R.

It is assumed that ionization occurs only as a result of direct inelastic electron impact on ground state atoms and that volume ionization such as by the Penning effect (Chapter 2, Section 2.4) is relatively not an important ion-producing process in the positive column. The salient points of the theory are as follows: If q is the number of ions produced per cm³ per second and is given by

$$q = \alpha n_e \qquad (3.5)$$

where n_e is the electron concentration and α is the number of ions produced per second per electron; then, if cylindrical geometry is assumed,

$$\frac{kT_e}{e}\left(\frac{d^2n_e}{dr^2} + \frac{1}{r}\frac{dn_e}{dr}\right) + \left(\frac{\alpha}{b^+}\right)n_e = 0 \qquad (3.6)$$

where T_e is the average electron temperature and b^+ is the ion mobility. The solution to this equation is a zero-order Bessel function having the boundary conditions $n_e = 0$ at the wall where $r = R$, leading to

$$\frac{\alpha}{b^+} = \frac{kT_e}{e}\left(\frac{2.405}{R}\right)^2 \qquad (3.7)$$

where the 2.405 is the first zero of $J_0(x) = 0$. Both α and b^+ depend on T_e, p, and the nature of the gas. The rate of ionization per electron α in a Maxwellian distribution of electrons of concentration n_e in a gas whose ionization potential is V_i is given by von Engel[10] as

$$\alpha \approx 600(2/\pi)^{1/2}(e/m)^{1/2} apV_i^{3/2}x^{-1/2}e^{-x} \qquad (3.8)$$

where $x = eV_i/kT_e$; the initial slope of the ionization mean free path curve a (in volts⁻¹) can be obtained from Fig. 3.3 or from Table 3.7 in von Engel,[10] and V_i and e/m are in esu. From eqn. (3.7)

$$\frac{kT_e}{e}\left(\frac{2.4}{R}\right)^2 = \frac{600(2/\pi)^{1/2}(e/m)^{1/2} ap^2V_i^{3/2}x^{-1/2}e^{-x}}{(b^+p)}. \qquad (3.9)$$

This reduces to

$$e^x/x^{1/2} = 1.2\times10^7 \, (cpR)^2 \qquad (3.10)$$

where c is a constant equal to $\left|\dfrac{aV_i^{1/2}}{b^+p}\right|^{1/2}$ with V_i in volts, a in cm, and p in torr; and pR is in torr-cm.

Equation (3.10) is a universal relationship between T_e (or x) and cpR for all gases. The nature of the gas is contained in the value of the constant c, which involves the efficiency of ionization (the coefficient a), the ionization potential V_i, and the product b^+p which is equal to a constant. Equation (3.10) is represented in Fig. 3.5 in which T_e/V_i is plotted as a function of cpR. In the figure caption approximate values of c are given for a few positive ions.

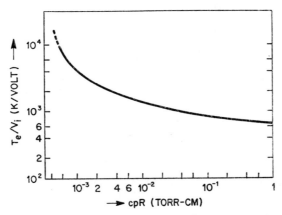

FIG. 3.5. Normalized electron temperature T_e/V_i as a function of the reduced tube radius Rp. Values for c: He, 4×10^{-3}; Ne, 6×10^{-3}; Ar, 4×10^{-2}; Hg, 7×10^{-2}; H$_2$, 1×10^{-2}; N$_2$, 4×10^{-2}. (After A. von Engel,[10] by courtesy of the Clarendon Press.)

From the figure it is clear that T_e can be varied over a wide range by varying the product cpR. Low values of cpR give a high electron temperature, and high values of cpR give a lower electron temperature and a value of T_e/V_i that approaches a limiting value of about 600 K per volt. The increase in electron temperature at low values of cpR (or pR) occurs because of the increased loss of ions by diffusion to the walls of the tube. The electron temperature has to increase in order to maintain the increased loss of positive ions. The opposite applies at high values of cpR (or pR), where the necessary electron temperature is reduced.

Figure 3.6, a plot somewhat similar to Fig. 3.5 restricted to the noble gases He, Ne, Ar, Kr, and Xe, gives the variation of electron temperature with the product of pressure and tube diameter (pD in torr-mm). These were calculated by using collected published values[21] of rates of ionization α and ion mobilities (He$^+$ in He, Ne$^+$ in Ne, Ar$^+$ in Ar, etc.) that are more recent than those used by von Engel to give the values of c given in the caption of Fig. 3.5.

It can be seen that there is an appreciable difference between the electron temperatures for helium and neon and those of argon, krypton, and xenon at all values of pD between 2 and 400 torr-mm. This is due to the relatively high ionization efficiency of electrons and low mobility of positive ions in argon, krypton, and xenon compared with the electron ionization efficiency and mobility of positive ions in helium and neon.

The relationship between the electric field E in the positive column and the electron temperature T_e can be obtained by equating the energy the electrons gain from the electric-

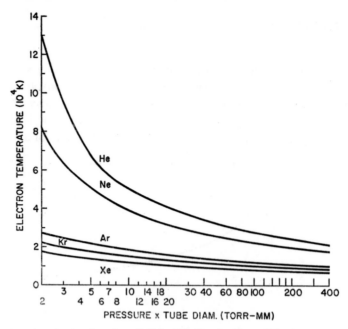

FIG. 3.6. Electron temperature as a function of pD for He, Ne, Ar, Kr and Xe. (By courtesy of R. T. Young, Jr., unpublished work.)

field per second to the energy they lose by collisions. The energy gained per second by a single electron in the electric field is given by the product of the force on the electron eE and the drift velocity v_d. If the electron loses in one collision on the average a fraction F of its energy, then the energy it loses per second is the product of F, the mean energy $\left(\frac{1}{2}mv^2\right)$, and the electron–atom collision frequency (f_e). Thus

$$eEv_d = F\frac{mv^2}{2}f_e = \frac{3}{2}FkT_ef_e. \tag{3.11}$$

Since the collision frequency $f_e = v/\lambda_e$ where λ_e is the mean free path of the electron, and the ratio of the drift velocity to the mean random velocity for a Maxwellian distribution of electron velocities[22] is equal to $(\pi/4)^{3/8}F^{1/2}$,

$$T_e = \frac{2}{3}(e/k)\frac{\lambda_e E}{F^{1/2}}(\pi/4)^{3/8}. \tag{3.12}$$

It is clear that

$$T_e \propto \frac{\lambda_e E}{F^{1/2}}. \tag{3.13}$$

Further, since λ_e is inversely proportional to the gas pressure p

$$T_e \propto \frac{E/p}{F^{1/2}}. \tag{3.14}$$

Expression (3.14) shows that the electron temperature is directly proportional to E/p, and that it is larger for a decreasing fractional collision loss factor F.

The above analysis applies to a single gas, but with suitable modifications it can be used also to determine the electron temperature in the positive column in a binary mixture. Essentially it means that the single-ionization rate and positive ion diffusion-rate terms, α and b^+, in (3.9), modify to

$$\left(\frac{\alpha_1}{b_1^+} + \frac{\alpha_2}{b_2^+}\right) \tag{3.15}$$

to account for the two gases in the mixture, instead of the (α/b^+) for the single gas. Young[23] has used a method of Dorgela, Alting, and Boers[24] in an extension of the elementary theory of the positive column of a single gas[25] to show that the electron temperature in a binary mixture is a function only of the pD product, the nature of the gases, and their relative concentrations, and that it is independent of the discharge current or the plasma density. The

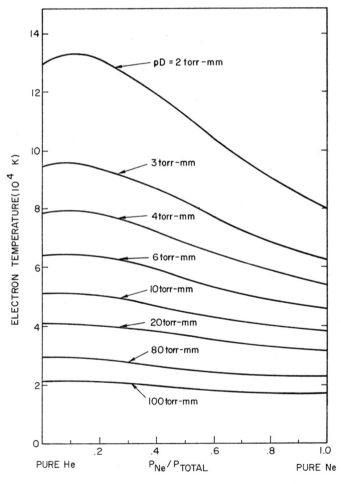

FIG. 3.7. Electron temperature as a function of the ratio of neon pressure to total pressure for various values of pD.

formula that Young derives, from which the electron temperature can be obtained, is

$$f_1 c_1^2 (pD)^2 (eV_1/kT_e)^{-1/2} [1 + \tfrac{1}{2}(eV_1/kT_e] \exp(-eV_1/kT_e)$$
$$+ f_2 c_2^2 (pD)^2 (eV_2/kT_e)^{-1/2} [1 + \tfrac{1}{2}(eV_2/kT_e)] \exp(-eV_2/kT_e)$$
$$= \left[(2/300\pi)^{1/2} \left(\frac{1}{2.4} \right)^2 (e/m)^{1/2} \right]^{-1} = 1.72 \times 10^{-7}, \quad V^{1/2} \text{ sec cm}^{-1}, \quad (3.16)$$

where $c_1^2 = (a_1 V_1^{1/2}/b_1^+ p)$ and $c_2^2 = (a_2 V_2^{1/2}/b_2^+ p)$ sec cm^{-3} torr^{-2}. The symbols and units are:
a_1, a_2, initial slopes of ionization efficiencies in ion pairs/cm-V/torr/primary electron; b_1^+
and b_2^+, mobilities of positive ions of gas 1 and 2 in cm/sec/V/cm; f_1, f_2, ratio of partial
pressures to total pressures; V_1 and V_2, ionization potentials in V; p, total pressure in
torr; D, diameter of tube in cm; eV, ionization energy in J; k, Boltzmann's constant in
J/K; T_e, electron temperature in K; e/m, electron charge to mass ratio in esu. The factor
300 arises from conversion of esu of potential to volts. The ion mobilities, $b_1^+ = 760\mu_1/p$
and $b_2^+ = 760\mu_2/p$, where μ_1 and μ_2 are the mobilities at STP. The μ's are evaluated by
application of Blanc's law:

$$1/\mu_1 = f_1/\mu_{11} + f_2/\mu_{12} \quad \text{and} \quad 1/\mu_2 = f_1/\mu_{21} + f_2/\mu_{22} \qquad (3.17)$$

where μ_{11} = mobility of ion 1 in gas 1; μ_{12}, of ion 1 in gas 2; μ_{21}, of ion 2 in gas 1; and
μ_{22}, of ion 2 in gas 2. This is given in detail as in Young[23] because some of the symbols
differ from those used in the earlier equations (3.5) to (3.10), and because the completeness
of the units makes it readily usable.

Results of calculations of T_e utilizing eqn. (3.16) for a mixture of helium and neon are
shown in Fig. 3.7 where T_e is plotted as a function of the ratio of neon to total pressure for
various products of pressure and tube diameter. Satisfactory agreement exists between the
calculated values derived from the simple theory and shown in Fig. 3.7 and experimentally
determined values of T_e in He–Ne mixtures.[26-28] The extent of the agreement is as indicated
in Table 3.2.

TABLE 3.2. *Comparison of calculated and
experimental values of electron temperature
in a 5:1, He–Ne mixture*

pD (torr-mm)	Electron temperature (K)	
	Calculated[28]	Experimental[26]
2	132,000	119,000
3	94,800	92,000
4	78,600	77,000
6	63,500	64,000

Curves of further calculated values of electron temperatures in noble-gas mixtures of
He–Ar, He–Kr, He–Xe; Ne–Ar, Ne–Kr, and Ne–Xe for various pD values are given as
Appendix Fig. A.1 to A.6. These figures, however, should be used with caution as it is possible that volume ionization by the Penning effect and associative ionization that are not
included in the calculation can be significant ion-producing mechanisms. This is likely to

be so at low partial pressures (less than 10 percent) of the gas that has the lower ionization potential of the two gases that comprise the mixture in all the stated mixtures in Figs. A.1 through A.6. It will be remembered that it was a requirement of the theory that direct ionization by electron impact from the ground state be the only ion-producing mechanism in the positive column. Actually, of the mixtures mentioned, it is only in the case of the He–Ne mixture that the metastable levels of the first gas are lower in potential than the ionization potential of the other so that the Penning reaction is not energetically possible.

In Fig. 3.7 and in Figs. A.1 through A.6 in the Appendix, calculations of T_e have not been included for values of pD below 2 torr-mm. This is because below a pD of about 2 torr-mm, the Schottky theory of ambipolar diffusion on which the calculations are based is not expected to be valid, since

(1) the electron mean free path can be comparable to the diameter of the discharge tube, so that diffusion is not applicable, and
(2) an extended negative space sheath, that develops on the inside wall of the discharge tube, reflects electrons back into the plasma and takes control of the radial motion of charged particles.

At high values of pD above about 20 torr-mm it has not been determined experimentally how accurate the calculated values of T_e actually are. In the calculation it is implicit that the ions be identified and furthermore that ions created in the discharge diffuse unchanged to the wall of the tube. At high values of pD, for small values of D giving pressures above 5 torr, even in a single noble gas, the relative percentages of molecular and atomic positive ions in the positive column of a glow discharge are not well known. Unfortunately, the calculations assume that the direct ionization is to the atomic state alone, and that mobilities used in the calculation are atomic mobilities. In gas mixtures other uncertainties arise. The ions of one gas can transfer charge to atoms of the other gas and so alter the effective ion mobilities.

Both relationships (3.10) and (3.16) for the single gas and binary mixture respectively, show that T_e is independent of the discharge current or the plasma density. This is true only as long as ionization does not proceed by stepwise processes and remains as direct ionization from the ground state. At high current densities appreciable excited and metastable atom densities can occur and ionization by stepwise electron-impact collisions from these excited states can become an appreciable fraction of the total ionization. The result is that T_e falls at higher current densities because of the facilitated ionization that has become current-density dependent. This effect is illustrated in Fig. 3.8 where T_e as a function of discharge current for various pD values in neon is shown.[29] When metastable concentrations can be reduced by the addition of another gas, such as the addition of neon to a discharge in helium, T_e does not decrease with increase in discharge current and remains constant over a considerable current range.

Finally, the assumption throughout the treatment of the positive column has been that the electron energy distribution is Maxwellian. That Maxwellian distribution occurs in typical neutral laser discharges in the noble gases at pD values that give, from the theory, average electron temperatures of less than 10 eV, is open to question. With the average electron temperature low in comparison with excitation and ionization potentials of the noble gases, the actual shape of the electron energy distribution in the high-energy tail of

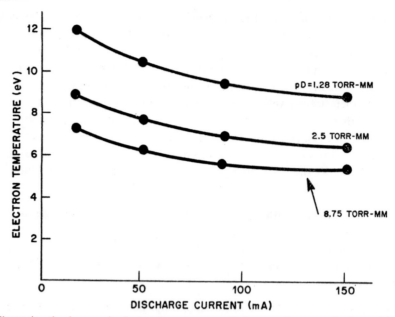

FIG. 3.8. Illustrating the decrease in electron temperature with increase in current in the positive column of a discharge in neon, at various pD values ($D = 25$ mm). (After Revald,[29] by courtesy of the Optical Society of America.)

the distribution can be extremely important in determining excitation and ionization processes that determine the properties of the plasma and its utilization as a gas laser medium. In some laser discharges it is apparent that the electron energy distribution cannot be characterized by a single Maxwellian distribution since it is depleted of high-energy electrons,[28] or is otherwise non-Maxwellian.[30, 31] In others, groups of high-energy electrons with well-defined energies have been found to exist,[32] similar to those observed in low pressure, non-laser discharges using refined probe techniques.[33, 34]

3.3. RF-Discharges

The early work on electrical excitation of gas lasers was concentrated on those systems that exhibited special advantages. The most widely used form of excitation was that of rf-excitation. It was chosen to avoid

(1) contamination and destruction of the internal cavity mirrors that were used prior to the introduction of Brewster's angled windows;
(2) gas contamination;
(3) gas clean-up that was known to occur with internal electrodes;
(4) cataphoretic effects.

A fifth reason was that rf-excitation of a discharge was believed to result in a higher electron temperature than would be realized using dc-excitation and that possibly the electron energy distribution in the rf-discharge was biased to higher energies than that in the positive-column region of the dc-excited discharge under the same conditions.

74

In the light of present-day knowledge the only advantage an rf-excited discharge has over a correctly designed dc-excited, positive-column discharge is that of a reduced cataphoretic effect when using gas mixtures. For the one advantage, it has more than a few disadvantages. The more important ones are: differential gas clean-up, difficulties in specifying the discharge parameters and maintaining a discharge in narrow-bore tubes of less than 3 mm i.d., and complexity of, and requirement for, specialized electrical circuitry.

3.3.1. GENERAL CONSIDERATIONS

A physical picture with maximum insight into the processes that occur in the rf-discharge with minimum recourse to diffusion theory might be as follows.

Ionization, the necessary condition for obtaining a discharge in a gas subjected to alternating electric fields, differs from ionization in steady dc-fields in the following ways:

(1) the field reverses periodically, thus charges may not be swept out of the volume on to the walls of electrodes, losses are reduced, and a sustained discharge can be maintained with low electric fields;

(2) secondary charged particles released from or reflected by the walls do not contribute to the growth of a discharge unless they are emitted when the direction of the field is favorable;

(3) alternating discharges can be maintained in insulating vessels with the static fields set up by the drifting of ions and electrons controlling the equilibrium density of ionization in the discharge volume;

(4) the maximum kinetic energy in the oscillatory motion of an electron, at the minimum field intensities needed for gas breakdown, corresponds to approximately 10^{-3} eV (Brown[35]).

This energy is insufficient to give breakdown, and sufficient energy can only be acquired by an electron experiencing collisions. A free electron in an oscillatory electric field oscillates in phase with the field about a mean position and the average power gained from the field is therefore zero. However, in a collision an electron can have its orderly oscillatory motion changed to random motion, and it can gain energy from the field. The average fractional kinetic energy lost by the electron per collision, F in eqn. (3.11) and equal to $2m/M$, where m is the mass of the electron and M is the mass of an atom or molecule, is more than made up by the gain in energy that can ensue by putting the electron out of phase with the oscillatory field and into the accelerating phase with the next half-cycle. The energy of the electron, therefore, can increase until it is high enough to enable the electron to excite or ionize in an inelastic collision when it loses a large fraction of its energy. This is in contrast with what happens in the dc-discharge, under conditions where plasma oscillations are not of consequence, where collisions result mainly in a loss of kinetic energy by the electrons.

Following the treatment of Francis,[36] the factors that influence the breakdown field and subsequent current and ion density of a fully developed plasma are:

(1) the gas pressure p (torr) and consequently the mean free path λ_e (cm) of the electrons and the collision frequency $f_c = v/\lambda_e$ sec^{-1} with which they hit atoms or molecules, where v again is the random velocity of the electron;

(2) the frequency f and wavelength λ of the applied electric field E;

(3) the dimension of the containing vessel; that is, the length l in the direction of the electric field and the perpendicular width or radius R.

3.3.2. PROPERTIES DETERMINED BY THE PRESSURE AND FREQUENCY OF THE FIELD

The following sets of conditions determine the physical processes occurring in the discharge:

Very low pressures ($< 10^{-2}$ *torr*), $\lambda_e > l, R$. The electrons hit the walls of the containing vessel more often than they hit atoms or molecules. The secondary effects at the walls then affect the breakdown of a gas or the maintenance of a discharge.

Medium or high pressures. (a) $\lambda_e < l, R$; low frequency, $v \ll f_c$. The electrons make many collisions for each cycle of the field and drift as a cloud in phase with the field.

If the frequency v is high, the amplitude of oscillation is low and can be less than l or R. Charged particles formed by ionizing collisions in the gas are then lost by diffusion to the walls.

If the frequency v is low, the amplitude of oscillation is high, and electrons are driven into the walls of the tube in each half-cycle. Secondary wall processes are then essential in the maintenance of a discharge.

(b) $\lambda_e < l, R$; high frequency $v \gg f_c$. The electrons make many collisions with atoms or molecules and make many small amplitude oscillations between collisions. They remain essentially stationary and spread out only by diffusion.

Very high frequencies, and ultra-high microwave frequencies, $\lambda_e < l, R$. The electrons are not under the influence of the oscillatory electric field but are subject to the standing oscillatory electromagnetic wave determined by the frequency of the field, the excitation, and the discharge tube geometry. That is to say, their behavior is set by the properties of the microwave cavity.

In a microwave discharge at 3000 MHz in helium, the average electron energies are much higher than in the positive column of a glow discharge for the same pressure and radius.[37] For a pD of approximately 220 torr-mm in a microwave discharge in helium, the electron temperature is about 47,000 K,[38] whereas according to theory (Fig. 3.6) the electron temperature in the positive column is 24,000 K or about one-half this. Excitation has been found also to depend on the frequency of the field.[39, 40]

Motornenko[39, 40] has shown that the excitation of lines with high excitation potentials increases with increase in frequency or decrease in wavelength of the oscillatory field as illustrated in Fig. 3.9. The excitation potential of the upper level of the Ne-I line at 337 nm is 20.03 V; for the Ne-II line at 336.7 nm, 34.5 V; and for the Ne-II line at 377.7 nm, 30.5 V.

In passing, it is worth noting that microwave discharges have been used successfully as gas laser excitation systems by Maksimov[38] for producing population inversion between the Ne $2s_2$ and $2p_4$ levels in a He–Ne mixture, and by Bloom and Goldsborough[41] for obtaining laser oscillation in the ion species of a number of noble gases and highly reactive gases.

The situation in (b) ($\lambda_e < l, R$; high frequency $v \gg f_c$) is applicable to He–Ne rf-excited laser discharges where commonly $R = 2$ to 5 mm, at a pD of 4 torr-mm. The pressures are

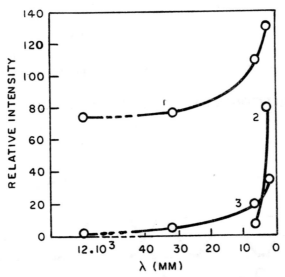

FIG. 3.9. Dependence of intensity of neon lines on wavelength of the excitation source. 1. Ne-I, 337.0-nm line; 2. Ne-II, 336.7-nm line; 3. Ne-II, 377.7-nm line. (After Motornenko,[39] by courtesy of the Optical Society of America.)

approximately 1 torr, and λ_e is about 0.8 mm for electrons with an average energy of 11 eV appropriate to a pD of 4 torr-mm (Brown[42]). As an item of interest, this is in the pD region where, in helium discharges, transitions from "weak" to "strong" discharges have been reported by Russian workers[43-45] (see Transition Discharges, Section 3.3.5).

3.3.3. LIMITS OF THE DIFFUSION THEORY

Certain basic assumptions are made in calculations of breakdown as a balance between ionization rate and loss of electrons by diffusion.[46]

At low frequency, dimensions of containing vessels are small compared with the wavelength of the electric field, and the uniform field assumption of the diffusion theory is very good.

At high frequency, there is a limit to the size of the discharge tube consistent with the uniform field assumption of the diffusion theory. The limit is when the size of the discharge tube is equal to a single loop of a standing wave of the electric field. The relation between this limiting length, L, the wavelength, λ, and the diffusion length, Λ, may be written[46] as

$$L = \lambda/2 = \pi\Lambda. \tag{3.18}$$

The diffusion length Λ of a cylindrical containment vessel is determined by the dimensions of the vessel and can readily be calculated from

$$1/\Lambda^2 = (\pi/l)^2 + (2.405/R)^2 \tag{3.19}$$

where R is the radius, and l is the length of the vessel. The first term on the right-hand side of the equation is the contribution due to diffusion to the end walls; and the second, diffusion to the side walls. Diffusion theory also does not apply when the electron mean free path λ_e becomes comparable to the size of the tube. In the limit, this occurs when $\lambda_e = \Lambda$.

77

Since the probability of collision P_m is equal to $1/p\lambda_e$

$$P_m = 1/p\Lambda, \quad \text{or} \quad p\Lambda = 1/P_m. \tag{3.20}$$

The value of P_m is not constant but depends on the electron energy. From measured values of P_m, therefore, it is possible to calculate the limit for the use of the diffusion theory.

3.3.4. AVERAGE ELECTRON ENERGY IN THE RF-DISCHARGE

The dependence of mean electron energy on frequency has been studied by Harries and von Engel.[47] Figure 3.10 taken from Harries and von Engel shows that at low frequency

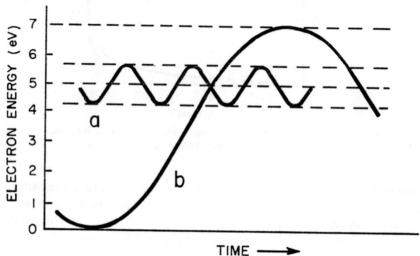

FIG. 3.10. Variation of average electron energy with time in an rf-excited discharge in helium. (a) High frequency; (b) low frequency. (After W. L. Harries and A. von Engel,[47] by courtesy of the Royal Society.)

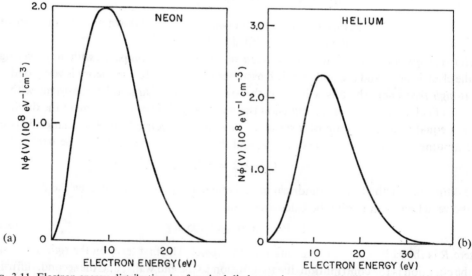

FIG. 3.11. Electron-energy distribution in rf-excited discharges in neon and helium. (a) $pD = 5.6$ torr-mm; (b) $pD = 10$ torr-mm, at a frequency of 21 MHz. (After R. S. Hemsworth and N. D. Twiddy,[48] by courtesy of the Secretariat, Öst. Studiengellschaft für Atomenergie, Austria.)

the modulation of the electron energy is 100 percent, and that the mean energy is low. At high frequency the mean energy is higher for the same field, with the modulation reduced to about 25 percent. This increase in mean energy for higher frequencies is consistent with the observed variation of intensity of spectral lines with change in frequency of the field noted in Section 3.2.2.[39, 40] Illustrations of electron energy distributions in rf-discharges in helium and neon are given in Fig. 3.11. These distributions are typical of those found in rf-discharges in the noble gases He, Ne, Ar, and Kr in the pD range 0.16 to 10 torr-mm in a 20-mm discharge tube. All the distributions are Druyvesteyn, having a higher mean energy than would be inferred if a Maxwellian distribution of electron energy was assumed (Hemsworth and Twiddy[48]).

3.3.5. TRANSITION DISCHARGES

The mechanism responsible for transitions from "weak" to "strong" plasma condition in rf-discharges mentioned as an "item of interest" in Section 3.2.2 is not clear.[43-45] It is discussed here because of its possible relevance to discharge mechanisms in He–Ne laser discharges as yet not understood.

At a certain pressure, as the pressure is increased over a narrow range for a particular discharge-tube diameter, a considerable increase in the intensity of a number of spectral lines in an rf-excited discharge occurs. Simultaneously the electron concentration increases by a factor of about 3.[44] The pronounced effect is not due to improved coupling-in of rf-power to the discharge or to a change in the electron temperature. According to Bochkova et al.[45] it could be due to the dissociation of molecular complexes formed in the discharge. Table 3.3 gives values of associated pressure and tube diameters for He, Ne and Ar in which transition discharges are observed. Such transition discharges in the positive column of a glow discharge have not been reported to the author's knowledge. (They have been looked for by the author without success in a dc-discharge in helium over a range of pD values encompassing that where a transition might be expected at a number of discharge currents.)

TABLE 3.3. *Pressures in torr at which transition discharges occur in rf-excited discharges*

Gas	Tube diameters			
	3 mm	12 mm	35 mm	60 mm
Helium	2.0–3.8	0.4	0.15	0.1
Neon	—	0.24	0.10	—
Argon	—	0.06	—	—

3.4. The Hollow-cathode Discharge (HCD)

The hollow-cathode discharge, although often called a Schuler discharge, was first developed and used as a spectroscopic source by Paschen.[49] Its first use as a laser discharge was reported by Smith.[50] Smith used it to obtain oscillation in a He–Ne mixture on Ne 2s–2p transitions in the near-infrared. Since this first report, population inversion or oscillation in HCD structures of various types has been reported in neutral species in neon–molecular gas mixtures and pure neon by Chebotayev,[51] and Chebotayev and Pokasov,[52] Znamenskii et al.,[53] Afanas'eva et al.;[32] in carbon dioxide and helium by Willett and Janney;[54] in ionized argon by Huchital and Rigden;[55] in metal–vapor ion lasers by Schuebel,[56, 57] and Collins et al.,[58] and by Russian[59] and Japanese workers.[60, 61]

The HCD is a particularly useful laser discharge. It can be operated at low sustaining voltages, and at high gas pressures, thus enabling high inversion densities to be realized. In addition it has certain properties that give it additional advantages over the positive column of the glow discharge for some gas laser systems. These are metal-vapor ion laser systems.[56–61]

3.4.1. GENERAL CONSIDERATIONS

An assortment of HCD anode–cathode configurations is shown in Figs. 3.12 and 3.13. Figure 3.12 shows the simplest configuration, an open-ended cylinder mounted in an enclosure that can be evacuated and filled with a gas with a pin anode a short distance away from it.[62] If such a tube is filled to a gas pressure between 0.1 and 20 torr, depending on the cathode diameter and the anode–cathode geometry, and if a few hundred volts is applied to

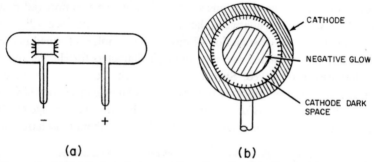

Fig. 3.12. Hollow cathode (or Schuler) discharge, (a) anode–cathode configuration, (b) luminous and dark-space regions in a cylindrical hollow cathode. (After S. Tolansky,[62] *High Resolution Spectroscopy* (1947), by courtesy of Methuen, London.)

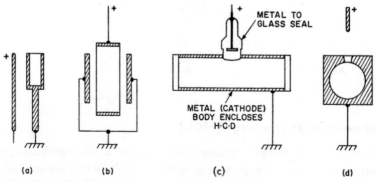

Fig. 3.13. Cross-sectional diagram of typical HCD anode–cathode configurations.

the electrodes, an intense discharge concentrates inside the cathode. The discharge inside the cathode is called a hollow-cathode discharge. As illustrated in Fig. 3.12b in the view along the axis of the cylindrical cathode, the discharge consists of an annular dark space close to the inner wall of the cathode and a very bright glow filling the rest of the interior of the cathode. The annular dark space constitutes the cathode dark space similar to that in the glow discharge. The bright glow is the negative-glow region. The same type of discharge occurs if the cylindrical cathode is replaced by two planar cathodes to which a common connection is made as in Fig. 3.13b. In this case the two planar cathodes must be separated by less than a certain distance which depends on the discharge-tube fill-pressure and the gas. This distance, or more exactly one-half this distance, corresponds to the reaching distance or range of high-energy beam-electrons that have been accelerated by the high potential existing across the cathode dark space. The distance or range can be calculated from eqn. (3.3) in the treatment of the negative glow of a glow discharge. The HCD configurations illustrated in Fig. 3.13 have been used: (a) in commercially available discharge tubes used as sources of line spectra for atomic absorption spectroscopy;[63] (b) as experimental arrangements for investigating the formation of the HCD;[64] (c) as laser structures;[51, 52] (d) as sources of electrons.[65] As far as laser structures are concerned, the most successful HCD configuration to date has been that used by Schuebel[56, 57] and illustrated in Fig. 5.6. In all the HCD structures mentioned, the glow has the same general appearance as that shown in Fig. 3.12b.

The brightness of the discharge depends in rather a critical manner on the carrier-gas pressure, the discharge current, and cathode diameter; or if two planar cathodes are used, the cathode separation. If the pressure is reduced below the approximate range of 1 to 3 torr for a noble gas, the brightness of the discharge falls until at a critical pressure the well-defined discharge quenches and a diffuse, weak-looking discharge forms inside the cathode. This discharge, called a suppressed mode discharge,[66] is characterized electrically by a high resistance. Two general rules for HCD operation are that

(1) it is not possible to run a normal HCD if the radius of the cathode (or half of the separation between two planar cathodes) is less than the width of the cathode dark space for a particular gas pressure. As a guide, for a neon–molybdenum HCD gas-cathode material combination with two parallel planar cathodes

$$1 \text{ torr} - \text{cm} < ap < 10 \text{ torr} - \text{cm} \tag{3.21}$$

where a is the cathode diameter or planar-cathode separation;[67]

(2) the ratio of the cathode length to diameter must not be greater than about 7 (von Engel, private communication, 1965).

These rules do not apply to the external-type HCD structure shown in Fig. 3.13c or the slotted HCD shown in Fig. 5.6. These structures will operate in the noble gases at low pressures.[51, 52]

In normal HCD-operation when negative-glow regions from opposite sides of the cathode or opposing planar cathodes have merged, the current density can be much higher than that obtained using a single planar cathode at the same pressure in a glow discharge. The quantity j/p^2 for a fully developed HCD where j is the discharge current surface density can be

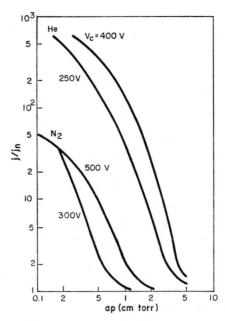

FIG. 3.14. Hollow cathode effect, ratio j/j_n of current density of a double planar cathode HCD to that of a single planar cathode as a function of the cathode separation and gas pressure product ap for various cathode falls in potential in nitrogen and helium. (After P. F. Little and A. von Engel,[64] by courtesy of the Royal Society.)

two to three orders of magnitude larger than that of a single planar cathode in helium for example, depending on the carrier gas, the pressure, and cathode diameter; and in a molecular gas an order of magnitude larger. This effect is shown clearly in Fig. 3.14 for a HCD of the coplanar-cathode type illustrated in Fig. 3.13b operating in helium and nitrogen, where j is the current density in the HCD and j_n in the normal current density for a single planar cathode.[64] The inclusion of results from two sources in Fig. 3.14 explains the intersection of the 300- and 500-V curves in N_2.

Consider the voltage versus discharge-current characteristic of a spherical-cavity-type HCD in neon (solid line) shown in Fig. 3.15.[67] The broken line shows a plot of the sustaining voltage versus current measured for a single planar cathode in neon. The gas pressure was 98 torr for the HCD and 60 torr for the planar-cathode discharge. When the current is a few mA, only a small area of the hollow cathode is covered with a glow. As the current increases, the glow extends to cover the interior of the cathode, and as it does so, the sustaining voltage decreases. At a discharge current of 20 mA, at the point at which the sustaining voltage starts to increase, the cathode is completely covered with a glow. Any further increase in current then requires an appreciable increase in sustaining voltage. With the single planar cathode, the sustaining voltage starts to increase at a discharge current of 8 mA when its surface is covered with a glow. Because of its construction, a glow remains on one side of the cathode only. When the physical dimensions of the HCD and the planar cathode given in ref. 67 are considered, j/p^2 for the HCD is 0.123 mA/cm²/torr²; for the planar cathode j_n/p^2 is 5.2 μA/cm²/torr². The ratio of j/j_n for the HCD and planar-cathode discharge is

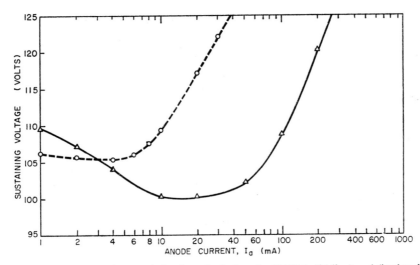

FIG. 3.15. Voltage vs. current characteristic of a spherical-cavity HCD (solid line) and (broken line) of a single planar cathode and anode discharge. (From J. W. Gewartowski and H. A. Watson's *Principles of Electron Tubes*, Copyright 1965, D. Van Nostrand Company, Inc., Princeton, New Jersey.)

therefore approximately 0.12 mA/5.2 μA or 230. This comparison of discharge current densities is more meaningful than the simple comparison of discharge currents obtainable from Fig. 3.15.

3.4.2. ELECTRON ENERGY DISTRIBUTION IN THE HCD

The HCD has always been considered to be a high-electron-energy plasma with the nature of its preferential excitation of highly excited neutral species or ionized species determined by direct electron impact by high-energy electrons. This assumption does not bear critical examination when one asks questions similar to, "Why is *this* state excited in a HCD, and not *that* one, when they are at the same potential energy, and both have good reason to be excited similarly?" In the light of recent quantitative work, it is evident that the HCD is characterized as much by a high concentration of low-energy electrons and a high thermal-energy positive-ion concentration as by any large concentration of high-energy electrons.

In the overlapping negative-glow regions that comprise the HCD, the energy of electrons would be expected to be similar in distribution to that in the normal negative glow in a glow discharge, though with perhaps a bias to low energies. In the negative-glow region of a glow discharge, the electrons can be divided into three groups of differing energies consisting of[11]

primary or fast electrons, with energies of about 20 eV; secondary electrons, of about 5–6 eV; and ultimate or slow electrons, at a low energy less than 1 eV or possibly of thermal energy,[68] the concentration of low-energy electrons being three orders of magnitude larger than that of primary or fast electrons (see Table 3.1).

From an **early** treatment of the negative glow in Section 3.2.1, it is evident that primary electrons **with considerably** more energy than a few tens of electron volts must exist to explain the extent of the negative glow transverse to the cathode surface. As shown in Section 3.2.1, the extent of the glow is due to high-energy electrons that enter the negative glow with an energy equal to, or almost equal to, that gained in crossing the cathode dark space without suffering inelastic collisions. The energy of the primary electrons, therefore, can be a few hundreds of electron volts. This is borne out by the results of work by Brewer and West-haver[16] mentioned earlier and those of Grimley and Emeleus.[69] Similarly, very high-energy beam-electrons need to exist in the HCD to account for the extent of the negative glow perpendicular to the surface of the cathode.

Figure 3.16, from Borodin and Kagan,[70] shows experimentally determined electron energy distributions in a cylindrical HCD in helium. For comparison purposes, included in the same figures are electron energy distributions in a positive-column discharge under (assumed) similar conditions. The figures clearly show that the HCD has more electrons with an energy over 18 eV than the positive column, yet simultaneously the HCD has a much

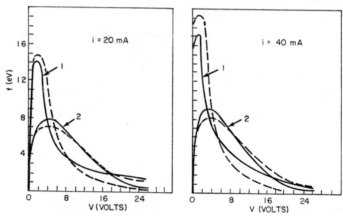

FIG. 3.16. Comparison of electron energy distributions in a cylindrical HCD and in a positive column in helium at two discharge currents. The broken lines denote Maxwellian distribution functions. Excitation conditions: $p = 0.9$ torr; 1. HCD (diam. 2 cm); 2. positive column (diam. 2 cm). (After V. S. Borodin and Yu. M. Kagan,[70] by courtesy of the Advisory Editor, Sov. Phys.-Tech. Phys., U.S.A.)

larger relative number of low-energy electrons in the range 0–4 eV than has the positive column. The broken curves in the figures denote Maxwellian distribution functions determined from the low-energy portion of probe curves. Comparison between them and the solid curves of the HCD and positive column shows that there is an excess of fast electrons in comparison with a Maxwellian energy distribution in a HCD, and a deficiency of fast electrons in a positive column. Afanas'eva et al.[71] have observed similar non-Maxwellian electron energy distributions in a HCD of the type shown in Fig. 3.13c.

On the basis of these quantitative measurements it is evident that the HCD is as much a low-energy plasma as it is a high-energy plasma. It can be characterized as much by an extremely high concentration of low-energy electrons and by its complementary, high positive-ion concentration as by a high concentration of high-energy electrons.

Spectroscopic examinations of a number of HCDs[72] and analyses of the literature indicate that the plasma of the HCD is similar to the high-electron-energy, beam-maintained, recombination-dominated plasma of the brush-cathode discharge of Persson[17] and that enhancements of particular spectrum lines in it are due principally to thermal-energy asymmetric charge-transfer reactions and not to any high-energy component of the electron energy distribution, nor to a high concentration of metastable atoms able to excite selectively by excitation transfer or the Penning effect.

3.4.3. EXCITATION THEORIES

A number of theories have been suggested to account for the interaction of the double-planar cathode and cylindrical hollow-cathode phenomenon, which results in the high-current, high-electron-density HCD. The HCD-phenomenon has been attributed to:

(1) a reduction of the positive space charge in front of the cathode, or an increase in ionization in the cathode dark space due to high-energy beam-electrons merging from opposite sides of the cavity;[64, 73, 74]

(2) an increase in ion density and radiation from high quantum states in the nearly field-free negative-glow region;[74, 75]

(3) more efficient geometric use of ions, photons, or metastable atoms in causing secondary emission of electrons from the cathode;[74, 76, 77]

(4) the presence in the cathode region of a larger number of sputtered metallic atoms with low ionization potentials which influence the secondary processes at the cathode and in the volume of the discharge.[65, 78]

It is an open question as to which of the four processes listed is the dominant process, if indeed there is a dominant process. The author favors the first process for the principal reason that properties peculiar to the HCD, such as the extremely high concentration relative to the negative-glow plasma of a single planar cathode, develop only when the negative glows start to overlap, and become more pronounced as the extent of the overlap increases as the applied voltage is increased.

3.4.4. SPUTTERING ACTION

When the light emitted by the HCD is examined, spectra of the material of which the cathode is made as well as of the inert or molecular carrier gas are found. Since the electric field in the plasma region of the HCD is small when operated under normal discharge conditions, the emission lines are almost free from Stark broadening. Further, if the cathode is cooled and as long as the discharge is not an abnormal discharge, the Doppler width of the spectrum lines can be reduced, and narrow linewidths obtained (Tolansky[62]). For example, linewidths of less than 0.01 cm^{-1} in a liquid-helium-cooled HCD have been reported by Roessler and de Noyer.[75]

The cathode material is sputtered by positive ions that impinge on the cathode with energies corresponding to the full cathode fall or a few hundred volts.[62] The majority of particles ejected from the cathode are atomic according to von Hippel.[78] The following proposals have been made to explain the sputtering process:

(1) momentum is transferred from the incident ion to the cathode surface and atoms, ions, and molecules are emitted from the cathode with varying velocities (Townes[79]);

(2) simple evaporation of atoms occurs under the ion bombardment.

Wehner and co-workers[80-83] have found that there is a periodic dependence of sputtering yield on atomic number similar to the reciprocals of the heat of sublimation of the elements. Peaks in sputtering yield occur at manganese, copper, silver and gold in the periodic table. The sputtering yield varies with the carrier gas used. Apparently the heavier the carrier gas, the higher is the sputter yield in atoms per ion incident on the cathode surface. Helium and neon produce considerably less sputtering than krypton, xenon, or argon.[84, 85]

Whatever the exact mechanism of the sputtering process might be, it has been found that the sputter yield $W \propto I^m$, where I is the discharge current and m is approximately equal to 3 for a molybdenum cathode, and 4.3 for a nickel cathode (Tokatsu and Toda[86]). Stocker[87] has reported that the sputter yield has a somewhat similar dependence on discharge current and that it is inversely proportional to the gas pressure to the 5/2 power, that is to say,

$$W \propto (I/p)^{5/2}. \tag{3.22}$$

3.4.5. USE AS A SPECTROSCOPIC SOURCE

The HCD has been used extensively as a source of line spectra of most of the nongaseous elements.[62] It is surpassed in the production of spectra having narrow linewidths only by the atomic beam, and laser devices. The main advantage the HCD has over other types of discharge sources of line spectra is in exciting high-melting-point metals. This is because the sputtering action can produce the requisite metal vapor in an atomic form without the necessity for having a high temperature to form the metal vapor by the normal means of evaporation.

The spectra in a HCD consist of that of both carrier gas and of the material of which the cathode is made. The intensity of each as a function of current differs significantly. The intensity of the spectra of the carrier gas is approximately proportional to the current I, and the intensity of the cathode-material spectra has been found to vary as I^m, where m can be between 2 and 3 depending on the cathode material and carrier gas combination.[65, 85, 88] These relationships could indicate that the carrier gas is excited by the single-step process of direct electron impact from the ground state, whereas the sputtered-cathode vapor is excited by stepwise processes. An alternative explanation for the I^m dependence of the intensity of the spectra of the cathode material could be that the intensity is simply following the concentration of sputtered-cathode material in the HCD that is available for excitation as it varies with discharge current. As noted in Section 3.4.4, the sputter yield also varies as I^m where m varies between 2 and 5. Subject to the concentration of sputtered metal vapor atoms not becoming appreciable enough to lower the effective ionization potential of the carrier gas and affecting the main ion-producing mechanisms of the discharge, the excitation of the sputtered atoms could in general be a single-step process and yet have a spectral intensity that varies as I^m. At high discharge currents, with the current depending on the diameter of the cathode, the spectra of the cathode has been found to predominate.[65]

The type of spectra, that is ion or neutral spectra, emitted by the HCD is affected considerably by the carrier gas used. Helium as a carrier gas produces cathode-material spectra in the HCD more ion-like than any of the other noble gases. This is because helium has high excitation potentials of its two lowest (metastable) $He^* 2^3S_1$ and $He^* 2^1S_0$ levels, a high ionization potential, and a high mobility of both neutral atoms and ions. These are all important parameters in Penning and charge-transfer reactions, both of which have large cross-sections and can produce preferential ionization and excitation of the sputtered cathode material in the HCD. Since helium is a less effective sputterer than other gases,[86, 87] when it is used as a carrier gas the vapor pressure of the sputtered cathode material is low, and Penning and charge-transfer ion-producing processes can predominate over ionization by direct electron impact. The high concentration of metastable atoms and positively charged ions and the low electric fields in the HCD enable these collisional processes to proceed readily and thus make it a favorable source for the excitation of ion spectra. Sawyer[72] has found that the extent of selective excitation and the level of simultaneous excitation and ionization is determined by asymmetric charge-transfer collisions of the 10/01' and 10'/1'0 type. As might be expected, if atom–atom, Penning, and charge-transfer reactions are important in a HCD, the type of excitation produced is not only a function of the excitation or ionization potentials of the carrier gas; it will depend on the coincidences of energy levels of the carrier gas and the sputtered cathode material. For example, Mitchell,[89] using an iron hollow cathode, found that xenon when used as a carrier gas gave more energetic ion spectra than argon, even though the excitation and ionization potentials of xenon are lower than those of argon.

Data given by Sawyer[72] show that the maximum level of excitation observed in a HCD can be attributed to the asymmetric (10/01') charge-transfer process. This process, with helium as a carrier gas, is apparently more probable than the Penning-like atom–atom 0'0'/01' process which should result in the excitation of more highly excited, singly ionized species than the 10/01' process.

3.4.6. EXCITED-STATES POPULATION IN THE HCD

Although the HCD has been used extensively as a source of spectra for over 50 years since its first use by Paschen in 1916, few studies have been directed at finding the excited-states populations responsible for the emission of the spectra or at finding the metastable concentrations. Even fewer systematic studies have been directed at determining the relationship between excited-state populations and experimental parameters. The systematic studies that have been done have been stimulated by interest in gas lasers since the early 1960s, and by the need to actually "get-a-grip" on processes that are responsible for the realization of population inversion in HCD lasers.

Information that one can use, and which is available at the time of writing, is restricted to results of studies of helium–neon HCD-laser discharges that operate in the near infrared and at 632.8 nm. Fortunately, because of the nature of helium–neon discharges in which helium constitutes the majority gas and the frequent use of helium as a carrier gas in other HCDs, this information is of direct relevance to other spectroscopic and HCD-laser systems. Such laser systems include the interesting and important He–Cd, He–Zn, and He–Hg singly ionized ion lasers that have discharge properties similar to He–Ne lasers in both hollow-cathode and positive-column discharges. The work of Znamenskii[90] has provided us with the majority of this much-needed basic information.

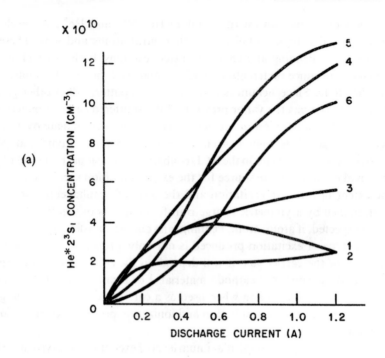

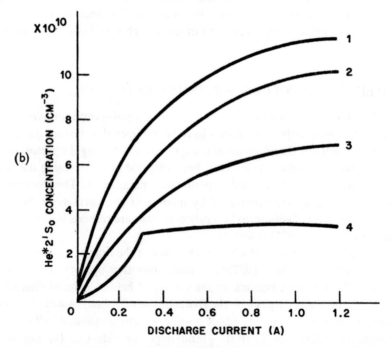

FIG. 3.17. Concentration of metastable He* 2^3S_1 atoms (a), and He* 2^1S_0 atoms (b), as a function of discharge current at various partial pressures of helium in a HCD (cathode diam. 8.5 mm, neon pressure 0.2 torr). Helium pressure (torr): 1. 0.7; 2. 1.4; 3. 2.8; 4. 5.6; 5. 8.5; 6. 11.3. (After V. B. Znamenskii,[90] by courtesy of the Optical Society of America.)

Znamenskii investigated the dependence of the concentration of both helium and neon metastables on the parameters of a HCD in a helium–neon mixture. The hollow-cathode-discharge system was basically the same as that illustrated in Fig. 3.13c that was developed and used by Chebotayev.[51, 52, 91] The metastable concentration measurements were made at values of pressure and cathode diameter product (pa) of 0.6 to 10 torr-cm, with He : Ne ratios of 2 to 80, and discharge-current surface density and cathode diameter product ja of 1 to 13 A-cm^{-1} in tubes of 0.85, 1.2, and 2.0-cm diameter. These parameters cover the conditions for which maximum laser output at 1.15 μm is realized in this type of HCD.[53]

Figures 3.17a and 3.17b and 3.18 illustrate the results obtained by Znamenskii. These three figures show the dependence of concentration of metastable helium atoms in the 2^3S_1 and 2^1S_0 states, and metastable neon atoms in the Ne* $1s_5$ state on discharge current at various partial pressures of helium in a HCD (8.5-mm diameter cathode, $p_{Ne} = 0.2$ torr). It can be seen Fig. 3.17a (curve 5) that: the concentration of He* 2^3S_1 metastables is a maximum at discharge currents above 0.6 A at a helium pressure of about 8.5 torr; the concentra-

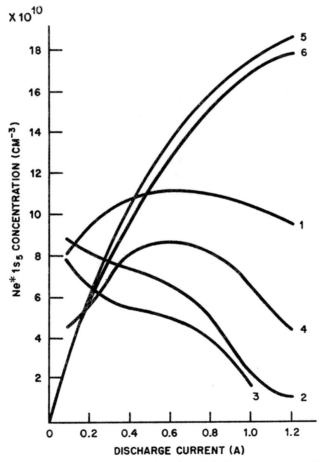

FIG. 3.18. Concentration of Ne* $1s_5$ metastables as a function of discharge current at various partial pressures of helium in a HCD (cathode diam. 8.5 mm, neon pressure 0.2 torr). Helium pressure (torr): 1. 0.7; 2. 1.4; 3. 2.8; 4. 5.6; 5. 8.5; 6. 11.3. (After V. B. Znamenskii,[90] by courtesy of the Optical Society of America.)

tion of He* 2^1S_0 metastables is still increasing at a helium pressure of less than 1 torr over all the current range; and the concentration of Ne* $1s_5$ metastables, Fig. 3.18, curve 5, maximizes above a discharge current of about 0.6 A at a helium pressure of about 8.5 torr within the range of pressures examined. As noted,[90] curves 1, 2, and 3 in Fig. 3.17b representing the concentration N of He* 2^1S_0 metastables with current are satisfactorily described by the expression

$$N(\text{He}^* \, 2\,^1S_0) = \frac{k_1 p_{\text{He}} I}{k_2 p_{\text{Ne}} + k_3 I}, \tag{3.23}$$

derived by Gordon and White[92] for the concentration N of He* 2^1S_0 metastables in the positive column of a glow discharge in a helium–neon mixture.

In the experiments it was found that the population of neon atoms in the quasi-metastable $1s_2$ state was equal to those in the metastable $1s_3$ state; that in the quasi-metastable $1s_4$ state the population was greater than in the $1s_2$ and $1s_3$ states; while the metastable $1s_5$ state, the lowest in potential energy of the four Ne-$1s$ levels, was more densely populated than any of the others. These results are in agreement with those of Bogdanova and Gi-Tkek[93] for the concentrations of excited Ne-$1s$ atoms in the positive column of a glow discharge, but differ from those in a HCD reported by them.

The measured values of the metastable concentrations of He* 2^3S_1, He* 2^1S_0, and Ne* $1s_5$ atoms plotted in Figs. 3.17 a and b and Fig. 3.18 lead to the extremely important conclusion that if the pa products of both helium and neon, and the ja product are maintained constant, then the metastable concentration N is inversely proportional to a, the internal diameter of the cathode.

At a constant pa, maintaining ja constant, if N is inversely proportional to a, this means that the metastable concentration is directly proportional to the pressure p. This direct relationship between the concentration of helium metastable atoms and pressure, and inverse dependence of the metastable concentration on cathode diameter can explain the observed relationship[53] that the output power and cathode-diameter product $Wa = $ a constant, in a HCD He–Ne laser operating at 1.15 μm. To do this it is necessary to assume that the process that dominates the production of population inversion is the upper-state selective excitation mechanism. This matter is treated more extensively in *Inverse radial dependence of gain*, in Section 4.1.1.6.

3.4.7. ELECTRICAL PROPERTIES OF THE HCD

The voltage versus current characteristic of the HCD is affected greatly by the cathode geometry.[85, 90] For cathodes of small diameter, and for high currents in cathodes of large diameter, the HCD exhibits a negative resistance characteristic.

Normal gas discharge similarity relations[11] are not obeyed in the HCD[65, 85] unless both pa, where a is the diameter of the cathode, and pj, where j is the discharge current density on the surface of the cathode, are maintained constant.

Little importance has been attached to the shape and position of the anode with respect to the cathode in a HCD. Though the shape and position of the anode do not appear to affect the spectral properties of the HCD, they can affect the electrical properties. It has been found that the HCD will exhibit electrical instability, as does a normal glow discharge, if an anode glow is present. Stability is not obtainable unless the anode glow is only a sheath-like glow over the anode surface, and it cannot be obtained by outside means.[94, 95] Under these conditions the anode current is supplied by diffusion of electrons from the negative glow inside the cathode.[65]

3.5. Pulsed Discharges

Pulsed gas discharges were first used as gas laser excitation sources at the Services Electronics Research Laboratory (S.E.R.L.), Baldock, England, by Boot and Clunie.[96] Boot and Clunie obtained oscillation in neon and carbon monoxide during the recombination period following a pulsed rf or dc discharge in helium mixed with either neon or carbon monoxide at wavelengths of 1.153 μm and 1.069 μm respectively. In later work, Boot, Clunie and Thorn[97] obtained oscillation in the afterglow of a high-current, pulsed discharge in a high-pressure mixture of helium and neon. Typical gas pressures were 5 torr of neon and 150 to 200 torr of helium, and the mixture was excited by means of high-voltage, microsecond pulses applied to external electrodes that capacitively coupled power into the gas mixture. Incidentally, as well as this work being the first to utilize the pulsed-discharge form of excitation, it was the first in which oscillation was realized in a gas or gas mixture that was at pressure that could be realistically given as a reasonably sized fraction of atmospheric pressure.

The same technique was used by Mathias and Parker, also of S.E.R.L., to give population inversion between electronic levels in molecular species for the first time.[98, 99] Basically the technique permits the introduction of much greater power into the discharge in a very short time than would be possible under CW conditions. The importance of this early pulsed work has not been sufficiently stressed. Pulsed excitation enabled possible laser systems to be rapidly examined for suitability of oscillation through a range of discharge conditions that cover initial gas breakdown; equivalent CW conditions; recombination; and afterglow conditions. Its use enabled the now commonplace, high-power CW, transient noble gas and molecular gas laser systems to be quickly developed and analysed. Prior to the innovation of pulsed excitation, rapid development of gas lasers was prevented by the absence of information of dominant collision-processes, excitation conditions, collision cross-sections of all kinds and effective transition probabilities appropriate to the complex medium of a gas discharge under the CW conditions that were first emphasized in laser feasibility and research studies.

It is convenient to consider the excitation processes in pulsed discharges in two parts: (1) at breakdown of the gas (Section 3.5.1), and (2) in the afterglow (Section 3.5.2).

3.5.1. EXCITATION PROCESSES AT BREAKDOWN

If a sufficiently high potential V is applied across a discharge tube containing a gas via internal electrodes in parallel with a capacitor, the gas will break down, and the current will try to become infinite at a particular voltage. This breakdown voltage (V_b) depends on the gas, the gas pressure, and the separation of the electrodes; i.e. on the E/p ratio and the gas. Plots of the variation of breakdown voltage with pressure p and separation d of the electrodes form the familiar Paschen breakdown curves (von Engel[100]). Provided the electric field is uniform and the pressure-electrode separation is less than 150 torr-cm, Paschen's law, which states that V_b is a function of the product pd only, is obeyed.[101]

At breakdown the current rises rapidly to a value determined by the inductance, the resistance in the circuit, and the energy available from the source. After the initial rapid rise in current, the electric field falls until the current reaches a particular value after which the maintaining potential is independent of the current. The current then decays exponentially according to the time constant of the circuit. In such a discharge, the current density can approach 300 A-cm^{-2} as a result of the considerable ionization that occurs. The reduction of the voltage gradient is probably due to a transition from ionization of ground-state atoms by direct electron impact to one of step-wise ionization by electron impact excitation of excited atoms.

The normal capacitor discharge is not a versatile one as far as analysis of a discharge is concerned since it is not possible to isolate processes in the afterglow of the initial breakdown pulse from those that occur during the decaying current pulse. However, by the use of a suitable delay line in conjunction with a thyratron switch and a pulse transformer, it is possible to produce quasi-CW rectangular discharge current pulses of up to a few hundred msec duration and so effectively isolate breakdown processes from the main discharge processes.

To the writer's knowledge, electron-energy-distribution determinations have not been made in pulsed discharges but it is believed that the average electron energy in them is high because of the large E/p, of the order of 200 V-cm^{-1} torr^{-1} at breakdown. The results of work by Cheo and Cooper[102] on pulsed, molecular gas laser discharges show that electrons are of high energy only during the time to complete breakdown of the gas. This is in agreement with the work of Breusova.[103] Breusova observed that beyond breakdown in all the noble gases she investigated, the electron temperature decreased with increase in current and the large electric field decayed at a rate determined by the diameter of the discharge tube, with the initial voltage drop being steeper in the tube of smaller diameter.

When the applied potential V exceeds the breakdown voltage V_b it is believed that the same ionization processes determine the early states of current growth on breakdown as those that occur when V is equal to V_b. In this case the electric field remains undistorted until the increasing current enhances the normal primary ionization and secondary ionization processes of the cathode, and so produces a collapse of the electric field. If the period of the current pulse can be reduced to a value ensuring that positive ions do not have time enough to move to the cathode and produce secondary ionization, it appears that a high E/p can be maintained throughout the period of the discharge without a transition occurring from a glow-like discharge to a low-voltage, high-current, arc-like discharge.

During the period of breakdown of the gas when the E/p can be high, the extent of excitation and ionization is determined principally by the E/p ratio. This ratio determines the energy gained by an electron in each mean free path, and this explains the dependence of excitation and ionization functions on the E/p ratio through its control of the mean electron energy.

3.5.2. EXCITATION PROCESSES IN THE AFTERGLOW

It is instructive to consider first of all afterglow processes in helium. Helium is commonly used as a so-called buffer gas in pulsed discharges, and it is principally with helium–noble gas mixtures that atom–atom, charge-transfer, and Penning reaction collisional energy-transfer processes appear to be responsible for producing population inversion long into the afterglow in a number of laser systems. In some of these systems it is only just becoming clear what role helium plays in the observed selective excitation that occurs. For this latter reason, afterglow processes in helium are discussed in detail in an attempt to show what excited species of helium are present in the afterglow in sufficient numbers to produce population inversion through various types of energy-transfer collisions. Unfortunately, opinions found in the literature on important processes in the afterglow in helium are highly contradictory.

3.5.2.1. Afterglow processes in helium

In the afterglow of a pulsed discharge in helium at low temperatures as well as at room temperatures, there is a sharp increase in intensity of atomic spectra 5 to 10 μsec after the cessation of the excitation pulse. This 5 to 10-μsec time interval up to the maximum of the afterglow represents the thermalization time of the electrons. The peak intensity varies with the particular level that is excited.[104] Biondi and Holstein[105] report that visible line-spectra from atomic species of helium are emitted in transitions from high-lying states $n = 3, 4, 5$, and 6 all within 1.5 eV of the helium atomic ionization potential at 24.58 eV above the ground state. The intensity of the atomic line-radiation follows closely the decay of the electron concentration n_e and is proportional to n_e^2. Since the positive-ion concentration is directly proportional to the electron concentration, it is clear that the early afterglow emission is associated with ion–electron recombination processes. This conclusion has been verified subsequently in microwave studies by Anderson.[106]

The most important deionization processes in the afterglow of an excitation pulse in helium are stated by Pakhomov et al.[104] to be: in the early afterglow, the collisional-radiative recombination process

$$He^+ + 2e^- \rightarrow He' + e^-, \tag{3.24}$$

followed in the later afterglow by the dissociative-recombination reaction

$$He_2^+ + e^- \rightarrow (He_2)' \begin{cases} \rightarrow He' + He \rightarrow 2He + h\nu, & (3.25a) \\ \rightarrow He_2' \rightarrow He_2'' + h\nu, & (3.25b) \end{cases}$$

in which He_2' and He_2'' are both excited molecular-states, and $(He_2)'$ is an unstable molecule (see Section 2.7.1).

The two-body deionization process $He^+ + e^- \rightarrow He'$, which might be expected on first sight to have a higher recombination than the process $He^+ + 2e^- \rightarrow He' + e^-$ of eqn. (3.24), has a much lower recombination than the three-body process.[104,107]

Processes in eqns. (3.24) and (3.25a) lead to deionization with the emission of spectra from atomic species, and process (3.25b) leads to deionization with the emission of molecular band-spectra. Pakhomov et al.[104] found that when the outside wall of their discharge tube was maintained at the temperature of liquid nitrogen (77 K) the intensity of the helium molecular band-spectra in the afterglow is low. This would indicate that reaction (3.25b) is gas temperature dependent and that it is not an important reaction at low temperatures.

Whereas it is clear that helium atomic ions will be present at the termination of the current pulse so that the three-body reaction involving a He^+ ion and two electrons can proceed in the early afterglow, the question arises as to how large numbers of molecular helium ions can be formed many milliseconds into the afterglow and so maintain an ionized plasma in the late afterglow. A number of proposals have been made to account for the production of this additional ionization and to account for the appearance of atomic spectra in the late afterglow that occurs after the thermalization of electrons on the cessation of the current pulse. They will be considered in some detail here as they are of relevance to a number of gas laser systems that operate in the late afterglow in axially excited discharges in conjunction with helium, sometimes at high helium pressures exceeding 50 torr. Examples of two such laser systems are the helium–mercury and helium–iodine singly ionized ion lasers covered in Chapter 5.

On the termination of the current pulse at pressures above 1 torr, energetic electrons initially present in the discharge are relatively quickly thermalized as a result of elastic collisions with atoms of the gas, and direct excitation and ionization cease. The time scale involved in this thermalization is between 1 to 5 μsec.[104, 108] After the thermalization, measured recombination coefficients indicate that the following ionization-producing reactions occur:

(1) below a pressure of 5 torr, the direct formation of molecular He_2^+ ions by pairs of metastable helium **atoms**

$$He^* \, 2\,^3S_1 + He^* 2\,^3S_1 \rightarrow He_2^+ + e^- + K.E. \ (\approx 15 \text{ eV}); \qquad (3.26)$$

(2) above a pressure of 5 torr, the formation of molecular He_2^+ ions from pairs of metastable helium atoms in the presence of ground state helium atoms

$$He^* \, 2\,^3S_1 + He^* \, 2\,^3S_1 + He \,^1S_0 \rightarrow He_2^+ + He + e^- + K.E. \ (\approx 15 \text{ eV}), \qquad (3.27)$$

and the collision of one metastable atom and two ground-state helium atoms

$$He^* \, 2\,^3S_1 + 2\,He \rightarrow He_2^* \, 2\,^3\Sigma_u^+ + He. \qquad (3.28)$$

The energetic electrons of approximately 15 eV energy produced in reaction (3.27), although incapable of exciting or ionizing helium atoms, are capable of ionizing the metastable $He_2^* \, 2\,^3\Sigma_u^+$ molecules produced in reation (3.28) by the following reaction:

$$He_2^* \, 2\,^3\Sigma_u^+ + (e^- + K.E. \approx 15 \text{ eV}) \left\{ \begin{array}{ll} \rightarrow He_2^+ + 2e^-., & (3.29) \\ \rightarrow He_2' + e^-, & (3.30) \end{array} \right.$$

Dissociative recombination of the He_2^+ molecular ions with slow electrons then continues via reactions (3.25a) and (3.25b) to give de-excitation, depending on the gas temperature, with the emission of atomic or molecular spectra respectively.

Since the concentration of helium metastable atoms, nearly all of which are in the 2^3S_1 state, can be high (greater than 10^{13} cm^{-3}) at the termination of the current pulse, processes (3.27) to (3.30) are of prime importance in the maintenance of ionization in the afterglow period. The measured time dependence of the population of atoms in the $He^* 2\,^3S_1$ state in the afterglow of a pulsed discharge in helium at various pressures is illustrated in Fig.

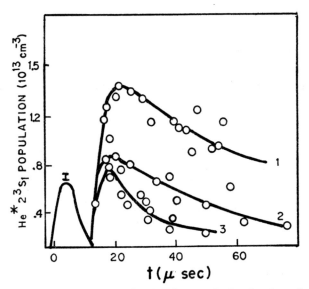

FIG. 3.19. Time dependence of the population of He* 2^3S_1 atoms in the afterglow of a pulsed discharge in helium at various pressure. *I*—current pulse of 100 A; pressure, 1—11.2 torr; 2—7.3 torr; 3—3.8 torr; $D = 14$ mm. (After V. S. Egorov, Yu. G. Kozlov, and A. M. Shukhtin,[109] by courtesy of the Optical Society of America.)

3.19.[109] It can be seen that the concentration of $He^* 2\,^3S_1$ metastables increases rapidly up to a maximum within 5 μsec of the end of the current pulse and falls away less rapidly at a rate that does not appear to be pressure dependent. The maximum concentration of $He^* 2\,^3S_1$ metastables was 2×10^{13} cm^{-3} at $p = 12$ torr of helium ($I_{max} = 100$ A, pulse length 15 μsec, and $D = 14$ mm). At high gas pressures and with high current pulses, extrapolation of Fig. 3.19 to longer afterglow times would indicate that sizeable populations of atoms in the 2^3S_1 state must occur out to a few hundred μsec after the end of the current pulse.

Strong emission in the visible region of the spectrum that is observed in the afterglow in helium is from highly excited atoms. These excited He′ atoms having quantum numbers of 3, 4, 5, and 6 are in states between 0.3 to 1.5 V of the atomic ionization potential of helium at 24.58 V above the ground state.[105] Because of the long duration of the atomic emission far in the afterglow at high helium pressures, it had been assumed that the dissociative-recombination reaction (3.25a) was responsible for the formation of the highly excited helium atoms. On account of the internuclear potential curves for He_2^+ molecules, in order to make plausible the production of an excited helium atom with a potential energy between 0.3 and 1.5 V of the helium atomic ionization potential in dissociative-recombination, it is necessary to postulate that the helium molecular ion in reaction (3.25a) is both metastable as well as vibrationally excited.[105, 107] The dissociative-recombination reaction would be, therefore,

$$He_2^{+v'} + e^- \rightarrow He' \,(> 23\ eV) + He. \tag{3.31}$$

According to Biondi and Holstein[110] this process is responsible for formation of highly excited helium atoms in states close to the helium atomic ion ground state "beyond any question". Although a metastable, vibrationally excited helium molecule with a potential energy close to the helium atomic-ion ground-state in helium in a mixture of a gas with helium would be an attractive candidate to explain energy transfer that occurs in a number of afterglow laser systems, it appears extremely unlikely that the presence of such a molecule in helium is required to explain the strong atomic emission from He' atoms.

Results by Ferguson, Fehsenfeld, and Schmeltekopf,[111] obtained using the powerful flowing-helium-afterglow technique, confirm the conclusion reached by Collins and Robertson[112] that the atomic emission is due, not to any dissociative-recombination reaction but to the simpler, collisional radiative-recombination reaction $He^+ + 2e^- \rightarrow He' + e^-$ of eqn. (3.24). Collins and Robertson[112] showed that, at extremes of high ionization and low neutral density, the collisional radiative-recombination process determined the rate of spontaneous emission from atomic species in the 50- to 250-μsec period in the afterglow. It was observed that the rate of spontaneous emission followed the concentration of the atomic He^+ ion and not the molecular He_2^+ ion even at pressures as high as 20 torr. Although it appears possible that the He^+ atomic ion could be formed progressively in the afterglow by the two-body reaction

$$He^* \ 2^3S_1 + He^* \ 2^3S_1 \rightarrow He^+ + He + e^- \qquad (3.32)$$

proposed by Rogers and Biondi,[108] recent experiments on charge-transfer-excited lasers show that helium atomic ions can have a lifetime long enough to account for the observed emission of atomic radiation long into the afterglow period. Figure 3.20 from Fite *et al.*,[113]

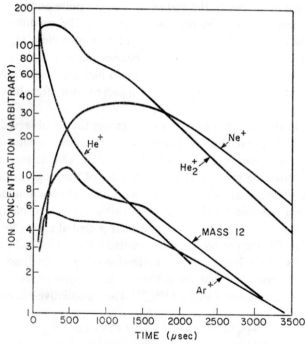

FIG. 3.20. Relative ion abundances in the afterglow of an rf-discharge in spectroscopically pure helium at 2.0 torr. (After W. L. Fite, J. A. Rutherford, W. R. Snow, and V. van Lint,[113] by courtesy of the Faraday Society.)

which illustrates the relative ion abundances in the afterglow of an rf-discharge in spectro-scopically pure helium at a pressure of 2.0 torr, shows that both atomic and molecular helium ions can exist well into the afterglow to more than 2000 μsec after the termination of the current pulse. The rapid rate of decay of the atomic He^+ ion compared with that of molecular He_2^+ ion throughout the afterglow period is clearly displayed. It is interesting to note that the ion of mass 12 noted in Fig. 3.20 is believed to be He_3^+ (and not C^+).

3.5.2.2. Afterglow processes in neon

Neutralization of charged particles in the initial afterglow period of a pulsed discharge in neon is due to dissociative-recombination following the formation of neon molecular ions from neon atomic ions and ground-state neon atoms.[114] The two-stage process involved is

$$Ne^+ + 2\,Ne \rightarrow Ne_2^+ + Ne \qquad (3.33)$$

followed by

$$Ne_2^+ + e^- \rightarrow (Ne_2)' \rightarrow Ne' + Ne \qquad (3.34)$$

where $(Ne_2)'$ is an unstable excited molecule and Ne' is a highly excited Ne atom. The dissociative-recombination reaction (3.34) is likely to be highly temperature dependent as is the similar reaction in helium.

3.5.2.3. Afterglow processes in a helium–neon mixture

The following triple-collision processes are believed to play a significant role in de-ionization in the afterglow of a pulsed discharge in a high-pressure helium–neon mix-ture.[115, 116]

$$Ne^+ + Ne + He \diagup\diagdown \begin{array}{l} \longrightarrow (HeNe)^+ + Ne, \\ \\ \longrightarrow Ne_2^+ + He, \end{array} \qquad (3.35a)$$

or

$$Ne^+ + 2\,He \qquad\qquad \longrightarrow (HeNe)^+ + He, \qquad (3.35b)$$

followed by

$$(HeNe)^+ + e^- \rightarrow (HeNe)' \rightarrow Ne' + He, \qquad (3.36)$$

or

$$Ne_2^+ + e^- \rightarrow (Ne_2)' \rightarrow Ne' + Ne, \qquad (3.37)$$

where $(HeNe)'$ and $(Ne_2)'$ are unstable excited molecules, and Ne' is an excited atom.

Table 3.4 taken from Egorov et al.[115] shows that there is a marked increase in the concentration of Ne^* $1s$ metastables in a pulsed discharge in a He–Ne mixture over that in pure Ne, whereas there is a decrease in the concentration of He^* 2^3S_1 metastables in a He–Ne mixture over that in a pulsed discharge in pure He. Although the gas pressures differ in the discharge and hence the excitation conditions will differ for the two gases and gas mixture intercompared, the clear increase in concentration by two orders of magnitude of Ne^* $1s$ metastables can be attributed to atom–atom excitation transfer from helium to neon via the reaction

$$He^*\ 2^3S_1 + Ne\ ^1S_0 \rightarrow He\ ^1S_0 + Ne\ 2s_{2-5} + \varDelta E_\infty \qquad (3.38)$$

and subsequent cascading $2s$–$2p$, and $2p$–$1s$ transitions into the metastable/quasi-metastable Ne $1s_{2-5}$ levels. The decrease in concentration of He* 2^3S_1 metastables in the He–Ne mixture by nearly two orders of magnitude of that in pure He follows from their de-excitation by neon atoms in reaction (3.38). It is also likely that deactivation of He* 2^3S_1 metastables in a high-pressure He–Ne mixture proceeds via the following associative-ionization process

$$\text{He* } 2^3S_1 + \text{Ne } {}^1S_0 \rightarrow (\text{HeNe})^+ + e^- \qquad (3.39)$$

that occurs in a pulsed high-pressure He–Ne laser.[117] Deionization of (HeNe)$^+$ ions would then follow via the dissociative-recombination reaction (3.36) in which the recombining electron may or may not have an energy differing from that possessed by the electron produced in the associative-ionization reaction of eqn. (3.39).

An examination of the literature referenced in the field will clearly indicate that a number of these afterglow processes are dependent on a variety of factors that include length and nature of the excitation current-pulse,[118, 119] the gas density, electron temperature and also the gas temperature.[118, 120]

TABLE 3.4. *Metastable atom concentrations in the afterglow of pulsed discharges in He, Ne, and He–Ne mixtures*[115]

Gas	Pressure (torr)	Current (amp)	Metastable specie	Metastable concentration
1. Pure Ne	0.5	70	Ne* $1s_5$	10^{11}
2. He–Ne mixture	0.5 Ne 4 He	60	Ne* $1s_5$ He* 2^3S_1	1.5×10^{13} 10^{11}
3. Pure He	4	80	He* 2^3S_1	7×10^{12}

Atom–atom, ion–atom resonant excitation transfer and charge-transfer reactions are likely to be of considerable importance in these afterglow processes (many are certainly sensitive to excitation and afterglow conditions). It seems inevitable that until more attention is paid to standardizing experimental arrangements and to the recording of more experimental parameters, differences in observation and interpretation of results by different researchers will continue to complicate analyses of afterglow processes.

References

1. FABRIKANT, V. A., Doctoral dissertation, Lebedev Institute, Acad. of Science, U.S.S.R. (1939).
2. *Soviet Gas Laser Research*, U.S. Government A.D. report 282253 (1962).
3. JAVAN, A., *Phys. Rev. Lett.* **3**, 86 (1959).
4. SANDERS, J. H., *Phys. Rev. Lett.* **3**, 87 (1959).
5. SANDERS, J. H., *Bell Telephone Lab.*, Case 38543 (1959).
6. BUTAYEVA, F. A. and FABRIKANT, V. A., *Academiya Nauk SSSR*, Lebedev Institute, JETP, *Studies of Experimental and Theoretical Physics*, pp. 62–70, U.S.S.R. Acad. Press, 1959.

7. BASOV, N. G. and KROKHIN, O. N., *Sov. Phys. JETP* **12**, 1240 (1961) and *Appl. Optics*, **1**, 213 (1962).
8. FABRIKANT, V. A., *Sov. Phys. JETP* **14**, 375 (1962).
9. JAVAN, A., BENNETT, W. R., Jr. and HERRIOTT, D. R., *Phys. Rev. Lett.* **6**, 106 (1961).
10. VON ENGEL, A., *Ionized Gases*, 2nd ed., pp. 217–57, Clarendon Press, Oxford, 1955.
11. FRANCIS, G., The glow discharge, in *Handbuch der Physik*, **22** (ed. S. FLUGGE), Springer-Verlag, Berlin (1956), pp. 53–203.
12. PARKER, P., *Electronics*, pp. 601–71, Arnold, 1953.
13. BROWN, S. C., *Basic Data of Plasma Physics*, pp. 275–301, Wiley, New York, 1959.
14. PRINGLE, D. H. and FARVIS, W. E. J., *Phys. Rev.* **96**, 536 (1954).
15. UDELSON, B. J. and CREEDON, J. E., *Phys. Rev.* **88**, 145 (1952).
16. BREWER, A. K. and WESTHAVER, J. W., *J. Appl. Phys.* **8**, 779 (1937).
17. PERSSON, K. B., *J. Appl. Phys.* **36**, 3086 (1965).
18. CARON, P. R. and RUSSO, F., *J. Appl. Phys.* **41**, 3547 (1970).
19. KNEWSTUBB, P. F. and TICKNER, A. W., *J. Chem. Phys.* **36**, 684 (1962).
20. HURT, B. H., Ph.D. Thesis, University Microfilm No. 11803 (1964).
21. YOUNG, R. T., Jr., unpublished paper, Calculation of average electron temperatures in rare gas mixtures (1968).
22. VON ENGEL, A., ref. 10, p. 123.
23. YOUNG, R. T., Jr., *J. Appl. Phys.* **36**, 2324 (1965).
24. DORGELA, H. B., ALTING, H. and BOERS, J., *Physica Haag* **2**, 959 (1935).
25. VON ENGEL, A. and STEENBECK, W., *Elektrische Gasentladungen*, Band II, p. 85, Julius Springer-Verlag, Berlin, 1932.
26. LABUDA, E. F. and GORDON, E. I., *J. Appl. Phys.* **35**, 1647 (1964).
27. MAMIKONYANTS, N. L. and MOLCHASHKIN, M. A., *Radio Engng Electron. Phys.* **3**, 518 (1967).
28. YOUNG, R. T., Jr., WILLETT, C. S. and MAUPIN, R. T., *J. Appl. Phys.* **41**, 2936 (1970).
29. REVALD, V. F., *Optics Spectrosc.* **18**, 318 (1965).
30. WADA, J. W. and HEIL, H., *IEEE J. Quant. Electr.* **QE-1**, 327 (1965).
31. WARD, R. C., Aerospace Report TDR-930 (2250-20) TN-1, "Preliminary calculations on the plasma of a helium–neon laser", Sept. 1961.
32. AFANAS'EVA, V. L., LUKIN, A. V. and MUSTAFIN, K. S., *Sov. Phys. Tech. Phys.* **12**, 233 (1967).
33. BOYD, R. L. F. and TWIDDY, N. D., *Proc. Roy. Soc.* **259**, 145 (1960).
34. RAYMENT, S. W. and TWIDDY, N. D., *Proceedings of the 8th International Conference on Phenomena in Gases*, Vienna (1967), p. 109.
35. BROWN, S. E., Breakdown in gases: alternating and high-frequency fields, in *Handbuch der Physik*, Band XXII, pp. 531–74 (ed. S. FLUGGE), Julius Springer-Verlag, Berlin, 1956.
36. FRANCIS, G., *Ionization Phenomena in Gases*, pp. 81–172, Academic Press, New York, 1960.
37. MAKSIMOV, A. I., *Sov. Phys. Tech. Phys.* **11**, 1316 (1967); and *Optics Spectrosc.* **26**, 369 (1969).
38. MAKSIMOV, A. I., *Optics Spectrosc.* **21**, 422 (1966).
39. MOTORNENKO, A. P., *Optics Spectrosc.* **18**, 603 (1965).
40. MOTORNENKO, A. P., *Optics Spectrosc.* **23**, 286 (1967).
41. BLOOM, A. L. and GOLDSBOROUGH, J. P., *Special Type Discharge Gas Lasers*, Tech. Report ECOM-02239-1, Feb. 1967.
42. BROWN, S. C., ref. 13, pp. 5 and 7.
43. BOCHKOVA, O. P., RAZUMOVSKAYA, L. P. and FRISH, S. E., *Optics Spectrosc.* **11**, 376 (1961).
44. BOCHKOVA, O. P. and RAZUMOVSKAYA, L. P., *Optics Spectrosc.* **17**, 8 (1964).
45. BOCHKOVA, O. P. and RAZUMOVSKAYA, L. P., *Optics Spectrosc.* **18**, 389 and 438 (1965).
46. BROWN, S. C., ref. 35, p. 537.
47. HARRIES, W. L. and VON ENGEL, A., *Proc. Roy. Soc.* A **222**, 490 (1954).
48. HEMSWORTH, R. S. and TWIDDY, N. D., *Proceedings of the 8th International Conference on Phenomena in Gases, Vienna* (1967), p. 486.
49. PASCHEN, F., *Ann. d. Physik* **50**, 901 (1916).
50. SMITH, J., *J. Appl. Phys.* **35**, 723 (1964).
51. CHEBOTAYEV, V. P., *Radio Engng Electron. Phys.* **10**, 314 and 316 (1965).
52. CHEBOTAYEV, V. P. and POKASOV, V. V., *Radio Engng Electron. Phys.* **10**, 817 (1965).
53. ZNAMENSKII, V. B., BUMOV, G. N. and BURAKOV, E. S., *Optics Spectrosc.* **20**, 292 (1966).
54. WILLETT, C. S. and JANNEY, G. M., *IEEE J. Quant. Electr.* **QE-6**, 568 (1970).
55. HUCHITAL, D. A. and RIGDEN, J. D., *IEEE J. Quant. Electr.* **QE-3**, 378 (1967).
56. SCHUEBEL, W. K., *Appl. Phys. Lett.* **16**, 470 (1970).
57. SCHUEBEL, W. K., *IEEE J. Quant. Electr.* **QE-6**, 574 and 655 (1970), and **QE-7**, 39 (1971).

58. COLLINS, G. J., JENSEN, R. C. and BENNETT, W. R., Jr., *Appl. Phys. Lett.* **18**, 50 and 282 (1971), and **19**, 125 (1971).
59. KARABUT, E. K., MIKHALEVSKII, V. S., PAPKIN, V. F. and SEM, M. F., *Sov. Phys. Tech. Phys.* **14**, 1447 (1970).
60. SUGAWARA, V. and TOKIWA, V., *Japan. J. Appl. Phys.* **9**, 588 (1970).
61. SUGAWARA, V., TOKIWA, V. and IIJIMA, T., *Digest of Tech. Papers*, QEC, Tokyo (1970), p. 320.
62. TOLANSKY, S., *High Resolution Spectroscopy*, p. 35, Methuen, London, 1947.
63. BURGER, J. C., GILLIES, W., and YAMASAKI, G., Westinghouse Product Engineering Memo ETD-6702, December (1967), presented at Soc. for Appl. Spectry, May 1967, Chicago, Illinois.
64. LITTLE, P. F. and VON ENGEL, A., *Proc. Roy. Soc.* A **224**, 209 (1954).
65. WHITE, A. D., *J. Appl. Phys.* **30**, 711 (1959).
66. STURGES, D. J. and OSKAM, H. J., *J. Appl. Phys.* **35**, 2887 (1964).
67. GEWARTOWSKI, J. W. and WATSON, H. A., *Principles of Electron Tubes*, chap. 15.3, p. 56, van Nostrand, 1965.
68. ANDERSON, J. M., *J. Appl. Phys.* **31**, 511 (1960).
69. GRIMLEY, H. M. and EMELEUS, K. G., *Brit. J. Appl. Phys.* **16**, 281 (1965).
70. BORODIN, V. S. and KAGAN, YU. M., *Sov. Phys. Tech. Phys.* **11**, 131 (1966).
71. AFANAS'EVA, V. L., LUKIN, A. V. and MUSTAFIN, K. S., *Sov. Phys. Tech. Phys.* **11**, 389 (1966).
72. SAWYER, R. A., *Phys. Rev.* **36**, 44 (1930).
73. BADAREU, E. and POPESCU, I., *Ann. Phys.* **5**, 508 (1960).
74. BADAREU, E. and POPESCU, I., *J. Electr. Control* **4**, 503 (1958).
75. ROESSLER, F. and DE NOYER, L., *Phys. Rev. Lett.* **12**, 396 (1964).
76. CIOBATARU, D., *J. Electr. and Control* **4**, 529 (1958).
77. DRUYVESTEYN, M. J. and PENNING, P. M., *Rev. Mod. Phys.* **12**, 87 (1940).
78. VON HIPPEL, A., *Ann. Phys.* **81**, 999 and 1043 (1926).
79. TOWNES, C. H., *Phys. Rev.* **65**, 319 (1964).
80. WEHNER, O. K., *Phys. Rev.* **108**, 35 (1957).
81. WEHNER, O. K., *Phys. Rev.* **112**, 1120 (1958).
82. LOEGREID, N. and WEHNER, O. K., *J. Appl. Phys.* **32**, 365 (1961).
83. WEHNER, O. K., *J. Appl. Phys.* **33**, 1842 (1962).
84. WEIJSENFELD, C. H., HOOGENDOORN, A. and KOEDAM, M., *Physica* **27**, 763 (1961).
85. MUSHA, T., *Japan. J. Phys. Soc.* **17**, 1440 (1962).
86. TOKATSU, K. and TODA, T., *Proceedings of the Conference on Ionization Phenomena in Gases*, vol. 1, pp. 96–105, Munich (1961).
87. STOCKER, B. J., *Brit. J. Appl. Phys.* **12**, 465 (1961).
88. CROSSWHITE, H. M., DIECKE, G. H. and LEGAGNEUR, C. S., *J. Opt. Soc. Am.* **45**, 270 (1955).
89. MITCHELL, K. B., *J. Opt. Soc. Am.* **51**, 846 (1961).
90. ZNAMENSKII, V. B., *Optics Spectrosc.* **15**, 7 (1968).
91. CHEBOTAYEV, V. P., *Radio Engng Electron. Phys.* **10**, 372 (1965).
92. GORDON, E. I. and WHITE, A. D., *Appl. Phys. Lett.* **3**, 199 (1963).
93. BOGDANOVA, I. P. and GI-TKEK, C., *Optika*; *Spectroskopiya* **2**, 681 (1957), translated by the National Lending Library for Science and Technology, Boston Spa, England, as RTS 2456 (1964).
94. LOMAX, R. W. and LYTALLIS, J., *Electr. Commun. (U.S.A.)* **37**, 367 (1962).
95. TOWNSEND, M. A. and DEPP, W. A., *Bell System Tech. J.* **32**, 1371 (1953).
96. BOOT, H. A. H. and CLUNIE, D. M., *Nature* **197**, 173 (1963).
97. BOOT, H. A. H., CLUNIE, D. M. and THORN, R. S. A., *Nature* **198**, 773 (1963).
98. MATHIAS, L. E. S. and PARKER, J. T., *Appl. Phys. Lett.* **3**, 16 (1963).
99. MATHIAS, L. E. S. and PARKER, J. T., *Phys. Lett.* **7**, 194 (1963).
100. VON ENGEL, A., ref. 10, p. 195.
101. LLEWELLYN-JONES, F., Ionization growth and breakdown, in *Handbuch der Physik*, **22**, pp. 1–52 (ed. S. FLUGGE), Julius Springer-Verlag, Berlin, 1956.
102. CHEO, P. K. and COOPER, H. G., *Appl. Phys. Lett.* **5**, 42 (1964).
103. BREUSOVA, L. N., *Radio Engng Electron. Phys.* **10**, 1595 (1965).
104. PAKHOMOV, P. L., REZNIKOV, G. P. and FUGAL, I., *Optics Spectrosc.* **20**, 5 (1966).
105. BIONDI, M. A. and HOLSTEIN, T., *Phys. Rev.* **82**, 962 (1951).
106. ANDERSON, J. M., *Phys. Rev.* **108**, 898 (1957).
107. LOEB, L. B., *Basic Processes of Gaseous Electronics*, p. 567, University of California Press, Berkeley, 1955.
108. ROGERS, W. A. and BIONDI, M. A., *Phys. Rev.* **134**, A1215 (1964).

109. EGOROV, V. S., KOZLOV, YU. G. and SHUKHTIN, A. M., *Optics Spectrosc.* **14,** 82 (1964).
110. BIONDI, M. A. and HOLSTEIN, T., *65th Conference on Gaseous Electronics, Washington, D.C.* (1953).
111. FERGUSON, E. E., FEHSENFELD, F. C. and SCHMELTEKOPF, A. L., *Phys. Rev.* **136,** A381 (1965).
112. COLLINS, C. B. and ROBERTSON, W. W., *J. Chem. Phys.* **40,** 2202 (1964).
113. FITE, W. L., RUTHERFORD, J. A., SNOW, W. R. and VAN LINT, V., *Disc. Faraday Soc.* **33,** 264 (1962).
114. EGOROV, V. S. and SHUKHTIN, A. M., *Optics Spectrosc.* **9,** 419 (1960).
115. EGOROV, V. S., KOZLOV, YU. G. and SHUKHTIN, A. M., *Optics Spectrosc.* **15,** 458 (1963).
116. LUKAC, P., *J. Phys.* D, **3,** 1689 (1970).
117. SHTYRKOV, E. I. and SUBBES, E. V., *Optics Spectrosc.* **21,** 143 (1966).
118. SUZUKI, N., *Japan J. Appl. Phys.* **4,** 642, Supplement 1 (1965) on Proc. Conf. Photographic and Spectroscopic Optics (1964).
119. HASTED, J. H., *Physics of Atomic Collisions*, p. 268, Butterworth, 1964.
120. SUZUKI, N., *Japan. J. Appl. Phys.* **4,** 285 (1965).

CHAPTER 4

Specific Neutral Laser Systems

Introduction

In this chapter particular neutral atomic laser systems are analysed to illustrate six of the selective excitation processes covered in Chapter 2:

Resonant Excitation-energy Transfer;
Dissociative Excitation Transfer;
Electron-impact Excitation;
Charge Neutralization;
Line Absorption and Molecular Photodissociation; and
Radiative Cascade-pumping.

Also considered are the ways in which population inversion is maintained in these systems in the presence of selective excitation of the upper laser level. Throughout the chapter, operating characteristics and discharge conditions are given of the specific laser systems discussed.

4.1. Resonant Excitation-energy Transfer Lasers

The neutral helium–neon and the 3.067-μm argon–chlorine lasers are two gas lasers in which population inversion is clearly produced by resonant excitation-energy transfer. Another laser exists, however, in which excitation transfer occurs in conjunction with other selective excitation mechanisms to give population inversion. This is the argon–oxygen laser that operates in neutral atomic species of oxygen at 0.8446 μm. The argon–oxygen laser is considered in Section 4.2, where it is more instructive to discuss its operation in conjunction with selective excitation mechanisms other than excitation energy-transfer.

4.1.1. THE HELIUM–NEON LASER

The helium–neon laser was the first gas laser in which oscillation was achieved in the near-infrared and visible regions of the spectrum, and of all present gas lasers is one of the most physically complicated and interesting as far as basic physical collision mechanisms are concerned. Even after a number of years of extensive investigations on it, there are still details of the basic physics of the excitation and de-excitation collisional processes that are not well understood.

Particular attention, therefore, is given to the He–Ne laser in this chapter to illustrate the important electronic, atomic, and molecular processes now known to be operative in it. Further, some aspects of it are emphasized in an attempt to correct a number of misunderstandings that exist about processes in it and that have been further related erroneously to other gas lasers in neutral species.

4.1.1.1. Continuous oscillation in the near-infrared

The first laser oscillation, on five transitions in the near-infrared at about 1.1 μm, was obtained by Javan, Bennett, and Herriott[1] at the Bell Telephone Laboratories in 1961. The five transitions occur between the $2s$ and $2p$ (Paschen notation) or $4s$ and $4p$ (Racah notation) levels in neutral neon.

Figure 4.1, a partial energy-level diagram of the lower levels of neutral helium and neon, together with well-known laser transitions, show that there is an energy-level coincidence

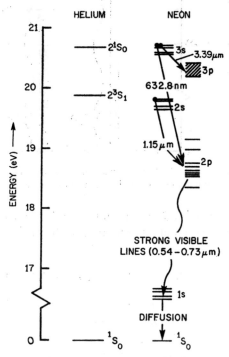

FIG. 4.1. Partial energy-level diagram for the He–Ne system, showing energy-level coincidences and well-known laser transitions.

between the He* 2^3S_1 level, which is metastable, and the four Ne-I levels of the $2s$ [or $4s$] group. Similar energy-level coincidences to be considered later occur for other energy levels.

According to the treatment in Section 2.2, large collision cross-sections are observed for collisions of the second kind between atoms in the reversible reaction

$$A' + B \rightleftarrows A + B' \pm \varDelta E_\infty, \tag{4.1}$$

where ΔE_∞ is less than a few kT. When the excited atom A' is metastable (A^*), and when B' is coupled radiatively to lower-lying energy levels, the forward process $A^* + B \rightarrow A + B'$ is very efficient.

Figure 4.2 shows in detail the energy coincidence for the He* 2^3S_1 level at 159,850 cm^{-1} and the Ne 2s levels, together with non-laser transitions of interest. All the four Ne 2s levels are within $+1247$ cm^{-1} of the He* 2^3S_1 levels so that excitation transfer might be

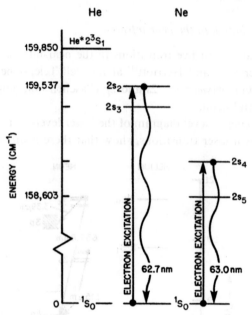

FIG. 4.2. Energy-level coincidence of the He* 2^3S_1 metastable level with Ne 2s levels, also showing strong vacuum-ultraviolet transitions to the Ne $^1S_0$ ground state and excitation paths.

expected between metastable He* 2^3S_1 atoms and ground state Ne $^1S_0$ atoms to give selective excitation of the Ne 2s levels and de-excitation of the He* 2^3S_1 atoms in the reaction (Paschen notation)

$$\text{He*} \ 2^3S_1 + \text{Ne} \ ^1S_0 \rightarrow \text{He} \ ^1S_0 + \text{He} \ (2s_{2,\,3,\,4,\,5}) + \Delta E_\infty \ (313\text{--}1247 \ \text{cm}^{-1}). \qquad (4.2)$$

The four Ne $2s_{2-5}$ levels in LS-coupling correspond to the $^1P_1$, $^3P_0$, $^3P_1$, and $^3P_2$ levels respectively, so that for the Wigner spin rule to be obeyed in the collision with the triplet He* 2^3S_1 metastables, the triplet $^3P_0$, $^3P_1$, $^3P_2$ levels should be the levels of the Ne 2s group to which excitation transfer occurs.[†] For reasons still not entirely clear the level that is preferentially excited is the Ne $2s_2$ level, which is a singlet level. This gives spin violation in eqn. (4.2) since on the left-hand side of the equation $\Sigma S = 1$, and on the right-hand side $\Sigma S = 0$ (following from the fact that the spin multiplicity is given by $2S+1$ for each state of the atom).

---

† Even though jl-coupling scheme is considered to give a better picture of the neon atom, in which case the s levels are not in fact singlets or triplets, for purposes of applying the Wigner spin rule they can be assigned a spin of $\frac{1}{2}$ or 1.

Lifetime considerations. The effective lifetimes of the Ne $2s$ states have been measured by Bennett and others,[2, 3] and of the Ne $2p$ state by Bennett *et al.*,[3] Oshirovich and Verolainen,[4] Klose,[5] Ladenburg,[6] and Griffiths.[7] Table 4.1 gives the effective lifetimes of these two states. The values for the Ne $2s$ states are those at a neon pressure of 1 torr.

TABLE 4.1. *Effective lifetimes for the Ne 2s and 2p (Paschen notation) states (in nsec)*

$2s_2$	$2s_3$	$2s_4$	$2s_5$	$2p_1$	$2p_2$	$2p_3$	$2p_4$	$2p_5$	$2p_6$	$2p_7$	$2p_8$	$2p_9$	$2p_{10}$	References
96	160	98	110	< 8	< 10	< 13	12	< 11	13	< 13	16	17	< 20	8
				14	20	18	24	23	21	22	25	24	26	4

The effective lifetimes are appreciably less than the pure radiative ones of isolated atoms given by the method of Bates and Damgaard[9] discussed in Chapter 2, since at a pressure of a few torr collisional de-excitation can become comparable to that of radiative decay. Since the Ne $2s$ lifetimes are longer than those of the $2p$ levels, it should be possible to obtain population inversion on all of the thirty allowed Ne $2s$–$2p$ transitions, given sufficient selective excitation of the $2s$ levels. Oscillation should thus be possible in a discharge in pure neon.

Javan[10] was the first to consider the near-infrared neon laser system in detail. Given a Maxwellian electron-energy distribution, Javan calculated that the reaction

$$\text{Ne}^* \, 1s_{2-5} + e^- \rightarrow \text{Ne} \, 2p_{1-10} + e^-, \qquad (4.3)$$

and radiation trapping of the strong (visible) lines on the $2p$–$1s$ transitions were second-order processes and that population inversion should be possible between the $2s$ and $2p$ states at high gas pressures (~ 10 torr). In subsequent experimental work in both pure neon and helium–neon mixtures it has been found that both of these processes are extremely important and determine the actual neon pressure and discharge current density at which the infrared laser can be operated. The neon pressure, which depends on the discharge-tube diameter, is low and is more like 0.1 torr than Javan's calculated value of about 10 torr.

Javan[10] suggested that the population of Ne* $1s$ metastables, whose presence reduces population inversion between the $2s$ and $2p$ levels by increasing the lower level $2p$ population by means of reaction (4.3), could be reduced by the addition of another gas such as argon. Argon did not prove to be a good choice, since it adversely affected the upper-state excitation rate by reducing the electron temperature or by altering the shape of the electron-energy distribution. The addition, however, of hydrogen or oxygen has been found to be beneficial in maintaining inversion between the Ne $2s$–$2p$ levels in certain types of laser discharges.[11]

Javan,[10] noting the resonant excitation-transfer reaction of eqn. (4.2), suggested that oscillation might be obtainable in a mixture of helium and neon in the afterglow period of a discharge. He deduced that (long-lived) metastable He* 2^3S_1 atoms would continue to excite the upper Ne $2s$ states selectively by excitation transfer while energetic electrons would not exist to excite the $2p$ levels from the metastable/quasimetastable atoms in the Ne $1s$ levels.

In the afterglow, electrons thermalize, depending on physical conditions, in times of the order of a few microseconds so that reaction (4.3), which reduces inversion between the Ne $2s$ and $2p$ levels, is not significant after a few microseconds in the afterglow. Experimental studies by Javan et al.[1] confirmed this. In the afterglow of a discharge in a helium–neon mixture the Ne $2s$ levels decayed with a time variation similar to that of the $He^* \, 2^3S_1$ metastables thus establishing that excitation transfer from the helium metastables to neon was occurring. Oscillation was subsequently observed in the afterglow at the five wavelengths listed below in Table 4.2.

TABLE 4.2. *Details of first laser transition in neon*

λ_{air} (μm)	Transition		Relative probability[a]
	Paschen	Racah	
1.118	$2s_5\text{--}2p_9$	$4s[3/2]_2^\circ\text{--}3p[5/2]_3$	7/9
1.152	$2s_2\text{--}2p_4$	$4s'[1/2]_1^\circ\text{--}3p'[3/2]_2$	5/9[b]
1.160	$2s_2\text{--}2p_3$	$4s'[1/2]_1^\circ\text{--}3p[1/2]_0$	0
1.199	$2s_3\text{--}2p_2$	$4s'[1/2]_0^\circ\text{--}3p'[1/2]_1$	1/9
1.207	$2s_5\text{--}2p_6$	$4s[3/2]_2^\circ\text{--}3p[3/2]_2$	1/2

[a] The relative probabilities for each transition are taken from Koster and Statz.[12]
[b] Indicates strongest laser transition.

Although strongest oscillation at 1.152 μm occurred on the $2s_2\text{--}2p_4$ transition that has the largest relative transition probability of any transitions from the Ne $2s_2$ level, strong oscillation at this wavelength was not predicted. The Ne $2s_2$ state was not expected to be the state that would be preferentially excited in excitation-transfer reaction (4.2) because of spin violation. Thus the spin-violated collision

$$He^* \, 2^3S_1(\uparrow\uparrow) + Ne \; {}^1S_0(\uparrow\downarrow) \rightarrow He \; {}^1S_0(\uparrow\downarrow) + Ne \; 2s_2(\uparrow\downarrow) + \Delta E_\infty \; (313 \text{ cm}^{-1}) \qquad (4.4)$$

is evidently the most important process in excitation transfer to the four Ne $2s$ states. Here apparently the collision cross-section depends more on energy discrepancy than spin conservation.[13]

Javan et al.[1] later obtained CW oscillation on four of the infrared transitions given in Table 4.2. They used an rf-excited discharge in a 10 : 1 mixture of helium and neon at a total pressure of 1.1 torr (in a tube of 15-mm bore and 80-cm length). The 1.1523-μm line exhibited the largest output power and gain.

In jl-coupling the Ne $2s_2$ $\left(4s'\left[\frac{1}{2}\right]_1\right)$ level is optically connected to the Ne ${}^1S_0$ ground state ($\Delta J = +1$) by a strong, vacuum uv-transition at 62.7 nm. It follows from (2.5) that it will have a large cross-section for direct electron-impact excitation from the ground state in the reaction

$$Ne \; {}^1S_0 + e^- \rightarrow Ne \; 2s_2 + e^- \, . \qquad (4.5)$$

The excitation of the $2s_2$ level under CW conditions then consists of both excitation transfer from $He^* \, 2^3S_1$ metastables and electron-impact excitation in a helium–neon mixture.

There is some disagreement on the relative importance of electron-impact excitation and excitation transfer to the Ne $2s$ levels. Bennett[8] states that not only are reactions (4.4) and (4.5) of equal importance in a CW He–Ne near-infrared laser, but that the direct electron-impact-excitation reaction (4.5) appears to be enhanced by the presence of helium, which increases the electron concentration without causing a decrease in the electron temperature. Later work has shown that excitation of the Ne $2s$ levels due to helium is the dominant mechanism in the CW He–Ne 1.15-μm laser discharge.[13] Figure 4.3, taken from Young,

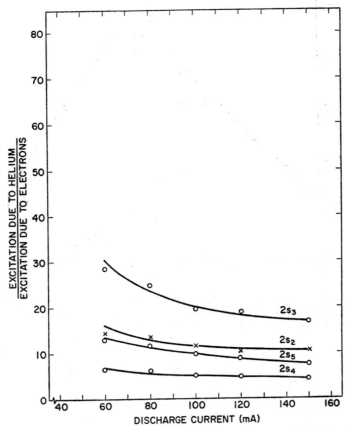

FIG. 4.3. Excitation ratios for the $2s$ levels of neon. Discharge conditions: 10 : 1 He–Ne mixture at a pD of 7.7 torr-mm, and tube diameter 10 mm. (After R. T. Young, Jr., C. S. Willett, and R. T. Maupin.[13])

Willett and Maupin,[13] shows that the ratio of excitation of the Ne $2s$ states due to helium to that due to electron impact is 20 for the $2s_3$ state, 13 for the $2s_2$ state, 11 for the $2s_5$ state; and 5 for the $2s_4$ state at a discharge current of 100 mA under discharge conditions characteristic of lasers operating at 1.15 μm. Theoretically, the average electron temperature should decrease, not increase or even remain constant, when helium is added to a discharge at a fixed pressure of neon. This follows from a consideration of Fig. 3.7.

Continuous oscillation in pure neon that can definitely be tied to oscillation from particular Ne $2s$ states and that results from direct electron-impact excitation of the Ne $2s_2$ state

107

has been reported by Bennett,[8, 14] and from excitation of the Ne $2s_4$ state by Petrash and Knyazev.[15] The Ne $2s_4$ $(4s[\frac{3}{2}]_1^\circ)$ level is optically connected to the ground state by a far-ultraviolet transition at 63.0 nm, so that, similar to the Ne $2s_2$ level, it has a large cross-section for direct electron-impact excitation from the ground state. Figure 4.4 taken from

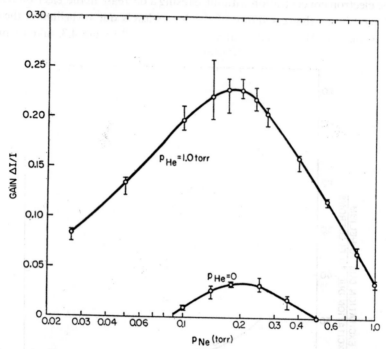

FIG. 4.4. Measured gain curves for the Ne $2s_2$–$2p_4$ transition at 1.1523 μm, in a 2-m long, 7-mm diameter discharge tube. The data shown gave the first proof that a practical amount of gain could be obtained in a pure gas by electron-impact excitation. (After W. R. Bennett, Jr.,[8] by courtesy of the Optical Society of America.)

Bennett[8] shows that the gain in a 1.15-μm pure-neon laser is about a factor of 7 less than that realized in a helium–neon laser mixture.

Table 4.3 gives the wavelengths, transitions, their relative probability, notes on, and references to lines on which oscillation has been reported of the thirty optically allowed transitions that exist between the Ne $2s$ and $2p$ states. On only the $2s_2$ and $2s_4$ to $2p$ transitions has oscillation been obtained as a result of electron-impact excitation alone.

Effects of additives. The addition of various gases in attempts to increase the atom–atom collisional decay rate of the Ne*-$1s$ metastables has been made by Bennett and others[8] and Chebotayev.[11, 29] Patel, Bennett, Faust, and McFarlane (unpublished work) were the first to try this technique by adding argon to the infrared He–Ne laser.[8] The results obtained were not encouraging. The addition of argon led to a decrease in gain, apparently because it adversely affected the excitation rate of He* 2^3S_1 metastables and the upper laser levels by reducing the electron temperature. The use of hydrogen or oxygen instead of argon, by Chebotayev[11, 29] in work after that of the group at the Bell Telephone Laboratory, was more successful.

TABLE 4.3. *Ne–I, 2s–2p laser transitions*

λ_{air} (μm)	Transition		Relative probability	Notes[a]	References
	Paschen	Racah			
1.5231	$2s_2$–$2p_1$	$4s'[3/2]_1^\circ$–$3p'[1/2]_0$	1/9	S	16–19
1.1767	–$2p_2$	–$3p'[1/2]_1$	2/9		11, 15, 16, 19, 20
1.1601	–$2p_3$	–$3p[1/2]_0$	0		16, 19, 21
1.1523	–$2p_4$	–$3p'[3/2]_2$	5/9	VS*	1, 11, 15, 19, 20–21, 23
1.1409	–$2p_5$	–$3p'[3/2]_1$	1/9		16, 19, 20
1.0844	–$2p_6$	–$3p[3/2]_2$	0		16, 20, 24
1.0621	–$2p_7$	–$3p[3/2]_1$	0		19, 22
1.0295	–$2p_8$	–$3p[5/2]_2$	0		19, 24
0.8865	–$2p_{10}$	–$3p[1/2]_1$	0		19, 24
1.1985	$2s_3$–$2p_2$	$4s'[1/2]_0^\circ$–$3p'[1/2]_1$	1/9		1, 19
1.1614	–$2p_5$	–$3p'[3/2]_1$	2/9		1, 19
1.0798	–$2p_7$	–$3p[3/2]_1$	0		16, 19
0.8989	–$2p_{10}$	–$3p[1/2]_1$	0		19, 24
1.7162	$2s_4$–$2p_1$	$4s[3/2]_1^\circ$–$3p'[1/2]_0$	0		19, 24
1.2887	–$2p_2$	–$3p'[1/2]_1$	0		24
1.2689	–$2p_3$	$4s[3/2]_1^\circ$–$3p[1/2]_0$	1/9	VS in HCD in Ne–H₂	19, 20, 25, 26
1.2594	$2s_4$–$2p_4$	–$3p'[3/2]_2$	0		20
1.2460	–$2p_5$	–$3p'[3/2]_1$	0		19, 20, 24
1.1789	–$2p_6$	–$3p[3/2]_2$	1/18		11, 19, 20, 24
1.1525	–$2p_7$	–$3p[3/2]_1$			1, 11, 20, 23, 27
1.1143	–$2p_8$	–$3p[5/2]_2$	1/2	VS in HCD in He–H₂	11, 15, 16, 19, 20
0.9486	–$2p_{10}$	–$3p[1/2]_1$	1/18		20
1.3912	$2s_5$–$2p_2$	$4s[3/2]_2^\circ$–$3p'[1/2]_1$	0		20
1.2912	–$2p_4$	–$3p'[3/2]_2$	0		11, 19, 20, 24, 26, 28
1.2767	–$2p_5$	–$3p'[3/2]_1$	0		19
1.2066	–$2p_6$	–$3p[3/2]_2$	1/2		1, 19, 20, 22
1.1790	–$2p_7$	–$3p[3/2]_1$	1/18		11, 19, 24, 20
1.1390	–$2p_8$	–$3p[5/2]_2$	1/18	S in HCD in Ne–H₂	16, 19, 20
1.1178	–$2p_9$	–$3p[5/2]_3$	7/9		1, 11, 19, 20, 22
0.9665	–$2p_{10}$	–$3p[1/2]_1$	5/18		20

[a] S indicates Strong; VS – Very Strong; VS* – Strongest line. HCD indicates hollow-cathode discharge.

It is known from the early optical-dispersion work of Ladenburg and Dorgela[30] that oxygen and hydrogen are effective in quenching Ne* $1s$ metastables in discharges in neon. On the basis of known cross-sections for de-activation of Ne*-$1s$ metastables by oxygen or hydrogen,[31, 32] oxygen should be more effective than hydrogen. A Ne–O₂ mixture, therefore, should give higher gain than a Ne–H₂ laser mixture. Rather surprisingly, hydrogen is found to be more effective than oxygen as an additive. There are two reasons for this. As well as acting to de-excite Ne* $1s$ metastables, the addition of hydrogen to neon:

(1) increases the electron-impact excitation of the Ne $2s$ upper laser levels by forming a pronounced peak in the electron-energy distribution at about 20 eV (Fig. 4.5). As indicated earlier in Fig. 4.1 the Ne $2s$ levels are situated at about 20 eV above the

ground state at a potential energy that corresponds to that of the observed peak in the electron-energy distribution;

(2) causes a depletion of electrons in the electron-energy distribution at about 16 eV so that production of Ne^* $1s$ metastables by direct electron-impact excitation from the ground states in electron-exchange collisions is reduced,[33] and two-step excitation of the $2p$ lower laser levels is reduced.

As shown in Fig. 4.5, the pronounced peak in the electron energy distribution is observed at high current densities (400-mA current) but not at low current densities (200-mA current).[33] The reason for its appearance at high current densities in a Ne–H_2 mixture is open to speculation. The discharge which exhibited these effects was a hollow-cathode discharge

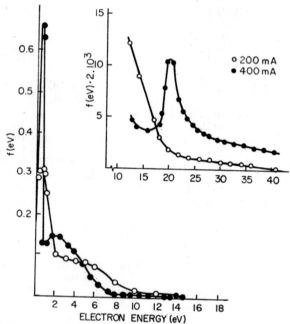

FIG. 4.5. Electron energy distribution in a hollow-cathode discharge in a Ne–H_2 laser mixture. Discharge conditions: $p(Ne)$, 0.8 torr; $p(H_2)$, 0.3 torr; cathode diameter, 1.2 cm; length 30 cm. (After V. L. Afanas'eva, A. V. Lukin, and K. S. Mustafin,[33] by courtesy of the Editor, *Soviet Physics-Technical Physics*.)

(HCD), and it is possible that the observed distribution is peculiar to it. However, whatever the discharge conditions might be, if the peak in the electron energy distribution is indeed responsible for enabling oscillation to be obtained at high current densities, its relative smallness (note the scale of the inset enlargement) would indicate that the extent and nature of the high-energy tail of the electron-energy distribution must be of considerable importance in this and other laser discharges.

Figure 4.6, after Chebotayev,[34] shows the variation of spontaneous intensity of the three Ne-I lines, at 1.1523 μm, 609.6 nm, and 632.8 nm as a function of helium pressure, at a fixed neon pressure of 0.09 torr in an external-type HCD laser structure of the type illustrated in Fig. 3.13(c). The spontaneous intensity of the 1.1523-μm line, $2s_2–2p_4$ transition, is directly proportional to the Ne $2s_2$ population, and the spontaneous intensity of the 609.6-nm line, $2p_4–1s_4$ transition (although it is probably subject to radiation trapping) gives an indication of the population of the Ne $2p_4$ state, the lower laser level of the 1.1523-μm line. The peak in intensity of the 1.1523-μm line at a helium pressure below 1 torr of helium can be ascribed to direct electron-impact excitation of the Ne $2s_2$ state from the ground state. At a helium pressure of 3 to 4 torr, the much wider peak in spontaneous intensity of the

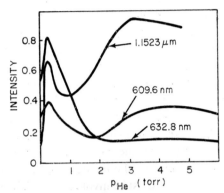

FIG. 4.6. Spontaneous intensity of Ne-I lines at 1.1523 μm, 609.6 nm and 632.8 nm as a function of helium pressure in a hollow-cathode discharge. Discharge conditions: fixed neon pressure of 0.09 torr; cathode diameter 12 mm. (After V. P. Chebotayev,[34] by courtesy of the Translation Editor, *Radio Engineering and Electronic Physics*.)

1.1523-μm line occurs as a result of excitation transfer from He* 2^3S_1 metastables to the Ne $2s_2$ state. Maximum laser output in this system occurs at 3 to 4 torr corresponding to the maximum difference in intensities between the 1.1523-μm and 609.6-nm lines. The additional curve of the 632.8-nm line intensity included in Fig. 4.6 shows that the Ne $3s_2$ state is excited more effectively at low pressures of helium (less than 0.4 torr) at the fixed neon pressure of 0.09 torr.

Overall results similar to these have been reported for helium–neon HCD lasers operating in the near infrared by Smith[23] and Znamenskii et al.[35] A very interesting aspect of the HCD laser, or of one operating in the negative-glow region of the glow discharge,[36] on the Ne-I $2s–2p$ transitions in the infrared, is that the optimum neon-to-helium ratio that gives maximum output power is about 0.02.[23, 34–36] This same ratio corresponds to that used in cold-cathode Dekatron discharge tubes, which have been found to give the shortest deionization time and the smallest sustaining voltage.[37] Whereas this effect could be ascribed to an optimum neon-to-helium ratio that maximizes destruction of He* 2^3S_1 metastables by neon thus making the maintenance of the discharge rely on direct ionization from the ground state, another mechanism is apparently of significance.

Desai and Kagan[38] have found that in addition to the electron-energy distribution differing substantially from a Maxwellian distribution at the center of a HCD in a helium–neon mixture, as the partial pressure of neon is increased at a fixed pressure of helium, the fraction of fast electrons increases to a maximum and then decreases. The maximum fraction of fast electrons occurs at a neon-to-helium ratio of between 0.01 and 0.04. This range clearly includes the neon-to-helium pressure-ratio of 0.02 for which the laser output power on the Ne $2s$–$2p$ transitions is a maximum using HCD excitation. The effect is observed at all discharge currents so that it is not related to any change in spatial variation of electron energies in the HCD. It is conjectured here that this surprising effect arises from an increased mobility of He^+ ions through a reduction in He^+-to-He symmetrical charge transfer at low partial pressures of neon, which causes the electron temperature to maximize at a particular neon-to-helium ratio. (See Fig. 3.7 and ref. 28, Chapter 3.)

The gain measurements illustrated in Fig. 4.4 on the Ne-I $2s_2$–$2p_4$ transitions proved for the first time that a practical amount of gain could be achieved in a pure neon discharge by direct electron-impact excitation of the Ne $2s_2$ state.[8] This result led naturally to investigations being carried out on other pure noble gases to see if CW oscillation could be obtained on similar transitions. As expected, oscillation was soon realized on a number of transitions in argon, krypton and xenon.[39] Again the addition of helium to these other noble-gas systems gives increased gain. This is particularly true in the xenon laser operating at 3.5 μm and 2.026 μm. In the case of xenon and in the other noble gases (excepting neon), there are reasons for believing that the increase in gain through the addition of helium is due to increased electron-impact excitation of the upper laser levels since excitation-transfer reactions with helium cannot account for the increased gain; nevertheless, it appears that some other selective-excitation process must be invoked to explain the effect. (This is covered in Section 4.3.1.1.)

4.1.1.2. Pulsed oscillation in the near infrared

Boot and Clunie[40] of the Services Electronics Research Laboratory were the first to use pulsed-discharge techniques to give population inversion and oscillation in a gas discharge. The technique enabled rapid advances to be made in the search for, and analysis of new laser systems, and resulted in the discovery of large numbers of CW as well as pulsed laser systems in the noble gases, and the discovery of the first molecular gas laser.[41] Boot et al.[42] in subsequent work, applied short-rise time, high-voltage, high-current pulses of short duration (30 to 60 kV, 90-A, 1-μsec pulse width) at repetition rates of about 1 kHz to a 150-cm long, 2-cm-bore tube containing a relatively high pressure (up to 240 torr) mixture of helium and neon. Oscillation at the (then) large peak-power of 84 W occurred in the afterglow on the four Ne-I $2s$–$2p$ transitions:

$$2s_5 - 2p_9, \quad \text{at} \quad 1.118 \ \mu m; \qquad 2s_2 - 2p_4, \quad \text{at} \quad 1.152 \ \mu m;$$
$$2s_3 - 2p_5, \quad \text{at} \quad 1.161 \ \mu m; \qquad \text{and} \quad 2s_5 - 2p_6, \quad \text{at} \quad 1.207 \ \mu m. \tag{4.6}$$

Population inversion in this system results from selective excitation of the Ne $2s$ levels by excitation-transfer collisions between ground-state neon atoms and the high concentration of He^* $2\,^3S_1$ metastables that are formed in the afterglow during the recombination of the well-ionized plasma, and likely by dissociative-recombination of $(HeNe)^+$ molecular ions (see Section 3.5.2.3). Using conditions differing little from those described above, Shtyrkov and Subbes[22] obtained oscillation during the excitation pulse, as well as in the afterglow in a high-pressure (above 200 torr) He–Ne mixture on the Ne-I $2s$–$2p$ transitions: $2s_2$–$2p_3$ at 1.160 μm; $2s_2$–$2p_4$ at 1.152 μm; $2s_2$–$2p_7$ at 1.062 μm; $2s_5$–$2p_6$ at 1.207 μm; and

$2s_5-2p_9$ at 1.118 μm. Oscillation during the excitation pulse is attributed to selective excitation by direct electron impact, and in the afterglow, to the dissociative-recombination reaction

$$(\text{HeNe})^+ + e^- \rightarrow \text{He} + \text{Ne}'. \tag{4.7}$$

Metastable $\text{He}^*\ 2^3S_1$ atoms play an essential role in the formation of the molecular $(\text{HeNe})^+$ ion, as oscillation disappears when an insignificant amount of argon is added to quench the metastable helium atoms.

At low pressures of neon, and helium–neon mixtures of about 1 torr, the selective-excitation process in pulsed discharges is clear. Oscillation in pure neon occurs only on transitions from the $2s_2$ and $2s_4$ levels, which are optically connected to the ground state. It follows that the excitation is by direct electron impact from the ground state. In the afterglow of a pulsed He–Ne discharge the selective excitation is by excitation transfer from $\text{He}^*\ 2^3S_1$ metastables. Figure 4.7, after Petrash and Knyazev,[15] shows very clearly the oscillation behavior of a number of Ne $2s-2p$ transitions under pulsed-excitation conditions at various pressures of neon and helium. This figure illustrates when, and on which tran-

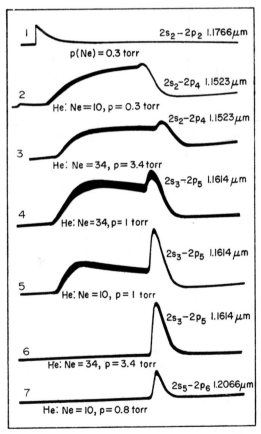

FIG. 4.7. Laser output in pure neon and He–Ne mixtures under pulsed conditions. In pure neon the duration of the excitation pulse was 60 μsec, in He–Ne mixtures it was 200 μsec. (After G. G. Petrash and I. N. Knyazev,[15] by courtesy of the Editor, *Soviet Physics-JETP*.)

sitions, electron-impact and excitation-transfer reactions are operating to produce population inversion. The initial sharp rise in the laser pulse in 1 is due to direct electron-impact excitation of the Ne $2s_2$ level, and the final peaks in traces 2 to 7 are due to excitation transfer in the afterglow from $He^* \, 2^3S_1$ metastables.

4.1.1.3. Oscillation in the visible

Shortly after CW oscillation was realized on the Ne-I $2s$–$2p$ transitions in the infrared by Javan, Bennett, and Herriott,[1] Rigden and White (ref. 8) of the Bell Telephone Laboratories observed that the visible Ne-I $3s_2$–$2p_4$ transition at 632.8 nm was strongly enhanced in an rf-excited discharge in a mixture of helium and neon compared with that in pure neon. They concluded that as the Ne $3s_2$ level was close to the $He^* \, 2^1S_0$ metastable level, selective excitation of neon atoms to the $3s_2$ state was occurring through the excitation-transfer

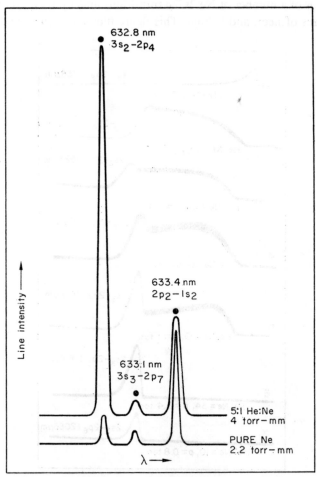

FIG. 4.8. Comparison of spontaneous intensities of the Ne-I lines at 632.8, 633.1, and 633.4 nm in a dc-discharge, in a 5:1 He : Ne mixture (upper) and in pure neon (lower) at the same average electron temperature, 10-mm-diameter discharge tube. (After C. S. Willett (unpublished work 1967).)

reaction[43]

$$\text{He* } 2^1S_0 + \text{Ne } {}^1S_0 \rightarrow \text{He } {}^1S_0 + \text{Ne } 3s_2 - \Delta E_\infty. \tag{4.8}$$

The extent of the intensification of the 632.8-nm line that occurs when helium is added to a discharge in neon is illustrated in Fig. 4.8. The figure shows the observed relative spontaneous intensities of the 632.8-nm line, the 633.1-nm line ($3s_3$–$2p_7$ transition), and the 633.-4-nm line ($2p_2$–$1s_2$ transition) in a pure neon discharge and in a discharge in a helium–neon mixture. The pressure X tube-diameter (pD) values chosen for each are such that the electron temperature is the same in both discharges.[13] Even though there is a factor of three difference in neon concentration between the pure neon and helium–neon discharge, it can be seen that the intensification of the 632.8-nm line is such as to make it considerably more intense than the 633.4-nm line, which is a strong line in a discharge in pure neon. It is also clear from Fig. 4.8 that the $3s_3$ level is not being selectively excited to any extent by helium atoms as the $3s_3$–$2p_7$ transition at 633.1 nm is not enhanced in a helium–neon mixture.

Actually, White and Rigden's observation of enhancement of a line from the Ne-I $3s_2$ state was not the first. Suga,[44] as early as 1938, reported enhancement of the strong vacuum-ultraviolet transition between the Ne $3s_2$ state and the Ne ${}^1S_0$ ground state at 60.004 nm in a pulsed discharge in helium, in which neon was present as an impurity, and concluded that the Ne $3s_2$ state was being selectively excited by He* 2^1S_0 metastables. At the same time, Suga observed that transitions between the Ne $2s_2$ and $2s_4$ levels and the neon ground state at about 63.0 nm were enhanced and concluded that the Ne $2s$ levels were being selectively excited by He* 2^3S_1 metastables. Apparently though, according to Suga,[44] the same observations were made, and similar conclusions were drawn by Dorgela and Abbink[45] even

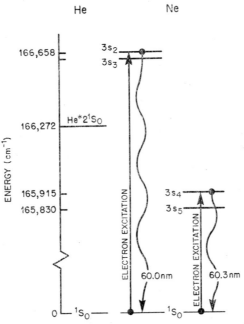

FIG. 4.9. Energy-level coincidence of the He* 2^1S_0 metastable level with Ne $3s$ levels, also showing strong vacuum ultraviolet transitions to the Ne ${}^1S_0$ ground state and excitation paths.

115

earlier in 1926. The energy coincidence for the He* 2^1S_0 level and the four Ne $3s_{2-5}$ levels, together with transitions referred to in the text, are shown in Fig. 4.9. The potential energies are taken from Moore.[46]

On the basis of the enhancement that they observed, White and Rigden attempted directly to obtain oscillation at 632.8 nm. Although they did not achieve oscillation in a discharge tube 1.5 cm in diameter, by going to smaller discharge-tube diameters and using pressures that differed from those used in the helium–neon near-infrared laser, White and Rigden were finally successful in obtaining oscillation at 632.8 nm.[43] The 632.8-nm He–Ne laser thus became the first gas laser to operate in the visible region of the spectrum.

Let us consider the energy coincidence between the metastable He* 2^1S_0 state and the Ne $3s$ states that gives rise to selective excitation of the Ne $3s_2$ state (Fig. 4.9).

The energy discrepancies ΔE_∞ in cm^{-1} between the metastable He* 2^1S_0 state and the Ne $3s$ states are as follows: $3s_2$, -386; $3s_3$, -336; $3s_4$, $+357$; and $3s_5$ $+442$. On the basis of the importance of energy discrepancy for excitation energy transfer (from the theory of Massey and Burhop discussed in Chapter 2) the order in which the four states would be expected to be selectively excited would be $3s_4$, $3s_5$, $3s_3$, and $3s_2$. This follows because of the magnitude of each of their respective ΔE_∞, and because, in the case of the $3s_3$ and $3s_2$ levels, energy has to be supplied during the collision to make up for the deficit in potential energy between the Ne $3s_2$ and $3s_3$ levels, and the lower-lying He* 2^1S_0 level. On this basis, the state that in practice is preferentially excited is the one least likely to be selectively excited. It is of course the $3s_2$ level, situated 386 cm^{-1} (approximately 2 kT at room temperature) *above* the He* 2^1S_0 level and further from it in energy than either the $3s_3$ or $3s_4$ levels. The reason for the preferential selective excitation of the Ne $3s_2$ level, as discussed in Section 2.2.1, appears to lie in the requirement for spin conservation. Only in the case of the excitation transfer reaction

$$\text{He}^* 2^1S_0(\uparrow\downarrow)+\text{Ne } ^1S_0(\uparrow\downarrow) \rightarrow \text{He } ^1S_0(\uparrow\downarrow)+\text{Ne } 3s_2(\uparrow\downarrow)-\Delta E_\infty \text{ (386 cm}^{-1}) \quad (4.9)$$

is spin conserved in collisions between He* 2^1S_0 metastables and neon ground state atoms that result in selective excitation of the Ne $3s$ states. Also, as discussed in Section 2.2.1, it is the actual shape of the molecular interaction potentials during the collision between two atoms that determines whether or not excitation transfer occurs, and the extent of the cross-section if it does occur. If an activation energy, or potential-energy barrier, occurs as a result of the interaction of the two colliding atoms, the end result is that the collision becomes a tunnelling phenomenon determined by statistical considerations in which the gas temperature of the discharge determines the effective cross-section for the reaction.

Although the addition of helium has so far been found to be essential for obtaining oscillation on the Ne $3s_2$–$2p_4$ transition at 632.8 nm and on the other $3s_2$–$2p$ transitions, some selective excitation of the Ne $3s_2$ state does occur by direct electron impact excitation from the neon ground state in the reaction

$$\text{Ne } ^1S_0+e^- \rightarrow \text{Ne } 3s_2+e^-. \quad (4.10)$$

Figure 4.9 shows that the Ne $3s_2$ level is connected to the ground state by the strong vacuum-ultraviolet 60.0-nm line observed by Suga[44] so that the optically allowed reaction (4.10) is likely to have a large cross-section for excitation by energetic electrons. Under

116

discharge conditions that ensure that the 60.0-nm line is radiation trapped, the Ne $3s_2$ level decays mainly through strong visible and infrared transitions to the lower-lying $2p$ and $3p$ states respectively. (Radiation trapping of the 60.0-nm line appears to be complete at $p_{Ne}D$ values greater than about 0.6 torr-mm.)

Reaction (4.10) is an important reaction in the selective excitation of the Ne $3s_2$ state, as CW oscillation on transitions between the $3s_2$ and $3p_{1-10}$ states has been reported in discharges in pure neon. The lifetime of He^* 2^1S_0 metastables, unlike that of He^* 2^3S_1 metastables, is very short in a He–Ne mixture at the pressures of about 1 torr used in CW, 632.8-nm He–Ne lasers, as clearly shown in Fig. 4.10. This lifetime of less than 8 μsec is comparable to the time it takes for electrons to thermalize in the afterglow of a pulsed discharge. Under the same conditions, the lifetime of He^* 2^3S_1 metastables, as evidenced by the decay of absorption on the 388.9-nm line, is more than an order of magnitude longer.[47]

The lifetime of He^* 2^1S_0 metastables is also reduced in the afterglow of a pulsed He–Ne discharge unless a short excitation pulse of less than 10 μsec is used.[48] Figures 4.11a and 4.11b after Suzuki[48] illustrate this, a result that is discussed further in a following section. Whereas it is possible to distinguish between excitation transfer and electron-impact excitation of the Ne $2s$ states in a He–Ne mixture by afterglow measurements because of the long lifetime of the He^* 2^3S_1 metastables, the time-resolved technique of Bennett et al.[49] cannot be used to determine the relative importance of reactions (4.9) and (4.10) in the selective excitation of the Ne $3s_2$ level since the lifetime of He^* 2^1S_0 metastables is comparable to the electron-thermalization time in a pulsed He–Ne discharge. It is clear, however, from the work of White and Gordon[50] and others[13] that selective excitation of the Ne $3s_2$ state is predominantly that of excitation transfer from He^* 2^1S_0 metastables in a discharge in a He–Ne mixture. In Fig. 4.12, after White and Gordon,[50] the spontaneous intensity of the 632.8-nm $3s_2$–$2p_4$ line and the He^* 2^1S_0 metastable density (circles) both measured along the axis of a discharge in a He–Ne mixture show an identical dependence on discharge

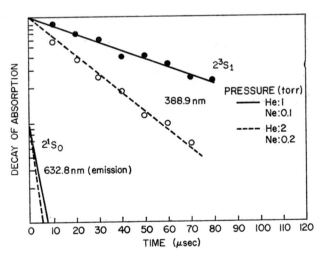

FIG. 4.10. Decay of absorption of the He-I line at 388.9 nm (He^* 2^3S_1 concentration), and emission of the 632.8-nm line (He^* 2^1S_0 concentration) in He–Ne mixtures. (After N. Suzuki,[48] by courtesy of the Editor, *Japanese Journal of Applied Physics*.)

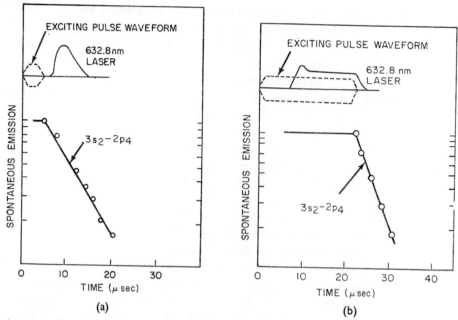

FIG. 4.11. Behavior of the laser output at 632.8 nm in the afterglow of a pulsed discharge, and decay of emission at 632.8 nm (which monitors the He* 2^1S_0 concentration). Discharge conditions: (a) short-pulse excitation (less that 10 μsec) showing strong afterglow oscillation and slower decay of He* 2^1S_0 metastables; (b) long-pulse-excitation. (After N. Suzuki,[48] by courtesy of the Editor, *Japanese Journal of Applied Physics*.)

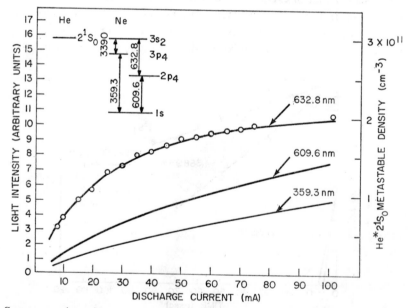

FIG. 4.12. Spontaneous intensities and He* 2^1S_0 metastable density in a He–Ne discharge versus discharge current. Circled points indicate the He* 2^1S_0 metastable density. Discharge conditions: 5:1 Ne : Ne ratio, $pD = 3.6$ torr-mm, $D = 6$ mm. (Adapted from a figure given by A. D. White and E. I. Gordon,[50] by courtesy of the American Institute of Physics.)

current. This indicates that (at least in a 0.5-torr[†] He, 0.1-torr Ne gas mixture) the significant excitation mechanism of the Ne $3s_2$ state is one in which $He^* 2^1S_0$ metastables are involved. A similar dependence of the 632.8-nm line intensity and $He^* 2^1S_0$ metastable concentration on discharge current has been reported by Chebotayev and Vasilenko.[20] Finally, quantitative determinations by Young et al.[13] of the relative importance of excitation transfer from helium to excitation by electron impact of the Ne $3s_2$ level have confirmed (Fig. 4.13) that

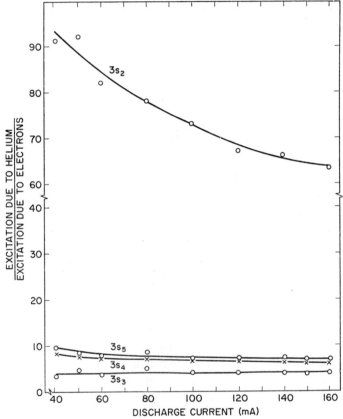

FIG. 4.13. Excitation ratios for the $3s$ levels of neon. Discharge conditions: 5:1 Ne : Ne mixture, $pD = 4$ torr-mm, tube diameter 10 mm. (After R. T. Young, Jr., C. S. Willett, and R. T. Maupin.[13])

excitation transfer from helium is by far the dominant selective excitation process of the Ne $3s_2$ level in a He–Ne 632.8-nm laser discharge. These results were obtained from comparisons made between spontaneous intensities of lines in a discharge in a He–Ne mixture and in a pure neon discharge, where the discharge conditions were chosen to give the same electron temperature in both discharges. Figure 4.13 shows that excitation transfer from helium to the Ne $3s_2$ state in a typical 5 : 1 He : Ne 632.8-nm laser discharge in a 10-mm-bore discharge tube is between 60 to 80 times more important than direct electron impact excitation from the Ne $^1S_0$ ground state. The predominance of the excitation due to helium to excitation due to electrons for the Ne $3s_2$ level compared with the other $3s_5$, $3s_4$ and $3s_3$ levels, also shown in Fig. 4.13, is in agreement with the data of Massey et al.[51]

---

† Due apparently to a misprint, this was given as 0.6 torr in ref. 50.

4.1.1.4. Additional laser transitions from the Ne 3s levels

There are four Ne-I $3s$ levels. In Paschen notation they are Ne $3s_2$, $3s_3$, $3s_4$ and $3s_5$. In Racah notation they are $5s'[\frac{1}{2}]_1^\circ$, $5s'[\frac{1}{2}]_0^\circ$, $5s[\frac{3}{2}]_1^\circ$, and $5s[\frac{3}{2}]_2^\circ$, in order of their decreasing potential energy.[46] All of these levels are radiatively connected to the lower-lying $2p$ and $3p$ groups of levels, each group of which constitutes ten levels. There are thirty optically allowed transitions between both groups of levels, $3s_{2-5}-2p_{1-10}$ and $3s_{2-5}-3p_{1-10}$. Transitions between the $3s$ and $2p$ levels give rise to the emission of radiation in the wavelength range 543.3 to 772.5 nm,[52] and between the $3s$ and $3p$ levels in the range 2.43 to 4.72 μm. Of the thirty allowed transitions between the $3s$ and $2p$ levels, oscillation has been observed on only nine. All nine of these laser transitions have the Ne $3s_2$ level as their upper laser state. Of the thirty allowed transitions between the $3s$ and $2p$ levels, oscillation has been reported on thirteen.[53] (See also Appendix Tables "Laser Transitions in Atomic Species".)

Oscillation on 3s–2p transitions. As stated above, laser oscillation has been reported on only nine of the thirty optically allowed transitions between the Ne $3s$ and $2p$ levels. In all nine transitions the $3s_2$ is the upper laser state. Oscillation on these nine transitions occurs only in He–Ne mixtures. Individual oscillation on seven of these transitions can be obtained only if the strong 632.8-nm ($3s_2-2p_4$) line, as well as the superradiant $3s_2-3p_4$ transition at 3.39 μm, are suppressed.[54-56] Under certain experimental and discharge conditions, oscillation on the $3s_2-2p_4$ and the $3s_2-2p_2$ transitions will occur simultaneously.[57, 58] Strongest oscillation was observed by White and Rigden[55] to occur, in order of their decreasing output power, on the following transitions: $3s_2-2p_4$ at 632.8 nm, $3s_2-2p_6$ at 611.8 nm, $3s_2-2p_2$ at 640.1 nm, $3s_2-2p_1$ at 730.5 nm, and $3s_2-2p_7$ at 604.6 nm. The order of the transitions is also in the ascending order of their expected relative transition probabilities (with the exception of the $3s_2-2p_6$ transition at 611.8 nm which has the lowest relative transition probability). The $3s_2-2p_{10}$ transition at 543.3 nm is the hardest transition on which to obtain oscillation; to do this it is necessary to use properly coated low-loss cavity mirrors peaked in the wavelength region 520 to 550 nm.[56]

The reason for oscillation not being observed on transitions between the $3s_{3, 4, 5}$, and the $2p$ levels in either He–Ne or pure neon discharges probably lies in the fact that these $3s$ levels are not sufficiently excited by He* 2^1S_0 metastables, electron impact from the ground state, or indirectly from the well-populated $3s_2$ level, to maintain population inversion over the $2p$ levels. Of course, the $2p$ levels are well populated in a He–Ne mixture via the strong $2s-2p$ laser transitions in the near infrared as a result of excitation transfer from He* 2^3S_1 metastables. In a discharge in pure neon the $2p$ levels are populated by the same transitions whose upper levels are selectively excited by direct electron impact.

Oscillation on 3s–3p transitions. Oscillation has been reported on thirteen of the thirty optically allowed transitions between the Ne $3s$ and $3p$ levels. Nearly one-half of these (six to be exact) have as an upper level the $3s_2$ level; one, the $3s_3$ level; four, the $3s_4$ level; and two, the $3s_5$ level.

Bloom, Bell, and Rempel[59] were the first to report oscillation between the Ne $3s$ and $3p$ levels, on the $3s_2-3p_4$ transition at 3.39 μm in a discharge in a He–Ne mixture. This transition has the largest relative transition probability of the $3s-3p$ transitions. The gain obtainable on the 3.39-μm line, which can be in excess of 20 dB m^{-1}, is sufficient to saturate the discharge medium, and can prevent and interfere with oscillation on the $3s_2$–$2p$ transitions in the visible by depleting the population of their common $3s_2$ upper state. Even in a short discharge tube the single-pass amplification of spontaneous emission at 3.39 μm can be sufficient to saturate the discharge medium.

The high gain on the $3s_2$–$3p_4$ transition, which is about two orders of magnitude larger than that on the visible $3s_2$–$2p_4$ transition, arises because of the large line strength of the 3.39-μm line (see Table 1.2). Relative to the 632.8-nm line, the line strength of the 3.39-μm line is a factor of 100 larger.[60] It is this large line strength that is responsible for the high gain, *and not any dependence of the gain on wavelength or line width* as has been commonly accepted (Section 1.5.1).

In addition to the superradiant $3s_2$–$3p_4$ transition, strong oscillation is also obtainable on other $3s_2$–$3p$ transitions, on the $3s_2$–$3p_2$ transition at 3.3903 μm and the $3s_2$–$3p_5$ transition at 3.3342 μm in He–He mixtures.

In the case of oscillation from the $3s_3$, $3s_4$, and $3s_5$ levels, the majority of the laser transitions, four in number are from the $3s_4$ level, and two are from the $3s_5$ level. The transitions on which oscillation occurs are those that have large relative transition probabilities. The excitation of the He $3s_3$, $3s_4$, and $3s_5$ levels is by excitation transfer from He* 2^1S_0 metastables; atom–atom collisional de-excitation of He $3s_2$ atoms (i.e. collisional mixing); direct electron-impact from the ground state; and cascading from higher-lying energy levels. Energy transfer to the $3s_3$, $3s_4$, and $3s_5$ levels from helium atoms is about an order of magnitude down on energy transfer to the Ne $3s_2$ level as shown in Fig. 4.13. It is probably of significance that ten of the thirteen reported $3s$–$3p$ laser transitions have the $3s_2$ or $3s_4$ levels as upper levels. These levels are optically connected to the ground state and thus have a higher probability of excitation by direct electron impact from the ground state than the $3s_3$ and $3s_5$ levels.

4.1.1.5. Additional neon levels populated by excitation transfer

Laser oscillation in neutral neon has been reported on more than 160 lines.[53] Most of the transitions responsible for the oscillation have been positively identified and their spectroscopy analysed extensively. Beyond a wavelength of about 1.0 μm the majority of the spectroscopic analysis has been carried out by Faust *et al.*[60–63]

Analysis shows that the laser lines fall into seven groups characterized by n and l for the initial and for the final levels (in Racah notation†):

$5s$–$4p$ between 2.8–4.0 μm; $6p$–$5d$ between 20.5–34.5 μm;

$5s$–$3p$ between 0.54–0.73 μm; $5p$–$4d$ between 12.9–22.8 μm;

$4s$–$3p$ between 0.88–1.55 μm; $4p$–$3d$ between 3.8–11 μm; and

$4f$–$3d$ between 1.83–1.86 μm.

---

† Paschen notation in the higher-lying states of Ne-I becomes cumbersome and no longer a convenient, easily remembered shorthand. Furthermore, it yields no information on the configuration of the state of concern, sometimes giving only the principal quantum number n. In this section for these reasons, Racah notation will be used extensively. When Paschen notation is included for continuity with earlier sections it will be obvious to the reader.

The relative line strength for a number of transitions in Ne-I from the various states n, l to other states m, $l–1$ have been calculated by Faust *et al.*[61] for $l = 1$, 2 and 3 using the $j–l$ coupling scheme of Racah. The coupling of the four angular momenta (spin and orbit of the core, spin and orbit of the excited electron) are:

$$[(l_c+s_c)+l_e]+s_e = J,$$

with

$$j_c = l_c+s_c, \quad \text{and}$$
$$k = (l_c+s_c)+l_e = j_c+l_e.$$

The rules for allowed transitions in $j–l$ coupling are $\Delta j_c = 0$; $\Delta l_e \pm 1$; $\Delta k = 0$; and $\Delta J = 0$, ± 1, with $J = 0$ to $J = 0$ forbidden. Examination of the calculated line strengths shows that of the allowed transitions, the strongest are those that satisfy $\Delta k = \Delta J$, particularly if $\Delta k = \Delta J = +\Delta l$, and more so for the higher J values. The effect is very strong for $f–d$ transitions, fairly strong for $d–p$ transitions, and rather weak for $s–p$ transitions.[61] Whereas most of the observed laser lines obey the $j–l$ selection rules, there are four groups of Ne-I $s–p$ laser lines in which strong oscillation is observed and in which the selection rules are violated. The reason for this lies in the fact that calculated line strengths are not too accurate for $s–p$ transitions, and that the s levels are efficiently pumped by excitation transfer from excited helium atoms in a discharge in a He–Ne mixture.

The four groups of s levels from which strong oscillation is observed are the $4s$, $5s$, $6s$, and $7s$ levels. The first two groups of levels, $4s$ and $5s$, are selectively excited by excitation transfer from He* 2^3S_1 and He* 2^1S_0 metastables respectively (as discussed in Sections 4.1.1.1 and 4.1.1.3); and the $6s$ and $7s$ levels are selectively excited by excitation transfer

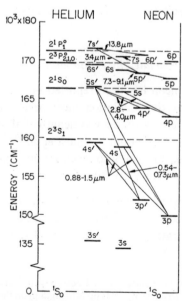

FIG. 4.14. Energy-level coincidences in the He–Ne system showing the four groups of s levels from which strong laser oscillation is observed. (After C. S. Willett and R. T. Young, Jr.[65])

from excited atoms in the well-populated He $2^3P_{0,1,2}$ and He 2^1P_1 states respectively. Figure 4.14, a partial energy-level diagram of helium and neon, shows the disposition of the four groups of neon s levels from which strong oscillation is observed in a He–Ne mixture with respect to He-I levels. The respective energy-level coincidences between helium and neon are given in Table 4.4.

TABLE 4.4. *Helium–neon energy coincidences*

Helium level		Neon group		Energy range
Symbol	Energy (cm^{-1})	Racah	Paschen	(cm^{-1})
He* 2^3S_1	159,850	4s	2s	158,600–159,540
He* 2^1S_0	166,272	5s	3s	165,830–166,660
		6s	4s	
He $2^3P_{0,1,2}$	169,082	4d	4d	167,013–170,291
		5d	5d	
He 2^1P_1	171,129	7s	5s	170,534–171,326

As well as showing that the Ne $6s$ levels ($4s$ in Paschen notation) are selectively excited by atoms in the He $2^3P_{0,1,2}$ levels, Colombo *et al.*[64] have shown that the $4d$ and $5d$ levels (notably the $5d$-level) are also excited by He 2^3P atoms. In the case of the energy coincidence between the He 2^1P_1 level and the Ne $7s$ levels, excitation transfer occurs mainly to the $7s'[\frac{1}{2}]_1^o$ level, which is the uppermost level of the $7s$ group at 171,326 cm^{-1} above the ground state. The particular excitation transfer that occurs can be attributed to the spin-conserved reaction (in Paschen notation)[65]

$$\text{He } 2^1P_1(\uparrow\downarrow) + \text{Ne } {}^1S_0(\uparrow\downarrow) \rightarrow \text{He } {}^1S_0(\uparrow\downarrow) + \text{Ne } 5s_2(\uparrow\downarrow) - \Delta E_\infty \text{ (197 cm}^{-1}). \quad (4.11)$$

The He 2^1P_1 level is optically connected to the ground state, and at the helium pressures used in typical He–Ne lasers of about 1 torr (at a pD product of 2 to 10 torr-mm) it is radiation trapped. Being the lowest of the helium p levels and having strong emission lines at 728.1 and 667.8 nm terminating on it, it has a large quasi-metastable population from which excitation transfer can occur. The excitation-transfer reaction (4.11) is similar to the excitation-transfer reaction (4.9) that occurs between He* 2^1S_0 metastables and neon atoms. It is endothermic by 197 cm^{-1} (or about 1 kT), and the total spin of zero is conserved.

Selective excitation of the Ne $4f$ levels also enables oscillation to be obtained in a He–Ne mixture on six $4f$–$4d$ transitions between 1.83 and 1.86 μm. The selective excitation is considered to be the result of excitation transfer from helium atoms in the $2^3P_{0,1,2}$ and the 2^1P_1 levels, and not, as previously supposed, from He* 2^1S_0 metastables.

For the interested reader, Table 4.5 gives a summary of collision cross-sections of processes pertinent to He–Ne and Ne laser systems.

4.1.1.6. *Operating mechanisms and characteristics*

Whereas the actual excitation process of the Ne $3s_2$ state in a He–Ne mixture is clear, it is not entirely clear just what process or processes are responsible for the variation of gain and output power of the 632.8-nm and 3.39-μm laser lines with current. Both of these lines have the same Ne $3s_2$ upper level, and have lower levels of similar configurations, $2p_4$ and

TABLE 4.5. *Collision cross-sections in He–Ne and Ne lasers*

Reaction (Paschen notation for neon)	Spin obeyed	ΔE_∞ (cm^{-1})	Cross-section (cm^2)	References
1. He* $2^1S_0 + Ne \rightarrow$				
(a) $He + Ne\ 3s_2$	yes	-386	10^{-15}–$10^{-17(a)}$	66, 20, 67, 70
(b) $He + Ne\ 3s_4$	no	$+357$	$< 10^{-18}$	68, 70
(c) $He + Ne\ 3s_5$	no	$+441$	$\sim 10^{-16}$–10^{-17}	68, 64, 70
2. He* $2^3S_1 + Ne \rightarrow$				
(d) $He + Ne\ 2s_{2-4}$	no	$+313$	10^{-16}–$10^{-19\ (a)}$	66, 1
(e) $He + Ne\ 2s_3$	yes	$+468$	1.4×10^{-17}	68
3. He $2^3P_{0,1,2} + Ne \rightarrow$				
(f) $He + Ne\ 5d_5$	—	-8	1.5×10^{-16}	64
(g) $He + Ne\ 7d_3$	—	$+2069$	$\sim 10^{-17}$	64, 70
(h) $He + Ne\ 4s_5$	—	$+55$	8.5×10^{-17}	64
4. He $2^1P_1 + Ne \rightarrow$				
(i) $He + Ne\ 5s_2$	yes	-197	$\sim 10^{-16}$	65
5. Ne $1s_{2,3,4} + H_2 \rightarrow$				
(j) $Ne + H + H'$	—	—	5×10^{-16}	69, 32
6. Ne* $1s_5 + H_2 \rightarrow$				
(k) $Ne + H + H'$	—	—	2×10^{-16}	69, 32

ᵃ Cross-section is dependent on gas temperature. See Figs. 2.7 and 2.8.

$3p_4$, respectively. We shall concern ourselves in this immediate section with the 3.39-μm and 632.8-nm laser lines.

White and Gordon[50] have shown that the observed current dependence of the concentration of He* 2^1S_0 metastables, and hence the population of the Ne $3s_2$ state in a discharge in a He–Ne mixture, may be written in the form

$$k_1 I/(k_2 + k_3 I),$$

where k_1 is proportional to the coefficient for excitation by direct electron impact from the ground state, k_2 is a constant involving loss by diffusion and excitation transfer, and k_3 involves loss by electron de-excitation. The electron de-excitation of He* 2^1S_0 metastables occurs in the super-elastic collision process

$$He*\ 2^1S_0 + e^- \text{ (slow)} \rightarrow He*\ 2^3S_1 + e^- \text{ (fast)}, \tag{4.12}$$

which has been shown by Phelps[71] to have the very large cross-section of 1.50×10^{-14} cm². White and Gordon[50] have also shown that the gain on the Ne $3s_2$–$3p_4$ transition at 3.39 μm, which is directly proportional to the difference in population of the Ne $3s_2$ and $3p_4$ levels, is given by the relationship

$$\text{gain (dB)} = \frac{2.14I}{(1 + k_3 I/k_2)} - 0.194I, \tag{4.13}$$

where I is the discharge current in mA, and k_3 and k_2 are constants. The first term in the expression for gain arises from the population density of the upper state divided by the statistical weight of the state, the second term $0.194I$ arises from the population density of the lower $3p_4$ laser state divided by its statistical weight. The ratio $k_3/k_2 = 4 \times 10^{-2}$ that is used in Fig. 4.15 is derived from the $He^* 2^1S_0$ metastable density versus discharge current (632.8-nm line) curve in Fig. 4.12, and the proportionality factor 0.194 in the second term is derived from the spontaneous intensity of the 359.3-nm $(3p_4-1s_2)$ Ne line. The curve for the 359.3-nm line represents the population density of the $3p_4$ state, and the curve of the 609.6-nm line represents the population density of the Ne $2p_4$ state (lower laser state of the 632.8-nm line). The intensities of these lines are sidelight intensities measured at right angles to the axis of discharge tube, and are not distorted by radiation trapping. As noted by White and Gordon,[50] the sidelight intensities of the 359.3-nm $(3p_4-1s_2)$ line and the 609.6-nm $(2p_4-1s_4)$ line do not exhibit the quadratic dependence on current that would be expected if the $3p_4$ and $2p_4$ states were being excited from the ground state in a two-step process via the intermediate metastable/quasi-metastable Ne $1s_{2-5}$ states. The almost linear dependence of the sidelight intensities of the 359.3-nm and 609.6-nm lines on discharge current, at least above a discharge current of 30 mA, indicates that both the $3p_4$ and $2p_4$ states are excited by a single-step electron-impact process and by some cascading from the strongly populated $3s_2$ and $2s_2$ levels.

Let us consider the relative importance of the Ne $3s_2$ upper-state population of the 3.39-μm and 632.8-nm lines, represented in (4.13) by the first term $2.14I/(1+0.04I)$, and the $3p_4$ lower-state population of the 3.39-μm line represented by the second term $0.194I$ divided by their respective statistical weights g_2 and g_1 (of 3 and 5) over the discharge-current range covered in Fig. 4.15. Representative calculated values of the $3s_2$ and $3p_4$ population densities divided by their statistical weights, together with the ratios of the population densities of the Ne $3s_2$ state to that of the $3p_4$ state at five discharge currents in the range 20 to 100 mA are tabulated in Table 4.6.

TABLE 4.6. *Representative population densities in the Ne $3s_2$ and $3p_4$ states in the 3.39-μm He–Ne laser*[a]

I Discharge current (mA)	$\dfrac{N_2(3s_2)}{g_2}$ represented by $\dfrac{2.14I}{1+0.04I}$	$\dfrac{N_1(3p_4)}{g_1}$ represented by $0.194I$	$\dfrac{N_2(3s_2)/g_2}{N_1(3p_4)/g_1}$	Gain (dB)
20	23.8	3.9	6.1	19.9
40	33.0	7.8	4.2	25.2
60	38.0	11.6	3.3	26.4
80	40.7	15.5	2.6	25.2
100	42.8	19.4	2.2	23.4

[a] Excitation conditions: 5 : 1 He : Ne mixture ratio, tube diameter 6 mm, at a pD of 3.6 torr-mm.

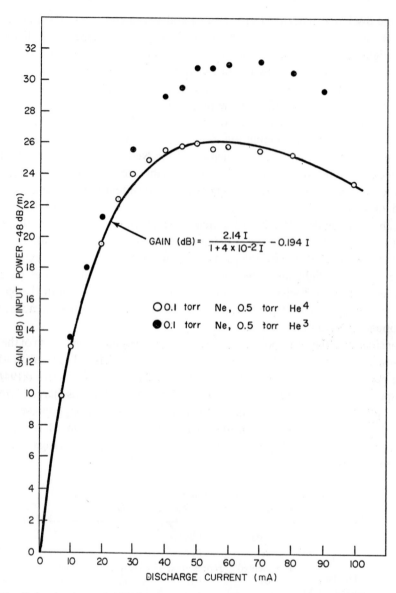

GAIN (dB) = $\dfrac{2.14\,I}{1+4\times10^{-2}\,I}$ − 0.194 I

○ 0.1 torr   Ne, 0.5 torr   He⁴

● 0.1 torr   Ne, 0.5 torr   He³

FIG. 4.15. Small signal gain on the Ne $3s_2$–$3p_4$ transition at 3.39 μm versus discharge current. Discharge conditions: 5:1 He:Ne mixture, $pD = 3.6$ torr-mm, $D = 6$ mm. The open circles are measured values, the solid curve was obtained by fitting the 632.8-nm and 359.3-nm data in Fig. 4.12. The gain for a He³–Ne fill is given by the solid circles. (From A. D. White and E. I. Gordon,[50] by courtesy of the American Institute of Physics.)

For all discharge currents, the table enables the very important conclusion to be drawn that the term $2.14I/(1+0.04I)$, which represents the $3s_2$-state population density divided by the statistical weight g_2, is the dominant term in the gain relationship (4.13) by a factor of between 2 and 6. Additional weight to this conclusion that the gain is determined primarily by the $3s_2$ upper laser level population (through excitation transfer from He* 2^1S_0 metastables) is given by gain measurements made on a similar discharge in a He–Ne mixture in which the lighter He^3 isotope is substituted for the normal He^4 gas filling. The solid circles plotted in Fig. 4.15 are gain values for a He^3 gas filling.[50] Most of the increased gain, which mounts to over 20 percent at the optimum discharge current of 50 to 60 mA, can be accounted for by the increased rate of excitation transfer from He* 2^1S_0 metastables. Because of the lighter mass of the He^3 metastable atoms, the increased rate of excitation transfer resulting from the increased collision frequency (considering the relatively heavier neon atoms to be stationary) would be expected to be proportional to the square root of the ratio of the mass of a He^4 atom to that of a He^3 atom. The excitation rate with He^3 should thus be approximately $\left(\frac{4}{3}\right)^{1/2}$ that with He^4 or about 115 percent.

The observed fact that the concentration of He* 2^1S_0 metastables saturates with increasing current, implies that the excitation rate of the Ne $3s_2$ level and optical gain on transitions originating on the $3s_2$ level will also saturate at high discharge currents. This is indeed the case with the 3.39-μm line. Figure 4.15 shows that gain on the 3.39-μm line is a maximum at a discharge current of between 50 to 60 mA. At these discharge currents the saturation of He* 2^1S_0 metastables is more than 90 percent complete, and a maximum difference occurs in intensities of the 632.8-nm and 359.3-nm lines originating on the Ne $3s_2$ and $3p_4$ levels, respectively.

Whereas differences in line intensities cannot normally be taken as absolute measures of differences in population (N_2–N_1) between two excited states because of the different frequencies and transition probabilities of the two appropriate transitions, it appears that the results of axial and sidelight intensities given by White and Gordon have been corrected to account for this and that the discharge current at which the maximum difference in intensity of the 632.8-nm and 359.3-nm lines occurs, corresponds to that at which the difference in population between the Ne $3s_2$ and $3p_4$ states, and the gain, are maxima.

Let us now consider the relative importance of the upper- and lower-state populations of the Ne $3s_2$–$2p_4$ transition at 632.8 nm. Unfortunately, in the case of the $3s_2$–$2p_4$ transition in spite of the extensive work and calculations that have been done on it over a period of years no one has apparently published combined measurements of spontaneous intensities of lines originating on the upper and lower states of the transition with actual gain measurements of the 632.8-nm line, as White and Gordon did for the $3s_2$–$3p_4$ transition at 3.39 μm. Although White and Gordon in Fig. 4.12 give measurements of the spontaneous sidelight intensity of the 609.6-nm ($2p_4$–$1s_4$) line as well as axial measurements of the 632.8-nm line, in their paper they do not give gain measurements of the 632.8-nm line taken with the same discharge tube under the discharge conditions used for the spontaneous intensity measurements. Nevertheless, other data[72] exists that we can use.

Labuda[72] has made measurements of the optical gain on the 632.8-nm line in a He–Ne discharge in tubes of differing internal diameters under discharge conditions very similar to those used by White and Gordon[50] in determining the 3.39-μm gain plotted in Fig. 4.15. Figure 4.16 shows values of gain on the 632.8-nm line as a function of discharge current for two capillary-tube diameters of 6 mm and 4 mm, at a pD product of 3.6 torr-mm. The curves showing the data points and indicated by the size of the capillary tube are those of Labuda.[72] The curve without data points has been obtained by fitting data given by White and Gordon[50] (Fig. 4.12) for the spontaneous intensities of the 632.8-nm and 609.6-nm lines to the gain values of the 632.8-nm line given in Fig. 4.16 for the 6-mm discharge tube. The basic assumption that has been made is that the intensities of the 632.8-nm and 609.6-nm lines in the 7 : 1 He : Ne mixture used by Labuda[72] vary in the same way with current as they do in the 5 : 1 He : Ne mixture used by White and Gordon.[50] The 7 : 1 He : Ne mixture ratio is slightly less favorable for producing population inversion between the Ne $3s_2$ and $2p_4$ states than the 5 : 1 ratio in a 6-mm diameter tube, at the common pD product of 3.6 torr-mm.[73] It is more nearly optimum than the 5 : 1 He : Ne mixture ratio for producing inversion on the Ne $2s$–$2p$ transitions in the near-infrared. This difference in the mixture ratios, it is maintained, is not likely to produce any significant changes in the intensities of the 632.8-nm and 609.6-nm lines relative to each other and thus affect adversely the analysis given, and the conclusions drawn here.

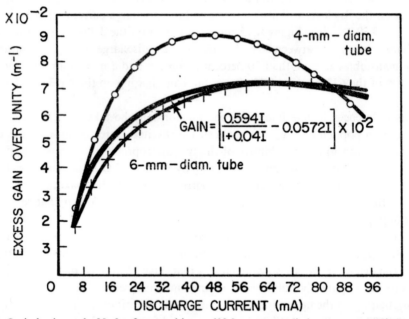

FIG. 4.16. Optical gain on the Ne $3s_2$–$2p_4$ transition at 632.8 nm versus discharge current. Discharge conditions: 7 : 1 He : Ne mixture, $pD = 3.6$ torr-mm; tube diameters of 6 and 4 mm. The curve without data points was obtained by fitting the 632.8- and 359.3-nm data in Fig. 4.12. (By courtesy of E. F. Labuda.[72])

The fitted curve in Fig. 4.16 is given by the relationship

$$\text{Gain} = \left[\frac{0.594I}{1+0.04I} - 0.0572I\right] \times 10^{-2} \qquad (4.14)$$

where the Gain is the excess gain over unity per meter, and I is the discharge current in mA. The relationship was derived by:

(1) letting the population of the Ne $3s_2$ state maintain the form $k_i I/(k_2+k_3 I)$ that was used by White and Gordon to give the first term in the gain expression $2.14I/(1+0.04I)$ $-0.194I$ for the 3.39-μm line (Fig. 4.15); and

(2) assuming that the population in the $2p_4$ state (upper state for the 609.6-nm line) varies

128

linearly with discharge current and that the slope of the 609.6-nm line in Fig. 4.12 above a current of about 35 mA will give the proportionality factor in the second term of the relationship.

Although the spontaneous sidelight intensity of the 609.6-nm line shows that the population of the $2p_4$ state does not vary as linearly with discharge current as does the population of the higher lying $3p_4$ state at low discharge currents, measurements by the author pubished in work by Young and Maupin[74] have shown that (2) is not an unreasonable assumption since the intensities of lines from Ne $2p$–$1s$ transitions in a He–Ne 632.8-nm laser discharge were found to be directly proportional to the discharge current up to even high current densities.

The fitted 632.8-nm gain curve in Fig. 4.16 is clearly not such a good fit to the measured gain curve, especially at low discharge currents, as that for the 3.39-μm laser line in Fig. 4.15. What is important, however, is that the saturated intensity-versus-current curve of the 632.8-nm line combined with an assumed intensity of the 609.9-nm line varying linearly with discharge current is a reasonable fit to the 632.8-nm gain curve.

As we did for the 3.39-μm laser line, let us consider the relative importance of the upper $3s_2$ state and the lower $2p_4$ state populations divided by their respective statistical weights represented by the first term $0.549I/(1+0.04I)$ and the second term $0.0572I$ in the fitted gain relationship (4.14) for the 632.8-nm line at various discharge currents. Representative values of the $3s_2$ and $2p_4$ populations and the ratios of these statistically weighted populations at five discharge currents from 20 to 100 mA are tabulated in Table 4.7.

TABLE 4.7. *Representative population densities in the Ne $3s_2$ and $2p_4$ states in the 632.8-nm He–Ne laser*[a]

I Discharge current (mA)	$\dfrac{N_2(3s_2)}{g_2}$ represented by $\dfrac{0.594I}{1+0.04I}$	$\dfrac{N_1(2p_4)}{g_1}$ represented by $0.0572I$	$\dfrac{N_2(3s_2)/g_2}{N_1(2p_4)/g_1}$	Gain (% m^{-1})
20	6.6	1.14	5.79	5
40	9.14	2.29	3.99	6.5
60	10.48	3.43	3.06	7.1
80	11.31	4.58	2.47	6.9
100	11.88	5.72	2.07	~ 6.5

[a] Excitation conditions: 7 : 1 He : Ne mixture ratio, tube diameter 6 mm, at a pD of 3.6 torr-mm.

The table enables the important conclusion to be drawn again that the first term in the gain expression, which represents the upper $3s_2$ state population together with its statistical weight g_2, is the dominant term. It dominates the second term, which is representative of the population in the lower $2p_4$ state together with its statistical weight g_1, by a factor of nearly 6 at lower discharge currents and by more than 2 at a discharge current of 100 mA.

At the optimum discharge current of 60 mA for laser oscillation at 632.8 nm in the 6-mm tube, the first term dominates the second by a factor of 3.

This is as might be expected. The $2p_4$ level is lower in potential energy than the $3p_4$ level and is therefore likely to be more highly populated than the $3p_4$ level. The ratios of the Ne $3s_2$ to Ne $2p_4$ population densities, given in Table 4.7 still incorporating their statistical weights (of 3 and 5 respectively), are less than the corresponding ratios of the Ne $3s_2$/Ne $3p_4$ states given in Table 4.6 over all the discharge currents covered. All the same, the ratios are not significantly different, indicating that the populations of the lower levels of both the 632.8-nm and the 3.39-μm laser lines are about the same. This is in agreement with the work of Faust and McFarlane[60] that shows that the inversion density inversion ($N_2/g_2 - N_1/g_1$ in atoms/cm³) is approximately 10^9 cm^{-3} for both the $3s_2$–$2p_4$ transition at 632.8 nm and the $3s_2$–$3p_4$ transition at 3.39 μm. This is the case even though the respective gains differ by more than two orders of magnitude.

The treatment of White and Gordon,[50] which is based on the assumption that the saturation behavior of He* 2^1S_0 metastables at high currents in a He–Ne mixture is due to their destruction in superelastic collisions with slow electrons [eqn. (4.12)], predicts completely the gain-versus-current behavior of the 3.39-μm laser line. The similar treatment given here also appears to predict satisfactorily the gain-versus-current behavior of the 632.8-nm He–Ne laser line. Wada and Heil,[75] however, on the basis of electron-energy-distribution measurements made in a He–Ne discharge and known excitation and de-excitation cross-sections, state that the de-excitation of He* 2^1S_0 metastables by slow electrons in the He–Ne 632.8-nm laser is definitely negligible. They conclude that some other process is responsible for the saturation of He* 2^1S_0 metastables at high discharge currents, and is thus responsible for the gain-versus-current behavior of the 632.8-nm and 3.39-μm laser lines in a He–Ne mixture. Although Wada and Heil do not suggest what this other de-excitation process minght be, there appears to be one reaction that it could be.

Suzuki[48] has postulated that the oscillation behavior of the 632.8-nm laser line with variation in excitation pulse-length in a pulsed He–Ne discharge is due to the temperature dependence of the reaction

$$He^* \; 2^1S_0 + He \; ^1S_0 \rightarrow He_2 \; A^1\Sigma_u^+ \tag{4.15}$$

in which a He* 2^1S_0 metastable has a collision with a ground-state helium atom to give a stable helium He$_2$ $A^1\Sigma_u^+$ molecule. This process clearly results in the destruction of He* 2^1S_0 metastables so that the rate of excitation transfer to neon ground-state atoms in the selective excitation of the Ne $3s_2$ state could be reduced if the discharge conditions were conducive to the reaction occurring. It appears possible that this process, which was observed by Suzuki[47, 48] to be of importance in a pulsed He–Ne laser operating at 632.8 nm, could be of importance under normal CW operating conditions, and could lead to the saturation with discharge current of He* 2^1S_0 metastables at high discharge currents. For this process to be important in the CW 632.8-nm He–Ne laser, it is necessary to invoke a nonlinear gas-temperature dependence on discharge current, and a large cros-section for reaction (4.15).

Suzuki[47, 48] has found that strong oscillation on the Ne $3s_2$–$2p_4$ transition at 632.8 nm in the afterglow of a pulsed discharge in a He–Ne mixture occurs only when short excitation pulses of less than 10 μsec are used (as shown in Figs. 4.11a and 4.11b). In addition, Suzuki found that strong oscillation at 632.8 nm is observed only when the vacuum-ultraviolet, 60.0-nm molecular-band emission that originates on the He_2 $A^1\Sigma_u^+$ level is low. Furthermore, there is both a nonlinear increase of 60.0-nm molecular-band emission and a nonlinear decrease in lifetime of He^* 2^1S_0 metastables with increase in excitation pulse-length of more than 10 μsec. Figures 4.17 and 4.18 show this clearly. In Fig. 4.18 the lifetime of He^* 2^1S_0 metastables is given by the absorption on the He^* $2s^1S_0$–$3p^1P_1^o$ transition at 501.6 nm. Presumably the longer the excitation pulse, the hotter the gas becomes, and the He^* 2^1S_0 metastables can acquire more kinetic energy to supply the energy necessary for reaction (4.15) to proceed more effectively.

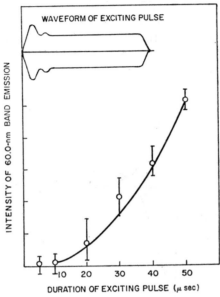

FIG. 4.17. Relation between intensity of the molecular-helium band $A^1\Sigma_u^+ - X^1\Sigma_g^+$ at 60.0 nm and duration of the excitation pulse. (After N. Suzuki,[48] by courtesy of the Editor, *Japanese Journal of Applied Physics*.)

The endothermic reaction (4.15) involving an atom–atom collision between an excited metastable He^* 2^1S_0 atom and a ground state helium atom is very similar to the endothermic spin-conserved, excitation-transfer reaction

$$He^* \ 2^1S_0(\uparrow\downarrow)+Ne \ ^1S_0(\uparrow\downarrow) \ \rightarrow \ He \ ^1S_0(\uparrow\downarrow)+Ne \ 3s_2(\uparrow\downarrow)-\Delta E_\infty \ (386 \ cm^{-1}),$$

which gives selective excitation of the 632.8-nm upper laser level. Both reactions involve He^* 2^1S_0 metastables and the ground-state He $^1S_0$ and Ne $^1S_0$ atoms have a similar closed-shell electronic configuration.

Buckingham and Dalgarno[76] have shown that the interaction potential of a collision between a metastable He^* 2^1S_0 atom and a ground-state helium atom that results in the formation of a stable He_2 $A^1\Sigma_u^+$ molecule has the form illustrated in Fig. 4.19. A maximum in the interaction potential occurs at an internuclear separation of about 2.0 Å. This maximum in the potential constitutes a barrier of approximately 0.26 eV[76] or less.[78] Because of the existence of this potential barrier, there is little chance of He_2 $A^1\Sigma_u^+$ molecu-

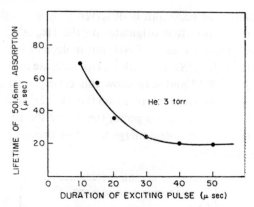

FIG. 4.18. Relation between lifetime of the He-I 501.6-nm line absorption (He* 2^1S_0 lifetime) and duration of the excitation pulse. (After N. Suzuki,[48] by courtesy of the Editor, *Japanese Journal of Applied Physics.*)

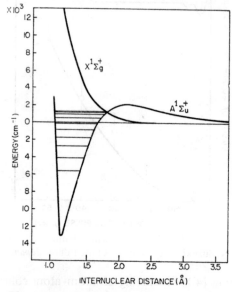

FIG. 4.19. Potential-energy curves of the $A^1\Sigma_u^+$ and $X^1\Sigma_g^+$ states of He_2. (After Y. Tanaka and K. Yoshino,[77] by courtesy of the American Institute of Physics.)

lar-helium formation at low relative collision velocities. At high relative collision velocities, or at high gas temperatures, formation can occur readily. Suzuki[48] has suggested that with excitation current pulses of less than 10-μsec duration the gas temperature is low, and the energy available in the discharge is too low for the potential barrier to be surmounted, and $He_2\ A^1\Sigma_u^+$ molecular formation is negligible. With excitation pulses longer than 10 μsec he suggests that the gas temperature increases sufficiently for $He_2\ A^1\Sigma_u^+$ molecules to form easily. This reduces the concentration of He* 2^1S_0 metastables enough to significantly reduce the population inversion between the Ne $3s_2$ and $2p_4$ levels.

Concluding, if the Wada and Heil analysis[75] is correct in showing that de-excitation of He* 2^1S_0 metastables by slow electrons is negligible in the 632.8-nm He–Ne laser discharge, it is suggested that there are indications that the current-saturation behavior of He* 2^1S_0

132

metastables, and hence the population density of the Ne $3s_2$ state, is due to a gas-temperature effect (related to the discharge current) on the $He_2\ A^1\Sigma_u^+$ molecular-helium-formation reaction of (4.15). However, any importance of this reaction in the 632.8-nm He–Ne laser is hard to reconcile with the small experimentally determined cross-sections of 3×10^{-20} cm² and 9.6×10^{-21} cm² reported for it by Phelps and Molnar[79] and Biondi,[80] respectively, or the theoretical one of 9×10^{-26} cm².[70, 81] It will be remembered that the superelastic collision-reaction, $He^*\ 2^1S_0 + e^-$ (slow) $\rightarrow He^*\ 2^3S_1 + e^-$ (fast), which White and Gordon[50] propose as being responsible for the saturation of $He^*\ 2^1S_0$ metastables at high discharge currents, has a cross-section[71, 79] of $\sim 10^{-14}$ cm² that is six orders of magnitude larger than that of the molecular $He_2\ A^1\Sigma_u^+$ formation reaction.

In spite of the reported large differences in cross-sections for the two reactions that compete for the $He^*\ 2^1S_0$ population, the observations by Suzuki[48] are even harder to reconcile with a cross-section as small as 10^{-20} cm² for the molecular-helium-formation reaction that apparently is large enough to compete effectively with the excitation-transfer reaction (4.9) which is responsible for the selective excitation of the Ne $3s_2$ state and has a cross-section of approximately 5×10^{-16} cm² at 300 K.

Radial spatial dependence of gain. We shall restrict ourselves in this discussion to 1.15-μm and 632.8-nm He–Ne laser discharges in dielectric tubes of circular cross-section, in which the internal diameter D is much less than the length of the tube l.

Both the 1.15-μm and 632.8-nm He–Ne lasers are four-level laser systems of the type shown in Fig. 4.20. In Fig. 4.20, F_2 and F_1 are the upper- and lower-state formation rates,

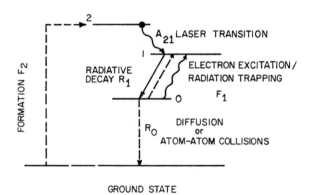

FIG. 4.20. Ideal four-level laser system showing excitation and de-excitation processes.

R_1 is the total relaxation rate of the lower level that occurs primarily through spontaneous radiative decay, and R_2 is the total relaxation rate of the upper laser level in the absence of stimulated emission (for $R_1 \gg R_2$). For the realization of steady-state inversion

$$R_1 > (g_2/g_1)(A_{21} + R_2 F_1/F_2). \qquad (4.16)$$

Clearly, the maximum output power realizable is fundamentally limited by saturation of the relaxation rate R_1 of the lower laser level, or by an increase in F_1/F_2 at high excitation rates. As pointed out by Bennett,[82] in the ideal four-level laser system shown in Fig. 4.20, the inversion density (N_2-N_1) would increase linearly with pump rate, and saturation of the output power would not occur. Level "0", however, is either metastable, or quasi-metastable, because of radiation trapping on transitions from it to the ground state in the far vacuum ultraviolet around 74.0 nm. Under these conditions, atoms in level '0' can only decay by diffusing to the walls where they are (assumed) de-excited, or are de-excited by atom–atom collisions. As these are slow processes, $R_0 \ll R_1$ or R_2, and a sizeable population of metastable atoms can build up in level 0. The effect of this is to increase the lower-laser level excitation rate F_1 to the detriment of the required inequality (4.16). This occurs by a combination of trapping between level 1 and the metastable 0, and electron-impact excitation of level 1 from atoms in level 0. Here, as level 0 is metastable and the population of the upper laser level 2 is determined by collisions with metastable helium (He* 2^3S_1 or He* 2^1S_0) atoms, it is assumed that the metastable excitation rate of level 0 is a related by a constant to that of the excitation F_2. Again as mentioned by Bennett,[82] saturation in the upper-state excitation rate R_2 can arise to limit the realizable output power at high excitation rates.

Experiments by Bennett and Knutson[83] in which radial profile and gain saturation characteristics were studied, have enabled the determination of the dominant processes that are responsible for these characteristics in the 1.1523-μm $(2s_2-2p_4)$ He–Ne laser. In the experiments, measurements were made of the variation of gain at 1.1523 μm (λ_1) across the diameter of a 3-m-long gain tube having a 15-mm bore while simultaneously monitoring the spontaneous intensity of another line that had the Ne $2s_2$ level as an upper state in common with the 1.1523-μm line. The spontaneous intensity of the second line at λ_2,

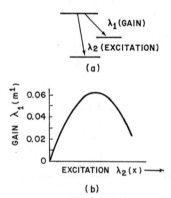

FIG. 4.21. (a) Relationship of gain (wavelength λ_1) and excitation rate (wavelength λ_2) for saturation determination on the 1.1523-μm He–Ne laser. (b) Variation of gain with excitation rate on the axis of a 1.1523-μm He–Ne laser. Discharge tube diameter 15 mm. (After W. R. Bennett, Jr.,[82] by courtesy of the Optical Society of America.)

which provided a direct measure of the Ne $2s_2$ upper-state population (Fig. 4.21a), was fed into the X-axis input of an XY recorder and the gain at λ_1 was fed into the Y-axis input. This arrangement permitted a direct plot to be made of the inversion saturation characteristics at discrete radial locations.[82] Data taken by Bennett and Knutson[82, 83] at a variety of gas pressures and radial locations for the variation of gain at 1.1523 μm with excitation rate on the axis of the 15-mm-bore gain tube, and the radial spatial variation of gain, are shown in Figs. 4.21b and 4.22, respectively.

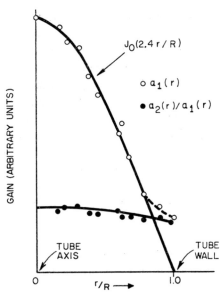

FIG. 4.22. Gain plotted as a function of radial position for the 1.1523-μm He–Ne laser transition, showing parameters in eqn. (4.17). (After W. R. Bennett, Jr.,[82] by courtesy of the Optical Society of America.)

As long as the excitation rate was varied in time intervals that were believed to be short enough to preclude the gas temperature changing, it was found that a typical plot of gain versus excitation rate (x) in Fig. 4.21b fitted the experimental points to within a few percent, and had the quadratic form

$$\text{Gain } (\lambda_1) = a_1 x - a_2 x^2 \ldots = a_1 x [1 - (a_2/a_1)x \ldots], \qquad (4.17)$$

where $(a_2/a_1)x$ is a saturation term.

The solid curve in Fig. 4.22 shows that the radial variation of gain $a_1(r)$ agrees closely with a J_0 Bessel function over nearly the whole tube diameter where a_1 is also the constant in eqn. (4.17). The departure of the experimental points near the tube wall is probably real and due to incomplete deactivation of both He^* 2^3S_1 metastables and Ne^* $1s$ metastables in collisions with the tube wall as is now known to occur.[179] The quantity $a_2(r)/a_1(r)$, however, which should also approximate a J_0 Bessel function if the saturation term $(a_2/a_1)x$ in (4.17) were in agreement with experiment, clearly does not have the shape of a Bessel function. From this observation, Bennett[82] deduces that the electron-metastable collision process in the 15-mm-bore discharge tube used is not the dominant saturation mechanism in contributing to the lower laser-level excitation rate F_1 (Fig. 4.20).

The plot of gain at 1.1523 μm versus excitation rate (x) in Fig. 4.21b exhibits a faster fall-off gain at higher excitation rates than the change in gain that occurs with excitation rate at low excitation rates. Whereas the population of the Ne $2s_2$ upper laser state continues to rise with discharge current, as indicated by the spontaneous emission at the wavelength λ_2, the population difference ($N_2 - N_1$), indicated by the gain at λ_1, falls sharply at a high excitation rate. This shows that the population of the lower laser level is increasing faster than that of the upper laser level and that the gain is being affected adversely by a disproportionate increase in F_1 relative to F_2 at high excitation rates. The result also enables

135

^the conclusion to be drawn that the gain-saturation mechanism in the 1.1523-μm He–Ne laser is due mainly to processes that affect the excitation rate F_1 of the $2p_4$ lower laser state. The decrease in gain is not due to a reduction of the upper-laser-state excitation rate F_2.

On the basis of these results, and the observed inverse variation of the gain in a number of four-level neutral-laser systems with the discharge-tube diameter, Bennett[82] has suggested that the inherent saturation mechanisms of a typical four-level gas laser system of the He–Ne type vary from resonance trapping on terminal laser transitions at the large tube-diameter extreme to electron-impact mechanisms involving the terminal-laser state 0 at small tube diameters. The connection of saturation mechanisms involving terminal laser states and the inverse-radial-gain dependence is unfortunate. It has been taken consistently in the laser literature to mean that the inverse radial dependence of the gain exhibited by four-level neutral-gas lasers follows from a predicted reduction in the population of their terminal states by increased diffusion losses in tubes of smaller diameter relative to that in tubes of larger diameter.

In the case of the 1.1523-μm He–Ne laser it is clear that the saturation mechanism is due primarily to the build-up of population in the lower $2p_4$ laser level. In the 3.39-μm and 0.6328-μm He–Ne lasers, Fig. 4.12 and the analysis given in the section prior to this show that the observed saturation in gain at high excitation rates follows from saturation in the *upper* $3s_2$ state population because of de-excitation of He* 2^1S_0 metastables in the super-elastic-collision reaction (4.12) at high electron densities, and not because of processes involving Ne $1s$ metastables that affect the lower laser level.

Inverse radial dependence of gain—the $1/D$ *gain relationship.* The gain in a number of neutral gas lasers, at optimum discharge conditions of gas pressure and discharge current varies approximately as the reciprocal of the diameter of the discharge tube.[8, 14, 39, 43, 82, 85–91] The lasers in which this gain relationship is observed include the 1.1523-, 3.39-, and 0.6328-μm He–Ne lasers; the He–Xe laser operating at 5.575, 3.507, and 2.025 μm; and the Ne–O₂ and Ar–O₂ lasers operating in atomic oxygen at 0.8446 μm. The Ne–O₂ and Ar–O₂ lasers, unlike the other lasers, are three-level laser systems and involve dissociation reactions.

The maximum unsaturated gain G_m in a He–Ne discharge optimized for gain at 632.8 nm is described by the relation[86]

$$G_m = 3.0 \times 10^{-2} \, l/D \text{ percent,} \tag{4.18}$$

where l is the discharge length in cm and D is the internal diameter of the tube in cm. This expression has been found to predict the measured gain to within 10 percent accuracy for a number of laser tubes with different fills that include 5 : 1 and 7 : 1 ratios of He⁴ : Ne, and a 7 : 1 ratio of He³ : Ne. Note that the relation only applies when competition with the high-gain $3s_2$–$3p_4$, 3.39-μm transition is eliminated. When competition with the 3.39-μm transition is not eliminated, a more complicated relationship,

$$G_m \cong \left[1 + 0.5 \left(\frac{D_0}{D} \right)^{1.4} \right]^{l/l_0}, \tag{4.19}$$

has been found to apply where $D_0 = 1$ mm and $l_0 = 100$ cm, and D and l, as defined above, are in mm and cm respectively.[87]

Optimum discharge conditions for obtaining maximum gain and output power on the

Ne-I 632.8-nm, $3s_2$–$2p_4$ transition in a He–Ne mixture when oscillation at 3.39 μm is suppressed are

1. He : Ne pressure ratio = 5 : 1,
2. pressure and tube-diameter product

$$pD = 3.6 \text{ to } 4.0 \text{ torr-mm,} \qquad\qquad (4.20)$$

where p is the total gas pressure.[73, 86, 92, 93]

The extremely important implication of the constant pD is that the average electron temperature has approximately the same value under optimum gain conditions for all 632.8-nm He–Ne laser discharges. This follows from eqns. (3.10) and (3.16), Fig. 3.7, and the treatment given in Section 3.2.2. Measurements of the average electron energy in a 5 : 1 He : Ne mixture in discharge tubes ranging from 2 to 6 mm diameter at a constant pD of 3.6 torr-mm by Labuda[72] and Labuda and Gordon[94] confirm this. Figure 4.23, which shows results of measurements of the average electron energy taken on four discharge tubes of 6-, 5-, 3-,

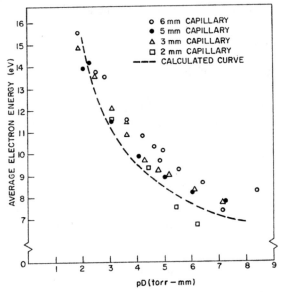

FIG. 4.23. Experimental values of average electron temperature as a function of the product of pressure and tube diameter. Discharge conditions: 5 : 1 He : Ne mixture for different tube diameters at discharge currents of less than 100 mA. (By courtesy of E. F. Labuda.[72])

and 2-mm bore, confirm that the average electron energy as a function of pD is a universal curve independent of the tube diameter.[72, 94] At a pD of 3.6 torr-mm the average electron energy is seen to be between 10 to 11 eV. This constancy of the average electron energy or temperature under optimum gain conditions simplifies considerably a discussion of the excitation mechanisms responsible for the gain of the 632.8-nm laser line since a precise knowledge of the energy dependence of various electron excitation or destruction mechanisms is not required.

The question that we wish to answer is "Does the $1/D$ gain relationship follow from a decrease in population of the terminal Ne $1s$, metastable/quasi-metastable states, or from an increase in upper $3s_2$-state population as the discharge-tube diameter is decreased at a constant pD?" Bennett[82] has related both the gain saturation at high excitation rates and inverse radial dependence of gain of four-level laser systems to mechanisms that involve the density of neon atoms in the metastable Ne $1s$ states, based on the assumption that in smaller bore tubes the Ne $1s$ metastable density is decreased through increased diffusion and de-activation at the wall of the tube. This assumption needs to be examined closely.

Let us consider how the concentration of metastable atoms varies with pressure and discharge-tube diameter in a plasma in a long cylindrical tube. Where diffusion to the walls is assumed to be the dominant loss mechanism, the number of atoms Q lost per second per unit length of plasma is given by[95]

$$Q = S\beta \, \Delta N \tag{4.21}$$

where S is the surface area of the plasma; ΔN is the difference in concentration of atoms on the axis of the discharge tube and on the inside wall of the tube; and $\beta = \mu_0^2 \mathcal{D}/R^2$ in which $\mu_0 = 2.405$ the first root of a Bessel function, $\mathcal{D}$ is the diffusion coefficient, and R is the tube radius. $\mathcal{D} \propto 1/p$ (and is directly proportional to the atom velocity and the gas tempera-ture).[96]

Substituting in eqn. (4.21) for β,

$$Q = \text{Area of the plasma column} \; \frac{(\mu_0)^2 \mathcal{D}}{R^2} \, \Delta N. \tag{4.22}$$

For a long, thin cylindrical plasma column of length L, and radius R, with R much less than L, the area of the ends of the column can be neglected in comparison with the surface area of the cylinder, and the diffusion loss per unit length is given by

$$Q \propto 2\pi R \frac{\mathcal{D}}{R^2} \, \Delta N. \tag{4.23}$$

Since $\mathcal{D} \propto 1/p$,

$$Q \propto 2\pi R \cdot \frac{\Delta N}{pR^2} \tag{4.24a}$$

making

$$Q \propto \frac{\Delta N}{pR} \quad \text{or} \quad \propto \frac{\Delta N}{pD}. \tag{4.24b}$$

At the constant product of pressure and tube radius pR (or pD) (observed for the neutral gas lasers mentioned) it follows immediately that the diffusion loss of metastables Q is independ-ent of the tube radius or diameter and depends *only* on ΔN. Only at constant pressure does the diffusion loss vary inversely as the tube radius R.

This simple treatment indicates, therefore, that at a constant pD and under optimum gain conditions, the population of metastable species that decay principally by diffusion to the walls of the discharge tube **remains constant** and does not decrease with decrease in tube diameter. Furthermore, at a constant electron concentration at a constant pD, with a de-crease in D, it is likely that ΔN and the metastable density on the axis of the tube **increases** due to the increase in gas pressure and an increased excitation rate. It seems clear then that the $1/D$ gain relationship in He–Ne or pure neon lasers cannot be correctly ascribed to an increased loss of Ne $1s$ metastables with decrease in tube diameter. In other gas lasers it also cannot be ascribed to a loss in population of terminal laser states.

Let us now consider the second part of the question previously asked, "Does the $1/D$ gain relationship follow from an increase in upper $3s_2$-state population as the discharge-tube diameter is decreased at a constant pD?"

Gordon and White[92] have shown that the $He^* \, 2^1S_0$ metastable density M on the tube axis is given by

$$M = \frac{k_1 n_e p_{He}}{k_2 p_{Ne} + k_3 n_e} \tag{4.25}$$

in which $k_1 p_{He}$ is the production rate by fast electrons and includes cascading from higher states; $k_2 p_{Ne}$ is the effective destruction rate by excitation transfer to neon ground-state atoms; and k_3 is the rate of destruction in superelastic collision with slow electrons. A similar relationship for the concentration of $He^* \, 2^1S_0$ metastables in $3.5:1$, $7:1$, and $14:1$ He : Ne mixture ratios in a hollow-cathode discharge has been reported by Znamenskii[97] (see Fig. 3.17b). Znamenskii also found that the concentration of $He^* \, 2^1S_0$ and 2^3S_1 metastables, as well as atoms in the metastable Ne $1s_5$ state varied inversely as the cathode diameter at a constant pD, and hence varied directly as the gas pressure. Using the fact that the population density of the Ne $3s_2$ upper laser level is mainly selectively excited by excitation transfer from $He^* \, 2^1S_0$ metastables, Gordon and White[92] have shown also that the maximum gain G_{max} is given by

$$G_{max} = [(pD)^2/(1+pD)/D] \, H(pD)/D^2, \tag{4.26}$$

in which H is expected to be a monotonically decreasing function of pD.

For pressures below 1 torr, keeping pD constant, G_{max} increases with p^2 or decreases with D^2. For increasing pressures above 1 torr, G_{max} increases less rapidly with pressure and eventually achieves a dependence proportional to p or inversely proportional to D in agreement with experimental results. It appears then that the treatment of Gordon and White, which is based on considerations of the way in which the upper $3s_2$ state population varies, increasing with pressure and decrease in tube diameter D, explains completely the $1/D$ gain relationship in the 0.6328-µm and 3.39-µm Ne–Ne lasers.

If the concentration of He 2^1S_0 and 2^3S_1 metastables and Ne metastables increase in the positive column of a glow discharge with decrease in tube diameter at a constant pD due to the increase in gas pressure (in the same way as they do in a hollow-cathode discharge) the inverse radial dependence of gain in the He–Ne laser at 1.1523 µm on the $2s_2$–$2p_4$ transition can also be attributed to an overall **increase** in population of both upper and lower state populations. If, as seems extremely likely, the population of both lower and upper laser states is directly proportional to the gas pressure by constants k_1 and k_2, respectively, then the gain, which is directly proportional to the difference in population of the upper and lower states, will also be directly proportional to the gas pressure by another constant k_3, which is equal to k_2-k_1.[98]

In systems such as the 3.39-µm and 0.6328-µm He–Ne lasers, in which the upper-state population is much larger than the lower-state populations and $N_2 > N_1 g_2/g_1$ in the gain expression (1.47) making $k_2 > k_1$, the gain is almost directly proportional to the upper-state population via the constant k_2 and hence to the gas pressure p. Where excitation transfer is the main selective-excitation process, the gain is then almost directly proportional to the population of the excited species from which energy is being transferred, the population of which is directly proportional to the gas pressure.

In summary:

1. Gain saturation at high excitation rates (high current densities) is due to processes that affect the excitation rate of the lower $2p_4$ laser state in the 1.1523-μm He–Ne laser. In the 3.39-μm and 0.6328-μm He–Ne lasers the gain saturation is due to processes that affect the population of $He^* \, 2^1S_0$ metastables, and the population of the $He^* \, 2^1S_0$ metastables in turn limit the realizable population of the upper $3s_2$ laser state. The processes that adversely affect the population of $He^* \, 2^1S_0$ metastables are the superelastic collision-process (4.12)

$$He^* \, 2^1S_0 + e^- \text{ (slow)} \rightarrow He^* \, 2^3S_1 + e \text{ (fast)};$$

and possibly the molecular-helium-formation reaction (4.15)

$$He^* \, 2^1S_0 + He \, ^1S_0 \rightarrow He_2 \, A^1\Sigma_u^+ .$$

2. The $1/D$ gain relationship is a pressure-gain relationship.[98] Under optimum laser conditions, where the pD product and hence the average electron temperature is maintained constant, the relationship arises because of increased excited-states populations in both upper and lower (and terminal) laser states as the tube diameter D is decreased, and p (which is effectively the ground-state population) is increased.

Discharge conditions. The output power and gain of the 632.8-nm and the 1.1523-μm He–Ne lasers is a function of a number of discharge parameters. The 632.8-nm laser is the better documented of the two. Important parameters for it include: total pressure p of helium and neon, discharge-tube diameter D, mixture ratio of helium to neon, electron concentration, which is related to the discharge current density, and gas temperature.

Figures 4.24 to 4.27 show the effect of total pressure, He : Ne mixture ratio, and tube diameter on the multimode output-power realizable on the Ne $3s_2$–$2p_4$ transition at 632.8 nm with oscillation on the $3s_2$–$2p_4$ transition at 3.39 μm suppressed.[73] Since the output power

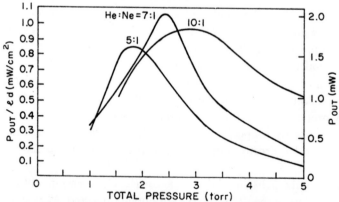

FIG. 4.24. Output power and normalized output power at 632.8 nm versus total pressure at various He : Ne pressure ratios (discharge-tube diameter 1.5 mm and plasma length 12.5 cm). (After R. L. Field, Jr.,[73] by courtesy of the American Institute of Physics.)

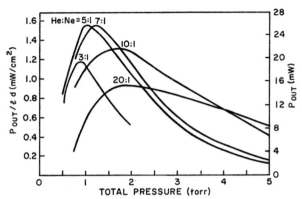

FIG. 4.25. Output power and normalized output power at 632.8 nm versus total pressure at various He : Ne pressure ratios (discharge-tube diameter 3 mm and plasma length 55 cm). (After R L. Field, Jr.,[73] by courtesy of the American Institute of Physics.)

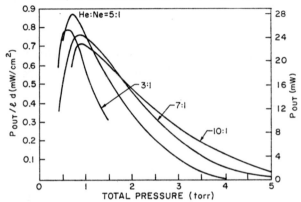

FIG. 4.26. Output power and normalized output power at 632.8 nm versus total pressure at various He : Ne pressure ratios (discharge-tube diameter 5 mm and plasma length 65 cm). (After R. L. Field, Jr.,[73] by courtesy of the American Institute of Physics.)

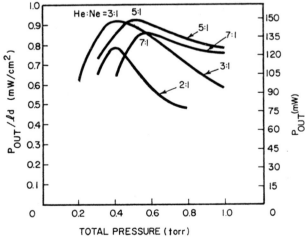

FIG. 4.27. Output power and normalized output power at 632.8 nm versus total pressure at various He : Ne pressure ratios (discharge-tube diameter 8 mm and plasma length 200 cm). (After R. L. Field, Jr.,[73] by courtesy of the American Institute of Physics.)

depends on the plasma length, laser cavity configuration, and spherical mirror reflectivities,[86, 99] the cavity and mirror data applicable to the results given in the figures (for a spherical-mirror and high-reflectivity-prism cavity-configuration) are supplied in Table 4.8.[100]

TABLE 4.8. *Mirror and cavity data applicable to Figs. 4.24 to 4.27*

Tube diam. D (mm)	Mirror radius (m)	Mirror transmission (percent)	Cavity length (cm)
1.5	0.5	1.1	22
3.0	2	1.0	70
5.0	2	1.0	80
8.0	10	1.1	215

The results given in Figs. 4.24 to 4.27 are summarized in Fig. 4.28; the pD is plotted as a function of capillary (tube) diameter D for various He : Ne pressure ratios and the optimum pD is also plotted as a function of the tube diameter. The optimum product $pD = 3.6$ torr-mm is in close agreement with results by Gordon and White,[92] Smith,[86] and Korolev *et al.*[93] As noted by Field,[73] the optimum He : Ne pressure ratio (i.e. the ratio that yields the

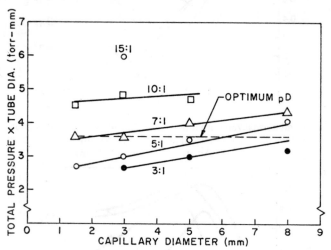

FIG. 4.28. Average optimum pD for all He–Ne mixture ratios (broken curve) and optimum pD for a given He : Ne ratio as a function of capillary (tube) diameter. (After R. L. Field, Jr.,[73] by courtesy of the American Institute of Physics.)

largest output power for a given tube diameter) tends to decrease with increasing tube diameter at the optimum pD of 3.6 torr-mm. Figure 4.29 shows this clearly.

There are some strong indications that the optimum He : Ne pressure ratio for each tube diameter is related to a maximization of the average electron energy given a sufficient neon pressure to radiation trap the Ne $3s_2$ to the ground state vacuum-ultraviolet transition. According to Fig. 4.29, the optimum He : Ne pressure ratio for a 5-mm-diameter tube is 5.5. This ratio corresponds exactly to the He : Ne ratio for a 5-mm discharge tube at which a maximum average electron energy has been observed experimentally in a 632.8-nm He–Ne

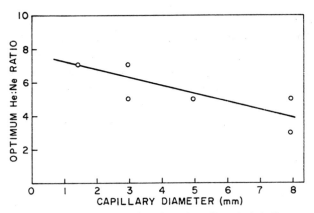

FIG. 4.29. Optimum He : Ne pressure ratio as a function of capillary (tube) diameter. (After R. L. Field, Jr.,[73] by courtesy of the American Institute of Physics.)

laser discharge by Labuda. Figure 4.30, after Labuda,[72] illustrates this. The broken curve shows the average electron energy as a function of He : Ne pressure ratio calculated using the theory of von Engel and Steenbeck,[101] similar to that by Young[102] given in Section 3.2.2. This He : Ne pressure ratio also corresponds to that at which the electron temperature is predicted to be a maximum in the positive column of glow discharge in a He–Ne mixture.

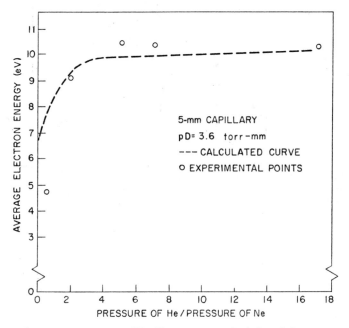

FIG. 4.30. Average electron energy versus He : Ne pressure ratio ($pD = 3.6$ torr-mm and $D = 5$ mm). (By courtesy of E. F. Labuda.[72])

The maximum is due to increased mobility of He^+ ions through a reduction in He^+ to He symmetrical charge transfer at low partial pressures of neon, Fig. 3.7 and ref. 13 (mentioned earlier in this chapter under *Effects of additives*).

143

Values of discharge current for giving maximum ouput power at 632.8 nm are shown plotted in Fig. 4.31. This figure is a modification of that given by Field.[73] Curve B exhibits a linear dependence of discharge current on tube diameter under optimum conditions found to apply to the 632.8-nm Ne–Ne laser when oscillation on the competing $3s_2$–$3p_4$ transition at 3.39 μm is suppressed. Note that B does not extrapolate to the origin, and the optimum current density (mA per mm²) decreases with increase in tube diameter, and that I/D is not a constant, in disagreement with the earlier observation of Gordon and White.[92] Curve A, which is an addition to the data given by Field, is a plot of the relationship

$$I \, (\text{mA}) = 3.5 + 1.5D^2 \tag{4.27}$$

between optimum discharge current (mA) and discharge-tube diameter in mm when oscillation at 3.39 μm is not suppressed.[103]

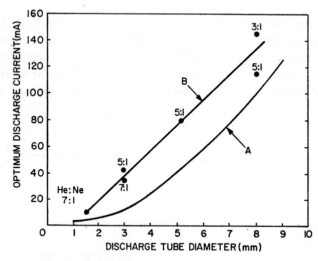

FIG. 4.31. Optimum discharge current as a function of discharge-tube diameter. Curve B is the observed linear relationship between optimum current and tube diameter when oscillation at 3.39 μm is suppressed.[73] Curve A is a curve of relationship (4.27) that applies when oscillation at 3.39 μm is not suppressed.[103] (After C. S. Willett, Neutral gas lasers, in *Handbook of Lasers*, by courtesy of the Chemical Rubber Co., Cleveland, Ohio.)

Curves A and B can be used to deduce that there is a universal optimum electron concentration in 632.8-nm He–Ne lasers under similar optimum discharge conditions. Figure 4.32 is basically a figure given by Labuda,[72] showing plots of the electron density versus discharge current for discharge tubes of differing diameters with a $pD = 3.6$ torr-mm and a He : Ne pressure ratio of 5 : 1. As might be expected, the figure shows that the electron density is directly proportional to the discharge current for each tube diameter. Rather surprisingly, it also shows that the electron density is not directly proportional to the current density independent of the discharge-tube diameter. The circled and crossed points plotted in the figure are values of the optimum discharge current for each of the 2-, 3-, 5-, and 6-mm discharge tubes used by Labuda, given by A and B in Fig. 4.31. These then are values of the opti-

144

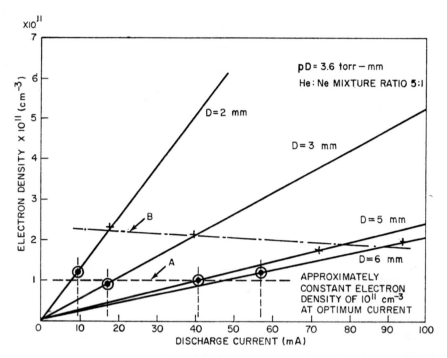

FIG. 4.32. Electron density versus discharge current for tubes of different internal diameter (pD = 3.6 torr-mm, 5 : 1 He : Ne mixture ratio). The circled and crossed points correspond to the values of optimum discharge current at the appropriate tube diameters given by A and B, respectively, in Fig. 4.31, and illustrate the constancy of electron density under comparable discharge conditions.

mum discharge current: A, when oscillation at 3.39 μm is not suppressed; and B when oscillation at 3.39 μm is suppressed. The two broken lines are drawn through each of these two sets of values of discharge current.

It can be seen from curve A that the electron density is approximately constant at 10^{11} cm^{-3} in the four discharge tubes when they are operated at their optimum discharge currents to give maximum output power at 632.8 nm with oscillation at 3.39 μm unsuppressed; from curve B that, with the 3.39-μm line suppressed, the laser can be run at discharge currents that give an electron concentration that is approximately 2×10^{11} cm^{-3} for all the four discharge tubes. A similar determination of the constancy of electron concentration in tubes of different diameters at optimum discharge currents, and in various He–Ne mixture ratios and pressures, has been reported by Fridrikhov.[104] The He–Ne laser operating at 632.8 nm appears, therefore, to be characterized by the same optimum values of T_e and n_e over a wide range of excitation conditions. Whereas a plausible argument could be given for the existence of an optimum electron temperature in the 632.8-nm He–Ne laser discharge, there does not appear to be any good reason why there should be an optimum electron concentration in such a discharge under comparable conditions.

The output power at 632.8 nm is affected by the gas temperature in the active region of the plasma.[105–7] This is illustrated in Fig. 4.33, where the output power at 632.8 nm is plotted against the concentration of unexcited atoms in a dc-excited laser discharge with the wall of

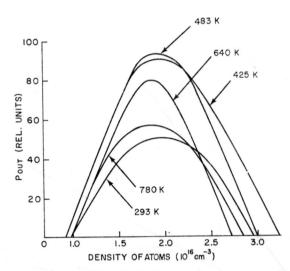

FIG. 4.33. Output power at 632.8 nm as a function of ground-state atom concentration for various discharge-tube wall temperatures. Discharge conditions: dc-discharge current 70 mA, He : Ne ratio 8 : 1, $D = 8$ mm. (After S. A. Gonchukov *et al.*,[105] by courtesy of the Optical Society of America.)

the discharge tube at various temperatures. The output power has a maximum at an atom concentration of 1.9×10^{16} cm^{-3} for all wall temperatures. The optimum wall temperature shown is 483 K.[108]

This optimum temperature is probably not a universal optimum wall temperature but rather is peculiar to the particular discharge tube used in the experiment. It does, however, illustrate the overall effect that gas temperature has on the numerous collisional and diffusive effects operating in the 632.8-nm He–Ne laser. These include the exponential variation of the cross-section with gas temperature for excitation transfer from He* 2^1S_0 metastables to the Ne $3s_2$ state [eqn. (2.11)]; collisional frequency for He*–Ne collisions; and possibly the molecular-helium-formation reaction [eqn. (4.15)]. A 30 to 40 percent increase in power at 632.8 nm and 1.1523 μm can be obtained by forced heating of the discharge tube for optimum partial pressures and total gas pressures of the He–Ne mixtures. The optimum gas temperatures T_g are approximately 700 and 750 K for the two laser systems, respectively.[107]

The detrimental effect of the high-gain $3s_2$–$3p_4$ transition at 3.39 μm on oscillation at 632.8 nm can be reduced drastically by Zeeman splitting the 3.39-μm line by a means of an inhomogeneous longitudinal magnetic field over the active length of the discharge.[109–110] The splitting reduces the gain and stimulated emission at 3.39 μm, and the reduced stimulated emission at 3.39-μm leads to an increase in population of the Ne $3s_2$ state and to an enhanced population inversion on the 632.8-nm $3s$–$2p_4$ transition. Prism wavelength selection can also be utilized to reduce the gain of the 3.39-μm line.[54, 55, 111] An approximate 50 percent increase in power at 632.8 nm over that in Figs. 4.24 to 4.27 can be achieved if a weak inhomogeneous magnetic field is used in conjunction with prism wavelength selection.[73] The inhomogeneous magnetic field can be generated simply by placing small magnets randomly along the discharge tube.[109] The variation of output power at 632.8 nm and 3.39 μm versus magnetic field is shown in Fig. 4.34. The figure clearly illustrates the competi-

tion that occurs between the 632.8-nm and 3.39-μm lines under the influence of the magnetic field. Curves 1 and 3, and 2 and 4 were recorded simultaneously with the rf-drive power
for 1 and 3 greater than for 2 and 4.

Differing excitation conditions between cathode and anode produced by cataphoretic
effects also cause the gain at both 0.6328 and 3.39 μm to increase appreciably toward the
cathode.[112]

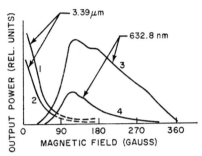

FIG. 4.34. Competition of emission at 3.39 μm (curves 1 and 2) and 632.8 nm (curves 3 and 4) with the plasma
in a longitudinal magnetic field. Discharge conditions: rf-excitation, He : Ne ratio 7 : 1, total pressure 0.8
torr, $D = 4$ mm. (After A. N. Alekseeva and D. V. Gordeev,[110] by courtesy of the Optical Society of
America.)

The 1.1523-μm He–Ne laser, the first gas laser, has not been as extensively investigated
as the 632.8-nm He–Ne laser. This has been because of the greater usefulness of the visible
632.8-nm laser and because of difficulties of spectroscopy in the near infrared region of the
spectrum where detector (photomultiplier) sensitivities are dropping off sharply. Most of
the transitions affecting the 1.1523-μm $2s_2$–$2p_4$ transition occur in the far-visible and near-
infrared region of the spectrum between about 0.8 to 1.5 μm where signal detection and
amplification is relatively hard. The majority of the early experiments conducted on the
1.1523-μm He–Ne laser utilized rf-discharge excitation. Because of the use of rf-excitation,
and the attendant difficulties of maintaining uniform discharges and estimating discharge
conditions, which are affected by electrode geometry, shape, coupling efficiency and rf-drive
power, coupled with the reluctance of researchers to duplicate their work using dc-excitation
(and sponsors to fund such work), it is not clear just what are the optimum discharge conditions for the 1.1523-μm He–Ne laser.

The "best" values of the product of gas pressure and tube diameter, and He : Ne mixture
ratio appear to be approximately

$$p_{Ne}D \simeq 1.2 \text{ torr-mm, } \text{and}$$

$$p_{He}D \simeq 12 \text{ to } 15 \text{ torr-mm,}$$

giving a He : Ne mixture ratio of between 8 and 10 : 1.[113–15] As can be seen from Fig. 4.35
the output power at 1.1523-μm is sensitive to the neon partial pressure but relatively insen-
sitive to the partial pressure of helium just as in the 632.8-nm He–Ne laser. The $p_{total}D$
product of approximately 14 to 17 torr-mm, at a mixture ratio of about 10 : 1, which coincides with a broad maximum in the curve of electron temperature versus mixture ratio (Fig.
3.7), gives an average electron temperature in the positive column of a glow discharge of

11   147

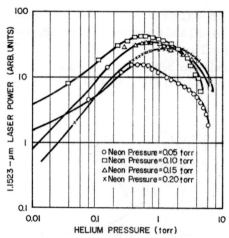

FIG. 4.35. Output at 1.1523 μm from an external confocal mirror He–Ne laser as a function of helium pressure for four different neon pressures. Discharge conditions: rf-excitation, $D = 15$ mm. (After C. K. N. Patel,[113] by courtesy of the American Institute of Physics.)

about 50,000 K (or 6 eV). This is in agreement with other work.[116] Axial electron concentrations in typical 1.1523-μm He–Ne laser discharges are similar to those at optimum conditions in 632.8-nm laser discharges. They are about 10^{11} cm^{-3} (Maximov,[114] Mamikonyants and Molchashkin,[116] and Belousova *et al.*).[117] No relationship between optimum discharge current for oscillation at 1.1523 μm and tube diameter appears to have been determined and published as it has for the 632.8-nm He–Ne laser.[73, 103] Though the electron concentrations in 1.1523-μm and 632.8-nm He–Ne lasers are about the same, the optimum pD products, He : Ne mixture ratios, and reflection coefficients for He* 2^3S_1 and He* 2^1S_0 metastables on glass surfaces[179] all differ. This means that the simple relationship (4.27) between optimum discharge current and tube diameter for the 632.8-nm He–Ne laser cannot be applied to the 1.1523-μm He–Ne laser. The optimum discharge current for the 1.1523-μm laser is probably larger than that for the 632.8-nm He–Ne laser for all tube diameters, and it probably depends on the reflectivity properties of the cavity at 1.1523, 3.39 and 0.6328 μm because of competition between transitions. If a factor is appropriate for a range of tube diameters of 3 to 15 mm, the optimum discharge currents for the infrared laser could be as much as a factor of 2 higher than those given by eqn. (4.27). For example, Rigden and White[115] using "hybrid" mirrors designed to give maxima of reflection at 1.1523 μm and 632.8 nm, find that the optimum currents for a 6-mm discharge tube are between 40 and 50 mA for the 632.8-nm line and approximately 75 and 90 mA for the 1.1523-μm line.

The $p_{Ne}D$ product of 1.2 torr-mm for the pure neon laser operating at 1.1523 μm is nearly the same as that of a He–Ne laser operating at the same wavelength.[113] The addition of helium to neon is responsible for a 50-times increase in output power.[113] In Fig. 4.4 a sixfold increase in gain is exhibited by a He–Ne 1.1523-μm laser discharge over that of a pure neon discharge at the same $p_{Ne}D$.[8] This significant increase in gain has been attributed to an increase in electron-impact excitation of the $2s_2$ upper laser state as a result of an increased electron concentration as well as to excitation transfer from He* 2^3S_1 metastables to neon atoms in the $2s_2$ state. According to theory, the electron temperature in

148

a discharge in pure neon at the $p_{Ne}D$ of 1.2 torr-mm used by Patel[113] is in excess of 90,000 K (Fig. 3.7). In a 8 : 1 He–Ne mixture at a pD of 13.5 torr-mm, the electron temperature is only about 50,000 K. This is, therefore, less by nearly one-half that in the pure neon laser discharge.[118] The theoretical decrease in average electron temperature together with an apparent *increase* in electron concentration that has been observed [8] when helium is added to neon discharge cannot be readily reconciled. Possible reasons are that helium alters the actual shape of the electron-energy distribution to a distribution that favors direct electron-impact excitation from the ground state or that the presence of $He^*\ 2\,^3S_1$ metastables leads to the production of $(HeNe)^+$ ions and additional electrons in the associative-ionization reaction

$$He^*\ 2\,^3S_1 + Ne\ ^1S_0 \rightarrow (HeNe)^+ + e^-,$$

which in some way produces additional electron-impact excitation of the upper Ne $2s$ states. This area is clearly still wide open for much-needed research to clarify just what is occurring.

4.1.2. THE 3.067-μm CHLORINE LASER

Oscillation at 3.067 μm on the $5p\ ^2D^\circ_{5/2}$–$5s\ ^2P_{3/2}$ transition in neutral atomic chlorine was reported by Dauger and Stafsudd[119] in 1970. Although there have been reports of oscillation in the near infrared in atomic chlorine in noble gas and chlorine discharges,[120] oscillation on the 3.067-μm transition had not been reported prior to 1970. In the case of the 3.067-μm line, oscillation occurs only in argon–chlorine discharge mixtures, and not in chlorine alone (Dauger and Stafsudd, private communication, October 1970).

4.1.2.1. Upper-level excitation

Selective excitation involved in populating the upper laser level is the excitation-transfer reaction

$$Ar^*\ 4s + Cl\ ^2P^\circ_{3/2} \rightarrow Ar\ ^1S_0 + Cl'\ 5p\ ^2D^\circ_{5/2} + \Delta E_\infty, \tag{4.28}$$

prior to which the ground state Cl $^2P^\circ_{3/2}$ atom has been produced by dissociation of Cl_2 in the discharge.

Figure 4.36, a partial energy-level diagram for Cl-I and Ar-I, illustrates the disposition of the $5p\ ^2D^\circ_{5/2}$ upper laser level and other Cl-I levels and the four lowest-lying excited levels of neutral argon. The Cl-I $5p\ ^2D^\circ_{5/2}$ level at 95,396 cm^{-1} is close to both the upper (quasi-metastable) $4s'[\frac{1}{2}]^\circ_1$ level at 95,400 cm^{-1} and the metastable $4s'[\frac{1}{2}]^\circ_0$ level at 94,554 cm^{-1}. In addition to the close energy coincidence between the $5p\ ^2D^\circ_{5/2}$ level and the Ar $4s$ levels, it can be seen from Fig. 4.36 that the $5p\ ^4D^\circ_{1/2,3/2,5/2,7/2}$, $5p\ ^2D^\circ_{3/2}$, $5p\ ^4P^\circ_{1/2,3/2,5/2}$, $5p\ ^4S^\circ_{3/2}$, and $5p\ ^2S^\circ_{1/2}$ levels are also in near coincidence with the Ar $4s$ levels. Dauger and Stafsudd[121] found when argon was added to discharge in chlorine and helium (0.2 torr Cl_2, 19 torr He, and 1.7 torr Ar)—the simple addition of which would be expected to lower considerably the average electron temperature of the discharge and thereby reduce the intensity of chlorine lines—the intensities of lines whose upper levels are listed above and which are in close

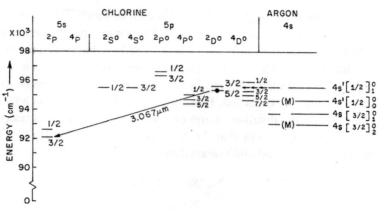

FIG. 4.36. Partial energy-level diagram for Cl-I and Ar-I showing the energy-level coincidences between the Ar-4s and Cl-I levels together with the 3.067-μm Cl-I laser transition. (After A. B. Dauger and O. M. Stafsudd,[119] by courtesy of the IEEE.)

energy coincidence with the Ar 4s levels, were increased in intensity. In the absence of evidence to the contrary, this shows that excitation transfer is occurring to all the Cl-I 5p states from atoms in the Ar 4s states. Table 4.9 lists several Cl-I lines for which the Cl' 5p levels listed are the upper states; their enhancements when argon was added, the nearest Ar 4s state, and the energy discrepancy ΔE_{∞} for excitation transfer between the two states.

TABLE 4.9. *Interaction between Cl' 5p and Ar 4s states*[121]

Wavelength of Cl-I line (nm)	Cl-I line upper level	Enhancement with Ar addition (percent)	Nearest Ar 4s state	ΔE_{∞} (cm^{-1})
436.3	$5p\ ^2D^{\circ}_{5/2}$ [a]	32	$4s'[1/2]^{\circ}_1$	+4
437.0	$5p\ ^2D^{\circ}_{3/2}$	24	$4s'[1/2]^{\circ}_0$	−302
438.0	$5p\ ^4D^{\circ}_{3/2}$	20	$4s'[1/2]^{\circ}_1$	+91
439.0	$5p\ ^4D^{\circ}_{7/2}$	41	$4s'[1/2]^{\circ}_0$	−164
443.8	$5p\ ^4P^{\circ}_{5/2}$	32	$4s'[1/2]^{\circ}_0$	+76
440.3	$5p\ ^4P^{\circ}_{3/2}$	40	$4s'[1/2]^{\circ}_0$	−105
432.3	$5p\ ^4S^{\circ}_{3/2}$	31	$4s'[1/2]^{\circ}_1$	−208

[a] Upper level of the 3.067-μm laser line.

Absorption studies of Ar-I lines that terminate on the Ar 4s levels confirmed that the Ar 4s states were significantly depopulated when additions of chlorine were made to a discharge in an argon–helium mixture (1.7 torr Ar, 19 torr He, tube diameter 6 mm). It should be noted that it is not known whether or not the spontaneous intensity of other Ar-I or He-I lines was reduced generally by the chlorine addition, which would reduce the average electron temperature or electron concentration. Such a reduction in electron temperature could also account for a reduction in population of the Ar 4s states. The effect of the addition of chlorine on the Ar 4s level population is given in Table 4.10.

TABLE 4.10. *Effect of chlorine on Ar 4s level population*[121]

Wavelength of Ar-I line (nm)	Terminal level	Change in absorption of line with Cl_2 addition (percent)
750.4	$4s'[1/2]_1^\circ$	-21
794.8	$4s'[1/2]_0^\circ$	-47
842.5	$4s[3/2]_1^\circ$	-9
811.5	$4s[3/2]_2^\circ$	-18

Comparison of the results given in Tables 4.9 and 4.10 shows the following. The Ar-I state that is closest to the Cl-I $5p\ ^4D_{7/2}^\circ$ level, which is the upper state for the chlorine line that is enhanced most by the addition of argon, is also that state to experience the most change in population when Cl_2 is added to an argon–helium discharge. The Ar-I $4s'\left[\frac{1}{2}\right]_0^\circ$ state (a metastable state) appears to be responsible for the largest enhancement of Cl-I lines, and itself is the most affected of the four Ar $4s$ states by the addition of chlorine to an argon–helium discharge (Table 4.10). This indicates that excitation transfer from the Ar $4s$ states to Cl $5p$ states is predominantly from the Ar $4s'\left[\frac{1}{2}\right]_0^\circ$ state to the $5p\ ^4D_{7/2}^\circ$, $5p\ ^4P_{5/2,3/2}^\circ$ and the $5p\ ^2D_{5/2}^\circ$ states. The Cl $5p\ ^2D_{5/2}^\circ$ level, however, upper level of the 3.067-μm laser line, is 842 cm^{-1} *above* the Ar $4s'\left[\frac{1}{2}\right]_0^\circ$ level. On the basis of what is known about atom–atom excitation transfer, this energy deficiency, which has to be acquired whatever the shape and interaction of the molecule potentials might be, is too much to acquire in kinetic energy in the discharge, and one is led to conclude that atoms in the Ar $4s'\left[\frac{1}{2}\right]_1^\circ$ level within 4 cm^{-1} of the $5p\ ^2D_{5/2}^\circ$ level, and not the $4s'\left[\frac{1}{2}\right]_0^\circ$ level, are responsible for its selective excitation. More specifically, eqn. (4.28) can then be written

$$\text{Ar } 4s'\left[\tfrac{1}{2}\right]_1^\circ + \text{Cl } ^2P_{3/2}^\circ \rightarrow \text{Ar } ^1S_0 + \text{Cl}'\ 5p\ ^2D_{5/2}^\circ + \Delta E_\infty \text{ (4 cm}^{-1}\text{)}. \qquad (4.29)$$

If, as a result of dissociation of Cl_2, atomic-chlorine atoms have been produced in the Cl $^2P_{1/2}^\circ$ quasi-ground-state level (at 881 cm^{-1} above the ground state) as well as in the Cl $^2P_{3/2}^\circ$ ground state (as assumed for reaction (4.29) throughout this analysis), excitation transfer to the Cl' $5p\ ^2D_{5/2}^\circ$ state from the Ar $4s'\left[\frac{1}{2}\right]_0^\circ$ metastable state does become energetically possible. The reaction would then be the exothermic one

$$\text{Ar}^*\ 4s'\left[\tfrac{1}{2}\right]_0^\circ + \text{Cl}'\ ^2P_{1/2}^\circ \rightarrow \text{Ar } ^1S_0 + \text{Cl}'\ 5p\ ^2D_{5/2}^\circ + \Delta E_\infty \text{ (39 cm}^{-1}\text{)}. \qquad (4.30)$$

Operating characteristics. The 3.067-μm laser output as a function of discharge current in two tubes of 2.5-cm and 1.0-cm diameter at the same pressures, Ar 2.1 torr and Cl 0.09 torr, is shown in Fig. 4.37. The gas pressures were not arbitrarily chosen to be the same for the two tubes but produced the maximum laser power in both laser tubes. A more accurate value of the optimum chlorine pressure is about 0.5 torr (Stafsudd, private communication, 1970).

The reduction in output power in the smaller 1.0-cm-bore discharge tube at the same optimum pressure as in the larger-bore tube is intriguing. One might have expected that the optimum pressure in the 1.0-cm-bore tube would have increased by the ratio of the larger-bore diameter to the smaller-bore diameter (that is by a factor of 2.5) if excitation transfer from atoms in the Ar $4s$ states dominated in producing population inversion between the

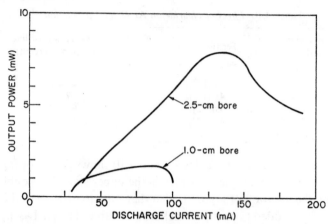

FIG. 4.37. 3.067-μm Cl-I laser output as a function of discharge current in different bore tubes. (After A. B. Dauger and O. M. Stafsudd,[119] by courtesy of the IEEE.)

$5p\ ^2D^o_{5/2}$ and $5s\ ^2P_{3/2}$ levels. That is to say, an approximate $1/D$ gain relationship might have been expected, and a higher output predicted for the smaller-bore tube as the population of atoms in the Ar $4s$ states would be anticipated to follow such a $1/D$ (or p) relationship.

4.2. Dissociative Excitation-transfer Lasers

A number of gas-laser systems utilize dissociative excitation transfer as an upper-laser-level, selective-excitation process. Lasers in which a collision of an excited atom with a molecule causes dissociation of the molecule and clearly result in the production of an atom in an excited state from which oscillation occurs include the following: the neon–oxygen laser, giving oscillation in atomic oxygen at 0.8446 μm;[90] the helium (or neon)–carbon monoxide (or carbon dioxide) laser, giving oscillation in carbon on a number of transitions in the infrared and near infrared;[40, 122-5] the helium (or neon)–nitrogen laser[123-5] oscillating in the near infrared in atomic nitrogen; various helium, neon or argon–sulfur-compound lasers giving oscillation in atomic sulfur, again in the near infrared;[125-6] the helium–fluorine and helium–CF_4, C_2F_6 or CF_6 lasers giving oscillation in the visible (near 0.7 μm) and near-infrared on transitions in atomic fluorine.[127-8]

In another laser system, the argon–oxygen laser, dissociative excitation transfer occurs in conjunction with two other processes to give selective excitation of particular states, and does not itself produce direct selective excitation of an upper laser state.

Only the neon–oxygen, argon–oxygen, and the helium–fluorine lasers are documented well enough to make a description of them illustrative of the selective excitation process of dissociative excitation transfer. For this reason, only these three laser systems will be considered here in any detail.

4.2.1. THE NEON–OXYGEN LASER

Historically, the neon–oxygen laser was the second gaseous system in which oscillation was achieved; CW oscillation in atomic oxygen at 0.8446 μm on the 0–1 $3p\ ^3P$–$3s\ ^3S_1$

transition being reported by Bennett *et al.* in early 1962.[90] Oscillation was observed in rf-excited mixtures of argon and oxygen as well as in neon–oxygen mixtures, with oxygen present in both systems at low pressure. In only the neon–oxygen laser does dissociative excitation transfer lead directly to selective excitation of the upper laser state.[129] The pertinent energy levels of neon, argon, krypton, molecular oxygen, and atomic oxygen are shown in Fig. 4.38. The noble-gas energy levels are defined in energy relative to the ground-state oxygen molecule, and only dissociation limits involving one ground-state $2p^4$ ^3P oxygen atom are shown (with the exception of the $O^- + O^+$ limit).[90] Laser oscillation occurs on the $3p$ ^3P–$3s$ ^3S transition of O-I at 0.8446 μm following selective excitation of the three states of the triplet-P state of oxygen, which is in close energy level coincidence with the four lowest excited levels of neutral neon. The dashed horizontal curves in Fig. 4.38 represent estimated repulsive curves that terminate at infinite nuclear separation on the excited

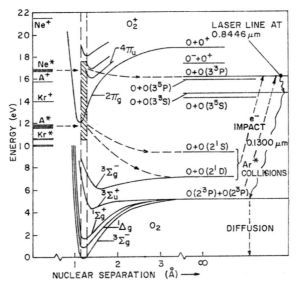

Fig. 4.38. Energy levels associated with laser action on the 0.8446-μm line of atomic oxygen in Ne–O_2, Ar–O_2, and Kr–O_2 mixtures. The excitation paths involving Ne* and Ar* atoms and oxygen atoms and excited states are shown by the arrows.

atomic-oxygen states specified. The dashed vertical lines indicate the region over which the Franck–Condon principle would predict excitation from the ground-state oxygen molecule to be probable, and the shaded area the region in which a number of excited molecular-oxygen "states" exist that can be repulsive and terminate on various excited states of atomic oxygen.

Dissociative excitation transfer occurs at the O-I $3p$ ^3P levels via the reaction

$$Ne' + O_2 \rightarrow Ne\ ^1S_0 + O + O'\ 3p^3P_{0,\,1,\,2} + \text{kinetic energy}, \tag{4.31}$$

where the excited Ne' atom is in any of the three lowest-lying, $1s_5$, $1s_4$, $1s_3$ metastable/quasi-metastable $1s$ states (Paschen notation) of neutral neon,[130] the O represents a ground-state $2p^4$ $^3P_2$ O-I atom, and the kinetic energy (of about 5000 cm^{-1}) is available to either or both of the dissociated O' or O atoms.

Although it was originally believed that only neon atoms in the $1s_4$ and $1s_3$ states took part in the dissociative excitation-transfer reaction (4.31),[8] it has been established that all of the three, closely situated, lowest-lying $1s_5$, $1s_4$ and $1s_3$ states of neon are involved in the reaction. This is made evident from the results given in Fig. 4.39 of experiments conducted by Kolpakova and Redko[130] on the variation of the concentration of Ne $1s$ atoms in a CW discharge in neon as oxygen was added. The neon pressure was maintained constant at 0.7 torr and the discharge-tube diameter was 16 mm. In order to maintain the partial pressure of oxygen at a set value in the presence of absorption on the walls of the discharge tube and clean-up at the electrodes, a steady flow of oxygen was allowed to enter the discharge tube. The oxygen pressure was varied in the range 2×10^{-3} to 4×10^{-2} torr, which encom passes the expected oxygen pressure for laser oscillation in a tube of 16 mm bore. Figure 4.39 shows that the concentrations of Ne $1s_5$, $1s_4$, and $1s_3$ atoms decrease, and all three change in exactly the same way when oxygen is added to the neon discharge. This reduction in Ne $1s_{3,4,5}$ atom concentration as oxygen is added to the discharge, alone is not enough to establish that this results from de-excitation collisions with oxygen atoms, as a reduction in electron temperature as the partial pressure of oxygen is increased could cause a reduction in the production of excited species of neon. Typical results, however, shown in Fig. 4.40 in which the intensity of the O-I laser line at 0.8446 μm and the concentration of Ne $1s$ atoms are shown plotted against discharge current, at a fixed partial pressure of oxygen of 2×10^{-2} torr, clearly show that the intensity of the 0.8446-μm line is indeed directly related to the concentration of Ne-I atoms in each of the $1s_5$, $1s_4$, and $1s_3$ states.

Figure 4.41 illustrates an interesting feature of the neon–oxygen 0.8446-μm laser discharge.[130] The variation of the intensity of the 0.8446-μm line with oxygen pressure for a fixed discharge current exhibits a maximum at an oxygen pressure of 1.3×10^{-2} torr. This maximum is seen to occur at the same oxygen pressure for different discharge currents, and corresponds to the pressure at which maximum laser power at 0.8446 μm (in a 7-mm-bore discharge tube) was obtained by Bennett et al.[90] Since the concentration of Ne $1s_{3,4,5}$ atoms does not exhibit a maximum (Fig. 4.39), as does the intensity of the 0.8446-μm O-I line (Fig. 4.41) at an oxygen pressure of 1.3×10^{-2} torr, and as the product of the concentration of Ne $1s$ atoms and the oxygen pressure does not maximize at that pressure as would be expected from reaction (4.31), it would appear that the dissociative excitation process is not the only process producing excitation of the upper state of the 0.8446-μm O-I line. An explanation could be that at low partial pressures of oxygen, below 2×10^{-2} torr, dissociative excitation transfer contributes significantly to the excitation, while above 2×10^{-2} torr, oxygen is determining the electron temperature of the discharge, and the excitation is by electron impact from the atomic oxygen ground state and is approximately directly proportional to the discharge current.

A strong resonance transition at 0.1300 μm connects the $3s\ {}^3S_1^{\circ}$ lower laser level to the atomix oxygen ${}^3P_2$ ground state so that in the absence of radiation trapping a good depopulation scheme exists for maintaining population inversion between the $3p\ {}^3P_{0,1,2}$ and the $3s\ {}^3S_1^{\circ}$ states. Unfortunately, the presence of the strong resonance transition connecting the lower laser level to the ground state means that the lower laser level is likely to be strongly excited by direct-electron impact from the O-I $2p^4\ {}^3P_2$ ground state if a sizeable concentration of ground-state atoms exists. Under actual lasing conditions, the concentration of ground-

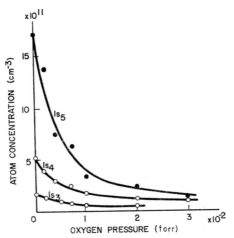

FIG. 4.39. Variation of the concentration of neon atoms in the $1s_{3,4,5}$ states with oxygen pressure at a fixed neon pressure ($p_{Ne} = 0.7$ torr, $D = 16$ mm, $I = 60$ mA). (After I. V. Kolpakova and T. P. Redko,[130] by courtesy of the Optical Society of America.)

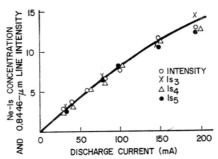

FIG. 4.40. Variation of the concentration of neon atoms in the $1s_{3,4,5}$ states and the intensity of the 0.8446-μm O-I line with discharge current ($P_{O_2} = 2.2 \times 10^{-2}$ torr, $D = 16$ mm). (After I. V. Kolpakova and T. P. Redko,[130] by courtesy of the Optical Society of America.)

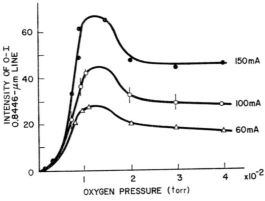

FIG. 4.41. Variation of intensity of the 0.8446-μm line with oxygen pressure at various discharge currents, $p_{Ne} = 0.7$ torr, $D = 16$ mm. (After I. V. Kolpakova and T. P. Redko,[130] by courtesy of the Optical Society of America.)

155

state atomic oxygen atoms is high enough so that radiation trapping on the O-I $3s\ ^3S_1^\circ-2p^4$ $^3P_2$ resonance transition can prevent population inversion occurring on the oxygen triplet transition at 0.8446 μm at line center.

As a result of the difference in potential energy of the excited Ne' atom and the sum of the potential energy of the excited O' oxygen atom and the dissociation energy of an oxygen molecule in reaction (4.31), the excess potential energy appears as kinetic energy of the dissociated oxygen atoms. An anomalously wide velocity distribution (as compared with that of atoms in the O-I ground state) is thereby given to atoms in the $3p\ ^3P_{0,1,2}$ upper laser states of the 0.8446-μm line.[131] Since atoms in the lower laser state are not similarly affected, population inversion can occur away from line center in the wings of the Doppler-broadened 0.8446-μm laser line. Subsequently overlapping of the components of the broadened O-I $3p\ ^3P_{0,1,2}-3s\ ^3S_1^\circ$ triplet transition causes oscillation to be possible only at one frequency. This frequency is shifted towards the violet from the fluorescent peak of the $3p\ ^3P_2-3s\ ^3S_1^\circ$ transition.[90, 129]

4.2.2. ARGON–OXYGEN LASER

Laser oscillation in the argon–oxygen laser occurs in atomic oxygen on four lines near 0.8446 μm. It occurs in the wings of the two ($3p\ ^3P_2-3s\ ^3S_1^\circ$ and $3p\ ^3P_1-3s\ ^3S_1^\circ$) transitions of the three $3p\ ^3P_{0,1,2}-3s\ ^3S_1^\circ$ O-I triplet transitions near 0.8446 μm.

Time-dependent studies of the 0.8446-μm emission line in argon–oxygen and krypton–oxygen mixtures by Bennett et al.[90] and Tunitskii and Cherkasov[132] have established that excitation of the $3p\ ^3P_{0,1,2}$ upper laser state occurs in the two-step processes:

$$\left.\begin{matrix} \mathrm{Ar}^* \\ \mathrm{Kr}^* \end{matrix}\right\} + \mathrm{O}_2 \rightarrow \left.\begin{matrix} \mathrm{Ar} \\ \mathrm{Kr} \end{matrix}\right\} + \mathrm{O}^*\ 2p^4\ ^1S_0,\ ^1D_2 + \mathrm{O} + \text{kinetic energy}, \qquad (4.32)$$

followed by[90]

$$\mathrm{O}^*\ 2p^4\ ^1S_0,\ ^1D_2 + e^- \rightarrow \mathrm{O}'\ 3p\ ^3P_{0,1,2} + e^-, \qquad (4.33)$$

where the O^* indicates an oxygen atom in a metastable $2p^4\ ^1S_0$ or $2p^4\ ^1D_2$ state, and O indicates an atom in the $2p^4\ ^3P_2$ O-I ground state, or[132]

$$\mathrm{Ar}^* + \mathrm{O}_2 \rightarrow \mathrm{Ar} + \mathrm{O}' + \mathrm{O} + \text{kinetic energy} \qquad (4.34)$$

followed by (4.33) together with

$$\mathrm{Ar}^* + \mathrm{O} \rightarrow \mathrm{Ar} + \mathrm{O}'\ 3p\ ^3P_{0,1,2} + \text{kinetic energy}. \qquad (4.35)$$

The excitation paths for these reactions are shown in Fig. 4.38.

Absence of laser oscillation in krypton–oxygen mixtures is due to the Kr^* metastable atoms having insufficient potential energy to excite the $3p\ ^3P$ upper laser state directly by excitation transfer from the oxygen-atomic ground state and to insufficient, indirect excitation by electron impact in reactions similar to (4.33).

Just as in the neon–oxygen laser, the spontaneous emission lines of the $3p\ ^3P_{0,1,2}-3s\ ^3S_1^\circ$ triplet at 0.8446 μm in an argon–oxygen laser are anomalously broadened. Oscillation though occurs at *four* frequencies displaced with respect to the maxima of the spontaneous emission of the $3p\ ^3P_1-3s\ ^3S_1^\circ$ and $3p\ ^3P_2-3s\ ^3S_1^\circ$ transitions. This displacement of the laser lines with respect to the maxima of the spontaneous emission lines is associated with the considerably different velocity distributions of atoms in the upper $3p\ ^3P$ and lower $3s\ ^3S_1^\circ$ states produced in the dissociative excitation-transfer reaction (4.32); the atom–atom excitation-transfer reaction (4.35); possibly electron-impact excitation from the O-I ground-state; and radiation trapping at line center on the strong resonance line at 0.1300 μm connecting the lower laser level with the O–I ground state.[134-5] The positions of the four laser lines

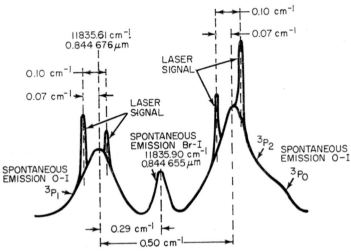

FIG. 4.42. Distribution of laser lines on the oxygen triplet at 0.8446 μm in argon–oxygen and argon–bromine lasers shown relative to spontaneous emission lines of O-I and Br-I. (After L. N. Tunitskii and E. M. Cherkasov,[133] by courtesy of the Optical Society of America.)

with respect to the maxima of the spontaneous-emission lines in the argon–oxygen laser, and the argon–bromine (oxygen) laser are illustrated in Fig. 4.42.[133] Oscillation at 0.8446 μm that has been reported in an argon–bromine mixture[49, 122] is due to the presence of oxygen as an impurity and oscillation does not occur on the Br-I line as once believed. This is also illustrated in Fig. 4.42.

4.2.2.1. *Operating characteristics of* Ne–O₂ *and* Ar–O₂ *lasers*

Provision must be made with neon–oxygen and argon–oxygen lasers to replenish oxygen that is rapidly lost to the walls of the discharge tube and to gas clean-up if internal electrodes are used. The strong cataphoretic effect that occurs in noble gas and oxygen mixtures in dc-discharges has meant that most studies of the 0.8446-μm oxygen laser have been carried out using rf-excited discharges. As mentioned in Chapter 3, this leads to uncertainties as to how discharge conditions such as current, or input-power actually coupled into the discharge can be specified or related between one laser system and another.

Optimum pressure. The argon–oxygen laser will operate over a wide range of pressures for widely differing partial pressures of argon and oxygen. Using rf-excitation, Tunitskii and

Cherkasov[129] operated an argon–oxygen laser over a total pressure range of 0.06 to 6 torr, with partial-pressure ratios of argon to oxygen of 10 : 1 to 200 : 1 in a 7-mm-bore tube. Their optimal conditions for oscillation were: total pressure, 0.44 torr; partial pressure ratio, 36 : 1; and optimum current, in excess of 550 mA; and $D = 7$ mm. Optimal conditions reported by Bennett *et al.*[90] for rf-excitation also using a 7-mm-bore tube were $p_{Ar} = 1.3$ torr and $p_{O_2} = 0.036$ torr. Though the partial pressure ratio is approximately the same at 36.1 : 1, the total pressure is higher than that used by Tunitskii and Cherkasov.[129]

Optimum discharge conditions for the rf-excited neon–oxygen laser have been given by Bennett *et al.*[90] as: $p_{Ne} = 0.35$ torr, $p_{O_2} = 0.014$ torr in a 7-mm-bore discharge tube, whereas in ref. 129 it is reported that the addition of argon is necessary to achieve sustained oscillation in a neon–oxygen laser.

It would appear from the results given by Kolpakova and Redko[130] for dc-excited, laser-like neon–oxygen discharges (Figs. 4.39 to 4.41), that the optimum conditions likely in a 16-mm-bore discharge tube are the following: p_{O_2}, 0.013 torr; p_{Ne}, less than 0.7 torr and discharge current in excess of 200 mA. The optimum pressure of neon is probably considerably less than 0.7 torr, and possibly as small as 0.15 torr if the results of Bennett *et al.*[90] for an rf-excited discharge can be related to the larger-bore discharge tube of 16 mm that was used by Kolpakova and Redko. The optimum current is suggested here as being in excess of 200 mA in a 16-mm-bore discharge tube, since Fig. 4.40 shows that the concentration of Ne $1s$ atoms, from which energy is transferred in dissociative excitation transfer to the O-I upper laser level, has not yet saturated at a discharge current of 200 mA.

The $1/D$ gain relationship. Measurements on rf-excited neon–oxygen and argon–oxygen lasers by Bennett *et al.*[90] have shown the gain on the 0.8446-μm transition under optimum discharge conditions varies roughly as the reciprocal of the tube diameter. Assuming that optimum conditions corresponded approximately to a constant total pressure and tube diameter product, it is conjectured that this relationship follows from another case in which the gain is determined primarily by the dominance of upper-laser-state selective excitation. Here it would be determined by the way the concentration of excited (quasi-metastable) Ne $1s$ and Ar $1s$ atoms varies with discharge-tube diameter under optimum laser conditions.

4.2.3. HELIUM–FLUORINE LASER

Pulsed laser oscillation in atomic fluorine in the visible at 0.7037, 0.7128, and 0.7204 μm, and, in addition, at 0.7800 μm and on twenty unidentified lines between 1.5900 to 9.34462 μm was reported in 1970 by Kovacs and Ultee,[128] and Jeffers and Wiswall[127] within a comparatively short period of time of each other. In the first reported work,[128] laser oscillation occurred in pulsed discharges in mixtures of CF_4, C_2F_6, or SF_6 (at a pressure of 0.03 to 0.1 torr) with helium at pressures in the range of 2 to 10 torr, and it was not clear what selective excitation processes were involved. Laser action, however, appeared at the onset of the current pulse and the behavior of the laser pulse with respect to changes in the length of the current pulse indicated that the selective excitation could be by direct electron-impact excitation of dissociated ground-state fluorine atoms.

158

Following this first report of oscillation in neutral atomic fluorine, Jeffers and Wiswall[127] obtained pulsed oscillation in flowing He–HF mixtures at partial pressures of 0.3 and 0.05 torr of He and HF, respectively, in a 10-cm-bore discharge tube. They determined that the presence of helium was required for oscillation to be obtained in such a mixture (and was also required with the SF_6(–He) mixture used by Kovacs and Ultee). Oscillation could not be obtained when helium was replaced by argon, nitrogen, oxygen, hydrogen, or carbon dioxide. The observation that helium was essential for obtaining oscillation tends to eliminate direct electron-impact excitation from the atomic fluorine ground state as the operative selective-excitation mechanism of the upper laser levels in both the SF_6–He and HF–He laser mixtures. This follows because at comparable electronic-excitation conditions of electron temperature and concentration, oscillation would be expected at least with argon as a carrier gas at some (lower) gas pressure than that of the helium pressure used.

4.2.3.1. Upper level excitation and lower level de-excitation

Table 4.11 lists the laser lines, their identification, and upper-level potential energy above the HF molecular ground state where D_0, the dissociation energy of HF, has been taken as 47,200 cm^{-1} (or 4.85 eV).

TABLE 4.11. *Laser transitions observed in a He–HF discharge and upper laser level potential energies*

Wavelenth in air (μm)	Identification in F-I	Upper state potential energy[a] above the HF ground state
0.703745	$3p\ ^2P^\circ_{3/2}$–$3s\ ^2P_{3/2}$	166,138
0.712788	$3p\ ^2P^\circ_{1/2}$–$3s\ ^2P_{1/2}$	166,283
0.720237	$3p\ ^2P^\circ_{3/2}$–$3s\ ^2P_{1/2}$	166,138
0.780022	$3p\ ^2D^\circ_{3/2}$–$3s\ ^2P_{1/2}$	165,074
1.5900	—	—
.	.	.
.	.	.
.	Twenty lines	.
.	unidentified	.
.		.
.		.
9.3462		—

[a] From C. E. Moore, *Atomic Energy Levels*, vol. 1 (ref. 46).

The selective excitation of the upper states of the identified F-I laser lines is believed to be due to dissociative excitation transfer between $He^*\ 2^1S_0$ metastables and HF molecules[127] in the reaction

$$He^*\ 2^1S_0 + HF\ (^1\Sigma^+,\ \nu = 0) \rightarrow He\ ^1S_0 + H + F'\ 3p\ ^2P^\circ_{3/2,\ 1/2}, {}^2D^\circ_{3/2} \pm \Delta E_\infty \quad (4.36)$$

where the H atom is in the ground state, and the ΔE_∞ is available as kinetic energy of either of the dissociated H or F' atoms when the reaction is exothermic (ΔE_∞ positive). The excitation energy of the $He^*\ 2^1S_0$ metastable state is 166,272 cm^{-1}, so that ΔE_∞ for the selective

159

excitation of the upper laser states is:

$$2p\ ^2P^\circ_{1/2},\ -11\ \mathrm{cm}^{-1};$$
$$3p\ ^2P^\circ_{3/2},\ +134\ \mathrm{cm}^{-1}_{\_};$$
$$3p\ ^2D^\circ_{3/2},\ +1198\ \mathrm{cm}^{-1}.$$

These energy discrepancies are much smaller than those found in other known dissociative collisions, in which ΔE_∞, for instance, is approximately 8000 cm^{-1} (or 1 eV) in the Ne–O$_2$, and about 16,000 cm^{-1} (2 eV) in the Ar–O$_2$ lasers operating at 0.8446 μm in atomic oxygen. With such small energy discrepancies, the collision is essentially resonant for the two upper states of the three strongest lines, and near-resonant for the $3p\ ^2D^\circ_{3/2}$ upper state of the weaker 0.7800-μm laser line.

The energy of approximately 166,000 cm^{-1} needed to provide simultaneously dissociation energy of HF and the excitation energy of the upper states of the identified F-I laser lines, suggests the reason why oscillation has been observed in He–HF mixtures only. Neon metastable/quasi-metastable 1s atoms with approximately 135,000 cm^{-1} of energy have insufficient energy to supply the necessary dissociation and excitation energy. Likewise, Ar 1s metastables with only approximately 94,000 cm^{-1} of energy, have insufficient energy to dissociate and excite the upper laser levels.

No selective excitation processes other than (4.36) appear to be able to account for the excitation in the fluorine laser. General excitation-transfer reactions such as

$$X^* + F \rightarrow X + F' \pm \Delta E_\infty \tag{4.37}$$

where the F atom is in the ground state and is assumed to be produced initially by other dissociative processes in the discharge, and X^* is a metastable atom of He, Ne, Ar, or Kr, are inappropriate as in none of these cases are the metastable energies such as to make the reactions resonant or mildly resonant as required.

Figure 4.43 shows the energy levels of atomic fluorine and helium (and H levels) relevant to the He–HF laser. The lower $3s\ ^2P$ laser levels are depopulated by the allowed transitions $3s\ ^2P$–$3p\ ^2P^\circ$ near 0.0950 μm. For partial pressures of a few torr and above, radiation trapping will limit the depopulation of the lower laser levels shortly after the F-I atomic ground state has become populated through dissociation processes that occur in the discharge. It could be that the time-delay, during which the F-I atomic-ground-state population is established and radiation trapping becomes complete, accounts for the limited duration (1 to 2 μsec) of the laser pulses in the CF$_4$–, CF$_6$–, or C$_2$F$_6$–He lasers.[128]

Fortunately, as noted by Jeffers and Wiswall,[127] there appears to be a possible second deactivation process in the He–HF laser for atoms in the lower laser states. The deactivation process involves ground-state hydrogen atoms which are produced, in addition to excited F' atoms, in the dissociative excitation-transfer reaction (4.36). The deactivation process is the atom–atom excitation-transfer reaction

$$F'\ 3s\ ^2P_{3/2,\ 1/2} + H\ 1s\ ^2S_{1/2}\ (n=1)$$
$$\rightarrow F\ 2p^5\ ^2P_0 + H\ (n=5) - \Delta E_\infty\ (560\ \text{or}\ 235\ \mathrm{cm}^{-1}), \tag{4.38}$$

where the principal quantum number $n = 5$ is used to designate the 5p, 5s, 5d, 5f and 5g levels of atomic hydrogen at 105,292 cm^{-1} above the ground state,[46] and the energy discre-

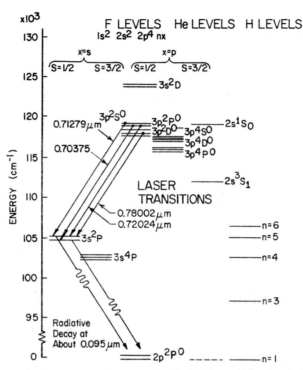

FIG. 4.43. Energy levels of F, He, and H relevant to the He–HF laser. (After W. Q. Jeffers and C. E. Wiswall,[127] by courtesy of the American Institute of Physics.)

pancies (ΔE_{∞}) have been calculated assuming that F atoms in the $3s\ ^2P_{3/2,\ 1/2}$ states have zero kinetic energy.

Whereas the energy discrepancy of -560 cm^{-1} (reaction (4.38) is endothermic) appears too large for efficient atom–atom excitation transfer, that of -235 cm^{-1} is less than the -386 cm^{-1} involved in excitation transfer from He$^*\ 2^1S_0$ metastables to the Ne $3s_2$ state in the 632.8-nm He–Ne laser, and (4.38) could have a cross-section that is approximately gas kinetic. If, however, the $3s\ ^2P_{2/3,\ 1/2}$ lower laser levels are also populated by dissociative excitation transfer as well as radiatively through the observed laser transitions, the available kinetic energy (~ 1.7 eV) will be more than sufficient to supply the required energy discrepancies or activation energy required for the reaction to proceed. It will be, of course, the shape of the interaction potential of the colliding F′ and H atoms (and any ensuing activation energy) that determines the effective cross-section of such a reaction at a particular gas temperature.

Operating characteristics. Unless the $3s\ ^2P_{3/2,\ 1/2}$ lower laser levels of F-I can be depopulated rapidly by collisions such as in (4.38) in a He–HF discharge by the presence of H$_2$ in another fluorine compound mixture or by cooling the discharge tube,[136] the resonant nature of the He*–HF upper-level selective-excitation reaction that must result in a narrow distribution of atom velocities in the upper laser levels, together with the radiative-decay processes of the lower laser levels, are such that the laser transitions discussed in the preceding section will continue to remain pulsed transitions incapable of CW operation.

Figure 4.44 shows a typical oscilloscope trace of the laser output of one of the visible F-I laser lines and its relationship to the discharge-current pulse in a He–HF mixture. Laser oscillation is seen to occur in the afterglow of the current pulse, and lasts for about 20 μsec. The 20-μsec duration of the laser pulse would appear to indicate that the lower laser levels *are* being depopulated more effectively in a pulsed Ne–HF mixture than in a CF_4^-, CF_6^-, or C_2F_2–He mixture in which the laser pulse-duration appears to be limited by lower-laser-level processes to 2 μsec.[128] Under the discharge conditions used ($p_{He} = 0.3$ torr, $p_{HF} = 0.05$ torr; discharge-current pulse less than 10-μsec duration at 600-A peak),[127] the 20-μsec duration of the laser pulse could be associated more with the relatively short lifetime expected

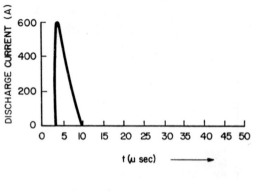

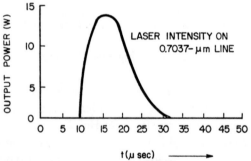

FIG. 4.44. Relationship between laser output of the 0.7037-μm F-I line and the discharge-current pulse in a He–HF mixture. Discharge conditions: capacitor discharge, 0.2 μF charged to 15 kV, He = 0.3 torr, HF = 0.05 torr, $D = 10$ cm. (After W. Q. Jeffers and C. E. Wiswall,[127] by courtesy of the American Institute of Physics.)

of He^* 2^1S_0 metastables in the afterglow of a pulsed discharge in the presence of a strong three-body de-excitation process such as (4.36) than with the build-up time of the concentration of atoms in the lower laser level or the F-I atomic ground state. If this is the case, it is conceivable that CW operation at low current densities in small-bore discharge tubes, similar to those used for 632.8-nm He–Ne lasers, is possible.

4.3. Electron-impact-excited Lasers

It was the gain results of Bennett et al.[8] on the $2s_2$–$2p_4$ transition in neon at 1.1523 μm in a discharge in pure neon that showed for the first time that a practical amount of gain was obtainable in a single-component gas discharge by the sole use of electron-impact excitation for selective excitation (Fig. 4.4). Subsequently, Patel et al.[39] produced oscillation on fifteen lines near 2 μm in helium, neon, argon, and krypton using electron-impact excitation in pure-gas discharges, and later produced oscillation on over 200 transitions in the noble gases and extended CW laser operation out to 133 μm. Most of these transitions are in neon (Appendix Table 2a).

Electron impact has now been used to give selective excitation in a number of neutral laser systems, which extend from low-pressure gases to liquids (in the noble gases). These neutral laser systems have distinct differences in their mode of operation. These differences lend themselves to a convenient classification into three broad groups:

(1) CW, noble-gas lasers;
(2) noble-gas and metal vapor lasers that are self-terminating or transient in operation; and
(3) dissociative-type, molecular metal–vapor lasers, in which population inversion occurs between neutral atomic states.

4.3.1. CW, NOBLE-GAS LASERS

If a strongly allowed optical transition connects the ground state of an atom or molecule with an excited state, it follows from the Born approximation[137] that where electron exchange can be neglected, selective excitation of that excited state can occur in inelastic electron-impact collisions.

The ground states in the noble gases neon, argon, krypton, and xenon are similar, and consist of closed-shell $(np)^6$ electron configurations in which n is respectively 2, 3, 4, and 5. In each there are strongly allowed optical transitions that connect the atom ground state with levels in the higher-lying $(np)^5$ ms and $(np)^5$ md electronic configurations where $m = n+1, n+2$, etc. As a result of these strong optical transitions, selective excitation in neon, argon, krypton, and xenon occurs through the electron-impact reactions

$$(np)^6 + e^- + \text{kinetic energy} \begin{cases} \longrightarrow (np)^5\,(n+1)s + e^-, \\ \\ \longrightarrow (np)^5\,(n+1)d + e^-, \end{cases} \qquad (4.39)$$

as illustrated in Fig. 4.45.[8] For pressures of a fraction of a torr (in discharge tubes of a few mm diameter) sufficient to produce radiation trapping of the strong optically allowed vacuum ultraviolet transitions between the $(np)^5\,(n+1)s$ and $(np)^5\,(n+1)d$ levels and the $(np)^6$ ground state, atoms in the excited ms and md states can only decay radiatively through transitions to the lower-lying p states marked LASER in the figure.

The radiative lifetimes of levels with the $(np)^5$ mp configuration are about 10^{-8} sec due to transitions of the type $mp \rightarrow (n+1)s$. Thus, given selective excitation of the $(np)^5$ ms and

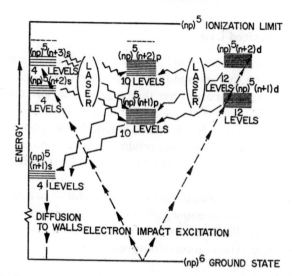

FIG. 4.45. Schematic indication of electron configurations of energy levels relevant to pure neon, argon, krypton, and xenon lasers. The dominant excitation and radiative-decay paths are shown by the arrows. (After W. R. Bennett, Jr.,[8] by courtesy of the Optical Society of America.)

$(np)^5\ mp$ levels and the occurrence of radiation trapping between them and the ground state, the $(np)^5\ mp$ levels are suitable lower laser levels for CW laser transitions of the type s–p and d–p. In practice, examination of the tables given in the Appendix will show that numerous CW laser transitions between high-lying energy levels in pure neon, argon, krypton, and xenon discharges are of the reverse type, i.e. p–s and p–d. These are understood to arise from excitation by electron–ion recombination and cascading and not from any preference for direct electron-impact excitation of the p states.

4.3.1.1. Effect of additives

The addition of large quantities of helium to certain low-pressure pure noble-gas lasers whose upper levels are believed to be selectively excited by direct electron impact from the atomic ground state (in accordance with the predictions in the preceding section) has been found to result in large increases in gain. The effect is particularly noticeable in the xenon laser operating at 2.026 and 3.508 μm, and in the argon laser operating at the four wavelengths 3.397, 2.208, 2.062, and 1.792 μm.

The high gain Xe and Xe–He lasers. A number of $5d$–$6p$ transitions in neutral xenon exhibit high gain. Three of them, the $5d[3/2]_1^\circ$–$6p[3/2]_1$ transition at 2.026 μm, the $5d[7/2]_3^\circ$–$6p[5/2]_2$ transition at 3.508 μm, and the $5d[7/2]_4^\circ$–$6p[5/2]_3$ at 5.575 μm exhibit extremely high gain. Optical gains of about 10^5 per meter are obtainable on the 3.508-μm line in a 7-mm-bore discharge tube in pure xenon at low pressure, or 10^6 per meter in a mixture of xenon and helium, with helium at pressures of tens of torr,[138-9] and in exceess of 10^8 per meter in isotopic xenon.[91] On the 2.026 μm line, optical gains in excess of 2×10^2 per meter in a He–Xe mixture in an rf-excited discharge in a 7-mm-bore tube have been realized.[88] An even higher optical gain of more than 10^{15} per meter has been measured on the 5.575-μm line.[91] With the 2.026-μm laser line, helium appears to be essential in producing high gain.

Population inversion occurs readily between the $5d$ and $6p$ levels because of favorable life-times of the $5d$ and $6p$ states. The lifetimes of the $5d$ states are approximately 30 times longer than those of the $6p$ states, so that even in the absence of selective excitation of the upper laser states for equal, or even unequal excitation of both upper and lower states, population inversion is easily achieved.[140-1, 149]

Strongly allowed vacuum-UV transitions connect the Xe-I $5p^6$ ground state with a number of levels in the $5p^5$ nd and $5p^5$ ns configurations for which the approximate selection rule $\Delta l = \pm 1$, and the rigorous selection rule for electric–dipole radiation $\Delta J = 0, \pm 1 (0 \rightarrow 0)$ are obeyed. As a consequence, these levels have the largest cross-sections for excitation by electron impact from the ground state. For the $5d[3/2]_1^\circ$ state, upper state of the 2.026-μm laser line, both selection rules are obeyed with $\Delta l = 1$ and $\Delta J = 1$. In the case, however, of the upper $5d[7/2]_3^\circ$ state of the 3.508-μm line, only the approximate selection rule is obeyed with the *rigorous* selection rule disobeyed (with $\Delta J = 2$).

In spite of the rigorous selection rule being disobeyed with excitation of the $5d[7/2]_3^\circ$ level, it is considered that excitation of both upper states of the 3.508-μm and 2.062-μm laser lines in pure xenon, or helium–xenon discharges, is by electron impact from the ground state. Studies of these two lines under pulsed and CW discharge conditions seem to confirm this.[142, 143, 149] Though the $5d$ levels would be expected to be excited in preference to the $6p$ levels, Moskalenko *et al.*[140] have shown that in fact the excitation rates of the $5d[7/2]_3^\circ$ and $5d[3/2]_1^\circ$ states (upper levels of the 3.508-μm and 2.026-μm lines respectively) are less than those of the appropriate $6p[5/2]_2$ and $6p[3/2]_1$ lower laser levels. The populations in the upper and lower level of each line, together with the excitation rate from the ground state (per unit volume of discharge) in the positive column of a helium–xenon discharge, are given below in Table 4.12.

TABLE 4.12. *Upper and lower state populations in the positive column of a He–Xe laser discharge*

Wavelength	Transition	Level population (cm^{-3})		Excitation rate $\times 10^{15}$ (cm^{-3} sec^{-1})	
		Upper	Lower	Upper	Lower
3.508	$5d[7/2]_3^\circ$–$6p[5/2]_2$	2×10^9	7×10^7	1	3
2.026	$5d[3/2]_1^\circ$–$6p[3/2]_1$	1.2×10^9	3×10^7	1	1

Discharge conditions: $p_{Xe} = 0.05$ torr, $p_{He} = 1.0$ torr, $D = 10$ mm; current $= 40$ mA. The electron temperature was 19,000 K, and electron concentration 5×10^{10} cm^{-3}.

Table 4.12 shows that the upper-level population in this representative laser discharge are about a factor of 30 to 40 times higher than the lower-state populations despite the fact that there are smaller (3.508-μm line) or equal (2.026-μm line) upper- and lower-state excitation rates.

12*

It is interesting to note that the measured electron temperature of 19,000 K given as a footnote to Table 4.12 is the same as the theoretical value given by the Appendix Fig. A.3 for the conditions under which the discharge was operated.

Figure 4.46, a partial energy-level diagram of neutral xenon and helium, shows mainly two things:

(1) energy-level coincidences do not exist between the lowest (He* 2^3S_1) helium metastable level and levels of neutral xenon so that atom–atom excitation transfer from helium to xenon cannot occur; and

(2) the He* 2^3S_1 metastable level is about 7.7 eV above the Xe$^+$ ground state.

Since atom–atom excitation transfer from helium to xenon cannot account for the significant increase in gain when helium is added to a pure xenon laser discharge, the increase

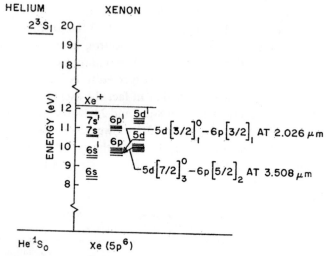

FIG. 4.46. Partial energy-level diagram of neutral xenon and helium, showing disposition of the two high-gain $5d$–$6p$ Xe-I laser transitions at 2.026 and 3.508 μm and the He* 2^3S_1 metastable level.

in gain in rf-excited discharges has been attributed to an increase in the average electron temperature and an increase in the electron density,[88] or an increase in electron density without an appreciable lowering of the average electron temperature.[8]

Bridges[138] has measured the gain on the 3.508-μm line in pure xenon and helium–xenon discharges using both rf and dc-excitation. He found that the addition of 1.9 torr of helium to 0.03 torr of xenon in a 7-mm-bore discharge tube with rf-excitation, increased the gain/meter by a factor of approximately 10 but at the expense of added exciting rf-power. The optimum rf-drive level was 6 W for the pure-xenon discharge, while the helium–xenon mixture optimized at 18 W. With the same rf-drive level of 6 W, the helium–xenon mixture gave only a small improvement in gain over the pure xenon discharge. This observation, as noted by Bridges,[138] is consistent with a factor of 3 increase in electron density reported by Aisenberg[144] for a helium–xenon mixture over pure xenon at constant rf input power, if electron-impact excitation is assumed to be the dominant excitation mechanism of the $6p$ and $5d$ levels. Aisenberg, using a multi-probe technique, observed an increase in electron density when helium was added to an rf-discharge in xenon, but did not notice any appreciable change in electron temperature.

It is hard to see how the addition of a gas to another could appreciably increase the average electron temperature postulated in ref. 88 to account for the increase in gain when helium is added to xenon. According to theory, under discharge conditions where a Maxwellian distribution of electron velocities might be expected, the addition of another gas, without a decrease in pressure of the first gas, can only decrease the electron temperature. The effect is particularly noticeable when the added gas has a much lower ionization potential than the first. At the typical pressures used in pure xenon lasers (of less than 0.03 torr in 7-mm-bore discharge tubes), the electron mean free path is much longer than the bore of the tube and it is questionable what the electron-energy distribution is like. It will be collisions of the electrons with the negatively charged sheath at the wall of the discharge tube, and not electron–atom or electron–electron collisions, that will determine the electron-energy distribution. While it is conceivable that the average electron temperature could remain the same (or even increase slightly) when *small* quantities of helium are added to a low-pressure xenon discharge by making the electron mean free path comparable to or just less than the bore of the discharge tube, the addition of *large* quantities of helium, making the pressure and diameter product as high as 30 torr-mm, *ensures* that the average electron temperature does not increase or remain the same, but must be lowered.

As the metastable levels of helium are higher in energy than the Xe^+ ground state, any observed increase in electron concentration when helium is added to a xenon discharge could possibly be accounted for by the (volume-ionization) Penning effect

$$He^* \ 2^3S_1 + Xe \rightarrow He + Xe^+ + (e^- + 7.7 \text{ eV kinetic energy}). \qquad (4.40)$$

The introduction of such a volume-ionization process that produces electrons in a well-defined energy range might then conceivably alter the electron-energy distribution to some optimum value for excitation of the $5d$ upper levels of the 3.507-μm and 2.067-μm Xe-I laser lines, and so increase the gain. If this process was important, one would expect that the amount of the helium addition would be such as to maximize the production rate of $He^* \ 2^3S_1$ metastables and so maximize the occurrence of the volume-ionization process. It follows from the optimum discharge conditions for 1.15-μm He–Ne, and 441.6-nm He–Cd lasers, that the $p_{He}D$ product in a helium–xenon laser should then be about 10 to 12 torr-mm. At this pD the $He^* \ 2^3S_1$ concentration apparently maximizes at helium pressures of a few torr, probably due to He_2 production at higher pressures via reaction (4.15) (W. T. Silfvast, private communication, 1972). As the $p_{He}D$ product in the 3.508-μm and 2.026-μm xenon lasers is considerably more than this, together with the fact that in the case of the 2.026-μm laser helium appears to be essential in producing high gain, the indication is that helium plays a more active role than just increasing the electron concentration and the extent of direct electron-impact excitation of the relevant laser levels. Recent unpublished work indicates that dissociation of (HeXe)' complexes into particular excited states of Xe-I could be responsible for the selective excitation in He–Xe laser discharges (G. J. Collins, private communication, 1972).

Operating characteristics and discharge conditions

Representative discharge conditions for the two high-gain 3.508-μm and 2.026-μm Xe-I laser lines are tabulated in Table 4.13.

The dependence of output power on the discharge current and pressure for a dc-excited, 3.508-μm laser operating in pure xenon is presented in Fig. 4.47. A curve of output power at 3.508 μm versus total (He–Xe) pressure using rf-excitation is given in Fig. 4.48.

Strong cataphoretic effects in dc-excited helium–xenon mixtures make it imperative to use rf-excitation. In addition, rapid clean-up of xenon in both pure xenon and helium–xenon mixtures makes it essential to use some form of pressure control.[145]

In the 3.508-μm pure xenon laser there is no optimum gas pressure in the range 10–150 mtorr. The optical gain increases monotonically as the pressure decreases. The observed

TABLE 4.13. *Neutral xenon CW laser discharge conditions*

Line (μm)	pD (torr-mm)	Mixture ratio He : Xe	Excitation	References
3.508	<0.93	pure Xe	dc	145
	0.21	pure Xe	dc	138
	<30	200 : 1	rf	146
	~13	60 : 1	rf	138
2.026	~15	100 : 1	rf	88
	~16	160 : 1	dc	147
	~20	500 : 1	rf	—[a]

[a] C. S. Willett, T. J. Gleason and J. S. Kruger (unpublished work, 1970).

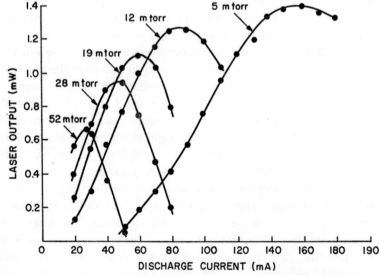

FIG. 4.47. 3.508-μm laser output power versus discharge current and gas pressure of a pure xenon laser (D = 6 mm). (After D. R. Armstrong,[145] by courtesy of the IEEE.)

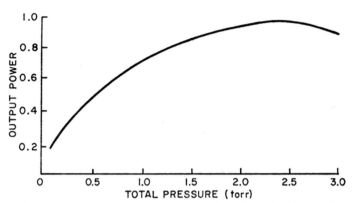

FIG. 4.48. 3.508-μm laser output power versus total pressure of a 200 : 1 He : Xe gas mixture. The output is relative to the maximum power at 2.5 torr. Discharge conditions: 30 MHz rf-excitation, input power approximately 100 W, $D = 12$ mm. (From A. A. Kuznetsov and D. I. Mash,[146] by courtesy of the Scripta Press.)

dependence between xenon pressure p and tube-diameter D is of the form

$$pD^n = C \qquad\qquad (4.41)$$

where $n \approx 3.2 \pm 0.2$, and C, a constant, depends on the gain.[148]

For large diameter tubes (i.d. $\geqslant 7$ mm) the gain is proportional to $1/D$.[148] For tube diameters of 5 to 26 mm, Aleksandrov et al.[91] find that the maximum gain per cm is equal to $0.08/D$, where D is in cm, for a Xe^{136} gas fill. For i.d. $\leqslant 3$ mm, the gain varies as D^{-n}, where $n > 1$ for xenon pressures > 50 mtorr.[148]

In the helium–xenon 2.026-μm laser, the gain is approximately proportional to $1/D$.[88] The gain on the 5.575-μm line is also inversely proportional to D, and is given by $0.15/D$ per cm.[91]

Although the 3.508-μm and 2.026-μm Xe-I lines exhibit very high gain, operating as CW laser lines, they have low-power capabilities of less than a mW.[148] However, operated under pulsed conditions using high-pressure helium–xenon mixtures with xenon partial pressures in the range 0.1 to 1.0 torr and transverse discharge excitation, pulses of 10^4 W peak power are obtainable.[150]

Neutral argon CW laser. The neutral argon CW infrared laser operates in the spectral region from 1.8 to 2.8 μm.[121] It has four principal laser lines in this wavelength region, at:

$$2.397 \text{ μm}, \quad (3d[\tfrac{1}{2}]_0^\circ - 4p'[\tfrac{1}{2}]_1 \text{ transition});$$
$$2.208 \text{ μm}, \quad (3d[\tfrac{1}{2}]_1^\circ - 4p'[\tfrac{3}{2}]_2 \text{ transition});$$
$$2.062 \text{ μm}, \quad (3d[\tfrac{3}{2}]_2^\circ - 4p'[\tfrac{3}{2}]_2 \text{ transition});$$
and
$$1.792 \text{ μm}, \quad (3d[\tfrac{1}{2}]_1^\circ - 4p[\tfrac{1}{2}]_1 \text{ transition}).$$

All four of the lines involve transitions from the $3d$ band to the $4p$ and $4p'$ band of levels (or $3d$ to $2p$ levels in Paschen notation). One notices that the levels involved in the transitions, though they do not include the directly comparable transitions that give the high-gain 5.575, 3.508 and 2.026-μm Xe-I laser lines, have the same configuration as those of the xenon lines. That is, they are $nd \rightarrow (n+1)p, (n+1)p'$ transitions, where $n = 5$ for xenon and 3 for argon. Having the same electronic configurations, the processes involved in their excitation are possibly similar in the two systems. In addition the ratio of the lifetimes of the $3d$ to $4p$ levels is probably abnormally high, as it is for $5d$–$6p$ levels in Xe-I, and so could account for easy population inversion in the presence of even equal excitation of the upper and lower laser levels.

Optimum operating conditions for the neutral argon CW laser are achieved by mixing helium with argon in a 10 : 1 ratio. In addition, when small amounts of chlorine gas are added to the helium–argon mixture, the output is considerably enhanced. The optimum discharge conditions for it are: 1.7 torr Ar, 19 torr He, and 0.2 torr Cl_2, under flowing conditions in a discharge tube with an optimum diameter of 6 mm.[121] Although oscillation on these argon lines has been observed under pulsed conditions at low pressures of argon,[39,61,125] at the higher pressures of argon used by Dauger and Stafsudd[121] helium was essential in obtaining positive gain in a pulsed discharge. Under pulsed conditions, helium reduces the effective lifetime of the lower laser levels by a factor of 4, compared to that with no helium additive.

The effect that the addition of helium has on the laser output of the CW argon laser has been attributed[121] to an *increase* in the electron temperature and electron concentration that increases the excitation rate of the upper laser levels. The proposed effect would thus be similar to that proposed to account for the increased gain when helium is added to a xenon laser operating at 3.508 and 2.062 μm.[88] According to theory, as in the He–Xe laser, the addition of helium does not lead to an increase but can result only in a *decrease* in electron temperature, unless the pressure of the gas to which helium is added is decreased on the addition of helium or if the helium addition is very small. The addition of helium to a pure-argon laser discharge under conditions used for the neutral argon CW laser reduces the theoretical value of the average electron temperature by as much as 36 percent[118] (Fig. A.1). In view of this, it appears that processes other than direct electron-impact excitation of the argon $3d$ levels from the ground state are involved when large quantities of helium are added to an argon discharge.

Shtyrkov and Subbes[22] have attributed selective excitation in a high-pressure pulsed helium–argon laser, in which oscillation is observed from the $3d$ states also, to the dissociative-recombination reaction

$$(HeAr)^+ + e^- \rightarrow He + Ar'\ 3d \tag{4.42}$$

in which the formation of the $(HeAr)^+$ composite ion required the presence of $He^*\ 2^3S_1$ metastables. Their studies showed that the addition of argon to a helium discharge quenched the $He^*\ 2^3S_1$ metastables, and then oscillation from the Ar-I $3d$ levels to the $4p$ levels at 1.21 and 1.28 μm was observed.

In addition to this process, ion–electron recombination

$$Ar^+ + e^- \rightarrow Ar' \tag{4.43}$$

following the Penning reaction

$$He^* + Ar \rightarrow He + Ar^+ + e^- + \text{kinetic energy} \tag{4.44}$$

could be responsible for increased excitation in a helium–argon discharge over that in pure argon. It also appears likely that the dissociative-recombination reaction (4.42) or the ion–electron recombination reactions (4.43) are responsible for selective excitation that is responsible for population inversion in helium–argon laser mixtures at atmospheric pressure using transverse (TEA-laser) excitation.[151] Oscillation under these excitation conditions occurred on three of the $3d$–$4p$ lines of Ar-I (at 2.397, 2.208, and 1.792 μm) that Dauger and Stafsudd[121] observed were enhanced by the addition of helium to a discharge in pure argon.

It has also been suggested that the addition of helium might alter the shape of the electron-energy distribution in a pure argon discharge to one more favorable to the electron-impact excitation process, and so increase the excitation rate of the $3d$ and $4p$ states[118] if (4.42) or (4.43) are not responsible for the selective excitation of Ar-I states in helium–argon discharge mixtures. This could also apply to the 3.508 and 2.062-μm helium–xenon lasers where, in addition, dissociative-recombination and ion–electron-recombination reactions similar to (4.42) and (4.43) might account for the increased excitation in the presence of helium.

There is little doubt that the addition of chlorine to an argon–helium CW laser discharge, that can increase the gain by a factor of almost 2 and the output power by a factor of 5, causes depopulation of the Ar $4s$ levels by at least one of the two reactions:

$$\text{Ar*}\ 4s + \text{Cl} \rightarrow \text{Ar} + \text{Cl}'\ 5p + \Delta E_\infty \tag{4.45}$$

and

$$\text{Ar*}\ 4s + \text{Cl}_2 \rightarrow \text{Ar} + \text{Cl} + \text{Cl}'\ 4s + \text{kinetic energy.} \tag{4.46}$$

Depopulation of the Ar $4s$ levels by either of these reactions causes a reduction in the rate of excitation of the $4p$ lower laser levels through the reaction

$$\text{Ar*}\ 4s + e^- \rightarrow \text{Ar}'\ 4p + e^- \tag{4.47}$$

and so increases population inversion between the $3d$ and $4p$ levels. It is a result of the resonant, atom–atom excitation-transfer reaction (4.45) in an argon–chlorine discharge that selective excitation occurs of the $5p\ ^2D^\circ_{5/2}$ upper laser level in the 3.067-μm chlorine laser (Section 4.1.2).

4.3.2. NOBLE-GAS AND METAL-VAPOR TRANSIENT LASERS

A number of gas-laser systems whose upper laser levels are selectively excited by electron impact operate by cyclic excitation and relaxation.[152-64] They are self-terminating or transient in operation. Systems that exhibit very high gain and are transient occur in neutral species in Pb, Cu, Au, Ca, Sr, and Mn, as well as in the noble gases He, Ar, Kr, and Xe.

Four criteria that define this class of pulsed lasers are given by Walter et al.[152]

(1) The upper laser level should be a resonance level connected to the ground state by a strong radiative transition.

(2) The lower laser level should not be optically connected to the ground state. This level will thus be metastable and the population inversion will be inherently of a transient nature.

These two conditions are fulfilled by the following atomic-level sequence: the ground level; a lower laser level with the same parity as the ground state; and an upper laser level having an opposite parity. Under discharge conditions where the electron energies are well above the excitation-energy threshold and the Born approximation applies (Section 2.6), the upper laser level will be preferentially excited.

(3) Radiation trapping of the resonance radiation should make the branching ratio into the laser transition approximately equal to one. Electric-dipole matrix elements for any other transitions out of the upper laser level should be much weaker than those of the laser transition and the resonance transition.

(4) The Einstein A-coefficient of the laser transition should be smaller than that of the excitation transition ($\sim 10^8$ sec^{-1}), but larger than that of the lower-laser-level relaxation transition (~ 1 sec^{-1}). A practical range could be

$$10^8 > A \text{ (laser transition)} > 10^4 \text{ sec}^{-1}.$$

If the radiative lifetime of the laser transition is shorter than the rise time of the excitation current-pulse, the spontaneous emission will depopulate the upper laser level before a sufficient population inversion can be achieved for oscillation to occur.

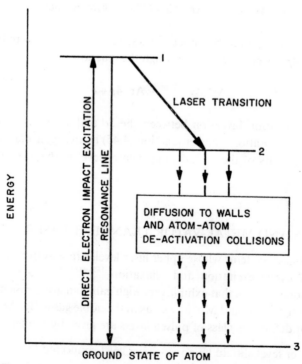

FIG. 4.49. Generalized energy-level diagram for a transient laser.

A generalized energy-level diagram for a typical transient laser is illustrated in Fig. 4.49. In such a three-level system, given rapid excitation of level 1, population inversion between levels 1 and 2 is only possible for a time T such that[82]

$$T \lesssim \tfrac{1}{2} A_{12} \tag{4.48}$$

where A_{12} is the Einstein A-coefficient for the laser transition that is typically 10^7 to 10^8 sec^{-1}. Typical transient laser pulse durations are thus of the order of 5 to 50 nsec.

4.3.2.1. Transient metal-vapor lasers

The copper-vapor laser that operates at two wavelengths in the visible at 0.51055 and 0.57821 μm on the $4p\ ^2P^{\circ}_{3/2}$–$4s\ ^2D_{5/2}$ and $4p\ ^2P^{\circ}_{1/2}$–$4s\ ^2D_{3/2}$ transitions respectively in neutral copper, and the manganese-vapor laser operating on six visible wavelengths between 0.5341 and 0.5537 μm, and on six wavelengths in the near infrared between 1.29 and 1.4 μm, satisfy the four criteria for transient lasers stated in Section 4.3.2. Figure 4.50 illustrates the excitation paths and laser transitions in the copper-vapor laser.

Pulsed gain coefficients of 58 dB/m on the green, 0.51055-μm line, and 42 dB/m on the yellow, 0.5782-μm line have been reported in the copper-vapor laser.[152] It is a laser that is capable of operation at high powers. For instance, operation at megawatt power levels has been achieved, with average power outputs approaching 1 W cm^{-3} of discharge.[154]

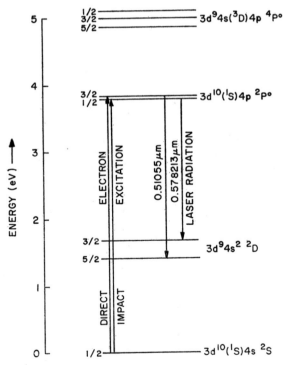

FIG. 4.50. Transient laser transitions and relevant excitation paths in the copper-vapor laser. (After W. T. Walter et al.,[152] by courtesy of the IEEE.)

The gold (0.7278-μm), calcium (5.546-μm), and strontium (6.456-μm) lasers also satisfy the four transient-laser criteria of Section 4.3.2. Their upper laser levels have large cross-sections for excitation by electron impact from the ground state, and their lower laser levels are metastable. Single-pass gains of 300 dB/m have been measured on the Ca-I and Sr-I laser transitions.[155]

Transient oscillation in lead vapor has been reported on at least four Pb-I transitions:[156-7]

$$6p7s\ ^3P_1^\circ-6p^2\ ^3P_1 \quad \text{at} \quad 0.36395\ \mu\text{m};$$

$$6p7s\ ^3P_1^\circ-6p^2\ ^3P_2 \quad \text{at} \quad 0.40578\ \mu\text{m};$$

$$6p6d\ ^3D_1^\circ-6p^2\ ^1D_2 \quad \text{at} \quad 0.40621\ \mu\text{m};$$

and $\quad 6p7s\ ^3P_1^\circ-6p^2\ ^1D_2 \quad$ at $\quad 0.7229\ \mu\text{m}.$

The 0.4062-μm and 0.7229-μm transitions are forbidden transitions for electric-dipole radiation in LS coupling and do not satisfy criterion (3). In spite of this, single-pass gains of 63 dB in a 10-cm effective length of plasma have been measured on the 0.7229-μm $6p7s\ ^3P_1^\circ-6p^2\ ^1D_2$ transition.[158] The $7s\ ^3P_1^\circ-6p^2\ ^3P_2$ transition at 0.40578 μm is said to exhibit even higher gain than the 0.7229-μm line.[157]

Operating conditions. High temperatures are necessary to obtain sufficient vapor pressure for operation of metal-vapor lasers. Figure 4.51 shows, schematically, apparatus used to provide vacuum-tight confinement of metal vapors up to temperatures of approximately 1500°C and yet have transparent, optical-quality windows. Deposition of metal vapor on the

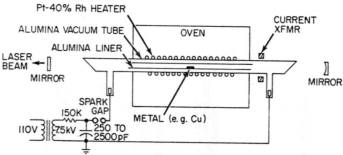

FIG. 4.51. Apparatus for pulsed metal-vapor lasers. (After W.T. Walter *et al.*,[152] by courtesy of the IEEE.)

Brewster's-angled windows is prevented by filling the tube with a few torr of helium. Helium also carries the discharge from electrodes outside the high-temperature region to the metal-vapor active region.

The output power can be increased considerably by utilizing low-inductance circuitry to decrease the risetime of the excitation current pulse to values of less than 200 nsec.

As most of the transient metal-vapor lasers can exhibit gains high enough to make them superradiant, and as their laser pulses are often limited in duration to about 10 nsec, multiple-pass resonant cavities are not necessary for their operation. To utilize some feedback, the resonant cavity normally consists of a highly reflecting mirror at one end of the cavity and a quartz flat, which is used as the output mirror, at the other end of the cavity.

The time relationship between the risetime of the current pulse in a copper vapor (and ionized-calcium) laser and the arrival and duration of the laser-output pulses is illustrated

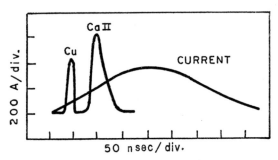

FIG. 4.52. 0.5106-μm Cu-I (and Ca-II) transient laser pulses superimposed on the discharge-current pulse. (After W. T. Walter *et al.*,[152] by courtesy of the IEEE.)

in Fig. 4.52. The Cu-I 0.5106-μm pulse is seen to be about 20 nsec long and the current pulse risetime about 200 nsec. Laser oscillation therefore occurs during the breakdown period of the discharge as the discharge current is rapidly increasing and while the electron temperature is still high.

4.3.2.2. Transient noble-gas lasers

A listing of transient noble-gas laser lines and transitions are given in Table 4.14.

TABLE 4.14. *Transient noble gas laser lines*[a]

Line (μm)	Specie	Transition	
		Racah	Paschen
2.0581	He	$2p\ {}^1P_1^{\circ}-2s\ {}^1S_0$	
0.54006	Ne	$3p'[1/2]_0-3s[3/2]_1^{\circ}$	$2p_1-1s_4$
0.58525	Ne	$3p'[1/2]_0-3s[1/2]_1^{\circ}$	$2p_1-1s_2$
0.59448	Ne	$3p'[3/2]_2-3s[3/2]_2^{\circ}$	$2p_4-1s_5$
0.61431	Ne	$3p\ [3/2]_2-3s[3/2]_2^{\circ}$	$3p_6-1s_5$
0.70672	Ar	$4p'[3/2]_2-4s[3/2]_2^{\circ}$	$2p_3-1s_5$
0.7503	Ar	$4p'[1/2]_0-4s'[1/2]_1^{\circ}$	$2p_1-1s_2$
0.81044	Kr	$5p\ [5/2]_2-5s[3/2]_2^{\circ}$	$2p_8-1s_5$
0.84092	Xe	$6p\ [3/2]_1-6s[3/2]_2^{\circ}$	$2p_7-1s_5$
0.90454	Xe	$6p\ [5/2]_2-6s[3/2]_2^{\circ}$	$2p_9-1s_5$
0.97997	Xe	$6p\ [1/2]_1-6s[3/2]_2^{\circ}$	$2p_{10}-1s_5$

[a] References to these laser lines will be found in the table in the Appendix.

Transient noble-gas laser lines are transient in nature only. They do not follow any of the four criteria stated in Section 4.3.2 to define the class of transient lasers. Their upper laser levels are not optically connected to the ground state, and in some cases it is their *lower* laser levels (the $1s_2$ and $1s_4$ levels) that are optically connected to the ground state. However,

175

in the presence of radiation trapping, none of these $1s_{2-5}$ lower laser levels can decay radiatively and they become effectively metastable and so satisfy the basic nature for metastability of criteria (3) in Section 4.3.2.

It is not clear which of the three possible excitation mechanisms

$$(1) \quad (np)^6 + e^- + \text{kinetic energy} \rightarrow np^5(n+1)p + e^-, \tag{4.49}$$

$$(2) \quad (np)^6 + e^- + \text{kinetic energy} \rightarrow np^5(n+1)s + e^-, \quad \text{followed by} \tag{4.50a}$$

$$np^5(n+1)s + e^- + \text{kinetic energy} \rightarrow np^5(n+1)p + e^-, \quad \text{or} \tag{4.50b}$$

(3)   Radiative cascading from higher-lying, excited s and d levels,

is or are responsible for excitation of the $np^5(n+1)p$ upper laser levels in the noble gas transient lasers, where $n = 1, 2, 3, 4$, and 5 for helium, neon, argon, krypton, or xenon, respectively. In eqn. (4.49) close to the threshold of excitation, it is possible that the much sharper maxima in the excitation functions of levels not optically connected to the ground state (Fig. 2.20) enable excitation of the $np^5(n+1)p$ states to occur in preference to excitation of the $np^5(n+1)s$ states. Reactions (4.50a) and (4.50b) would be more in keeping with theoretical predictions of selective excitation in electron-impact collisions. The excitation would then be a two-step process, involving direct excitation of the metastable/quasi-metastable $np^5(n+1)s$ levels from the ground state, and then indirect excitation of the $np^5(n+1)p$ states from the $np^5(n+1)s$ states by electron impact. The cross-section for eqn. (4.50b) is expected to be large and so could make up for a limited population of atoms in the $np^5(n+1)s$ states as compared with the $(np)^6$ ground-state population.

Radiative cascading from higher-lying s and d levels, and two-step (4.50a) and (4.50b) processes would appear to be too slow to account for transient population inversion that occurs in the 0.5401-μm transient neon laser. This laser operates during the risetime of the excitation current pulse for periods of about 5 nsec in relatively high-pressure discharges of pure neon.[159-60] The requirement for a low-impurity concentration in the neon fill-gas in the 0.5401-μm neon laser as compared with that in transient molecular N_2 lasers, however, argues for the importance of the presence of neon metastables in the upper-level selective-excitation process (4.50b). It is also conceivable that pure gases are required for obtaining electrons of high energy for direct electron-impact excitation of the $np^5(n+1)p$ levels to proceed effectively during the risetime of the discharge current pulse in the breakdown-period of the gas discharge, though that requirement for the production of energetic electrons does not seem to apply to the transient N_2 laser (Chapter 6).

4.3.3. DISSOCIATIVE-TYPE, MOLECULAR METAL-VAPOR LASERS

Just as a "collision of the second kind" between an excited atom and a molecule can lead to dissociation and selective excitation of one or both of the dissociated atoms, so can a "collision of the first kind"

$$e^- + \text{kinetic energy} + AB \begin{cases} \longrightarrow e^- + A' + B + \Delta E_\infty, \tag{4.51a} \\ \longrightarrow e^- + A + B + \Delta E_\infty, \tag{4.51b} \end{cases}$$

where ΔE_{∞} can appear as kinetic energy of the dissociated A', B', A or B atoms. Whereas excitation of a molecule from the ground state, in accordance with the Franck–Condon principle, normally leads to the formation of population inversion between excited electronic states of the molecule (as in the 0.3371 and near-infrared N_2 laser, Section 6.2.3), Isaev and Petrash[161] realized that when potential-energy curves of the molecule were suitable, electron-impact collisions with molecules in the ground state could lead to dissociation into one of the excited states of the atom.

An example of such an arrangement of the potential-energy curves of a suitable hypothetical molecule is shown schematically in Fig. 4.53. In accordance with the Franck–Condon

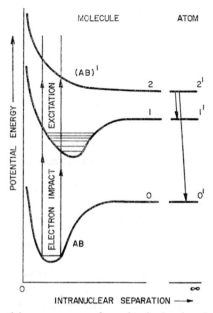

FIG. 4.53. Arrangement of potential-energy curves of a molecule showing electron-impact collisions leading to dissociation and population inversion between atomic levels. (After A. A. Isaev and G. G. Petrash,[161] by courtesy of the American Institute of Physics.)

principle, excitation can occur without any change of internuclear separation to the molecular levels 1 or 2. Excitation of molecular level 1 leads only to the production of vibrationally excited electronic states of the molecule, whereas excitation to the repulsive state 2 of the molecule leads to excitation of the atomic state 2' as the repulsive state of the molecule AB' dissociates. In principle, inversion is possible between the excited states 2' and 1' of the atom, and between 2' and the atomic ground state 0'. Advantages of this type of selective excitation process include: operation over a wide range of electron energies as long as they exceed the sum of the dissociation energy of the molecule and the excitation energy of the atomic state 2'; ability to excite high-lying electronic states of the atom that would not be accessible by photodissociation, thus making it possible to obtain inversion in transitions in the visible and ultraviolet regions of the spectrum.[161]

Superradiance under pulsed conditions in thallium vapor on the $7s\,^2S_{1/2}$–$6p\,^2P^\circ_{3/2}$ transition in atomic thallium at 0.5350 μm is believed to occur as a result of dissociative electron-

177

impact excitation. Superradiance was observed by Isaev and Petrash[161] in thallium vapor in the presence of the carrier gases helium, neon, argon, and air in the temperature range 370° to 410°C with the intensity of the superradiance increasing with different carrier gases in the order given. The duration of the superradiant pulse did not exceed 20 nsec. The observation when superradiance of the 0.5350-μm Tl-I was present in the temperature range 370° to 410°C that the main lines were in singly ionized and triply ionized species of thallium, while strong lines of neutral atomic Tl-I appeared and superradiance disappeared at temperatures above 410°C, indicates that direct electron-impact excitation of ground-state thallium atoms to the $7s\ ^2S_{1/2}$ state was *not* a strong excitation process. It follows that the superradiance is more likely produced as a result of excitation of the thallium molecules in the dissociation reaction (4.51).

4.4. Miscellaneous Lasers

4.4.1. CHARGE NEUTRALIZATION LASERS

A number of gas lasers operate in pulsed discharges where charge neutralization occurs in the form of the two-body ion–ion recombination reaction

$$A^+ + B^- \rightarrow A'(\text{or } A) + B(\text{or } B') + \text{kinetic energy}; \tag{4.52}$$

or in the form of the three-body ion-electron dissociative-recombination reaction

$$(AB)^+ + e^- \rightarrow A + B + \text{kinetic energy}. \tag{4.53}$$

The dissociated atoms A and B may, or may not, be excited.

In the pulsed sodium–hydrogen and potassium–hydrogen lasers[162–4] that oscillate in neutral species of sodium and potassium at low pressures of the metal vapor with a few torr of hydrogen, selective excitation of the upper laser levels occurs in the ion–ion recombination reactions

$$\text{Na}^+ + \text{H}^- \rightarrow \text{Na}'\ 4s\ ^2S_{1/2} + \text{H}, \tag{4.54}$$

and

$$\text{K}^+ + \text{H}^- \rightarrow \text{K}'\ 5s\ ^2S_{1/2} + \text{H}, \tag{4.55}$$

with the negatively charged hydrogen atom H^- being produced in the electron-molecule dissociative reaction

$$e^- + \text{H}_2 \quad\begin{array}{l} \longrightarrow \text{H}^- + \text{H}, \\ \longrightarrow \text{H}^- + \text{H}', \\ \longrightarrow \text{H}^+ + \text{H}^- + e^-. \end{array}\qquad\begin{array}{l}(4.56a)\\(4.56b)\\(4.56c)\end{array}$$

The $\text{Na}'\ 4s\ ^2S_{1/2}$ atoms have been determined to have more kinetic energy directed towards the cathode so they are clearly formed from Na^+ positive ions and not as a result of electron-impact excitation.

In a gaseous discharge above a pressure of a few torr, recombination of positive and negative ions is believed to be via dissociative-recombination reactions.[165] Selective excitation in three laser systems in the afterglow of a pulsed discharge appears to occur via the ion–electron dissociative-recombination, charge-neutralization reactions:

$$\text{(1a)} \quad O_2^+ + e^- \rightarrow O'\ 3p + O + \text{kinetic energy}, \tag{4.57}$$

$$\text{(1b)} \quad O_3^+ + e^- \rightarrow O'\ 3p + O_2 + \text{kinetic energy} \tag{4.58}$$

to give laser oscillation at 0.8446 μm in atomic oxygen;[136]

$$\text{(2)} \quad (HeNe)^+ + e^- \rightarrow He' + Ne'\ 2s; \tag{4.59}$$

and

$$\text{(3)} \quad (HeAr)^+ + e^- \rightarrow He' + Ar'\ 3d,\ 3d'' \tag{4.60}$$

giving oscillation in neon or argon in pulsed, high-pressure discharges of helium and neon mixtures at 1.15 μm, and in helium–argon mixtures at 1.21 μm and 1.28 μm, respectively.[22]

4.4.2. LINE-ABSORPTION AND MOLECULAR-PHOTODISSOCIATION LASERS

Line absorption (or optical pumping) has had limited utilization as a means of producing population inversion in gas lasers. It has been used exclusively to excite the $8p\ ^2P_{1/2}^{\circ}$ upper level of the 3.204-μm and 7.182-μm cesium laser lines selectively in Cs-I;[166–8] and to a limited extent to produce oscillation indirectly at approximately 1.8 μm in the 4f–3d transitions,[169] and at 1.64 μm and 1.258 μm in Ne-I on other transitions.[170]

Figure 4.54 shows energy levels pertinent to the cesium laser. This laser is optically pumped by the strong $3p\ ^3P_{2,1}^{\circ} - 2s\ ^3S_1$ transitions in He-I at 0.3888 μm. The center of this line falls within the Doppler half-width of the Cs-I $6s\ ^2S_{1/2} - 8p\ ^2P_{1/2}^{\circ}$ resonance transition at 0.388851 μm. The coincidence of the line shapes of the 0.3888-μm He-I line in spontaneous emission and the Cs-I resonance transition is illustrated in Fig. 4.55. Selective excitation of the $8p\ ^2P_{1/2}^{\circ}$ state leads to population inversion being realized between the $8p\ ^2P_{1/2}^{\circ}$ state and the $8s\ ^2S_{1/2}$ and $6d\ ^2D_{3/2}$ states. The coincidence of the Cs-I, 0.388851-μm line away from the line center of the spontaneous emission line shape of the He-I 0.3888-μm line, and the narrowness of the cesium line, severely limits the amount of selective excitation of the $8p\ ^2P_{1/2}^{\circ}$ state that can occur through absorption.

Optical pumping, has been used to increase the excitation rates of the 4f states of Ne-I in comparison with the excitation rate of the 3d states of neutral neon. The processes responsible are

$$He^*\ 2s\ ^3S_1 + h\nu \rightarrow He\ 2p\ ^3P_{2,1,0}^{\circ} \tag{4.61}$$

and

$$He\ 2p\ ^3P_{2,1,0}^{\circ} + Ne\ ^1S_0 \rightarrow He\ ^1S_0 + Ne\ (6s, 5d, \text{etc.}) + \Delta E_\infty \tag{4.62}$$

followed by radiative cascading into the 4f levels in preference to the 3d levels.[169]

Molecular photodissociation that occurs during flash photolysis in the reaction

$$CX_3I + h\nu \rightarrow I' + CX_3 \tag{4.63}$$

where $X = $ H or F produces iodine atoms mainly in the first excited $^2P_{1/2}^{\circ}$ state of atomic

13

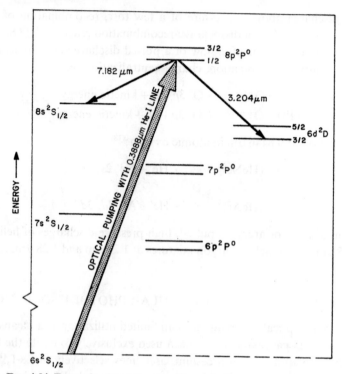

FIG. 4.54. Energy levels pertinent to the optically pumped cesium laser.

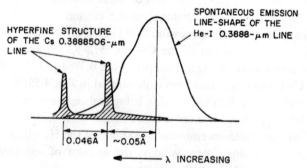

FIG. 4.55. Coincidence of the line-shapes of the Cs-I resonance line at 0.388851 μm and the He-I ($3p\ ^3P^\circ_{2,1}$–$2s\ ^3S_1$) line at 0.3888 μm. (After P. Rabinowitz and S. Jacobs,[167] by courtesy of the Columbia University Press.)

iodine in sufficient excess of those produced in the $^2P^\circ_{3/2}$ atomic ground state to produce stimulated emission on the $^2P_{1/2}$–$^2P_{3/2}$ magnetic-dipole transition at 1.315 μm.[171–4] High output powers at extremely high pulse energies are realizable on the 1.315-μm transition.[174]

4.4.3. RADIATIVE CASCADE-PUMPING

As stated in Chapter 2 (Section 2.9), cascading through strong transitions (in particular laser transitions) provides selective excitation of levels having the same parity as the ground state that are not accessible by direct electron-impact excitation from the ground state. In a clear example, in the He–Ne laser radiative cascade-pumping occurs via strong s–p transitions from the well-populated Ne $3s_2$ level (Paschen notation) to give selective excitation of the $3p$ levels of Ne-I.[175-7] Oscillation then occurs on a number of $3p$–$2s$ transitions at approximately 2 μm. As illustrated in Fig. 4.56, cascade-pumping by $3p$–$2s$ laser transitions

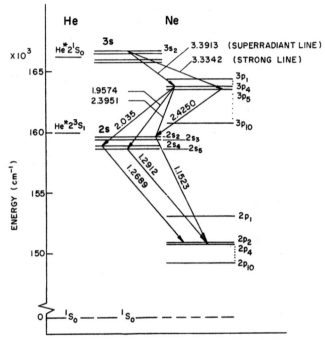

FIG. 4.56. Ne-I $3p$–$2s$ and $2s$–$2p$ laser lines (wavelengths in μm) observed in cascade following selective excitation of the $3p_4$ and $3p_5$ levels by the strong 3.3913-μm and 3.3342-μm lines. (After C. S. Willett, Neutral gas lasers, in *Handbook of Lasers*, by courtesy of the Chemical Rubber Company, Cleveland, Ohio.)

of the $2s_{2-4}$ levels can subsequently cause population inversion to occur between the $2s$ and $2p$ levels and enable oscillation to be obtainable on the well-known near-infrared laser lines at about 1.2 μm whose upper levels are normally excited by atom–atom excitation transfer from He* 2^3S_1 metastables or by electron-impact excitation from the $2p^6$ ground state of Ne-I.

Radiative cascade-pumping also occurs in laser discharges in pure neon at low pressure, where again transitions from the $3s$ levels initiate the cascade laser transitions.[175, 178]

References

1. JAVAN, A., BENNETT, W. R., Jr. and HERRIOTT, D. R., *Phys. Rev. Lett.* **6**, 106 (1961).
2. BENNETT, W. R., Jr., *Advances in Quantum Electronics* (ed. J. INGER), pp. 28–43, Columbia University Press, New York, 1961.
3. BENNETT, W. R., Jr., JAVAN, A. and BALLIK, E. A., *Bull. Am. Phys. Soc.* **5**, 496 (1960).
4. OSHIROVICH, A. L. and VEROLAINEN, YA. F., *Optics Spectrosc.* **22**, 181 (1967).
5. KLOSE, J. S., *Bull. Am. Phys. Soc.* **9**, 488 (1964).

6. LADENBURG, R., *Rev. Mod. Phys.* **5**, 243 (1933).
7. GRIFFITHS, J. H., *Proc. Roy. Soc.* (London) **143**, 588 (1934).
8. BENNETT, W. R., Jr., *Appl. Optics*, Supplement 1: *Optical Masers* (1962), pp. 24–61.
9. BATES, D. R. and DAMGAARD, A., *Phil. Trans.* A **242**, 10 (1950).
10. JAVAN, A., *Phys. Rev. Lett.* **3**, 87 (1959).
11. CHEBOTAYEV, V. P., *Radio Engng Electron. Phys.* **10**, 316 (1965).
12. KOSTER, G. F. and STATZ, H. J., *Appl. Phys.* **32**, 2054 (1961).
13. YOUNG, R. T., Jr., WILLETT, C. S. and MAUPIN, R. T., *J. Appl. Phys.* **41**, 2936 (1970).
14. BENNETT, W. R. Jr., *Bull. Am. Phys. Soc.* **7**, 15 (1962).
15. PETRASH, G. G. and KNYAZEV, I. N., *Sov. Phys. JETP* **18**, 571 (1964).
16. MCFARLANE, R. A., PATEL, C. K. N., BENNETT, W. R., Jr. and FAUST, W. L., *Proc. IRE* **50**, 2111 (1962).
17. RIGDEN, J. D. and WHITE, A. D., *Nature* **198**, 774 (1931).
18. PAANANEN, R. A., TANG, C. L., HORRIGAN, F. A., and STATZ, H. J., *Appl. Phys.* **34**, 3148 (1963).
19. ZITTER, R. N., *J. Appl. Phys.* **35**, 3070 (1964).
20. CHEBOTAYEV, V. P. and VASILENKO, L. S., *Sov. Phys. JETP* **21**, 515 (1965).
21. WHITE, A. D. and RIGDEN, J. D., *Proc. IRE* **50**, 2366 (1962).
22. SHTYRKOV, E. I. and SUBBES, E. V., *Optics Spectrosc.* **21**, 143 (1966).
23. SMITH, J., *J. Appl. Phys.* **35**, 723 (1964).
24. ZITTER, R. N., *Bull. Am. Phys. Soc.* **9**, 500 (1964).
25. AGOBIAN, R. DER, OTTO, J. L., CAGNARD, R. and ECHARD, R., *Compt. Rend.* **257**, 1044 (1963).
26. AGOBIAN, R. DER, OTTO, J. L., ECHARD, R. and CAGNARD, R., *Compt. Rend.* **257**, 3844 (1963).
27. BENNETT, W. R., Jr. and KNUTSON, J. W., Jr., *Proc. IEEE* **52**, 861 (1964).
28. AGOBIAN, R. DER, CAGNARD, R., ECHARD, R. and OTTO, J. L., *Compt. Rend.* **258**, 3661 (1964).
29. CHEBOTAYEV, V. P. and POKASOV, V. V., *Radio Engng Electron. Phys.* **10**, 817 (1965).
30. LADENBURG, R. and DORGELA, H. B., *Physica* **5**, 90 (1925).
31. BENTON, E. E., *Phys. Rev.* **206**, 128 (1962).
32. BENTON, E. E. and ROBERTSON, W. W., *Bull. Am. Phys. Soc.* **7**, 114 (1962).
33. AFANAS'EVA, V. L., LUKIN, A. V. and MUSTAFIN, K. S., *Sov. Phys. Tech. Phys.* **12**, 233 (1967).
34. CHEBOTAYEV, V. P., *Radio Engng Electron. Phys.* **10**, 314 (1965).
35. ZNAMENSKII, V. B., BUINOV, G. N. and BURSAKOV, E. S., *Optics Spectrosc.* **20**, 292 (1966).
36. MUSTAFIN, K. S., SELEZNEV, V. A. and SHTYRKOV, E. I., *Optics Spectrosc.* **21**, 429 (1966).
37. ACTON, J. R. and SWIFT, J. D., *Cold Cathode Discharge Tubes*, p. 226, Academic Press, New York, 1963.
38. DESAI, SH. K. and KAGAN, YU. M., *Optics Spectrosc.* **28**, 234 (1970).
39. PATEL, C. K. N., BENNETT, W. R., Jr., FAUST, W. L. and MCFARLANE, R. A., *Phys. Rev. Lett.* **9**, 102 (1962).
40. BOOT, H. A. H. and CLUNIE, D. M., *Nature* **197**, 173 (1963).
41. MATHIAS, L. E. S. and PARKER, J. T., *Appl. Phys. Lett.* **3**, 16 (1963).
42. BOOT, H. A. H., CLUNIE, D. M. and THORN, R. S. A., *Nature* **198**, 773 (1963).
43. WHITE, A. D. and RIGDEN, J. D., *Proc. IRE* **50**, 1697 (1962).
44. SUGA, T., *Sci. Papers Inst. Phys. Chem. Research* (Tokyo) **34**, 7 (1938).
45. DORGELA, H. B. and ABBINK, A., *Z. Phys.* **37**, 667 (1926).
46. MOORE, C. E., *Atomic Energy Levels*, vol. 1 (1949), NBS Circular 467, U.S. Government Printing Office, Washington, D.C. 20402.
47. SUZUKI, N., *Japan. J. Appl. Phys.* **4**, Supplement I (1965), pp. 642–7.
48. SUZUKI, N., *Japan. J. Appl. Phys.* **4**, 285 (1965).
49. BENNETT, W. R., Jr., KINDLMANN, P. J. and MERCER, G. N., *Appl. Optics*, Supplement 2: *Chemical Lasers* (1965), pp. 34–57.
50. WHITE, A. D. and GORDON, E. I., *Appl. Phys. Lett.* **3**, 197 (1963).
51. MASSEY, J. T., SCHULZ, A. G., HOCHHEIMER, B. F. and CANNON, S. M., *J. Appl. Phys.* **36**, 658 (1965).
52. YOUNG, R. T., Jr., Spectrum transitions and their wavelengths from 2000 to 10,000 Å for rare gas atoms, Harry Diamond Laboratories, TM-68-27, published as AD 690198.
53. WILLETT, C. S., *Laser Lines in Atomic Species*, in vol. 1 of *Progress in Quantum Electronics* (ed. J. H. SANDERS and K. W. H. STEVENS), pp. 273–357, Pergamon Press, 1971.
54. BLOOM, A. L., *Appl. Phys. Lett.* **2**, 101 (1963).
55. WHITE, A. D. and RIGDEN, J. D., *Appl. Phys. Lett.* **2**, 211 (1963).
56. PERRY, D. L., *IEEE J. Quant. Electr.* **QE-7**, 102 (1971).
57. BLOOM, A. L. and HARDWICK, D. L., *Phys. Lett.* **20**, 373 (1966).

58. Tobias, I. and Strouse, W. M., *Appl. Phys. Lett.* **10**, 342 (1967).
59. Bloom, A. L., Bell, W. E. and Rempel, R. C., *Appl. Optics* **2**, 317 (1963).
60. Faust, W. L. and McFarlane, R. A., *J. Appl. Phys.* **35**, 2010 (1964).
61. Faust, W. L., McFarlane, R. A., Patel, C. K. N. and Garrett, C. G. B., *Phys. Rev.* **133**, A1476 (1964).
62. McFarlane, R. A., Faust, W. L. and Patel, C. K. N., *Proc. IEEE* **51**, 468 (1963).
63. Patel, C. K. N., Faust, W. L., McFarlane, R. A. and Garrett, C. G. B., *Proc. IEEE* **52**, 713 (1964).
64. Colombo, L., Markovic, B., Pavlovic, Z. and Persin, A., *J. Opt. Soc. Am.* **56**, 890 (1966).
65. Willett, C. S. and Young, R. T., Jr., *J. Appl. Phys.* **43**, 725 (1972).
66. Jones, C. R. and Robertson, W. W., *Bull. Am. Phys. Soc.* **13**, 198 (1968). See also Chapter 2, ref. 26.
67. Benton, E. E., Ferguson, E. E., Matsen, F. A. and Robertson, W. W., *Phys. Rev.* **128**, 206 (1962).
68. Massey, J. T., Schulz, A. G., Hochheimer, B. F. and Cannon, S. M., *J. Appl. Phys.* **36**, 658 (1965).
69. Chebotayev, V. P., *Optics Spectrosc.* **1**, 10 (1966).
70. Ionikh, Yu. H. and Penkin, N. P., *Optics Spectrosc.* **31**, 453 (1971).
71. Phelps, A. V., *Phys. Rev.* **99**, 1307 (1955).
72. Labuda, E. F., Ph.D. Dissertation, Polytechnic Institute of Brooklyn, New York, 1967 (obtainable from University Microfilms, 313N First Street, Ann Arbor, Michigan 48108).
73. Field, R. L., Jr., *Rev. Sci. Instr.* **38**, 1720 (1967).
74. Young, R. T., Jr. and Maupin, R. T., *J. Appl. Phys.* **40**, 3881 (1969).
75. Wada, J. Y. and Heil, H., *IEEE J. Quant. Electr.* **QE-1**, 327 (1965).
76. Buckingham, R. A. and Dalgarno, A., *Proc. Roy. Soc.* (London) A **213**, 327 (1952).
77. Tanaka, V. and Yoshino, K., *J. Chem. Phys.* **39**, 3081 (1963).
78. Brigman, G. H., Brient, S. J. and Matsen, F. A., *J. Chem. Phys.* **34**, 958 (1961). Recent experiments have indicated that the height of this barrier is considerably less than the 0.25 eV calculated value of Buckingham and Dalgarno and less than the experimentally determined value of 0.16 eV by Brigman *et al.*[78] (C. R. Jones, Private Communication, May 1968). See also Ph.D. Dissertation, "Temperature and pressure dependence of the lifetime of the singlet metastable of helium", by Sister J. Hungerman, University of Texas, August 1972.
79. Phelps, A. V. and Molnar, J. P., *Phys. Rev.* **89**, 1203 (1953).
80. Biondi, M. A., *Phys. Rev.* **82**, 543 (1951).
81. Hasted, J. B., *Physics of Atomic Collisions*, p. 463, Butterworth, Washington, 1964.
82. Bennett, W. R., Jr., Inversion mechanisms in gas lasers, in *Applied Optics*, Supplement 2: *Chemical Lasers* (1965), pp. 3–33.
83. Bennett, W. R., Jr. and Knutson, J. W., *Bull. Am. Phys. Soc.* **9**, 500 (1964). See also ref. 82.
84. Lisitsyn, V. N., Provorov, A. S. and Chebotayev, V. P., *Optics Spectrosc.* **29**, 119 (1970).
85. McFarlane, R. A., Patel, C. K. N., Bennett, W. R., Jr. and Faust, W. L., *Proc. Inst. Radio Engrs.* **50**, 2111 (1962).
86. Smith, P. W., *IEEE J. Quant. Electr.* **QE-2**, 77 (1966).
87. Herziger, G., Holzapfel, W. and Seelig, W., *Z. für Physik* **189**, 385 (1966).
88. Patel, C. K. N., Faust, W. L. and McFarlane, R. A., *Appl. Phys. Lett.* **1**, 84 (1962).
89. Clarke, P. O., *IEEE J. Quant. Electr.* **QE-1**, 109 (1965).
90. Bennett, W. R., Jr., Faust, W. L., McFarlane, R. A. and Patel, C. K. N., *Phys. Rev. Lett.* **8**, 470 (1962).
91. Aleksandrov, E. G., Kulyasov, V. N. and Mamyrin, A. B., *Optics Spectrosc.* **31**, 170 (1971).
92. Gordon, E. I. and White, A. D., *Appl. Phys. Lett.* **3**, 199 (1963).
93. Korolev, F. A., Odintsov, A. I. and Misai, V. N., *Optics Spectrosc.* **19**, 36 (1965).
94. Labuda, E. F. and Gordon, E. I., *J. Appl. Phys.* **35**, 1647 (1964).
95. Mandelstam, S. L. and Nedler, V. V., *Optics Spectrosc.* **10**, 196 (1961).
96. Brown, S. C., *Handbuch der Physik*, vol. 22 (ed. S. Flugge), p. 531, Springer-Verlag, Berlin, 1956.
97. Znamenskii, V. B., *Optics Spectrosc.* **15**, 7 (1968).
98. Willett, C. S. and Gleason, T. J., *Laser Focus* (June 1971), pp. 30–33.
99. Smith, P. W., *IEEE J. Quant. Electr.* **QE-2**, 62 (1966).
100. The prism-mirror cavity-configuration is used to suppress oscillation on the 3.39-μm line (see ref. 55). The high output powers per unit length of discharge are average (not best possible) values that can be achieved.
101. von Engel, A. and Steenbeck, W., *Electrische Gasentladungen*, Band II, p. 85, Springer-Verlag, Berlin, 1932.
102. Young, R. T., Jr., *J. Appl. Phys.* **36**, 2324 (1965).
103. Ramsay, J. V. and Tanaka, K., *Japan. J. Appl. Phys.* **5**, 918 (1966).

104. Fridrikhov
105. Gonchukov, S. A., Ermakov, G. A., Mikhenko, G. A. and Protsenko, E. D., *Optics Spectrosc.* **20**, 601 (1966).
106. Belousova, I. M., Danilov, O. B., El'kina, I. A. and Kiselyov, V. M., *Optics Spectrosc.* **16**, 44 (1969).
107. Belousova, I. M., Danilov, O. B. and Kiselyov, V. M., *Sov. Phys. Tech. Phys.* **13**, 363 (1968).
108. The gas temperature is of course higher than the wall temperature. It is probably between 100 to 200 K higher than the outside wall temperature for commonly used discharge tubes.
109. Bell, W. E. and Bloom, A. L., *Appl. Optics* **3**, 413 (1964).
110. Alekseeva, A. N. and Gordeev, D. V., *Optics Spectrosc.* **23**, 520 (1967).
111. White, A. D., *Appl. Optics* **3**, 431 (1964).
112. Mazanko, I. P., Malchanov, M. I., Ogurok, N.-D. D. and Sviridov, M. V., *Optics Spectrosc.* **30**, 495 (1971).
113. Patel, C. K. N., *J. Appl. Phys.* **33**, 3194 (1962).
114. Maximov, A. I., *Optics Spectrosc.* **22**, 100 (1967).
115. Rigden, J. D. and White, A. D., *Proc. IRE* **50**, 2366 (1962).
116. Mamikonyants, N. L. and Molchashkin, M. A., *Radio Engng Electr. Phys.* **3**, 518 (1967).
117. Belousova, I. M., Danilov, O. B. and Elkina, I. A., *Sov. Phys. Tech. Phys.* **12**, 1229 (1968).
118. Willett, C. S., *Appl. Optics* **11**, 1429 (1972).
119. Dauger, A. B. and Stafsudd, O. M., *IEEE J. Quant. Electr.* **QE-6**, 572 (1970).
120. Paananen, R. A. and Horrigan, F. A., *Proc. IEEE* **52**, 1261 (1964).
121. Dauger, A. B. and Stafsudd, O. M., *Appl. Optics* **10**, 2690 (1971).
122. Patel, C. K. N., McFarlane, R. A. and Faust, W. L., *Phys. Rev.* **133**, A1244 (1964).
123. Atkinson, J. B. and Sanders, J. H., *J. Phys. B (Proc. Phys. Soc.)*, Ser. 2, **1**, 1171 (1968).
124. Tunitskii, L. N. and Cherkasov, E. M., *Sov. Phys. Tech. Phys.* **13**, 1696 (1969).
125. Patel, C. K. N., McFarlane, R. A. and Faust, W. L., Further infrared spectroscopy using stimulated emission techniques, in *Quantum Electronics*, vol. III (P. Grivet and N. Bloembergen, eds.), pp. 561–72, Columbia University Press, New York, 1964.
126. Martinelli, R. U. and Gerritsen, H. J., *J. Appl. Phys.* **37**, 444 (1966).
127. Jeffers, W. Q. and Wiswall, C. E., *Appl. Phys. Lett.* **17**, 444 (1970).
128. Kovacs, M. A. and Ultee, C. J., *Appl. Phys. Lett.* **17**, 39 (1970).
129. Tunitskii, L. N. and Cherkasov, E. M., *Optics Spectrosc.* **23**, 154 (1967).
130. Kolpakova, I. V. and Redko, T. P., *Optics Spectrosc.* **23**, 351 (1971).
131. Rautian, S. G. and Rubin, P. L., *Optics Spectrosc.* **18**, 180 (1965).
132. Tunitskii, L. N. and Cherkasov, E. M., *Optics Spectrosc.* **26**, 344 (1969).
133. Tunitskii, L. N. and Cherkasov, E. M., *J. Opt. Soc. Am.* **56**, 1783 (1966).
134. Feld, M. S. and Javan, A., *Phys. Rev.* **177**, 540 (1969).
135. Domash, L., Feld, M. S., Feldman, B. J. and Javan, A., *Bull. Am. Phys. Soc.* **16**, 593 (1971).
136. Tunitskii, L. N. and Cherkasov, E. M., *Sov. Phys. Tech. Phys.* **13**, 993 (1969).
137. Massey, H. S. W. and Burhop, E. H. S., in *Electonic and Ionic Impact Phenomena*, p. 137, Clarendon Press, Oxford, 1952.
138. Bridges, W. B., *Appl. Phys. Lett.* **3**, 45 (1963).
139. Paananen, R. A. and Bobroff, D. L., *Appl. Phys. Lett.* **2**, 99 (1963).
140. Moskakenko, V. F., Ostapchenko, E. P. and Pugnin, V. I., *Optics Spectrosc.* **23**, 94 (1967).
141. Allen, L., Jones, D. G. C. and Schofield, D. G., *J. Opt. Soc. Am.* **59**, 842 (1969).
142. Walter, W. T. and Jarrett, S. M., *Appl. Optics* **3**, 789 (1964).
143. Clark, P. O., *Phys. Lett.* **17**, 190 (1965).
144. Aisenberg, S., *Appl. Phys. Lett.* **2**, 187 (1963).
145. Armstrong, D. R., *IEEE J. Quant. Electr.* **QE-4**, 968 (1968).
146. Kuznetsov, A. A. and Mash, D. I., *Radio Engng Electr. Phys.* **10**, 319 (1965).
147. Smiley, V. N., Forbes, D. K. and Lewis, A. L., *Appl. Phys. Lett.* **7**, 1 (1965).
148. Clark, P. O., *IEEE J. Quant. Electr.* **QE-1**, 109 (1965).
149. Davis, C. C. and King, T. A., *Phys. Lett.* **39A**, 186 (1972).
150. Targ, R. and Sasnett, M. W., *IEEE J. Quant. Electr.* **QE-8**, 166 (1971).
151. Wood, O. R., Burkhardt, E. G., Pollack, M. A. and Bridges, T. J., *Appl. Phys. Lett.* **18**, 261 (1971).
152. Walter, W. T., Solimene, N., Piltch, M. and Gould, G., *IEEE J. Quant. Electr.* **QE-2**, 474 (1966).
153. Piltch, M., Walter, W. T., Solimene, N., Gould, G. and Bennett, W. R., Jr., *Appl. Phys. Lett.* **7**, 309 (1965).
154. Leonard, D. A., *IEEE J. Quant. Electr.* **QE-3**, 380 (1967).
155. Deech, J. S. and Sanders, J. H., *IEEE J. Quant. Electr.* **QE-4**, 474 (1968).

184

156. FOWLES, G. R. and SILFVAST, W. T., *Appl. Phys. Lett.* **6,** 236 (1965).
157. ISAEV, A. A. and PETRASH, G. G., *Sov. Phys. JETP-Lett.* **10,** 119 (1969).
158. SILFVAST, W. T. and DEECH, J. S., *Appl. Phys. Lett.* **11,** 97 (1967).
159. CLUNIE, D. M., THORN, R. S. A. and TREZISE, K. E., *Phys. Lett.* **14,** 28 (1965).
160. LEONARD, D. A., *IEEE J. Quant. Electr.* **QE-3,** 133 (1967).
161. ISAEV, A. A. and PETRASH, G. G., *Sov. Phys. JETP Lett.* **7,** 156 (1968).
162. TIBILOV, A. S. and SHUKHTIN, A. M., *Optics Spectrosc.* **21,** 69 (1966).
163. POGORELYI, P. A. and TIBILOV, A. S., *Optics Spectrosc.* **25,** 301 (1968).
164. TIBILOV, A. S. and SHUKHTIN, A. M., *Optics Spectrosc.* **25,** 221 (1968).
165. FOWLER, R. G., Radiation from low pressure discharges, in *Handbuch der Physik* (ed. S. FLUGGE), *Gas Discharges II*, vol. 22, pp. 209–53, Springer-Verlag, Berlin, 1956, p. 240.
166. JACOBS, S., RABINOWITZ, P. and GOULD, G., *Phys. Rev. Lett.* **7,** 415 (1965).
167. RABINOWITZ, P. and JACOBS, S., The optically pumped cesium laser, in *Quantum Electronics*, vol. III (eds. P. GRIVET and N. BLOEMBERGEN), pp. 489–98, Columbia University Press, New York 1964.
168. RABINOWITZ, P., JACOBS, S. and GOULD, G., *Appl. Optics* **1,** 511 (1962).
169. LISITSYN, V. N. and CHEBOTAYEV, V. P., *Optics Spectrosc.* **20,** 603 (1966).
170. LISITSYN, V. N., FECHENKO, A. I. and CHEBOTAYEV, V. P., *Optics Spectrosc.* **27,** 157 (1969).
171. KASPER, J. V. and PIMENTAL, G. C., *Appl. Phys. Lett.* **5,** 231 (1964).
172. POLLACK, M. A., *Appl. Phys. Lett.* **8,** 36 (1966).
173. O'BRIEN, D. E. and BOWEN, J. R., *J. Appl. Phys.* **42,** 1010 (1971).
174. KOMPA, K. L. and HOHLA, K., Postdeadline Paper, Q.E.C. Montreal (1972). 30-J pulses in less than 5-nsec pulses are predicted with this system; 30-J pulses have already been obtained (*Laser Focus*, 26 July 1972).
175. AGOBIAN, R. DER, OTTO, J. L., ECHARD, R. and CAGNARD, R., *Compt. Rend.* **257,** 3344 (1963).
176. AGOBIAN, R. DER, CAGNARD, R., ECHARD, R. and OTTO, J. L., *Compt. Rend.* **258,** 3661 (1964).
177. SMILEY, V. N., *Appl. Phys. Lett.* **4,** 123 (1964).
178. AGOBIAN, R. DER, OTTO, J. L., CAGNARD, R. and ECHARD, R., *Compt. Rend.* **259,** 323 (1964).
179. LISITSYN, V. N., PROVOROV, A. S. and CHEBOTAYEV, V. P., *Optic Spectrosc.* **29,** 119 (1970).

CHAPTER 5

Specific Ionized Laser Systems

Introduction

In this chapter, particular ionized atomic laser systems are considered to illustrate four of the inelastic collision processes discussed in Chapter 2:

5.1. Resonant Excitation-energy Transfer;
5.2. Charge Transfer;
5.3. Penning Reactions; and
5.4. Electron-impact Excitation

As in Chapter 4, consideration is given in this chapter to the ways in which population inversion is maintained in the presence of selective excitation of the upper laser level.

5.1. Resonant Excitation-energy Transfer Lasers

There are two ion lasers that rely on a special type of resonant excitation transfer between an atom and an ion for the selective excitation of their upper laser levels. They are the helium–krypton and neon–xenon ion lasers in which oscillation occurs on singly ionized transitions of krypton and xenon. The selective excitation mechanism operative in both takes the general form

$$A^* + B^+(0) \rightarrow A + B^{+\prime} + \Delta E_\infty \tag{5.1}$$

where A^* is a metastable atom, $B^+(0)$ is a ground state ion, and $B^{+\prime}$ is an excited ion state. The energy-level coincidence for reaction (5.1) is not as stringent as that for atom–atom resonant energy-excitation transfer, with an allowable maximum ΔE_∞ being at least 0.25 eV and possibly as much as 0.67 eV, while maintaining a cross-section that is still probably gas kinetic. Since the reaction has been observed to occur at thermal energies in the afterglow of pulsed discharge, it is one that probably involves "curve-crossing" of the intermolecular potentials of the incoming $A^* + B^+(0)$, and outgoing $A + B^{+\prime}$ atoms during the collision. As such, it would appear likely that the reaction is not restricted to the He–Kr and Ne–Xe laser systems described here, and might be observable in discharges in other gas or vapor mixtures.

5.1.1. THE HELIUM–KRYPTON LASER

Oscillation that occurs in the visible region of the spectrum in the afterglow of a pulsed discharge in a mixture of helium and krypton was first reported by Laures, Dana and Frapard[1] of the Compagnie Générale d'Électricité. Laures *et al.* used 1-μsec, 4–8-kV excitation pulses in a discharge tube 1 m long and 4 mm diameter to produce laser oscillation on five lines between 512.8 and 431.9 nm. The optimum gas-mixture pressures were 0.001 to 0.003 torr of krypton and about 10 torr of helium; and the discharge current was varied between 40 to 150 A. The wavelengths at which oscillation was observed and probable transitions are given in Table 5.1

TABLE 5.1. *Wavelengths of singly ionized krypton laser lines in pulsed He–Kr discharges*

λ_{vac} (nm)	Transition	Intensity
512.714	$6s\ ^4P_{3/2}$–$5p\ ^4D^\circ_{3/2}$	Weak
469.574	$6s\ ^4P_{5/2}$–$5p\ ^4D^\circ_{7/2}$	Strong
458.413	$6s\ ^4P_{3/2}$–$5p\ ^4D^\circ_{5/2}$	Moderate
438.777[a]	$6s\ ^4P_{5/2}$–$5p\ ^4P^\circ_{5/2}$	Weak
431.902	$6s\ ^4P_{5/2}$–$5p\ ^4D^\circ_{5/2}$	Strong

[a] Observed only using prism wavelength-selection.

All the five laser lines originate from the two excited singly ionized krypton states $6s\ ^4P_{3/2}$ and $6s\ ^4P_{5/2}$. Figure 5.1, which illustrates the laser transitions and the potential-energy coincidence that occurs between the sum of the potential energy of a $He^*\ 2^3S_1$ atom and a ground-state Kr^+ ion, shows that there are near potential-energy coincidences of about 0.25 eV ($+1965\ cm^{-1}$ and $+1971\ cm^{-1}$, respectively) with the $6s\ ^4P_{3/2}$ and $6s\ ^4P_{5/2}$ states of singly ionized krypton. Except for the $5p\ ^4D^\circ_{7/2}$ lower laser level of the strong afterglow 469.6-nm line, the lower laser levels are upper p levels of laser lines that have been observed in low-pressure, pure krypton discharges[2] where strong p–s laser transitions are favored. A good radiative lower laser level depopulation mechanism exists, therefore, for the afterglow lines, and the lifetimes of the upper laser levels are such that population inversion is able to build up in the afterglow after electron thermalization.

In later work, Dana and Laures[3, 4] summarized their earlier results and examined the population inversion mechanisms operative in the pulsed He–Kr ion laser. They conclude that the selective excitation responsible for the oscillation that occurs in the afterglow is the exothermic reaction

$$He^*\ 2^3S_1(\uparrow\uparrow) + Kr^+\ ^2P(\uparrow) \rightarrow He\ ^1S_0(\uparrow\downarrow) + Kr^+\ 6s\ ^4P_{3/2,\ 5/2}(\uparrow\uparrow\uparrow) + \Delta E_\infty . \tag{5.2}$$

In this reaction, in which it is observed that the total spin of 3/2 is conserved in the collision, energy is transferred from a $He^*\ 2^3S_1$ metastable to a singly ionized Kr^+ ion in the $^2P^\circ_{3/2}$

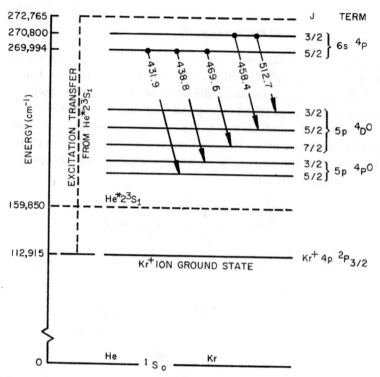

FIG. 5.1. Partial energy-level diagram of Kr-II showing afterglow laser transitions, and illustrating the potential-energy coincidence between the Kr-II $6s$ $^4P_{3/2, 5/2}$ levels and the sum of the potential energies of the He* 2^3S_1 state and the Kr$^+$ ground-state ion. Wavelengths are in nm.

ground state to give selective excitation of the two excited $6s$ $^4P_{3/2}$ and $6s$ $^4P_{5/2}$ states of singly ionized krypton.

The selective excitation process occurs in two steps:

(1) Ionization of the krypton "impurity" during the current pulse and in the afterglow.
(2) Excitation of these ground-state ions through the atom–ion excitation transfer reaction (5.2).

In the afterglow of the pulsed discharge, after thermalization of high-energy electrons that would be capable of ionizing has occurred, ionization of krypton at low partial pressures in the presence of excited atoms of helium is believed to occur via the Penning reaction

$$\text{He}^* \ 2^3S_1 + \text{Kr} \ ^1S_0 \rightarrow \text{He} \ ^1S_0 + \text{Kr}^+ + e^- + \text{kinetic energy.} \tag{5.3}$$

The relatively high pressure of helium of 10 torr used by Dana and Laures enables a high concentration of long-lived metastable He* 2^3S_1 atoms to be produced during, and at the termination of, the current pulse and inhibits the diffusion of ions and metastables to the wall of the discharge tube to the betterment of the excitation transfer process (5.2). It is perhaps only a fortuitous coincidence but a pressure of 10 torr of helium has been shown by Egorov et al.[5] to be approximately the optimum pressure for the production of He* 2^3S_1

188

metastables in the afterglow of a capacitor discharge in helium in a discharge tube of 1.4-cm bore (Fig. 3.19). The low partial pressure of krypton of $\sim 10^{-3}$ torr, and the strong oscillation observed in the afterglow, indicates that reaction (5.2) must have a large cross-section.

5.1.2. THE NEON–XENON LASER

Oscillation in the visible, in the afterglow of pulsed discharge in a mixture of neon and xenon, in addition to that observed in a helium–krypton mixture, was also reported by Dana and Laures.[3, 4] The selective excitation mechanism responsible for the afterglow oscillation is similar to that responsible for selective excitation in the pulsed He–Kr ion laser. It involves the transfer of potential energy from an excited (metastable) atom to a singly ionized ion in the ground state. It occurs in the resonant exothermic reaction

$$\text{Ne}^* \; 1s_5(\uparrow\uparrow) + \text{Xe}^+ \; {}^2P^\circ_{3/2}(\uparrow) \rightarrow \text{Ne} \; {}^1S_0(\uparrow\downarrow) + \text{Xe}^+ \; 7s \; {}^4P_{3/2,\,5/2}(\uparrow\uparrow\uparrow) + \varDelta E_\infty \quad (5.4)$$

in which the total spin of 3/2 is again conserved and where the Ne^* $1s_5$ state (in Paschen notation) is the lowest excited (and metastable) state of neon. Wavelengths at which oscillation occurred, probable transitions, and laser intensities are given in Table 5.2.

TABLE 5.2. *Wavelengths of singly ionized xenon laser lines in pulsed Ne–Xe discharges*

λ_{vac} (nm)	Transition	Intensity
609.529	$7s \; {}^4P_{3/2}–6p \; {}^4D^\circ_{3/2}$	Weak
572.855	$5d' \; {}^2S_{1/2}–6p \; {}^4P^\circ_{3/2}$	Weak
531.536	$7s \; {}^4P_{5/2}–6p \; {}^4D^\circ_{7/2}$	Strong
486.383	$7s \; {}^4P_{5/2}–6p \; {}^4P^\circ_{5/2}$	Moderate

Figure 5.2, giving the laser transitions and potential-energy coincidences of excited singly ionized states of xenon and the sum of the potential energies of a Ne^* $1s_5$ metastable and a ground-state Xe^+ ion, shows that there is a near potential-energy coincidence between the Xe^+ $7s \; {}^4P_{3/2}$, $7s \; {}^4P_{5/2}$ and $5d' \; {}^2S_{1/2}$ states and the total potential energy of 231,877 cm^{-1} of the Ne^* $1s_5$ atom and the Xe^+ ground-state ion. The energy discrepancy $\varDelta E_\infty$ for excitation transfer to the Xe^+ $7s \; {}^4P_{3/2}$ state is $+854 \; \text{cm}^{-1}$ (or about 0.1 eV) and 1524 cm^{-1} (about 0.2 eV) for the Xe^+ $7s \; {}^4P_{5/2}$ state. The $5d' \; {}^2S_{1/2}$ state, which is also selectively excited, is more than 0.6 eV below the sum of the potential energies of the Ne* atom and Xe^+ ion. It is possible in the case of this state that the observed selective excitation is due to the buffering reaction

$$\text{Xe}^+ \; 7s \; {}^4P_{3/2,\,5/2} + \text{Ne} \rightarrow \text{Xe}^+ \; 5d' \; {}^2S_{1/2} + \text{Ne} \quad (5.5)$$

and not to the excitation-transfer reaction (5.4) since spin is not conserved in energy transfer to the $5d' \; {}^2S_{1/2}$ state.

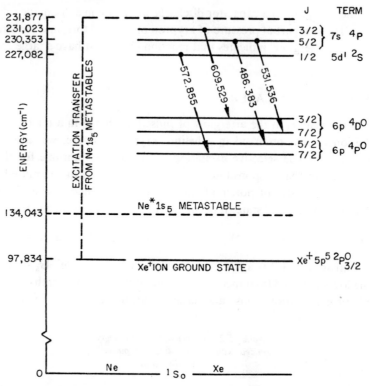

FIG. 5.2. Partial energy-level diagram of Xe-II showing afterglow laser transitions, and illustrating the potential-energy coincidence between the Xe-II $7s$ $^4P_{3/2, 5/2}$ and $5d'$ $^2S_{1/2}$ levels and the sum of the potential energies of the Ne* $1s_5$ state and the Xe$^+$ ground-state ion. Wavelengths are in nm.

As in the He–Kr ion laser, the lower levels of the afterglow singly ionized xenon laser lines tabulated in Table 5.2 include upper levels of strong p–s laser lines that occur in low-pressure, pure xenon discharges[2, 6] so that a good radiative depopulation scheme exists in the same way for the lower laser levels.

Selective excitation in the Ne–Xe ion laser also apparently occurs in two steps. The first step involves ionization of xenon, which is present at low partial pressures of 10^{-3} torr compared with the neon pressure of 4 torr, by the Penning reaction

$$\text{Ne } 1s_{2-5} + \text{Xe } ^1S_0 \rightarrow \text{Ne } ^1S_0 + \text{Xe}^+ + e^- + \text{kinetic energy} \qquad (5.6)$$

where the Xe$^+$ ion is probably in the ground state but could be in an excited state. The first step is followed by excitation transfer from Ne* $1s_5$ metastables to the ground-state Xe$^+$ $^2P^{\circ}_{3/2}$ ion giving selective excitation to the Xe$^+$ $7s$ $^4P_{3/2, 5/2}$ and $5d'$ $^2S_{1/2}$ states. These processes have a long life-time relative to electron-thermalization times and last around 25 to 30 μsec. This is evidenced by the spontaneous-emission behavior of the laser lines exhibited in Fig. 5.3.

Figure 5.3 shows that laser oscillation begins after the maximum of spontaneous emission of the 531.5-nm Xe-II laser line, about 2 μsec after the termination of the current pulse. The delay and duration of the laser pulse varies with the strength of the lines and the cavity

190

losses. The time-delays from the commencement of the current pulse are about 3 μsec for the strongest lines (e.g. the Xe-II 531.5-nm line in Fig. 5.3) to 13 μsec for the weakest Xe-II 572.9-nm line. Laser oscillation of the strongest lines occurs for a period of 15 μsec under good experimental conditions.[3] The variation of delay of the laser pulse with respect to the current pulse and the maximum of the spontaneous emission is associated with the

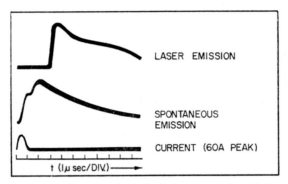

FIG. 5.3. Illustrating the time relation between the excitation-current pulse, spontaneous and laser emission of the Xe-II 531.5-nm afterglow laser line. (A composite of figures given by L. Dana and P. Laures.[3])

decay of excited ions in the lower laser states that have been populated by electron impact during the current pulse, and the time taken for gain to build up in the cavity and overcome cavity losses.

All the krypton and xenon lines obey the selection rules $\Delta S = 0$, and all the strongest lines have $\Delta J = \Delta L$. The only exception is the Xe-II 572.9-nm line starting from the $5d'\,{}^2S_{1/2}$ state with a transition $5d'^2 \rightarrow 6p^4$ giving $\Delta S = 1$ but with $\Delta J = \Delta L$. This change of spin of $\Delta S = 1$ differs from that of the other three observed laser lines and shows departure from L–S coupling.

5.2. Charge-transfer Lasers

It has been established that thermal-energy asymmetric charge transfer of the general form

$$A^+ + B \rightarrow A + B^{+\prime} \pm \Delta E_\infty \tag{5.7}$$

is responsible *with some degree of certainty* for selective excitation that occurs in the following singly ionized metal-vapor ion lasers:

He–Hg, He–Cd, and He–Zn, and very likely for selective excitation in the He–I₂, He–Se, and He–Te ion lasers. In all these lasers the presence of helium is essential to their operation, since it is the atomic-helium ground-state ion that provides the potential energy needed to simultaneously ionize and excite atom B in (5.7) to $B^{+\prime}$.

It is not yet clear how the thermal-energy charge-transfer reactions can be treated theoretically. And it is not known if the collisions follow the Massey adiabatic hypothesis (see Charge Transfer in Section 2.3 and the "near-adiabatic hypothesis" mentioned in Section 2.2.2). Indications are that the collisions can occur nonadiabatically through "curve-crossings" between the different intermolecular potentials that describe the temporary molecule formed during the collision. If this is the case, then it is likely that the cross-sections for asymmetric charge-transfer reactions exhibit an exponential dependence on the energy discrepancy ΔE_∞, similar to that illustrated in Figs. 2.14a and 2.14b for the thermal-energy, asymmetric charge-transfer reaction $Ne^+ + Pb \rightarrow Ne + Pb^{+\prime}$; and that there is a preference for total spin conservation in the collision, as in (2.21) to (2.24), and (5.2) and (5.4).

5.2.1. THE HELIUM–MERCURY LASER

Laser oscillation in the ion species of a gas was first reported in 1964 by Bell.[7] Laser oscillation on singly ionized mercury transitions in the afterglow of a pulsed discharge in a helium–mercury mixture was observed by Bell on red (615.0-nm) and green (567.7-nm) lines, and on two other lines at 734.6 nm and 1.0583 μm. In later work Bloom, Bell and Lopez[8] extended the laser oscillation to nineteen lines, nine of which were singly ionized ion lines; six were identified as neutral mercury transitions; and four were unidentified. All the ion lines reported and identified as Hg-II lines by these researchers, except for the 1.1179-μm line ($7g$–$6f$ transition), are shown in the partial energy-level diagram of ionized mercury and neutral helium of Fig. 5.4. The figure also shows the transition $7p\ ^2P^\circ_{1/2}$–$7s\ ^2S_{1/2}$ (at 741.8 nm) on which pulsed oscillation has been reported and the $7p\ ^2P^\circ_{3/2}$–$7s\ ^2S_{1/2}$ transition (at 794.5 nm) on which pulsed and CW-laser oscillation, together with CW oscillation of the 615.0-nm line, has subsequently been reported.[9–11] Pulsed oscillation has been reported also in doubly ionized mercury on the $5d^8\ 6s^2\ (J = 4)$–$5d^9(^2D_{5/2})\ 6p_{1/2}(J = 3)$ transition in the visible at 479.7 nm.[12] This report, incidentally, was the first of oscillation in a doubly ionized species.

The two visible lines at 615.0 nm ($7p\ ^2P^\circ_{3/2}$–$7s\ ^2S_{1/2}$) and 567.7 nm ($5f\ ^2F^\circ_{7/2}$–$6d\ ^2D_{5/2}$) exhibit high gain. A peak output power of 40 W at 615.0 nm by discharging a 250-μF capacitor charged to 10 kV through a 15-mm-bore, 2.25-m-long discharge tube has been obtained by Bell.[7] The optimum pressures for the green line were about 0.001 torr of mercury and 0.5 torr of helium. The use of discharge tubes of differing bores showed that the gain was not dependent on the tube diameter.

A better form of discharge structure than the common axially pulsed glow-discharge tube for producing pulsed oscillation at 615.0 nm on the $7p\ ^2P^\circ_{3/2}$–$7s\ ^2S_{1/2}$ Hg-II transition has proved to be the metal, hollow-cathode discharge laser structure illustrated in Fig. 5.5, or that developed by Wieder et al.,[13] which consists of a metal, hollow cylindrical cathode and a pair of simple ring anodes (one at each end of and spaced away from the cathode) sealed into a glass enclosure. The considerable advantage with both of these discharge structures is that they can be operated as sources with wide numerical apertures as high gain can still be obtained with large tube diameters. Cathodes ranging from 2.5 to 7.5 cm in diameter and 7.5 to 25 cm in length have been used by Wieder et al.[13] The discharge in these structures fills the inside of the cathode and it is essentially a transverse discharge. It extends but a short distance beyond the cathode, and constitutes an extended negative-glow region of a glow discharge. The extent of the discharge region in the cathode depends on the discharge current, the inside surface area of the cathode and the gas pressure.

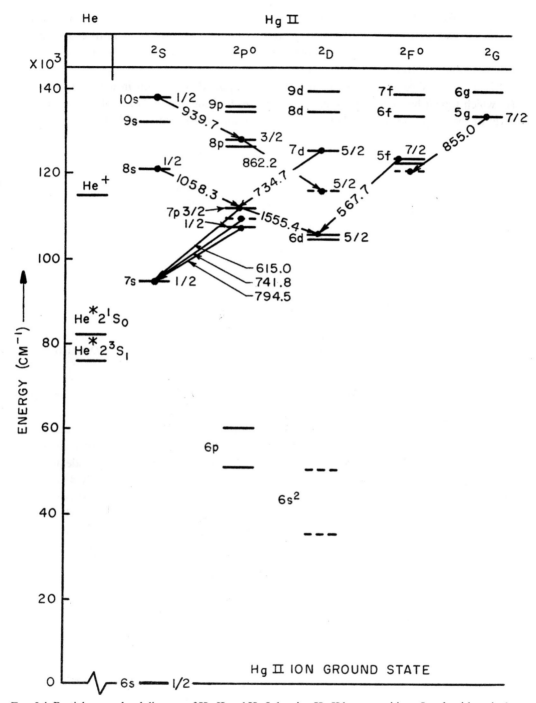

FIG. 5.4. Partial energy-level diagram of Hg-II and He-I showing Hg-II laser transitions. Levels with excited *d*-electrons, when included, are shown as dashed lines. Wavelengths are in nm.

The 615.0-nm He–Hg laser has proved to be different from noble-gas ion lasers. Unlike singly ionized noble-gas lasers, which operate at high current densities in discharge tubes of a few mm bore, the He–Hg 615.0-nm laser under pulsed conditions favors the use of discharge tubes with internal diameters of a few cm.[14]

Six and a half years after Bell's first report of pulsed oscillation in a He–Hg mixture (in which time it had begun to look as if processes limiting depopulation of the lower laser level of the 615.0-nm line made CW operation impossible), Schuebel[11] was successful in

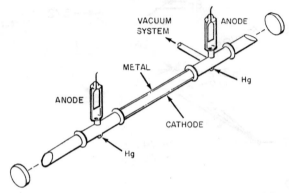

FIG. 5.5. Pulsed hollow-cathode laser structure. The cathode is a Kovar tube 2.5 cm in diameter and 30 cm long. (After R. L. Byer, W. E. Bell, E. Hodges, and A. L. Bloom,[10] by courtesy of the Optical Society of America.)

achieving continuous oscillation at 615.0 nm (and also 794.5 nm). The laser structure developed by Schuebel for CW oscillation is illustrated schematically in Fig. 5.6.[15] This structure is a particularly useful one for producing oscillation in metal vapors of elements such as Hg, Cd, Zn, and Se and elements that have low vapor pressures at room temperatures.[16–19] The description and important dimensions are as follows. The cathode and anode are made of Kovar (stainless-steel) tubes of 5.6 and 9.4 mm i.d., both are approximately 45 cm long. The cathode is slotted along its whole length with a slot about 2 mm wide. It is mounted inside the anode. The Kovar cathode can be suitably water-cooled. The elevated temperatures necessary to give the required vapor pressure of material placed inside the anode or a side-arm reservoir is maintained by heater tapes wrapped around the outside of the tube.

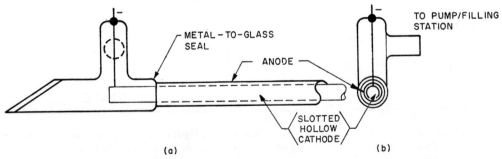

FIG. 5.6. Schematic diagram of slotted hollow-cathode laser tube. (a) Side view. (b) View along the tube axis. (After W. K. Schuebel,[15] by courtesy of the IEEE.)

194

The elevated temperature is also necessary to prevent condensation of metal vapor on the Brewster's angled windows. The device operates at low voltages of a few hundred volts; provides a uniform axial distribution of laser media; does not suffer from cataphoretic effects as do mixtures of gases in normal glow-discharge lasers; and is electrically less noisy than glow-discharge metal-vapor lasers.[16]

Continuous oscillation from the doublet Hg-II $7p\ ^2P^\circ_{3/2}$ and $7p\ ^2P^\circ_{1/2}$ levels to the $7s\ ^2S_{1/2}$ level at 615.0 and 794.5 nm respectively was obtained by Schuebel[11] at helium pressures from 4 to 9 torr, with the anode tube at temperatures of 60° to 80°C. This corresponds to a partial pressure of mercury of 0.02 to 0.1 torr in the hollow-cathode part of the discharge tube and is considered to be a lower limit because of a temperature differential between the inside and the water-cooled outside of the cathode. Threshold discharge currents under these conditions were 700 and 240 mA for the 615.0-nm and 794.5-nm lines, respectively, with no saturation exhibited up to the limit of the power supply of 1 A.

Two important points about this structure are:

(1) the smallness of the internal diameter of the cathode;
(2) the small radial separation between the cathode and the anode, in addition to the cathode being slotted.

Since an optimum product of pressure p and hollow-cathode diameter a appears to be a condition for maximizing the ionization, the small diameter of the cathode ensures that the gas pressure is high, and the excited-state populations are also correspondingly high. As shown in Chapter 3 (Section 3.4.6) in a HCD at a constant pa, the concentration of meta-stables (and probably other excited species and ions) is directly proportional to the pressure. This means that energy transfer reactions that rely on the presence of excited or ion species are enhanced in the HCD as compared with those in a normal glow-discharge tube. As the HCD is a negative-glow plasma, high-pressure discharges are readily maintained by the high-energy beam electrons that exist in this type of plasma.

The smallness of the radial separation between cathode and anode and their concentric configuration tends to suppress the formation of a discharge on the outside of the cathode; causes the anode fall to disappear; increases the cathode-fall potential,[20] and gives a good anode-collection efficiency for electrons that diffuse out of the cathode slot to maintain the discharge. The slotted cathode enables a plasma to be obtained that can be long and uni-form, and axially field-free. As this type of discharge structure is likely to prove important in the technological development of certain types of gas lasers (viz. charge-transfer-excited lasers), and in research on atomic-collisions processes, same attention is paid to it here.

Three mechanisms have been suggested as being responsible for the selective excitation of the singly ionized mercury doublet $7p\ ^2P^\circ_{3/2}$ and $7p\ ^2P^\circ_{1/2}$ that occurs in a discharge in helium–mercury mixture. Figure 5.4 shows that these are the upper states of the high-gain laser lines at 615.0 nm and 1.5554 μm, and 794.5 nm, respectively. The 615.0-nm and 794.5-nm lines have the same $7s\ ^2S_{1/2}$ lower state. Depending upon the type of discharge and the discharge geometry employed, the three mechanisms that may play a role in the selective excitation of the $7p\ ^2P_{3/2,\,1/2}$ doublet are: single, direct electron-impact excitation from the Hg-I ground state;[8, 9] charge transfer;[14, 21, 22] and a Penning-like reaction.[23] Each of these mechanisms will be considered separately. Bloom, Bell, and Lopez,[8] using a 15-mm-bore discharge tube, and a pulsing network that gave peak pulse currents of 10 to 50 A with a duration of about 3 μsec, and helium pressures of 0.8 to 1.2 torr and a mercury partial pressure of approximately 0.01 torr, observed that a number of the Hg-II laser lines exhibited a hollow-beam oscillation behavior. This phenomenon is displayed in the frontispiece. When minor changes are made to the excitation-pulse risetime and peak current, significant changes occur in the spatial variation of the laser power output and wavelength of the laser emission across the laser beam. At low pulsed input energy, oscillation occurs in a central spot on the orange 615.0-nm line, and at a higher input energy on the green 567.7-nm line. At an intermediate input energy, oscillation occurs on the green line in a central spot and is surrounded by an annulus of oscillation on the orange line. In addition to this, when the 615.0-nm line oscillates in an annular mode the 1.5554-μm line ($7p\ ^2P^\circ_{3/2}$–$6d$ $^2D_{5/2}$ transition), which, as can be seen from Fig. 5.4 shares the same upper state as the 615.0-nm line and at the same time has a lower state that is common to the green 567.7-nm line ($5f\ ^2F^\circ_{7/2}$–$6d\ ^2D_{5/2}$), also oscillates in an annulus, while the 567.7-nm line continues to oscillate in the central spot. Further, while the 734.7-nm and 1.0583-μm lines, which operate in cascade into the $7p\ ^2P^\circ_{3/2}$ upper state of the 615.0-nm orange line, lase in the central spot, the 615.0-nm line continues to oscillate in a "mode" that is annular in cross-section. This behavior should not be confused with the output-mode characteristics of low-gain laser lines in which the cavity can determine the output profile of the laser emission. The annular mode observed here is not a TEM_{02} mode determined by any cavity properties.

To account for this behavior, Bloom, Bell, and Lopez proposed that the process involved in the selective excitation of the upper states of a number of the Hg-II laser lines was one of direct electron-impact excitation from the Hg-I ground state. Since the green laser line at 567.7 nm and other infrared laser lines (with the exception of the 1.5554-μm line), which have upper states that are higher in potential energy than that of the 615.0-nm line, would only oscillate in the central spot while the orange line continued to oscillate with an annular output profile, it was presumed that electrons near the walls of the discharge tube are less energetic that those in the center, and that the population of excited ion states is a fairly sensitive function of electron energy.

The two other mechanisms of charge transfer and Penning-like collisions that could play a role in selective excitation of the Hg-II $7p\ ^2P^\circ_{3/2,\,1/2}$ doublet in a discharge in a helium–mercury mixture may be written, respectively, as

$$He^+\ ^2S_{1/2}+Hg\ ^1S_0 \rightarrow He\ ^1S_0+Hg^{+\prime}\ 7p\ ^2P^\circ_{2/2,\,1/2}+\Delta E_\infty\ (0.26,\ 0.72\ eV), \quad (5.8)$$

and

$$He^*\ 2^3S_1+Hg^*\ 6p\ ^3P_0 \rightarrow He\ ^1S_0+Hg^{+\prime}\ 7p\ ^2P^\circ_{3/2,\,1/2}+e^-\ (0.16,\ 0.62\ eV). \quad (5.9)$$

The energy differences in parentheses (5.8) and (5.9) refer to the excited Hg-II doublet states $7p\ ^2P^\circ_{3/2}$ and $7p\ ^2P^\circ_{1/2}$, respectively. In the Penning-like reaction (5.9), a reaction considered earlier by Paschen[24] and Takahashi[25] in the late 1920s, a collision occurs between two excited and metastable species of helium and mercury to give simultaneous excitation and ionization of the mercury atom and de-excitation of the helium atom to the neutral ground state. The energy differences involved in both reactions are positive, and so allow both to occur effectively in either a dc-discharge, or in the afterglow of a pulsed discharge when ion

kinetic energies are low. Figure 5.7 illustrates the energy-level coincidences that occur between excited states of neutral mercury and the He^* 2^3S_1 and He^* 2^1S_0 metastable states,
and shows their relationship to the upper states of the singly ionized mercury laser lines at
615.0, 794.5 nm, 1.5554 μm, and 567.7 nm. The dashed lines indicate the He^* 2^3S_1 level, and
the path of the excitation transfer reaction He^* $2^3S_1 + Hg^*$ $6p$ $^2P_0^\circ$. The broken lines show
those of other Penning-like reactions involving singlet helium metastables and the two

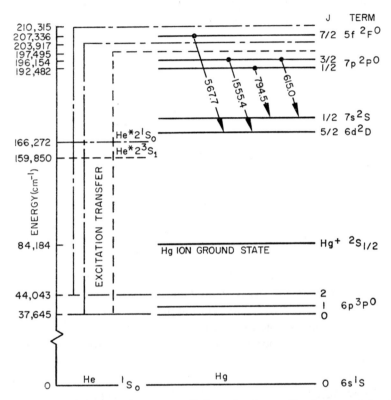

FIG. 5.7. Partial energy-level diagram of mercury and helium showing combined energy relationships between
the He^* 2^3S_1 and He^* 2^1S_0 and the metastable Hg-I $6p$ $^3P_0^\circ$ and $6p$ $^3P_2^\circ$ levels and the excited Hg-II $7p$ $^2P_{3/2, 1/2}^\circ$
and $5f$ $^2F_{7/2}^\circ$ upper laser levels involved in the Penning-like reaction (5.9).

$6p$ $^3P_0^\circ$ and $6p$ $^3P_2^\circ$ metastable states of mercury. The singly ionized helium ground state, not
shown in the figure, at 198,305 cm^{-1} above the gound state, would be situated in energy
above that of the He^* $2^3S_1 + Hg^*$ $6p$ $^3P_0^*$ reaction, and lower than the reactions involving
He^* 2^1S_0 metastables and metastable mercury atoms. The asymmetric charge transfer
reaction (5.8) and both Penning-like reactions with the Hg^* $6p$ $^3P_0^\circ$ metastable state are
energetically suited for the selective excitation of the $7p$ $^2P_{3/2, 1/2}^\circ$ doublet. The He^* 2^1S_0
$+ Hg^*$ $6p$ $^3P_2^\circ$ reaction is the only reaction that is, or likely to be, energetically capable of
selectively exciting the higher-lying $5f$ $^2F_{7/2}^\circ$ upper state of the high-gain, green 567.7-nm
line.

Using a pulsed, positive-column discharge in a mixture of helium and mercury vapor,
Dyson[14] has shown conclusively that the only selective excitation process that can explain

the afterglow decay of spontaneous emission of the 615.0-nm Hg-II laser line in his discharge is the asymmetric charge-transfer reaction (5.8).

Figure 5.8 shows the spontaneous-emission intensity as a function of time, of some mercury lines from a helium–mercury vapor mixture in a 2.3-cm-bore, 1-m-long tube following the discharge through it of a capacitor of a few hundred picofarads charged to 5–10 kV. (These conditions are favorable for laser oscillation at 615.0 nm.) The intensity of the 546.1-nm Hg-I line shows a time variation typical of the majority of Hg-I lines; the 567.7-nm line has a time variation typical of the majority of the singly ionized mercury lines; whereas

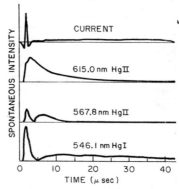

FIG. 5.8. Spontaneous-emission pulse-shapes of Hg-II and Hg-I lines from a pulsed helium–mercury discharge compared with the discharge current pulse. (After D. J. Dyson,[14] by courtesy of *Nature*.)

the 615.0-nm line, and others resulting from cascade transitions from the lower level of the 615.0-nm line, exhibit a smooth decay of spontaneous emission following an initial rapid rise in intensity during the relatively short current pulse. Dyson found that the decay rate of the 615.0-nm spontaneous emission was sensitive to the temperature of the discharge tube that determined the mercury vapor pressure, but was not sensitive to the partial pressure of helium. He also found that the decay rate of the 615.0-nm line was directly proportional to the neutral mercury ground-state population, and that it was not associated with excited mercury species produced in the discharge. The only excitation process energetically capable of selectively exciting the $7p\ ^2P^{\circ}_{3/2}$ level, consistent with the observations, was concluded to be the asymmetric charge-transfer reaction (5.8).

The cross-section derived by Dyson for this reaction is the very large one of 1.3×10^{-14} cm^2, and as such is almost an order of magnitude larger than that of 1.4×10^{-15} cm^2 for the corresponding Penning ionization cross-sections between He* 2^3S_1 and He* 2^1S_0 metastables and ground-state mercury atoms (Table 2.2).

Also using a pulsed positive-column discharge in a mixture of helium and mercury vapor together with a time-resolved gating technique, Suzuki[23] has shown that the intensity of the spontaneous emission of the Hg-II 615.0-nm laser line was proportional to the product of the concentrations of metastable He* 2^3S_1 atoms and metastable Hg* $6p\ ^3P^{\circ}_0$ atoms, and the decay rate of the 615.0-nm afterglow was proportional to the product of the decay rate of the metastable states of helium and mercury in the afterglow of the pulsed discharge he investigated. Suzuki concluded that selective excitation process of the upper state of the 615.0-nm laser line was via the Penning-like reaction (5.9).

198

The conclusions of Dyson and Suzuki are not necessarily contradictory.[21] Dyson investigated discharges in wide-bore discharge tubes of 2.3-cm bore and Suzuki used narrow-bore discharge tubes of 6-mm bore, and differences in excitation conditions could account for the seemingly contradictory conclusions regarding the principal selective-excitation mechanisms in their 615.0-nm helium–mercury lasers. Another possibility exists that the intensity of the 615.0-nm line in spontaneous emission in the observations of Suzuki is distorted due to optical gain in the long, 1-m tube that he used. Whereas Dyson observed only side-light emission from a region where optical gain should not affect the intensity, Suzuki apparently observed end-on emission along the axis of the discharge tube where distortion could occur.

Additional support for the conclusion that the principal selective excitation of the $7p\,^2P^\circ_{3/2}$ Hg-II state in a helium–mercury discharge in asymmetric charge transfer involving the He$^+$ ion, is provided by the results of a simple, elegant experiment by Aleinikov.[26] Aleinikov used an experimental technique that is essentially the same as that used in studying excitation functions of spectral lines. He measured the changes in the relative intensities and shapes of the ionization functions of Hg-II lines as helium, neon or argon was added to the electron-gun chamber of his apparatus. All the experiments were performed at a mercury vapor pressure of 0.002 torr.

Figure 5.9 shows the dependence of the relative intensities of the mercury lines on the helium pressure in the electron gun. The relative intensity of the singly ionized mercury lines at 615.0, 284.7, and 398.4 nm in the cascade sequence $7p\,^2P^\circ_{3/2} \to 7s\,^2S_{1/2} \to 6p\,^2P^\circ_{3/2} \to 6s^2\,^2D_{5/2}$ from the upper level of the 615.0-nm line is seen to increase linearly with helium pressure, and has increased by two orders of magnitude when the pressure reaches 0.1 to 0.2 torr. The increase in intensity of the Hg-II lines became detectable even with the addition of extremely small amounts of helium (about 0.005 torr) to the mercury vapor, while the addition of neon or argon (0.001–0.01 torr) to the electron-gun chamber did not cause any significant changes in the relative intensities of Hg-I or Hg-II lines. For helium pressures in

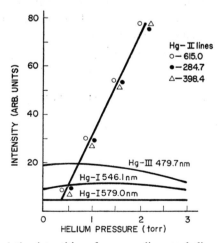

FIG. 5.9. Dependence of the relative intensities of mercury lines on helium pressure at a mercury vapor pressure of 0.002 torr in an electron-gun collision chamber. (After V. S. Aleinikov,[26] by courtesy of the Optical Society of America.)

excess of 0.01 to 0.015 torr the intensities of the 615.0-nm and the 284.7-nm Hg-II lines were clearly determined by collisions of the second kind with helium and were no longer determined by electron impact excitation from the Hg-I ground state. The threshold for excitation of the 615.0-nm line in a helium–mercury mixture was found to be close to 24.5 eV, and therefore in the vicinity of the ionization energy of the He^+ ion. Aleinikov[26] found that the excitation functions of the 615.0-nm lines (and the 284.7-nm and 398.4-nm lines in cascade from it) have practically the same shape as the curve representing the dependence of the helium ionization cross-section on electron energy[27] as illustrated in Fig. 5.10. Illustrated

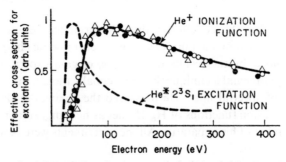

Fig. 5.10. Excitation function of Hg-II lines in the cascade $7p\ ^2P^\circ_{3/2}-7s\ ^2S_{1/2}$ (at 1615.01 nm); $-7s\ ^2S_{1/2}-6p\ ^2P^\circ_{3/2}$ (at 284.7 nm); and $6p\ ^2P^\circ_{4/2}-6s^2\ ^2D_{5/2}$ (at 398.4 nm) in He–Hg mixtures, ionization function of helium and excitation function of the $He^*\ 2^3S_1$ metastable state. The symbols denote experimental points obtained in determining the ionization functions of the 615.0, 284.7, and 398.4-nm Hg-II lines in He–Hg mixtures. (After V. S. Aleinikov,[26] by courtesy of the Optical Society of America.)

also in Fig. 5.10 (broken curve) is the excitation function for $He^*\ 2^3S_1$ metastables.[28] If the upper state of the 615.0-nm line was being excited in the Penning-like collision process (5.9) suggested by Paschen[24] and investigated by Suzuki[23] the product of the excitation functions of the $He^*\ 2^3S_1$ and $Hg^*\ 6p\ ^3P_0$ metastables would be expected to decrease sharply after reaching the 30–40-eV range (the maximum of the function describing the production of $He^*\ 2^3S_1$ metastables). This was not observed in the experiments indicating that the Penning-like reaction (5.9) was not a significant process.

These results show conclusively that the $7p\ ^2P^\circ_{3/2}$ state is selectively excited in the asymmetric charge transfer reaction

$$He^+\ ^2S_{1/2}(\uparrow)+Hg\ ^1S_0(\uparrow\downarrow) \rightarrow He\ ^1S_0(\uparrow\downarrow)+Hg^{+\prime}\ 7p\ ^2P^\circ_{3/2}(\uparrow)+\Delta E_\infty\ (0.26\ eV). \qquad (5.10)$$

This reaction is one in which the Wigner spin rule is obeyed with a spin of 1/2 conserved in the collision.

Aleinikov[26] does not give results for the 794.5-nm Hg-II line $(7p\ ^2P^\circ_{1/2}-7s\ ^2S_{1/2})$ so that it is not possible to conclude whether or not the $7p\ ^2P^\circ_{1/2}$ state, doublet-state to the upper level of the 615.0-nm line, is also selectively excited via an asymmetric charge-transfer reaction with the He^+ ion in which the ΔE_∞ is large, as charge-transfer reactions go at thermal energies, at 0.72 eV. If the $7p\ ^2P^{1/2}$ state is not selectively excited directly via charge transfer as suggested by Willett,[21] it is possibly excited in the buffering reaction

$$Hg^{+\prime}\ 7p\ ^2P^\circ_{3/2}+He \rightarrow Hg^{+\prime}\ 7p\ ^2P_{1/2}+He+kinetic\ energy \qquad (5.11)$$

proposed by Goldsborough and Bloom,[9] or in the collisional-mixing reaction

$$Hg^{+\prime}\ 7p\ ^2P^{\circ}_{3/2} + e^- \rightarrow Hg^{+\prime}\ 7p\ ^2P^{\circ}_{1/2} + e^- + \text{kinetic energy} \qquad (5.12)$$

suggested by Aleinikov.[26] This reaction could be significant at high electron and ion concentrations in high-pressure pulsed[9] and CW hollow-cathode discharge lasers.[11]

Even though calculations show that there is a significant probability for simultaneous ionization and excitation of mercury by direct electron impact from the Hg-I ground state,[29] that a selective excitation process of the $7p\ ^2P^{\circ}_{3/2, 1/2}$ doublet involving ionized or excited species of helium is the dominant selective excitation process in a discharge in a helium–mercury mixture is strongly indicated by the results of Dyson[14] and Aleinikov[26] and the following:

(1) high operating pressure of helium in helium–mercury ion lasers (of a few to tens of torr);[9, 11, 14, 23]

(2) the requirement for helium (oscillation from the $7p\ ^2P^{\circ}$ doublet has not been reported for any mixture other than a helium–mercury mixture);

(3) oscillation at 615.0 nm in the afterglow of a current pulse for times considerably longer than electron-thermalization times, decay-times of radiative-cascade processes into the upper laser level, and mildness of the excitation pulse of a few amperes necessary for oscillation;[8, 13, 14]

(4) preference for oscillation in wide-bore discharge tubes, unlike noble-gas ion lasers in singly ionized species[14] that are excited by electron impact and which work best using small-bore tubes;

(5) high gain at 615.0 nm, and absence of oscillation on the 567.7-nm line that is easily obtainable in a normal pulsed discharge from singly ionized levels other than the Hg-II $7p\ ^3P_{3/2, 1/2}$ doublet levels and the $5d^9\ ^2P^{\circ}_{3/2}$ level in pulsed hollow-cathode discharges at high helium pressures,[9, 10] where the electron energy distribution would not be expected to differ appreciably from a pulsed glow discharge under the same conditions; and

(6) a spontaneous emission linewidth at 615.0 nm appropriate to a gas temperature of only 370 K that is considerably smaller than that of noble gas ion lasers excited by electron impact (1850 K for Xe^+ or 3000 K for Ar^+).[30]

As high positive-ion concentrations occur in hollow-cathode discharges in what are essentially strongly abnormal negative glow discharges, it seems entirely in accord with the above observations to attribute the dominant selective-excitation mechanism of the Hg-II $7p\ ^2P^{\circ}_{3/2, 1/2}$ doublet level in a helium–mercury mixture in hollow-cathode discharges to be that of asymmetric charge transfer involving the ground-state He^+ ion.

The excitation and ionization process that is responsible for the observation of strong oscillation on the Hg-II, $5f\ ^2F^{\circ}_{7/2}$–$6d\ ^2D_{5/2}$ transition at 567.7 nm in pulsed positive-column discharges in helium–mercury mixtures at pressures of 0.001 torr of mercury with a trace to 0.5 torr of helium is not known. The disposition in energy of the $5f\ ^2F^{\circ}_{7/2}$ level and the He^+ ion state, together with the observation that oscillation at 567.7 nm also occurs, in what must be assumed were helium-free Ar–Hg, and Ne–Ar mixtures, certainly precludes charge transfer as being the selective excitation mechanism.

Likewise, it is not known what mechanism (or mechanisms) is responsible for selective excitation of the $5d^8\,6s^2$ ($J = 4$) state, upper state of the Hg-III laser line at 479.7 nm that is observed in pulsed helium–mercury discharges at low partial pressures of helium.[12] This line can be as intense as singly ionized mercury lines in a high-frequency discharge in mercury at low pressure,[31] and thus it appears likely that the excitation mechanism of its upper state is direct electron-impact excitation from the Hg-I ground state.

As mentioned earlier, the cross-section for the asymmetric charge-transfer reaction (5.8) that produces selective excitation of the $7p\,^2P^\circ_{3/2}$ level in mercury–helium mixtures is a relatively large one in excess of 10^{-14} cm^2 (Dyson[14] and Cotman et al.[32]).

5.2.2. THE HELIUM–CADMIUM LASER

Laser oscillation in cadmium under pulsed discharge conditions in cadmium–helium and cadmium–neon mixtures was first reported by Fowles and Silfvast[33] in 1965. Oscillation occurred in the visible spectra of singly ionized cadmium on the doublet transitions $4f\,^2F^\circ_{5/2}-5d\,^2D^\circ_{3/2}$ and $4f\,^2F^\circ_{7/2}-5d\,^2D_{5/2}$ at 533.7 nm and 537.8 nm, respectively. Subsequently pulsed laser oscillation in cadmium–helium mixtures only, was observed at 441.6 nm in Cd-II by Silfvast, Fowles and Hopkins.[34] Strong oscillation occurred at cadmium vapor pressures of 0.001 to 0.01 torr and helium pressures of 1 to 2 torr in a 1-m-long discharge tube of 6-mm bore. The pulsed discharge was obtained with a 100-mA, 12-kV neon-sign transformer paralleled with a 0.003-μF capacitor. A spark gap in series with the discharge tube controlled the breakdown voltage across the tube.

Rather surprisingly, this report of oscillation in a mixture of cadmium and a rare gas (helium) at pressures that were indicative of a selective excitation mechanism occurring that had a cross-section comparable to excitation transfer reactions in the He–Ne laser, did not appear to arouse any noticeable interest. Whereas the presence of helium was essential for operation of the Cd-II line at 441.6 nm, examination of possible energy-level coincidences between He$^*\,2^3S_1$ and He$^*\,2^1S_0$ metastables or the He$^+$ ion ground state, and the $5s^2\,^2D_{5/2}$ upper state of the 441.6-nm Cd-II laser line, revealed that the energy discrepancies ΔE_∞ of more than $+6$ eV involved in both the Penning-like reaction

$$\text{He}^* + \text{Cd}^* \rightarrow \text{He} + \text{Cd}^{+\prime} + (e^- + \text{K.E.}) \tag{5.13}$$

and the asymmetric charge-transfer reaction

$$\text{He}^+ + \text{Cd} \rightarrow \text{He} + \text{Cd}^{+\prime} + \Delta E_\infty \tag{5.14}$$

were too large to expect these reactions to be of significance[35] unless cascading into the $5s^2\,^2D_{5/2}$ level was occurring following some other selective-excitation mechanism. However, extensive cascading from higher-lying ion levels did not appear to be occurring as oscillation at other (longer) wavelengths, in addition to that at 533.7 and 537.8 nm, would have been expected. The observation that oscillation occurred at 533.7 and 537.8 nm in the presence of neon, in what were presumably helium-free discharges, further complicated the search for the selective excitation mechanism that was occurring. In the light of these observations, the only reasonable mechanism that could be invoked was that of electron-impact excitation. At the time of the report of oscillation at 441.6 nm in Cd-II by Silfvast et al.[34] it was not

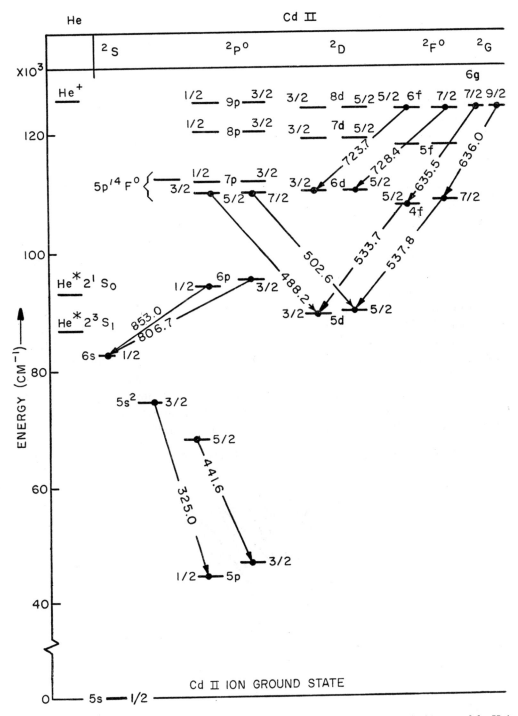

FIG. 5.11. Partial energy-level diagram of Cd-II and He-I, illustrating energy-level coincidences of the He* 2^3S_1 and He* 2^1S_0 metastable levels and the He$^+$ ion ground state with excited Cd-II states, and including identified Cd-II laser transitions (wavelengths in nm).

appreciated that of the Penning reactions

$$A^* + B \Big\langle \begin{array}{l} \longrightarrow A + B^+ + e^- + \text{K.E.}, \qquad (5.15a) \\[2ex] \longrightarrow A + B^{+\prime} + e^- + \text{K.E.}, \qquad (5.15b) \end{array}$$

reaction (5.15b), in which $B^{+\prime}$ is an excited state of the ion (and B^+ in (5.15a) is an ion in the ground state), could be a significant selective-excitation reaction for the $5s^2\ ^2D_{5/2}$ state. And it was considered that the Penning reaction occurred through channel (5.15a) alone to produce ground-state ions of B, and not excitation as well as ionization as in (5.15b). In all fairness to earlier workers, it must be mentioned that such a possibility had been considered and proposed to explain certain spectrum-line enhancements that had been observed in negative-glow discharges in cadmium–helium mixtures.[25, 36, 37]

It was not until low-current, quasi-CW laser operation for periods of about 5 msec was observed on the 441.6-nm Cd-II line at helium pressures of about 4 torr in a 5-mm-bore discharge tube by Fowles and Hopkins,[38] and Hopkins,[39] and true CW operation by Silfvast,[40] at 0.02 percent efficiencies comparable to that of typical of He–Ne lasers, that extensive interest was shown in the (helium) cadmium ion laser.

The exact role that helium plays in exciting Cd-II levels from which laser oscillation has been reported in He–Cd mixtures is now known to be both by asymmetric charge transfer and the Penning reaction.[41, 42] Figure 5.11, a partial energy-level diagram of Cd-II and He-I adapted from Bridges and Chester,[43] shows the energy-level coincidences between the $He^*\ 2^3S_1$, $He^*\ 2^1S_0$ metastable levels; and the He^+ ion ground-state, and various Cd-II states from which laser oscillation occurs at the wavelengths indicated. The potential energy of the helium levels shown are measured from the Cd-II ground state which is 72,540 cm^{-1} above the Cd-I and He-I ground states. Asymmetric charge transfer of the type

$$He^+\ ^2S_{1/2}(\uparrow) + Cd\ ^1S_0(\uparrow\downarrow) \rightarrow He\ ^1S_0(\uparrow\downarrow) + Cd^{+\prime}(\uparrow) + \Delta E_\infty \ (512 \text{ to } 2045 \text{ cm}^{-1}) \quad (5.16)$$

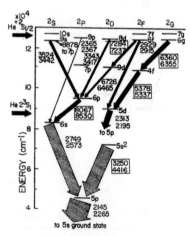

FIG. 5.12. Partial energy-level diagram of Cd-II. Transitions excited by charge transfer from He* ions are indicated by solid arrows. Transitions excited by Penning reactions are indicated by cross-hatched arrows. Boxed wavelengths (in Å) are CW laser transitions. (After C. E. Webb, A. R. Turner-Smith, and J. M. Green,[41] by courtesy of the Institute of Physics.)

provides a majority of the excitation to Cd-II doublet terms with the electronic configurations $4d^{10} 6g$, $4d^{10} 6f$, $4d^{10} 8d$, $4d^{10} 9p$, and $4d^{10} 9s$ in which only the outer electron is excited.[41, 42] The selective excitation gives rise to CW laser transitons at 635.5, 636.0 nm $(4d^{10} 6g)$; and 723.7, 728.4 nm $(4d^{10} 6f)$, and to the observation of other Cd-II lines strong in spontaneous emission in helium–cadmium discharges. Figure 5.12 (after Webb *et al.*[41]) shows graphically the dominant paths for laser, non-laser, and cascade transitions following selective excitation of the 9s, 9p, 8d, 6f, and 6g states by asymmetric charge transfer from the He^+ $^2S_{1/2}$ ground-state ion. The photon-emission rate, derived from observed line intensities in a flowing afterglow of cadmium and helium is indicated on a crude logarithmic scale by the width of the arrow connecting initial and final levels on the partial term diagram. Included also in the figure are those transitions selectively excited by Penning reactions. Transitions excited by charge transfer are indicated by solid arrows; those excited by Penning reactions by the cross-hatched arrows. Discrimination between Penning reaction and charge-transfer excitation was made by applying a pulsed rf-field to the afterglow

TABLE 5.3. *CW laser transitions in Cd-II in a slotted hollow-cathode discharge structure that are excited by charge transfer*[a]

Wavelength (nm)	Transition	ΔE_∞ (cm^{-1})*
728.4	$6f\ ^2F^\circ_{7/2}-6d\ ^2D^\circ_{5/2}$	1794
723.7	$6f\ ^2F^\circ_{5/2}-6d\ ^2D^\circ_{3/2}$	1778
636.0	$6g\ ^2G^\circ_{9/2}-4f\ ^2F^\circ_{7/2}$	1615
537.8[b]	$4f\ ^2F^\circ_{7/2}-5d\ ^2D^\circ_{5/2}$	†
635.4	$6g\ ^2G^\circ_{7/2}-4f\ ^2F^\circ_{5/2}$	1615
533.8[b]	$4f\ ^2F^\circ_{5/2}-5d\ ^2D^\circ_{3/2}$	†

[a] In order of increasing threshold current, observed in a 45-cm long, 5.6-mm-i.d. cathode, over a pressure range of helium 3 to 8 torr, and anode temperature of 260° to 340°C; optimum helium pressure of 5 torr, anode temperature of 290° to 310°C at a discharge current of 400 to 480 mA. Threshold currents are between 130 and 180 mA, with the threshold current increasing in the order in which the lines are given.[17]

[b] Weak oscillation on this line has been observed in a Ne–Cd mixture.[33, 95] Although the $-\Delta E_\infty$ is large, selective excitation is possibly by the asymmetric charge-transfer reaction

$$Ne^+ + Cd \rightarrow Ne + Cd^{+\prime}\ 4f\ ^2F^\circ_{7/2} - \Delta E_\infty\ (7009\ cm^{-1}),$$

or by cascading from $Cd^{+\prime}$ levels excited in the reaction

$$Ne^+ + Cd^* \rightarrow Ne + Cd^{+\prime} + \Delta E_\infty.$$

* ΔE_∞ is the difference in potential energy between the He ionization energy and the appropriate excited ion species of Cd-II.

† Indicates cascade transitions.

205

in the flow region upstream of the helium–cadmium reaction zone. The He* 2^3S_1 metastable density is unaffected by the rf-field applied to a pulsed microwave sustaining discharge that provides the excited atoms and ions, whereas during the rf-pulse the He$^+$ $^2S_{1/2}$ ion density is reduced. Emission from transitions excited by charge transfer then shows a characteristic notch, corresponding to the reduction in ion density, while emission from transitions excited by Penning or Penning-type reactions appears as a sequence of identical pulses.

Webb et al.[41] noted that any correlation between cross-section and the energy discrepancies ΔE_∞, which vary between about 0.06 and 0.25 eV in the asymmetric charge-transfer reaction (5.16), is conspicuously absent. The total velocity-averaged cross-section of thermal energy He$^+$ $^2S_{1/2}$ ions by cadmium ground-state atoms in (5.16) has been determined by Collins et al.[42] to be the relatively large one of 3.7×10^{-15} cm^2.

Information on characteristics of the helium–cadmium ion laser operating under discharge conditions *where charge-transfer excitation of Cd-II levels predominates* is limited to that given by Schuebel[15, 17, 18] on slotted, hollow-cathode-discharge structures in which CW operation in Cd-II was first achieved,[15] and by Sugawara et al.[95] on another type of hollow-cathode-discharge structure that requires high discharge currents. This data is included as a footnote to Table 5.3.

5.2.3. THE HELIUM–ZINC AND HELIUM–NEON–ZINC LASERS

5.2.3.1. The helium–zinc laser

Fowles and Silfvast[33] were the first again to report laser oscillation in the ionic spectra of zinc using pulsed discharges in zinc vapor mixed with helium. Oscillation was realized in the visible on the $4f\ ^2F^\circ_{7/2}$–$4d\ ^2D_{5/2}$ transition at 492.4 nm, and on the $6s\ ^2S_{1/2}$–$5p\ ^2P^\circ_{3/2}$ transition at 775.7 nm using the same apparatus in which pulsed oscillation in Cd-II was observed. Whereas oscillation could be obtained in Cd-II in the presence of helium or neon carrier gases, oscillation in zinc occurred only in the presence of helium, at a pressure of 1 to 2 torr with zinc at a vapor pressure of 0.001 to 0.1 torr (in a 5-mm-bore discharge tube).

Possible excitation mechanisms that could have been involved in the selective excitation of the $4f\ ^2F^\circ_{7/2}$ state, upper level of the 492.4-nm Zn-II line in the presence of helium include:

(1) the asymmetric thermal-energy, charge-transfer reaction

$$\text{He}^+\ ^2S_{1/2}(\uparrow) + \text{Zn}\ ^1S_0(\uparrow\downarrow) \rightarrow \text{He}\ ^1S_0(\uparrow\downarrow) + \text{Zn}^{+\prime}\ 4f\ ^2F^\circ_{7/2}(\uparrow) + \Delta E_\infty\ (5275\ \text{cm}^{-1}) \quad (5.17)$$

in which spin is conserved; and

(2) the metastable-metastable Penning-like reaction

$$\text{He}^*\ 2^3S_1 + \text{Zn}^*\ 4p\ ^3P^\circ_0 \rightarrow \text{He}\ ^1S_0 + \text{Zn}^{+\prime}\ 4f\ ^2F^\circ_{7/2} - \Delta E_\infty\ (869\ \text{cm}^{-1}). \quad (5.18)$$

Although an energy discrepancy of 5275 cm^{-1} (or about 0.63 eV) seemed in 1965 too large to ascribe to a reaction that clearly had a large cross-section, earlier published work in 1964 by Fowles and Jensen[44] on the helium–iodine ion laser indicated large cross-sections occurred with asymmetric charge-transfer reactions involving energy discrepancies of 0.23 to 0.51 eV, reaction (5.17) appeared to be a more likely reaction than the endothermic reaction (5.18) to account for the selective excitation of the $4f$ state.[35]

It has now been established that laser oscillation from the $3d^{10}\,4f$ state and the $4d^{10}\,5d$ and $4d^{10}\,6s$ states which occurs in the presence of helium[18, 34, 45, 46] is due to selective excitation by asymmetric charge transfer, the respective energy discrepancies ΔE_∞ for the $5d$ and $6s$ states being 4545 and 8039 cm^{-1} (nearly 1 eV).[19, 40, 46, 48] Figure 5.13, a partial-

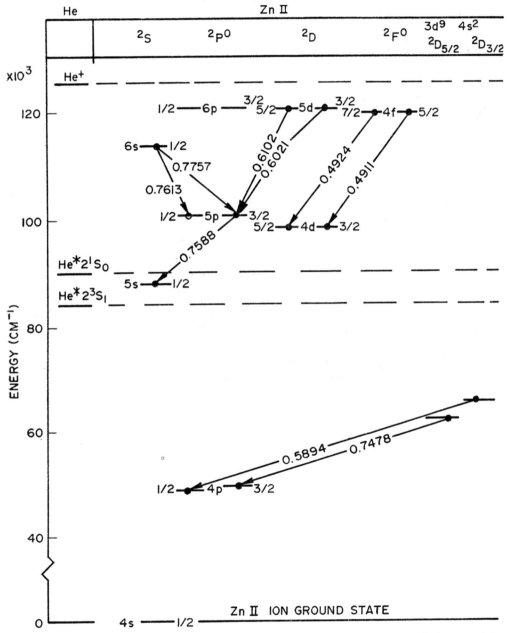

FIG. 5.13. Partial energy-level diagram of Zn-II and He-I illustrating the energy-level coincidences of the He* 2^3S_1 and He* 2^1S_0 metastable states and the He$^+$ ion ground state with excited Zn-II states, and including Zn-II laser transitions (wavelengths in μm).

energy-level diagram of Zn-II and He-I, shows laser transitions that have been reported in Zn-II, and illustrates the energy-level coincidences between the He$^+$ $^2S_{1/2}$ ground state and the upper-lying Zn-II, $3d^{10}$ $4f$, $5d$, $6p$, and $6s$ states. Also shown is the disposition of the He* 2^3S_1 and He* 2^1S_0 metastable states with respect to the charge-transfer excited Zn-II lines and two other Zn-II CW laser lines at 589.4 nm and 747.8 nm. These two latter lines, though once ascribed to charge-transfer excitation from molecular He$_2^+$ ions,[46] are selectively excited by the Penning reaction[19, 41, 48] (as discussed in Section 5.3.2). Population inversion on the $5p$ $^2P_{3/2}^\circ$–$5s$ $^2S_{1/2}$ transition at 758.8 nm occurs as a result of cascading transitions from the well-populated $6s$ $^2S_{1/2}$ and $5d$ $^2D_{5/2, 3/2}$ states.

The cross-section for thermal-energy, asymmetric charge transfer to the upper-lying Zn-II states from the He$^+$ $^2S_{1/2}$ ground state has been found to be 2.1×10^{-15} cm^2 (Jensen et al.[46] and Riseberg and Schearer[48]).

As in helium–cadmium ion lasers that utilize hollow-cathode-discharge excitation[17] charge-transfer selective excitation in the helium–zinc laser predominates in hollow-cathode-discharge laser structures.[19] Also as with the helium–cadmium ion laser, published output characteristics of charge-transfer excited Zn-II lines as a function of helium pressure, zinc temperature or vapor pressure, and discharge current in either positive-column or hollow-cathode discharges, are very limited. What there are, are more conveniently given in Section 5.3 on Penning-reaction lasers.

5.2.3.2. The helium–neon–zinc laser

By utilizing He–Ne–Zn mixtures in a dc, ion-drift "cataphoresis" discharge tube similar to that illustrated in a later figure (Fig. 5.24a), Collins[49] obtained CW laser oscillation at 632.8 nm and 3.39 μm from the $3s_2$ level of Ne-I (Paschen notation) as well as on the

TABLE 5.4. *CW laser transitions in the helium–neon–zinc laser*

Wavelength (nm)	Upper state classification	Atomic species	Excitation mechanism
492.4	$3d^{10}$ $4f$ $^2F_{7/2}$	Zn-II	Charge transfer with He$^+$ $^2S_{1/2}$ ions
589.4	$3d^9$ $4s^2$ $^2D_{3/2}$	Zn-II	Penning reaction with He* 2^3S_1 metastables
747.8	$3d^9$ $4s^2$ $^2D_{5/2}$	Zn-II	Penning reaction with He* 2^3S_1 metastables
632.8	$3s_2$	Ne-I	Excitation transfer with He* 2^1S_0 metastables
3391.3	$3s_2$	Ne-I	Excitation transfer with He* 2^1S_0 metastables

Zn-II lines listed in Table 5.4. The operating conditions and output characteristics for the red 632.8-nm Ne-I laser line, and the yellow 589.4-nm and blue 492.4-nm lines are given in Fig. 5.14. For discharge currents below 10 mA, more than 5-mW output power is realized on the 632.8-nm Ne-I line. Oscillation ceased entirely at intermediate current levels, but at

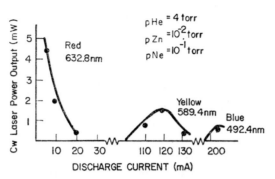

FIG. 5.14. Output characteristics of the He–Ne–Zn laser as a function of discharge current. The optimum helium, neon, and zinc pressures are indicated, $D = 3$ mm. (After C. J. Collins,[49] by courtesy of the American Institute of Physics.)

100 mA the yellow 589.4-nm Zn-II line reaches threshold and continues to oscillate until the current exceeds 150 mA. At 200 mA the 492.4-nm Zn-II begins to oscillate.

This three-stage behavior illustrates the current ranges in which, respectively, atom–atom excitation from $He^* \, 2^1S_0$ metastables, Penning excitation from $He^* \, 2^3S_1$ metastables, and charge transfer from the $He^+ \, {}^2S_{1/2}$ ion dominates the selective excitation of the appropriate upper laser levels listed in Table 5.4. Collins[49] found, however, that the output power from the 492.4-nm Zn-II transition was very unstable, and when the neon partial pressure exceeded 0.1 torr, even by 0.02 torr, a strong quenching of the 492.4-nm line occurred. Subsequent spontaneous-emission studies on lines from the $4d \, {}^2D_{5/2, \, 3/2}$ doublet states, lower states of the 492.4 and 491.1-nm Zn-II laser lines, established that the $4d \, {}^2D$ states were selectively excited in a He–Ne–Zn discharge in an exothermic, thermal-energy, charge-transfer reaction between Ne^+ ground-state ions and ground-state Zn atoms

$$Ne^+ + Zn \, {}^1S_0 \rightarrow Ne \, {}^1S_0 + Zn^{+\prime} \, 4d \, {}^2D_{5/2, \, 3/2} + \Delta E_\infty \, (1205 \text{ and } 1256 \text{ cm}^{-1}). \quad (5.19)$$

The total velocity averaged cross-section for this reaction was found to be 2.3×10^{-15} cm². It is thus directly comparable to that of 2.1×10^{-15} cm² for charge transfer between $He^+ \, {}^2S_{1/2}$ ground-state ions and ground-state Zn atoms in eqn. (5.17).[46]

5.2.4. THE HELIUM–IODINE LASER

Apparently Fowles and Jensen of the University of Utah were the first to realize and propose that asymmetric charge-transfer reactions of the type

$$A^+ + B \rightarrow A + B^{+\prime} + \Delta E_\infty$$

could produce population inversion in the ionic species of a gas or vapor. Fowles and Jensen,[22] following observation of oscillation in a pulsed helium–iodine discharge on two transitions in singly ionized atomic iodine from two states within 0.18 V of the helium ionization potential of 24.58 V, suggested that the main selective excitation process of the upper laser

209

levels was the asymmetric charge-transfer reaction

$$He^+ + I \rightarrow He + I^+ \text{ (excited)}. \tag{5.20a}$$

In addition to suggesting this, they related a similar type of excitation process to the mercury ion laser in which oscillation at 615.0 nm in pulsed helium–mercury mixtures had in 1964 been reported by Bloom *et al.*[50]

Fowles and Jensen first reported oscillation on the two singly ionized atomic-iodine transitions $6p'\ ^3D_2$–$6s'\ ^3D_1^\circ$ at 576.1 nm, and $6p'\ ^3D_1$–$6s'\ ^3D_2^\circ$ at 612.7 nm, where the notation used here, and throughout the treatment of the helium–iodine laser, is that of Martin and Corliss.[51] In subsequent work, oscillation in the afterglow of pulsed discharges of various types, sometimes in combination with rf-sustaining discharges, in discharge tubes that varied in bore from 5 to 8 mm at helium pressures of a few torr and iodine vapor pressures of about 0.1 torr, was reported at the seventeen wavelengths listed in Table 5.5. Oscillation on some of the lines as indicated in the table was also observed when the helium was replaced with neon or krypton. Laser action was not obtained on any of the reported lines using argon or xenon in conjunction with iodine.

Additional pulsed laser oscillation in singly ionized iodine at 606.9 nm ($6p'\ ^3F_2$–$6s'\ ^3D_1^\circ$); 651.6 nm ($6p'\ ^3F_2$–$5p^5\ ^1P_1^\circ$); 682.5 nm ($6p'\ ^3F_2$–$6s'\ ^3D_3$); and 713.9 nm ($6p'\ ^3D_2$–$5d'\ ^3D_2$) has been reported by the author.[35, 51, 55] CW oscillation on the 651.6-nm and the 713.9-nm lines has also been reported (footnote (a) of Table 5.5).

TABLE 5.5. *Laser transitions observed in noble gas–iodine vapor mixtures (in order of increasing wavelength)* [52, 53]

Wavelength (nm)	Transition	ΔE_∞ (eV)	Carrier gas
498.7	$6p'\ ^3D_2$–$5d\ ^3D_1^\circ$	0.23	He
521.6	$6p'\ ^3F_2$–$5d\ ^3D_1^\circ$	0.34	He
540.7[a]	$6p'\ ^3D_2$–$6s'\ ^3D_2^\circ$	0.23	He
541.9	A krypton impurity line		
562.6	$6p\ ^3P_1$–$6s\ ^3S_1^\circ$	2.0	He or Ne
567.8[a]	$6p'\ ^3F_2$–$6s'\ ^3D_2$	0.34	He
576.1[a]	$6p'\ ^3D_2$–$6s'\ 3D_1^\circ$	0.23	He
612.7[a, b]	$6p'\ ^3D_1$–$6s'\ ^3D_2^\circ$	0.51	He
658.5[a]	$6p'\ ^3D_1$–$6s'\ ^3D_1^\circ$	0.51	He
690.5	$6p'\ ^3D_2$–$6s'\ ^1D_2^\circ$	0.23	He, Ne, or Kr
703.3[a]	$6p'\ ^3F_2$–$5d'\ ^3D_2^\circ$	0.34	He
825.4[a]	$6p'\ ^3D_1$–$5d'\ ^3P_1^\circ$	0.51	He
880.4[a]	$6p'\ ^3F_2$–$5d'\ ^3F_3$	0.34	He or Ne
980.0	Neutral lines(?)	—	He or Ne
1010.0	Neutral lines(?)	—	He or Ne
1030.0	Neutral lines(?)	—	He or Ne
1060.0	Neutral lines(?)	—	He or Ne

[a] CW oscillation has now been observed on this line in a HCD laser structure by J. A. Piper, G. J. Collins, and C. E. Webb (see ref. 293 of Appendix Tables 1 to 46).

[b] CW oscillation has been reported on thus line by G. J. Collins, H. Kuno, S. Hattori, K. Tokutome, M. Ishikaaw, and N. Kamide in *IEEE J. Quant. Electr.* **QE-8**, 679 (1972).

With the exception of the strong 562.6-nm line that can be obtained in Ne, or krypton–iodine mixtures and whose upper state is about 2 volts below the helium ionization potential, all the upper states of the identified singly ionized laser transitions are within a fraction of an electron volt of the He^+ ground state at 24.58 eV. All the identified lines, with the exception of the 690.5-nm line, have upper states that are triplet states, even though there are singlet levels (not selectively excited) that are closer to the He^+ ground state than some of the triplet levels from which oscillation occurs, and that have transitions isoelectronic with laser transitions in doubly ionized xenon. (The requirement of the 690.5-nm iodine line for a combination of rf and pulsed excitation, together with its observation of oscillation in Ne or Kr and iodine vapor mixtures as well as in helium–iodine vapor mixtures, strongly suggests that it is excited in a two-step process that involves electron impact excitation from the I^+ ground state or some well-populated metastable state. A similar process is suggested as being responsible for oscillation of the 562.6-nm line, which is a "usual" p–s laser transition.)

Because of the closeness of the upper laser levels to the He^+ ground state and the preference for selective excitation of triplet levels, Fowles and Jensen[53] proposed that the selective excitation process in a discharge in helium–iodine vapor was

$$He^+ \, ^2S_{1/2} + I \, ^2P_{1/2} \rightarrow He \, ^1S_0 + I^{+\prime}(X). \tag{5.20b}$$

If the $I^{+\prime}(X)$ excited level of ionized iodine is a triplet level, spin is conserved, but it is not conserved if it is a singlet level. Figure 5.15, a partial energy-level diagram of singly ionized

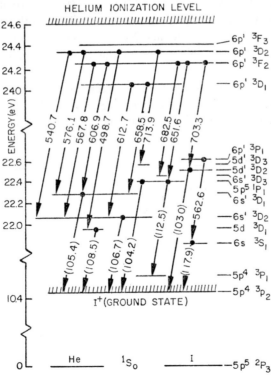

FIG. 5.15. Partial energy-level diagram of singly ionized iodine giving laser transitions and showing the energy-level coincidence of the upper-laser levels with the helium ionization level. Bracketed wavelengths indicate lower-laser-level transitions to or near the I^+ ion ground state. (After C. S. Willett.[35])

atomic iodine, shows most of the laser transitions that have been observed,[35, 52, 55] and the closeness of the upper laser states to the He^+ ionization level. It can be seen that apart from the $6p'\ ^3P_1$ state, upper state of the 562.6-nm line that we can neglect for reasons given earlier, oscillation is observed from just the three triplet levels $6p'\ ^3D_2$, $6p'\ ^3F_2$ and $6p'\ ^3D_1$ so that eqn. (5.20b) becomes

$$He^+\ ^2S_{1/2}(\uparrow)+I\ ^2P_{1/2}(\uparrow) \rightarrow He\ ^1S_0(\uparrow\downarrow)+I^+\ 6p'(^3X)\,(\uparrow\uparrow)+\Delta E_\infty , \qquad (5.21)$$

where ^3X indicates $^3D_2$, $^3F_2$ or $^3D_1$, and ΔE_∞ is 0.23, 0.34, and 0.51 eV, respectively.

As well as the asymmetric charge-transfer selective-excitation process, other excitation processes are capable of exciting the iodine $6p'$ states in discharges in helium, or neon–iodine vapor mixtures. Both helium and neon metastables have sufficient energy to excite a ground-state iodine $I^+\ ^3P_2$ ion to the $6p'$ levels by processes analogous to those reported by Dana and Laures for the He–Kr and Ne–Xe afterglow ion lasers[3, 4] (Sections 5.1.1 and 5.1.2)

$$He^*\ 2^3S_1+I^+\ ^3P_2 \rightarrow He\ ^1S_0+I^+\ 6p'+\Delta E_\infty \qquad (5.22)$$

and

$$Ne^*\ 1s+I^+\ ^3P_2 \rightarrow Ne\ ^1S_0+I^+\ 6p'+\Delta E_\infty . \qquad (5.23)$$

The $I^+\ ^3P_2$ ground-state iodine ion *could* be formed by a prior collision of an iodine molecule or atom with another helium or neon metastable atom, or by photodissociation in the discharge. The energy discrepancies in eqns. (5.22) and (5.23) are approximately 5 eV and 2.5 eV respectively. The reactions, therefore, are not even mildly exothermic, and certainly not near-resonant as they are in the He–Kr and Ne–Xe ion lasers. They thus do not seem likely selective-excitation mechanisms in the case of this laser and are only mentioned for the sake of completeness.

Figure 5.16 illustrates the time relationship between the spontaneous emission of a typical strong ionized iodine line at 576.1 nm (upper trace) and the discharge-current pulse (lower trace) under lasing conditions. The figure shows that maximum spontaneous emission occurs

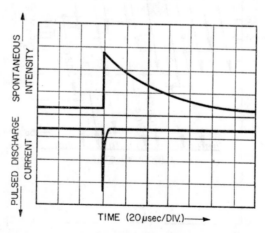

TIME (20 μsec/DIV.)——➤

FIG. 5.16. Time relationship of the spontaneous emission of the singly ionized iodine 576.1-nm laser line (upper trace) and discharge-current pulse (lower trace) under lasing conditions. (Redrawn from oscilloscope trace.) (After C. S. Willett.[35])

when the current pulse is practically zero, and that the emission is strong for more than 140 μsec after the termination of the current pulse. Under ideal discharge conditions the duration of oscillation can be in excess of 250 μsec. The duration depends on the vapor pressure of iodine, the helium pressure, and rise time of the excitation current pulse. Oscillation can be obtained at helium pressures of up to approximately 60 torr, with an optimum at about 6 torr, at about 0.2 torr vapor pressure of iodine in discharge tubes from 7- to 50-mm bore.

The conclusion that the main selective excitation in the helium–iodine laser is that of asymmetric charge transfer, as originally proposed by Fowles and Jensen,[22] is evidenced as follows:

(1) the strong spontaneous emission for a long period in the afterglow of a pulsed high-pressure helium–iodine mixture;

(2) strong enhancement of singly ionized iodine lines on which laser oscillation has been observed in relatively high-pressure helium–iodine mixtures in the high current-density plasma of hollow-cathode discharges[35, 56, 57] where charge-transfer reactions are enhanced;

(3) a decay of spontaneous emission unlike the time variation of the concentration of molecular He^+ ions but similar to that exhibited by He^+ atomic ions in the presence of impurity atoms in a pulsed discharge in helium (Fig. 3.20);

(4) similarity with the 615.0-nm helium–mercury laser in which charge transfer is the main excitation mechanism; and

(5) the adherence to spin conservation in (5.21) as in other charge-transfer reactions.

5.2.5. THE HELIUM–SELENIUM LASER

One of the most promising laser systems for multi-wavelength CW oscillation at a few tens of mW output power is the helium–selenium ion laser discovered by Silvast and Klein[58] in 1970. CW oscillation on at least forty transitions in the visible spectrum of singly ionized selenium between 446.8 and 779.6 nm, together with six from 783.9 nm out to 1.26 μm, has subsequently been reported.[58, 59] Output powers varying from 3 to 50 mW have been obtained over the entire spectral range of operation in a 2-m-long discharge tube of 3 mm bore, with a combined output power of 250 mW on the six strongest blue-green transitions at 497.6, 499.3, 506.9, 517.6, 522.8, and 530.5 nm operating simultaneously. The forty-six laser transitions reported by Silvast[58] and Klein[59] are listed in Table 5.6. Included in the table are the identifications according to Martin;[60] the relative intensities of the lines listed as strong (S), moderate (M), or weak (W) as they were observed in (1) a 4-mm-bore, 1-m-long discharge tube at a helium pressure of 8 torr and a discharge current of 200 mA (marked by an asterisk) and (2) a 3-mm-bore, 2-m-long discharge tube at a helium pressure of 7 torr and a temperature of 270 °C for each selenium reservoir side-arm; and the measured gain. The optimum partial pressures of the two gases are approximately 6 to 8 torr of helium and 5×10^{-3} torr of selenium. The gain per meter was found to be higher on most transitions in the 3-mm-bore tube. The maximum gain measured was 11 percent/meter on the 522.7-nm line at a discharge current of 500 mA.

TABLE 5.6. *Helium–selenium ion laser transitions*

Wavelength (nm)	Identification	Relative intensity	Measured gain
0.446760	$5p\ ^2P^\circ_{1/2}-5s\ ^2P_{1/2}$	M	1.4
0.460434	$5p\ ^2D^\circ_{5/2}-5s\ ^4P_{5/2}$	M*	—
0.461877	$5p\ ^4P^\circ_{5/2}-5s\ ^4P_{3/2}$	W	< 1.0
0.464844	$5p\ ^4P^\circ_{3/2}-5s\ ^4P_{1/2}$	W*	—
0.471823	$5p\ ^4S^\circ_{3/2}-3^a$	W	< 1.0
0.474097	$5p\ ^2P^\circ_{3/2}-4s\ 4p^4\ ^2P_{3/2}$	W	< 1.0
0.476365	$5p\ ^2D^\circ_{3/2}-5s\ ^4P_{3/2}$	W*	—
0.476552	$5p\ ^2P^\circ_{1/2}-5s\ ^2P_{3/2}$	M	1.6
0.484063	$5p\ ^2S^\circ_{1/2}-5s\ ^4P_{3/2}$	W*	—
0.484496	$5p\ ^4S^\circ_{3/2}-5s\ ^4P_{5/2}$	M*	2.3
0.497566	$5p\ ^2D^\circ_{5/2}-4s\ 4p^4\ ^2P_{3/2}$	S*	3.3
0.499275	$5p\ ^4P^\circ_{3/2}-5s\ ^4P_{3/2}$	S*	3.3
0.506865	$5p\ ^4P^\circ_{5/2}-5s\ ^4P_{5/2}$	S*	3.3
0.509650	$5p\ ^4D^\circ_{7/2}-4d\ ^4F_{9/2}{}^b$	W*	—
0.514214	$5p\ ^4D^\circ_{3/2}-5s\ ^4P_{1/2}$	M*	1.3
0.517598	$5p\ ^4D^\circ_{5/2}-5s\ ^4P_{3/2}$	S*	4.6
0.522751	$5p\ ^4D^\circ_{7/2}-5s\ ^4P_{5/2}$	S*	5.4
0.525307	$5p\ ^2D^\circ_{5/2}-5s\ ^2P_{3/2}$	M*	1.7
0.525363	$5p\ ^4D^\circ_{1/2}-5s\ ^4P_{1/2}$	M*	1.7
0.527111	$5p\ ^4D^\circ_{5/2}-4d\ ^4F_{7/2}{}^b$	W*	—
0.530535	$5p\ ^2D^\circ_{3/2}-5s\ ^2P_{1/2}$	S*	2.6
0.552242 {	$5p\ ^4P^\circ_{3/2}-5s\ ^4P_{5/2}(?)$	M*	—
	$5p\ ^4P^\circ_{5/2}-4s4p^4\ ^2P_{3/2}(?)$	M*	—
0.556693	$5p\ ^4D^\circ_{3/2}-5s\ ^4P_{3/2}$	W	< 1.0
0.559116	$5p\ ^4P^\circ_{3/2}-5s\ ^2P_{1/2}$	W*	—
0.562313	$5p\ ^4P^\circ_{1/2}-5s\ ^4P_{1/2}$	W	< 1.0
0.569788	$5p\ ^4D^\circ_{1/2}-5s\ ^4P_{3/2}$	W*	—
0.574762	$5p\ ^4D^\circ_{5/2}-5s\ ^4P_{5/2}$	W*	—
0.584268	$5p\ ^2S^\circ_{1/2}-5s\ ^2P_{3/2}$	M	1.6
0.586627	$5p\ ^4P^\circ_{5/2}-5s\ ^2P_{3/2}$	M	1.6
0.605596	$5p\ ^2P^\circ_{3/2}-7$	M*	1.3
0.606583	$5p\ ^4P^\circ_{3/2}-5s\ ^2P_{3/2}$	M	1.6
0.610196	$5p\ ^2D^\circ_{3/2}-5s\ ^2P_{3/2}$	W	< 1.0
0.644425	$5p\ ^2D^\circ_{5/2}-7$	M*	—
0.649048	$5p\ ^4D^\circ_{1/2}-5s\ ^2P_{1/2}$	M*	—
0.653495	$5p\ ^2P^\circ_{1/2}-7$	W*	—
0.706389^c	$5p\ ^4P^\circ_{1/2}-5s\ ^2P_{1/2}$	M	1.8
0.739199	$5p\ ^4P^\circ_{5/2}-7$	M	1.4
0.767482	$5p\ ^2P^\circ_{3/2}-5s'\ ^2D_{5/2}$	W	1.0
0.772404	$5p\ ^4D^\circ_{1/2}-5s\ ^2P_{3/2}$	M	1.1
0.779615	$5p\ ^2P^\circ_{1/2}-5s'\ ^2D_{3/2}$	M	1.3
0.783881	$5p\ ^4P^\circ_{1/2}-4s4p^4\ ^2P_{3/2}$	M	1.9
0.830952	$5p\ ^2D^\circ_{3/2}-5s'\ ^2D_{5/2}$	M	1.9
0.92493^d	Unknown	W	< 1.0
0.995515^c	$5p\ ^4P^\circ_{5/2}-5s'\ ^2D_{5/2}$	M	1.1
1.040881^c	$5p\ ^4D^\circ_{1/2}-6$	S	2.8
1.258678^c	$5p\ ^4P^\circ_{3/2}-10$	S	4.3

^{a} Incorrect identification in ref. 60.
^{b} Identification: S. G. Krishnamurty and K. R. Rao, *Proc. Roy. Soc.* (London) A **149**, 56 (1935).
^{c} Calculated from energy-level differences.
^{d} Measured wavelength (± 0.1 nm).

5.2.5.1. Upper level excitation

All the laser transitions originate from each of the thirteen $5p$ levels in Se-II with an excited electronic configuration $4s^2\,4p^2\,5p$, that are energetically close (at the ion energies expected in a glow discharge) to the He^+ ion ground state. The $5p$ levels are uniformly distributed from approximately 0.2 eV above to 0.8 eV below the $He^+\,{}^2S_{1/2}$ ground state. The close, exothermic energy coincidences within 0.8 eV, and the endothermic energy coincidences within 0.2 eV of the He^+ ion ground state, together with a linear dependence of sidelight intensities of the selectively excited laser lines on discharge currents at moderate currents,[59] and the high helium pressure for optimum output, indicates that the selective excitation mechanism that occurs in helium–selenium discharges is again that of asymmetric charge transfer

$$He^+\,{}^2S_{1/2}(\uparrow)+Se\,{}^3P_2(\uparrow\uparrow) \rightarrow He\,{}^1S_0(\uparrow\downarrow)+Se^{+\prime}\,5p\,{}^4S,\,{}^4P,\,{}^4D,\,{}^2S,\,{}^2P,\,{}^2D$$
$$+\Delta E_\infty\;(6599\;\text{to}\;-1735\,\text{cm}^{-1}). \qquad (5.24)$$

If electron impact excitation was responsible for the selective excitation of the $5p$ states the optimum helium pressure would be much lower in order to provide the necessary higher electron temperature. The discharge conditions would correspond to the conditions under which pulsed laser action was first observed in Se-II at 509.7 and 522.7 nm by Bell, Bloom, and Goldsborough,[61] in which the optimum carrier gas was neon at a pressure of 100 mtorr.

Although there appears to be a preference for strong laser lines to originate on quartet states (^4D and ^4P) in the selective excitation process, which would give spin conservation in eqn. (5.24), there appears to be no preference for charge transfer to those states with a minimum ΔE_∞. The closeness (to within approximately kT) of the numerous Se-II $5p$ states from which oscillation is observed means that collisional mixing with electrons or helium atoms probably plays an important role in distributing any charge transfer that could be occurring to a limited number of $5p$ states.

The absence of oscillation in Se-II on strong transitions below 400.0 nm that originate also from the $5p$ levels, and terminate on three levels that lie in energies between the $He^*\,2^1S_0$ and $He^*\,2^3S_1$ metastable levels, has been attributed to possible Penning collisions.[59] Such collisions would inhibit the attainment of population inversion by populating the lower levels.

5.2.5.2. Operating characteristics and discharge conditions

Figures 5.17, 5.18, and 5.19 show the variation of laser power with current of the strongest transition at 522.7 nm for a 3-mm-bore, 50-cm-long discharge tube; and the variation of laser power with helium pressure and with reservoir side-arm temperature for a 4-mm-bore, 1-m discharge tube at a current of 200 mA. The discharge tube had a centrally mounted cathode and an anode at each end. Selenium vapor was introduced into the discharge from an independently heated sidearm attached to the discharge tube near each anode. Figure 5.17 shows that the output power increases approximately linearly with current until it saturates at about 400 mA. It peaks at 500 mA and then falls off to one-third of the maximum at a discharge current of 1 A. A threshold current of 40 mA is observed for the 522.7-nm

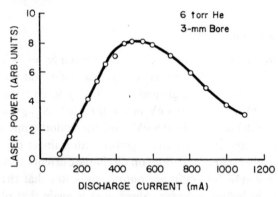

FIG. 5.17. Se-II 522.7-nm laser power versus discharge current under optimum conditions. (After W. T. Silfvast and M. B. Klein,[58] by courtesy of the American Institute of Physics.)

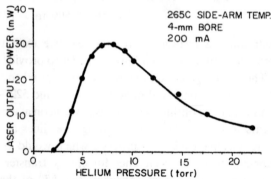

FIG. 5.18. Combined Se-II laser power at 497.6, 499.3, 506.9, 517.6, 522.7, and 530.5 nm versus helium pressure. (After W. T. Silfvast and M. B. Klein,[58] by courtesy of the American Institute of Physics.)

line in the 4-mm-bore tube. The variation of output with helium pressure in Fig. 5.18 at a side-arm temperature of 265°C has a rather broad maximum over the region 6 to 10 torr. The low-pressure threshold is at 2 torr and oscillation can be observed up to pressures as high as 22 torr. The 3-mm-bore tube showed a similar pressure variation, with an optimum at 6 torr. Unlike the He–Cd ion laser operating under conditions for optimizing Penning reaction energy transfer, the He–Se ion laser appears to have neither an optimum pressure and tube diameter product, nor any discharge-current density that is common to all tube diameters. Rather surprisingly, the optimum helium pressure of about 8 torr in the 4-mm-bore tube is higher than the 6 torr that is the optimum for the smaller-diameter 3-mm-bore tube. Likewise, the optimum discharge of about 500 mA in the smaller-bore tube is surprisingly more than twice the 200 mA that is presumably the optimum discharge current in the 4-mm bore laser tube.[58] One wonders if these unexpected results arise from some requirement of the discharge to provide atomic selenium vapor from Se_2, Se_4, Se_6, or Se_8 molecules, as well as in the maintenance of an optimum electron temperature that is normally achieved by maintaining the pressure and tube-diameter product constant. Possibly it is associated with the necessity of maintaining a temperature high enough to prevent condensation of the selenium on the tube walls.

The variation of output power with side-arm temperature is shown in Fig. 5.19 at a helium pressure of 8 torr. The half-width of the curve occurs over a temperature range of 30°C with the maximum occurring at 265°C. This temperature corresponds to a selenium vapor pressure of about 0.005 torr. The narrow vapor-pressure region over which laser action occurs is characterized by a relatively white discharge. This is contrasted with the pinkish discharge associated with pure helium on the low-temperature side of the discharge tube, and the blue discharge of selenium on the high-temperature side.[58]

In order to obtain more information on the saturation behavior of the laser output power as the discharge is increased above an optimum value (as shown in Fig. 5.17), Klein and Silfvast[59] measured the variation of gain, upper-level spontaneous emission, and laser power as a function of discharge current on several of the strong laser transitions. Their results for a 3-mm-bore tube are plotted in Fig. 5.20 for the 522.7-nm line. (All of the other

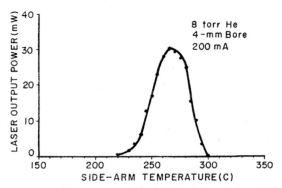

Fig. 5.19. Combined Se-II laser power at 497.6, 499.3, 506.9, 517.6, 522.7, and 530.5 nm versus selenium sidearm reservoir temperature. (After W. T. Silfvast and M. B. Klein,[58] by courtesy of the American Institute of Physics.)

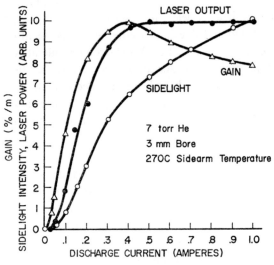

Fig. 5.20. He–Se laser variation of gain, upper-laser-level spontaneous emission, and laser power at 522.7 nm with discharge current under optimum laser conditions. (After M. B. Klein and W. T. Silfast,[59] by courtesy of the American Institute of Physics.)

lines examined behaved similarly.) Both the gain and the laser power reach a maximum at approximately 400 mA, whereas the sidelight emission continues to increase with current (to the highest currents measured). The sidelight emission increases faster than linearly from 75 to 300 mA, and sublinearly at higher currents. The rapid variation at low currents is attributed to cataphoretic effects and to the collisional dissociation of selenium molecules. The linear variation at the intermediate-current range is consistent with charge-transfer excitation from He$^+$ ions. At higher currents the positive ion charge density will still be increasing, but selenium ions constitute an increasing fraction of it and thus reduce the increase of the He$^+$ ions to a sublinear rate. The saturation of the laser power and the gain (which begins at approximately 150 mA) is due apparently to a large increase in the population of the lower laser level. It is not clear if this arises through electron impact excitation of selenium ions in the ground state or through the alternative process of radiation trapping on the terminal transition to the selenium-ion ground state.[59]

5.2.6. THE HELIUM– OR NEON–TELLURIUM ION LASER

Another laser in which one of the main selective mechanisms involved is charge transfer is the CW, helium–, or neon–tellurium ion laser. Following a lapse of nearly 6 years from the first report by Bell, Bloom, and Goldsborough[61] of (pulsed) laser action on three transitions in tellurium using neon as a carrier gas and almost 4 years after the report by Webb[62] of additional oscillation on other transitions under pulsed conditions in Ne–Te and He–Te mixtures, Silfvast and Klein[63] reported laser action on thirty-one transitions in singly ionized tellurium using both helium and neon as carrier gases with tellurium vapor. Using helium, a total of sixteen lines ranging in wavelength from 484.3 to 873.4 nm were observed. With neon, twenty laser lines were observed from 545.0 to 937.8 nm with eleven lines lasing simultaneously. Only five lines were common to both gas–vapor mixtures. The discharge tube by means of which the laser lines were obtained was a 2-m, 3-mm-bore quartz tube similar to that used for the helium–selenium laser described in the preceding section. The thirty-one laser lines are listed in Table 5.7, together with their identification (when known), carrier gas used, relative intensities (S, M or W for strong, moderate or weak) and measured gains if more than 0.5 percent/meter. As spectroscopic identification of the Te-II energy levels in incomplete and excited states have been classified only in terms of their J-value and energy, the levels in Table 5.7 are identified by writing their energy above the Te$^+$ ion ground state in units of 10^3 cm^{-1} with the J-value, as usual, given as a subscript.

5.2.6.1. Upper-level excitation

Figure 5.21, a partial energy-level diagram for Te-II and neutral helium and neon, shows the energy-level coincidences between the Te-II laser levels and the metastable, and ground-state ion levels in He and Ne. As noted by Silfvast and Klein[63] most of the upper laser levels lie in between the He$^+$ and Ne$^+$ ion ground states in energy, with a few levels extending slightly below that of the Ne$^+$ ground state. A more detailed diagram of the energy levels involved in the laser transitions is given in Fig. 5.22. Denoted also in this figure are those levels that act as upper laser levels with helium as a carrier gas, and those involved when neon is used.

TABLE 5.7. *Helium– and neon–tellurium ion laser transitions*

Wavelength (nm)	Identification	Buffer gas	Relative intensity	Measured gain (percent/ meter)
484.29	$122'_{5/2}-102_{3/2}$	He	M	0.7
502.04	Unknown	He	M	0.8
525.64	Unknown	He	W	—
544.98	$103_{3/2}-85'_{5/2}$	Ne	S	2.1
547.91	$105_{3/2}-86_{3/2}$	Ne	S	1.1
557.63	$112_{7/2}-94_{5/2}$	He	S	1.5
566.62	$101_{3/2}-83_{1/2}$	He	S	1.3
570.81	$103_{7/2}-85'_{5/2}$	He	S	4.3
574.16	$112_{3/2}-94_{5/2}$	Ne	M	0.6
575.59	$100_{5/2}-82_{3/2}$	He	W	—
		Ne	S	2.5
576.53	$112_{5/2}-95_{3/2}$	He	W	—
585.11	$111_{5/2}-94_{5/2}$	He	W	—
		Ne	M	1.0
593.61	$99_{3/2}-82_{3/2}$	Ne	S	1.7
597.26	$111_{5/2}-95_{3/2}$	He	W	—
		Ne	M	0.9
597.47	$102_{5/2}-85'_{5/2}$	He	S	1.3
601.45	$105_{5/2}-88_{3/2}$	Ne	W	—
608.23	Unknown	He	W	—
623.07	$105_{3/2}-88_{3/2}$	Ne	S	2.1
624.54	$99_{3/2}-83_{1/2}$	He	W	—
		Ne	S	1.3
658.51	Unknown	He	W	—
664.86	$97_{1/2}-82_{3/2}$	Ne	M	0.7
667.61	$103_{3/2}-88_{3/2}$	Ne	W	—
688.51	$100_{5/2}-85'_{5/2}$	Ne	W	—
703.91	$97_{1/2}-83_{1/2}$	Ne	S	1.7
780.17	$105_{3/2}-92_{3/2}$	Ne	M	0.7
792.17	$97_{1/2}-85_{3/2}$	Ne	M	0.7
860.46	Unknown	He	S	1.3
873.38	Unknown	He	M	1.0
897.21	$103_{3/2}-92_{5/2}$	Ne	W	—
899.82[a]	Unknown	Ne	W	—
937.85[b]	$99_{3/2}-88_{5/2}$[b]	Ne	M	0.7

[a] Measured wavelength (± 0.06 nm).
[b] Obtained from energy-level differences.

It is easier to tie down the selective excitation mechanisms in the Ne–Te laser than these in the He–Te laser. In the Ne–Te laser the majority of the upper laser levels are situated in a group about the Ne$^+$ ion ground state and within it by $\pm 4\,kT$, where T is about 1000 K. The upper laser levels have the configuration given by Charlotte Moore[64] as $5s^2\,5p^2(^3P)\,6p$. This configuration is thus similar to the $4s^2\,4p^2(^3P)\,5p$ configuration of the Se-II states excited by charge transfer involving He$^+$ ground-state ions. The energy-level coincidences, together with an observed linear variation of laser power and sidelight intensity of all the laser lines with discharge current, and the high optimum neon pressure of 2 torr in a 3-mm-

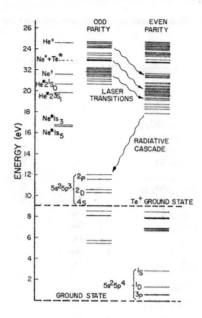

FIG. 5.21. Partial energy-level diagram showing relationship between Te-II laser levels and the long-lifetime (metastable and ground-state ion) levels in He and Ne. (After W. T. Silfvast and M. B. Klein,[63] by courtesy of the American Institute of Physics.)

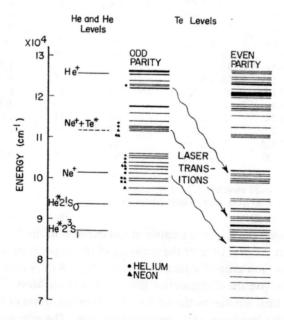

FIG. 5.22. Detailed diagram of Te-II laser levels showing specific upper laser levels when He or Ne are used as carrier gases together with energy coincidences between Ne+, Ne+ + Te*, and He+ and Te-II levels. (After W. T. Silfvast and M. B. Klein,[63] by courtesy of the American Institute of Physics.)

bore tube, indicate that the selective excitation occurs in two, asymmetric charge-transfer reactions. The first reaction

$$\text{Ne}^+ \, {}^2P^\circ_{3/2} + \text{Te} \, {}^3P_2 \rightarrow \text{Ne} \, {}^1S_0 + \text{Te}^{+\prime} \pm \varDelta E_\infty(4kT) \qquad (5.25)$$

would be responsible for the selective excitation of the states close to the Ne^+ ion ground-state, with the kinetic energy of the Ne^+ ions in the discharge able to supply the energy discrepancy of $-4 \, kT$ between the potential energy of the excited $\text{Te}^{+\prime}$ ion and the ground state $\text{Ne}^+ \, {}^2P^\circ_{3/2}$ ion where the gas temperature is about 1000 K (Silvast[65]). The second reaction

$$\text{Ne}^+ \, {}^2P^\circ_{3/2} + \text{Te}^* \, {}^1D_2 \rightarrow \text{Ne} \, {}^1S_0 + \text{Te}^{+\prime} - \varDelta E_\infty(0.03 \text{ eV}) \qquad (5.26)$$

involves a charge-transfer collision between a ground-state Ne^+ ion and a tellurium atom in the lowest-lying $5p^4 \, {}^1D_2$ metastable state, and provides a near energy-coincidence for exciting the two uppermost Te-II states in the presence of neon. Penning-like $\text{Ne}^* + \text{Te}^*$ collisions do not have enough energy to both ionize and excite the $\text{Te}^{+\prime}$ states. The atom–ion collision process $\text{Ne}^* + \text{Te}^+$, similar to that operative in the He–Kr ion laser described in Section 5.1, would put the excitation close to the Te^{++} ion ground state at 150,000 cm^{-1} above the Te^+ ground state, so that it cannot account for the selective excitation.

In the He–Te laser, only one of the nine upper-laser levels is close to the He^+–ion ground state. This level is approximately $3 \, kT$ (at the temperature of the discharge) below it, while the others lie at 1.5 to 3 eV below the He^+ ion ground-state potential energy. Even with the limited information that is available, an energy discrepancy in excess of 1.5 eV makes the probability of direct charge transfer extremely remote, so that it has been suggested by Silfvast and Klein that the excitation of these lower-lying laser levels is derived from radiative cascade transitions from levels more nearly resonant with the He^+ ion ground state. Penning-type collisions between metastable helium atoms and the various metastable levels of neutral tellurium, as in the Ne–Te system, do not have enough energy to excite the uppermost upper laser levels of the observed lines. Likewise, no other known excitation processes can account for the selective excitation that occurs in the He–Te ion-laser system.

5.2.6.2. Operating characteristics and discharge conditions

Laser output on the strongest transitions for discharges in both He–Te and Ne–Te mixtures occurred at side-arm temperatures from 380° to 440°C. The optimum temperature of approximately 420°C corresponds to a tellurium vapor pressure of 0.05 torr.[66] The behavior of the laser output with current was similar for both helium and neon, and was similar to that observed for the He–Se ion laser as discussed earlier. The sidelight of the laser lines in spontaneous emission continues to increase linearly with discharge current up to 500 to 550 mA, although the laser output power has levelled off at 400 to 450 mA and dropped slightly at 600 mA. The laser-power saturation, therefore, can be attributed to processes that affect the population of the lower laser levels.

The variation of laser power with neon or helium gas pressure is shown in Fig. 5.23. In neon, the pressure range for laser action is from 1 to 7 torr with an optimum of 2 torr, giving a $p_{\text{Ne}} D$ product of 6 torr-mm. In helium the pressure range is from 2.5 to 11 torr, with an optimum of 5 torr giving a $p_{\text{He}} D$ of 15 torr-mm. Though it is hard to see any reason why

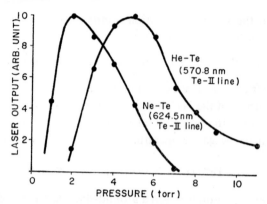

FIG. 5.23. Variation of laser power of two Te-II lines in Ne–Te and He–Te mixtures versus Ne or He carrier-gas pressure, $D = 3$ mm, tellurium sidearm temperature 420°C. (After W. T. Silfvast and M. B. Klein,[63] by courtesy of the American Institute of Physics.)

it should be so, the electron temperature that corresponds to these pD products is the same[63] (assuming that the small partial-pressure of tellurium does not significantly alter the electron temperature from that of pure neon or helium discharges)[63] at approximately 47,000 K (from Fig. 3.7). In the absence of any known reason why this should be so, one could conjecture that the occurrence of the same optimum electron temperature in both Ne–Te and He–Te mixtures is related to some preferential kinetic energy of Ne^+ and He^+ ground-state ions that is produced in acceleration to the negatively charged wall of the discharge tube. This could affect the excitation because of the resonant nature of the asymmetric charge-transfer reactions. The magnitude of the negative charge on the inside wall of the discharge tube presumably would be directly related to the average electron temperature.

5.3. Penning Reaction Lasers

Penning reactions of the general form

$$A^* + B \rightarrow A + B^{+\prime} + e^- + \text{kinetic energy} \qquad (5.27)$$

are definitely responsible for selective excitation of upper-laser levels in at least the five metal-vapor ion lasers: He–Cd, He–Zn, He–Sr, He–Sn, He–Pb, and possibly in the He–Mg ion laser that operates in the near infrared. In another laser, the He–Se ion laser, Section (5.2.4), it has been suggested that the Penning reaction is responsible for the selective excitation of *lower* levels of transitions whose upper levels are excited by asymmetric charge transfer from the He^+ ion ground state.[59] This selective excitation in the positive column of a glow discharge (but probably not in a hollow-cathode discharge) is sufficient to prevent oscillation on the charge-transfer-excited transitions in Se-II.

The lasers to be discussed here will be restricted to the He–Cd, He–Zn and He–Mg laser systems. It is considered that a description of these three systems is sufficient to illustrate generally the properties of Penning-reaction selective excitation and the behavior of Penning-reaction-excited lasers.

5.3.1. THE HELIUM–CADMIUM LASER

Aspects of the Penning-reaction-excited helium–cadmium ion laser have already been discussed in Section 5.2.2, which is concerned with charge-transfer-excited laser transitions in discharges in helium–cadmium mixtures. Figures used in Section 5.2.2 that are relevant to Penning-reaction excitation will be referred to in this section.

As stated in Section 5.2.2, selective-excitation processes in the helium–cadmium ion laser, in which the presence of helium is essential for operation, include both Penning and asymmetric charge-transfer reactions. In low-current dc-discharges such as the positive column of a glow discharge, Penning collisions are the dominant source of both ionization, and selective ionization and excitation of singly ionized Cd-II levels.[67] In the high-electron and ion-concentration plasma of dc and pulsed hollow-cathode discharges, charge-transfer collisions predominate in the simultaneous ionization and excitation of $Cd^{+'}$ states.[17]

5.3.1.1. Upper level excitation

Figure 5.11 (Section 5.2.2), which illustrates the energy-level coincidences of the metastable $He^* \; 2^3S_1$ and $He^* \; 2^1S_0$ states, and the He^+ ion ground state with Cd-II levels, together with the disposition of reported Cd-II laser lines, shows two doublet states $6p \; ^2P^o_{1/2, \, 3/2}$ and $5d \; ^2D_{3/2, \, 5,2}$, and one singlet state $6s \; ^2S_{1/2}$ close in potential energy to the two helium metastable states; and two $5s^2 \; ^2D_{3/2}$ and $^2D_{5/2}$ states situated between 10,000 and 20,000 cm^{-1} below the $He^* \; 2^3S_1$ and $He^* \; 2^1S_0$ metastable levels. Although the $6s$, $6p$, and $5d$ levels of Cd-II are closer to the helium metastable levels than are the $5s^2 \; ^2D$ levels of Cd-II, preferential excitation transfer occurs from helium metastables to the two $5s^2 \; ^2D$ states in the Penning reaction

$$He^* \; 2^3S_1(\uparrow\uparrow) + Cd \; 5s^2 \; ^1S_0(\uparrow\downarrow) \; \rightarrow \; He \; ^1S_0(\uparrow\downarrow) + Cd^{+'} \; 5s^2 \; ^2D_{3/2, \, 5/2}(\uparrow)$$
$$+ e^-(\uparrow) + \text{kinetic energy}, \qquad (5.28)$$

in which spin is conserved.[68] Here, since the ground-state configuration of neutral cadmium is that of $4d^{10} \; 5s^2$, and the configuration of the final $d^{+'}$ state is $4d^9 \; 5s^2$, the collision has resulted in the removal of an inner d-shell electron. Radiative decay then occurs between the $4d^9 \; 5s^2 \; ^2D$ states and the $4d^{10} \; 5p$ states to give the two strong (laser) lines at 441.6 and 325.0 nm, and a nonlaser line at 353.6 nm. The two-electron transitions of the type $nd^9(n+1)s^2 \rightarrow nd^{10}(n+1)p$, where $n = 4$ for cadmium, are actually forbidden, but a strong admixture of an "orbitally degenerate" component $nd^{10}(n+1)p^2$ makes them strong transitions.[69, 70] The Einstein A-coefficient for the $4d^9 \; 5s^2 \; ^2D_{5/2} - 4d^{10} \; 5p \; ^2P_{3/2}$ transition at 441.6 nm is about $1.5 \times 10^6 \; sec^{-1}$; $3.33 \times 10^6 \; sec^{-1}$ for the 325.0-nm transition; and $0.52 \times 10^6 \; sec^{-1}$ for the $5s^2 \; ^2D_{3/2} - 5p \; ^2P_{3/2}$ transition at 353.6 nm.[71]

Webb et al.[41] have shown that the $6s \; ^2S_{1/2}$ state of Cd-II, about 5000 cm^{-1} below the $He^* \; 2^3S_1$ level, is also selectively excited via the Penning reaction. An earlier figure, Fig. 5.12, gives their results. It shows the $5s^2 \; ^2D$ levels to be entirely pumped by the Penning reaction, together with the majority of excitation of the $6s \; ^2S_{1/2}$ state. The rest of the excitation of the $6s \; ^2S_{1/2}$ state is provided by cascade transitions from the charge-transfer-excited $9s$, $8d$, and $6f$ states. These results are, of course, applicable to flowing-afterglow conditions that could differ from the conditions in the positive column of a glow-discharge ion laser, or hollow-cathode-laser discharges. Laser oscillation on the $6s\ ^2S_{1/2}-5p\ ^2P^o_{1/2,\ 3/2}$ transitions at 274.9 and 257.3 nm is not to be expected, as population inversion does not exist between the $6s$ and $5p$ states.[71]

The cross-section for the Penning reaction in the excitation of the $Cd^{+\prime}$ states is approximately 50×10^{-16} cm^2 (Schearer and Padovani[72] and Collins et al.[42]).

5.3.1.2. Operating characteristics and discharge conditions of the 441.6-nm He–Cd ion laser

Most of the work on the 441.6-nm and 325.0-nm helium–cadmium ion lasers and other metal-vapor ion lasers has been carried out with discharge-tube structures that use a single, heated cadmium reservoir near the anode and utilize the cataphoretic effect to distribute the cadmium vapor along the active bore of the discharge tube,[73–79] and rely on the discharge current to provide enough self-heating along the bore of the tube to prevent condensation of the cadmium vapor in the bore. A commonly used laser-tube design that relies on self-

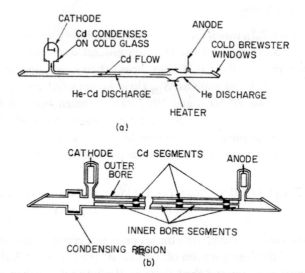

FIG. 5.24. (a) Single reservoir, cataphoresis He–Cd (or metal-carrier gas) laser tube.[78] (b) Segmented-bore He–Cd laser tube.[81] (By courtesy of the Editor of *Laser Focus*, and the American Institute of Physics, respectively.)

heating of the active bore of the tube is given in Fig. 5.24a. The rapid loss of helium in some sealed-off laser tubes has led to the development of improved designs[80] to minimize this effect. Nonuniform distribution of cadmium vapor in the bore region, which is inherent in the design of the tube shown in Fig. 5.24a, has been improved in the segmented-bore, helium–cadmium laser tube of Silfvast and Szeto.[81] This tube design can lead to an increase in gain of 50 percent over that using a single cadmium source. Its construction is fully described in ref. 81 and will not be discussed here.

Operating characteristics. Figure 5.25 shows plots of the laser power at 441.6 nm as a function of current for various helium- and cadmium-reservoir temperatures.[79] The laser tube used to give these results was 85-cm long, and had a 4-mm bore. It relied on the cataphoretic

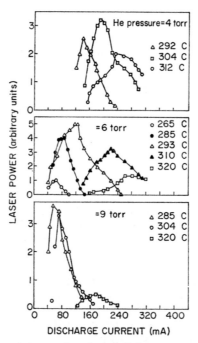

FIG. 5.25. 441.6-nm laser output power as a function of discharge current for various cadmium-reservoir temperatures and helium pressures, $D = 4$ mm, active region of tube maintained at 350°C. (After T. G₁ Giallorenzi and S. A. Ahmed,[79] by the courtesy of the IEEE.)

effect to transport cadmium vapor between the anode and cathode, and had the temperature of the active region maintained by heater tapes at 350°C ± 5 percent. This temperature was always above that of the cadmium reservoir.

At a given helium pressure, the laser power at 441.6 nm and the spontaneous emission of the 441.6-nm line in sidelight, peaked at higher currents as the cadmium-reservoir tempera- ture, and hence the cadmium-pressure, is increased. It is important to note that, though not shown in the figure, the 441.6-nm sidelight spontaneous emission is found to follow the same behavior as the laser output.[79] Maximum laser output occurs at a helium pressure of about 4 to 5 torr, and a cadmium-reservoir temperature of about 295°C. In Fig. 5.26 the spontane- ous emission from several lines in He-I, Cd-I and Cd-II is plotted as a function of discharge current for a fixed helium pressure (5 torr) and cadmium-reservoir temperature (294°C), for the same tube-diameter of 4 mm. Crucial to the following analysis is the behavior of the He-I $4s\,^3S_1$–$2p\,^3P_{0,1,2}$ transition (Racah notation) at 471.3 nm with discharge current, and its assumed relationship to the variation of the He* 2^3S_1 metastable concentration with discharge current. The He-I $4s\,^3S_1$ state can be assumed to be populated by one or both of the following processes: direct electron-impact excitation from the ground state, cascading from higher-lying excited states that can be excited directly from the ground state, or elec- tron impact excitation from the He* 2^3S_1 metastable state. In Fig. 5.26, above 140 mA the 471.3-nm line increases nearly linearly with current, while below 140 mA, it increases faster than linearly with current. The faster than linear increase of the 471.3-nm line intensity below 140 mA is attributed to the fact that the He* 2^3S_1 metastable population is not yet saturated

225

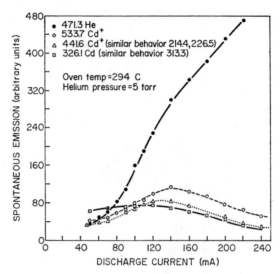

FIG. 5.26. Spontaneous emission of several He-I, Cd-I, and Cd-II lines as a function of discharge current, $D = 4$ mm. (After T. G. Giallorenzi and S. A. Ahmed,[79] by courtesy of the IEEE.)

and that the contribution of excitation due to it will be a minimum of a two-step process from cascade transitions into the $4s\,{}^3S_1$ state from levels excited from the He* 2^3S_1 metastable state, with an I^2 dependence. At higher currents, it is assumed that the quadratic dependence changes to one linear with I as the He* 2^3S_1 concentration saturates. The linearity of the 471.3-nm line intensity for currents more than 140 mA, combined with a relatively flat voltage–current characteristic of the discharge above 140 mA, implies that the electron temperature is roughly independent of current over this range. At a constant electron temperature, the electron concentration is directly proportional to the discharge current, and the intensity of the neutral cadmium line at 326.1 nm is expected to be directly proportional to the product of current and cadmium concentration. The spontaneous emission of the 326.1-nm line maximizes around 100 to 120 mA, and indicates that the neutral cadmium concentration maximizes at this discharge current, and decreases with further increase in current. Since the intensity of the 441.6-nm Cd-II laser line in spontaneous sidelight emission is expected to be directly proportional to the product of the He* 2^3S_1 triplet-metastable concentration $M_T(2^3S_1)$ and the neutral cadmium-atom concentration N_{Cd}, the intensity of the 441.6-nm should reflect the (assumed) saturation of He* 2^3S_1 metastables above 120 mA, and the decrease in neutral cadmium-atom concentration beyond about 100 to 120mA. This is approximately the case. Evidence that supports this explanation for the spontaneous-intensity behavior of the 441.6-nm line has been supplied by Silfvast.[67] Silfvast monitored the intensity of the 441.6 nm in sidelight in a 20-cm-long, 10-mm-bore discharge tube, measured the He* 2^3S_1 triplet-metastable concentration $M_T(2^3S_1)$ and that of the He* 2^1S_0 singlet metastables $M_S(2^1S_0)$, and determined the neutral cadmium concentration N_{Cd} from the temperature of the cadmium metal in the sidearms mounted along the active region of the discharge tube. The measured values of M_T and M_S versus N_{Cd} at the (low) discharge current of 10 mA are shown in Fig. 5.27. The figure also shows the normalized plot of $(M_T + M_S)N_{Cd}$ compared with the experimental values of the sidelight intensity of the

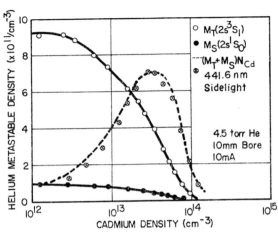

FIG. 5.27. Correlation of helium metastable densities with 441.6-mm sidelight intensity. (After W. T. Silfvast,[67] by courtesy of the American Institute of Physics.)

441.6-nm Cd-II laser line. The agreement of the latter is very good and shows that Penning collisions of both $He^* 2^3S_1$ and $He^* 2^1S_0$ metastables are responsible for the excitation of the $5s^2 {}^2D_{5/2}$ upper state of the 441.6-nm line (and also the $6s {}^2S$, $5d {}^2D$, and $5p {}^2P$ Cd-II levels, lines originating from which levels have the same intensity variation as the 441.6-nm line). At more laser-like discharge currents than the 10 mA (in a 10-mm-bore discharge tube) used by Silfvast, the relative concentration of $He^* 2^1S_0$ metastables in the total metastable concentration $(M_T + M_S)$ will decrease appreciably with a concomitant increase in the relative concentration of $He^* 2^3S_1$ metastables because of the superelastic collision process

$$He^* 2^1S_0 + e^- \text{ (slow)} \rightarrow He^* 2^3S_1 + e^- \text{ (faster)} \qquad (5.29)$$

that has a very large cross-section of 1.40×10^{-14} cm^2 for electrons of thermal energy. The result will be that Penning ionization from the $He^* 2^3S_1$ triplet level will play the major role in the selective excitation of the Cd-II, $5s^2 {}^2D$; $6s {}^2S$, $5d {}^2D$, and $5p {}^2P$ states.

As the laser output at 441.6 nm has the same behavior as the spontaneous-emission intensity of the 441.6-nm line, it follows immediately that the saturation of output power is due to processes that affect the $5s^2 {}^2D_{3/2}$ upper-state population and not to processes that affect the lower-state population. This saturation of output power with discharge current under optimum discharge conditions of pressure and cadmium concentration is due primarily to the saturation of $He^* 2^3S_1$ triplet metastables with discharge current.

Electron temperature and concentration, and ion concentration. The electron temperature in a helium–cadmium laser decreases very rapidly with increasing cadmium pressure. Results of double-probe determinations of the electron temperature over the range of typical helium–cadmium 441.6-nm laser-discharge conditions in a 5.6-mm-bore tube by Goto et al.[82] are shown plotted in Fig. 5.28. Although the electron temperature hardly changes with discharge current over the range 15 to 80 mA at both 1 torr and 2.5 torr pressures of helium, it decreases significantly with increasing cadmium pressure. Over the cadmium vapor range of 10^{-4} to 10^{-2} torr, which covers the pressure at which maximum laser output at 441.6 nm

227

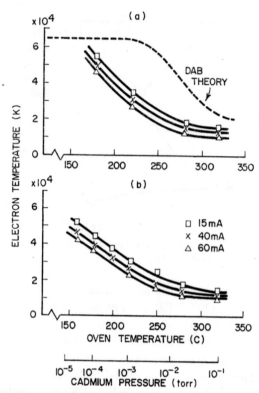

FIG. 5.28. Average electron temperature in the positive column of a He–Cd discharge as a function of cadmium pressure (and oven temperature). Helium pressure: (a) 1.0 torr, (b) 2.5 torr; $D = 5.6$ mm. (After T. Goto, A. Kawahara, G. J. Collins, and S. Hattori,[82] by courtesy of the American Institute of Physics.)

occurs, the electron temperature decreases by a factor of 2.5. For cadmium partial pressures above 10^{-1} torr the electron temperature is determined by the cadmium pressure alone, and approaches that characteristic of a pure cadmium discharge. The broken line in Fig. 5.28a (marked DAB) is a plot of the theoretical electron temperature calculated using the method of Dorgela, Alting, and Boers,[83] which is applicable to the positive column in a binary mixture (Section 3.2.2). Three fundamental assumptions in the DAB method are that ionization occurs only via single-step electron impact from the ground state and that volume ionization by excited species (such as by the Penning effect) does not occur, and also that two-step ionization from metastable states, that would make the electron temperature current-dependent, does not occur. The helium–cadmium ion laser clearly does not satisfy the first fundamental assumption as significant ionization in it is produced in Penning reactions. It is this ionization that is probably the cause for the large discrepancy between the experimental and theoretical curves of electron temperature versus oven temperature of Fig. 5.28a.

The results of probe measurements and calculations of the electron concentration in the helium–cadmium plasma versus cadmium-vapor pressure at a helium pressure of 2.5 torr at various currents covering the current region in which oscillation can occur, are shown in Fig. 5.29 on a log-linear scale.[82] It can be seen that the electron concentration increases approximately linearly with increasing discharge current over the whole range of cadmium (vapor) pressure, but slightly more rapidly than linearly at higher cadmium pressures than at lower pressures. In addition, the electron density exhibits a broad minimum at cadmium-vapor pressures, depending on the discharge current, between approximately 10^{-3} to 10^{-2} torr. The electron density at the minimum is only about 50 percent of what it is at either end

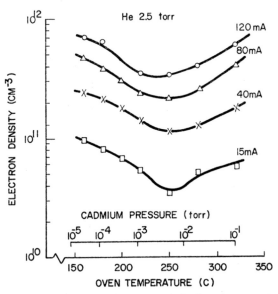

FIG. 5.29. Electron density as a function of cadmium pressure (and oven temperature) in a helium–cadmium laser discharge, $D = 5.6$ mm. (After T. Goto, A. Kawahara, G. J. Collins, and S. Hattori,[82] by courtesy of the American Institute of Physics.)

of the cadmium-vapor-pressure range. This region corresponds to that in which the electron temperature is starting to flatten out (Fig. 5.28b), between 25,000 and 20,000 K, and is the region in which the largest discrepancy exists between the experimental and calculated values of electron temperature in Fig. 5.28a. The electron temperature at the minimum is less than one-half of what it is in a pure helium discharge at a pD of 14 torr-mm.

It seems clear from these results that the Penning reaction between helium metastables and ground-state cadmium atoms that results in volume ionization is responsible for the reduction in electron temperature as the cadmium-vapor pressure increases. The increase in volume ionization means that the electron temperature can decrease in order to maintain the ionization rate needed to sustain the discharge at the same discharge current. Possibly the supply by the Penning process of energetic electrons, that have an energy that can be as much as the 11.6 eV difference in potential energy between $He^* \, 2^1S_0$ atoms and the ionization energy of neutral cadmium, to an electron-energy distribution that has an average electron temperature of only about 20,000 to 28,000 K (or 3 to 4 eV) enables more ionization of cadmium to occur by direct electron impact. The ionization energy of cadmium is only 8.99 eV, so that this process is energetically possible. Also, electrons having an energy of about 11 eV are capable of producing additional ionization by electron impact on helium metastables similar to those that produced them.

Silfvast[67] has shown that where the 441.6-nm line peaks in intensity with variation in cadmium density, the Penning ionization rate is a factor of about 2 greater than the electron-impact-ionization rate of cadmium. Penning ionization, therefore, appears to be the dominant ionization source of cadmium in the cadmium-vapor-pressure region where the output power at 441.6 nm is maximized. Figure 5.30, which is part of a figure given in ref. 67, shows this. The solid curve marked Cd$^+$ is a plot of the ion density in the $5s\ ^2S_{1/2}$ Cd$^+$ ion ground

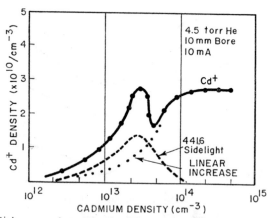

FIG. 5.30. Variation of Cd$^+$ ion ground-state density and 441.6-nm sidelight intensity, with cadmium density; also showing the expected (linear) increase of Cd$^+$ ground-state ions with cadmium density in electron-impact excitation. (After W. T. Silfvast,[67] by courtesy of the American Institute of Physics.)

state. It is shown together with the variation with cadmium density of the sidelight intensity of the 441.6-nm Cd-II laser line. Clearly, the Cd$^+$ concentration goes through a peak at the same value of cadmium density as the peak in the 441.6-nm sidelight intensity. The pronounced peak of Cd$^+$, and its association with the maximum in the 441.6-nm sidelight intensity can be attributed to the Penning process. The dotted curve is a plot of the estimated cadmium ion density that is produced by electron impact in the absence of the Penning effect. The pronounced peak in the Cd$^+$-ion ground-state density reported in Silfvast's work is hard to reconcile with the reduction in electron density in the laser-discharge region reported by Goto et al.[82] (shown here in Fig. 5.28b) unless there is a corresponding large reduction in the concentration of helium ions.

Discharge conditions. As yet, it is not clear just what actually are the optimum discharge conditions for the 441.6-nm helium–cadmium ion laser. A variety of "optimum" helium-pressure and tube-diameter products, cadmium-reservoir temperatures, tube temperatures, and discharge currents have been reported in the literature. Differences in individual "optimum comditions" probably arise as a result of the use of differing constructional detail, such as laser tubes with different wall thicknesses, and depend on whether or not external heating of the active region of the laser tube is used in conjunction with self-heating. Table 5.8 lists a number of reported optimum discharge conditions, together with constructional notes and references. The data presented indicates that the most popular value for the optimum helium-pressure and tube-diameter product is about 12 torr-mm for laser tubes that have an unheated (self-heated) active region. The optimum discharge current given here for

TABLE 5.8. *441.6-nm laser optimum discharge conditions*

Tube diam. D (mm)	Helium pressure p (torr)	Temperature (°C) of tube (T) or reservoir (R)	pD product (torr-mm)	Discharge current (mA)	Cadmium type (natl. or isotopic 114)	Bore type	References
5.0	2.0	200(T)	10	> 125	Natl.	Heated	40
2.4	3.4	233(R)	8.2	110	Natl.	Unheated	75
2.6	3.0	205(R)	7.8	140	114	Unheated	137
4.0	5.0	294(R)	20	130	Natl.	Heated	79
1.6	< 8.0	—	< 12.8	50	114	Unheated	85
2	6.0	—	12	67	Natl.	Self-heated segmented	81
3	4.0	—	12	140	Natl.	Unheated	71

a variety of tubes does not appear to scale with tube diameter, though Goldsborough and Hodges[75] report that it does in their work. As the partial pressure of cadmium vapor is quite critical with the output falling off rapidly at 0.001 and 0.003 torr from an optimum of 0.002 torr[75] corresponding to the vapor pressure of cadmium of 233°C, variation in "optimum" discharge currents for laser tubes of approximately the same internal diameter is likely to be associated with the exact physical construction, constituents, and external dimensions of the active-bore region of the tube. The self-heated, segmented-bore laser tube[81] illustrated in Fig. 5.24b requires a discharge current of 67 mA (for a tube of 2-mm i.d.) that is significantly lower than others having a comparable internal diameter.

Theoretically, the maximum available laser power on the 441.6-nm transition using cadmium of natural isotopic abundance in a 3-mm-bore tube is 12 mW/cm³. For the same discharge conditions 10 mW/cm³ has been reported,[84] while approximately 13 mW/cm³ has been obtained in a laser tube of 2-mm bore.[81] More than 30 mW/cm³ has been reported from a 1.6-mm-bore tube using an isotopic He³–Cd¹¹⁴ mixture.[84]

Inverse radial dependence of gain—the $1/D$ gain relationship. Just as in certain neutral gas lasers (in particular the 632.8-nm and 1.152-μm He–Ne lasers), at optimum discharge conditions of helium pressure, cadmium-vapor pressure, and discharge current, the gain of the 441.6-nm helium–cadmium laser varies inversely as the tube diameter.[40, 74, 78, 84, 86] Since for optimum discharge conditions, the helium pressure and tube-diameter product is a constant (Table 5.8), the inverse radial dependence of gain means that the gain is directly proportional to the helium pressure. The direct dependence of the gain on helium pressure follows from the mechanism that is responsible for the saturation in the population of the 441.6-nm upper laser level, and not from a lower laser level depopulating mechanism. The saturation of 441.6-nm laser power with discharge current is due to the saturation of the He* 2^3S_1 triplet metastable population (as shown earlier). It follows then, since the gain is determined primarily by the upper laser state population, which is directly proportional to the He* 2^3S_1 metastable population, that the $1/D$ gain relationship (or pressure-gain relationship) is caused by the variation of the He* 2^3S_1 metastable population with helium pressure p_{He}. This would be so if the $He^* \, 2^3S_1$ metastable population M_T on the tube axis at high current densities where saturation occurs, followed the relationship

$$M_T = \frac{k_1 \, n_e \, p_{He}}{k_2 \, p_{Cd} + k_3 \, n_e + \gamma} \tag{5.30}$$

in which $k_1 \, n_e \, p_{He}$ is the production rate of $He^* \, 2^3S_1$ metastables by fast electrons, $k_2 \, p_{Cd}$ is the effective destruction rate of $He^* \, 2^3S_1$ metastables by ground-state cadmium atoms, $k_3 \, n_e$ is a current-saturation term for the deactivation rate by electrons in excitation to higher levels, and γ is a diffusion-loss term (which could be included in k_2).

This relationship between M_T and the helium pressure is similar to that derived by:

(1) Silfvast[67] for the concentration of $He^* \, 2^3S_1$ metastables in the a helium–cadmium discharge;

(2) Gordon and White[87] for the concentration of $He^* \, 2^1S_0$ metastables in a 3.39-μm He–Ne laser discharge;

(3) Herziger et al.[88] for the concentration of helium metastables in the presence of neon atoms, derived from a general differential equation for the formation and destruction of metastable atoms established by Fabrikant,[89] and

(4) deduced by Young et al.[90] to apply to the concentration of $He^* \, 2^3S_1$ metastables in a 1.152-μm He–Ne laser mixture in the positive column of a glow discharge.

The validity of relationship (5.30) is likely to be restricted in positive column discharges to helium pressures of less than 5 torr. Above a pressure of 5 torr, helium metastable loss becomes appreciable through the two-body, molecular-helium-formation reactions

$$He^* \, 2^3S_1 + He \, ^1S_0 \rightarrow He_2 \, ^3\Sigma_u^+ \tag{5.31}$$

and

$$He^* \, 2^1S_0 + He \, ^1S_0 \rightarrow He_2 \, A \, ^1\Sigma_u^+, \tag{5.32}$$

and three-body reactions that involve metastable helium atoms and ground-state helium atoms (3.27) and (3.28).

To the author's knowledge no direct experimental measurements have been published of helium metastable concentrations in a glow discharge as a function of helium pressure at constant electron concentration while the electron temperature is maintained constant (by keeping the pD constant) that could confirm the validity of the general form of relationship (5.30). The only published work that comes close to doing this is restricted to measurements on helium–neon mixtures in hollow-cathode discharges (referred to in Section 3.4.6, "Excited-states population in the HCD"). In this work by Znamenskii,[91] in which $p_{He} \, a$, $p_{Ne} a$, and ja were kept constant (j is the discharge-current surface density on the inside of the HCD, and a is the hollow-cathode diameter), it was found that the concentration of $He^* \, 2^3S_1$, $He^* \, 2^1S_0$, as well as $Ne^* \, 1s_5$ metastables was inversely proportional to the cathode diameter a. Since pa was maintained constant, the metastable concentrations are thus directly proportional to the gas pressure.

All indications from laser-orientated experiments is that the concentration of metastables (and possibly most excited-state populations) in the positive column of a glow discharge, or in rf-discharges, is directly proportional to the pressure (at a constant pD) up to a few tortorr of helium just as it is in hollow-cathode discharges at higher helium pressures. One looks forward to an early experimental verification of this prediction of a facet of gas discharge physics that is basic to the understanding of gas laser processes. Silfvast (private communication, June 1972) has found that the metastable concentration in the positive column of glow discharge in helium maximizes at optimum discharge currents at about 3 torr (in a 4-mm-bore discharge tube) and at the same helium pressure in a helium–cadmium laser discharge.[92] This gives a pD of 12 torr-mm that corresponds to the optimum $p_{He}D$ product of the 441.6-nm helium-cadmium laser (Table 5.8), and which is close to the optimum $p_{He}D$ product of the 1.15-μm He–Ne laser (Chapter 4, p. 147). This result might have been anticipated as He* 2^3S_1 metastables provide the main upper level selective excitation in both laser systems.

5.3.2. THE HELIUM–ZINC LASER

Laser oscillation in the helium–zinc ion laser has been reported on the nine lines listed in Table 5.9. As well as listing the wavelengths, and upper states of the lines, the mechanism by which their upper states are excited is specified.

TABLE 5.9. *Helium–zinc ion laser lines*

Wavelength (nm)	Identification	Upper state excitation
589.4	$3d^9 4s^2\ ^2D_{3/2}$	Penning reaction
747.8	$3d^9 4s^2\ ^2D_{5/2}$	Penning reaction
491.1	$3d^{10}4f\ ^2F_{5/2}$	Charge transfer
492.4	$3d^{10}4f\ ^2F_{7/2}$	Charge transfer
602.1	$3d^{10}5d\ ^2D_{3/2}$	Charge transfer
610.2	$3d^{10}5d\ ^2D_{5/2}$	Charge transfer
758.8	$3d^{10}5p\ ^2P_{3/2}$	Cascade processes
761.3[a]	$3d^{10}6s\ ^2S_{1/2}$	Charge transfer
775.7[a]	$3d^{10}6s\ ^2S_{1/2}$	Charge transfer

[a] Observed in pulsed oscillation only.

5.3.2.1. *Upper level excitation*

As indicated in Table 5.9 the majority of the Zn-II laser lines are excited via charge transfer. Of the nine lines listed, only two, the 589.4-nm line ($3d^9\ 4s^2\ ^2D_{3/2}$–$3d^{10}\ 4p\ ^2P^\circ_{1/2}$) and the 747.8-nm line ($3d^9\ 4s^2\ ^2D_{5/2}$–$3d^9\ 4p^2\ P^\circ_{3/2}$), are not selectively excited by charge transfer from ground-state He$^+$ ions.[19, 41, 46] Selective excitation of the upper states of the 589.4-nm and 747.8-nm lines occurs in the Penning reaction

$$He^*\ 2^3S_1(\uparrow\uparrow) + Zn\ 4s^2\ ^1S_0(\uparrow\downarrow) \rightarrow He\ ^1S_0(\uparrow\downarrow) + Zn^{+\prime}\ 4s^2\ ^2D_{3/2,\ 5/7}(\uparrow)$$

$$+ e^-(\uparrow) + \text{kinetic energy} \qquad (5.33)$$

in which spin is conserved.[68] The reaction, as in the Penning reactions in the helium–cadmium laser, results in the removal of an inner d-shell electron from the Zn-I $3d^{10}\,4s^2$ ground-state atom to give selective ionization and excitation of the $3d^9\,4s^2\,{}^2D$ doublet-state of Zn-II. (The two transitions of the 589.4-nm and 747.8-nm lines are similarly "forbidden" but are strong transitions[69, 70] as are the two strong 441.6-nm and 325.0-nm laser lines in the helium–cadmium laser.) The kinetic energy of the electron emitted as a result of the reaction is about 2.5 eV. The energy disposition of the He* 2^3S_1 state and the Zn-II $4s^2\,{}^2D$ states is shown in the earlier figure (Fig. 5.13).

By means of fluorescence-decay measurements in pulsed helium–zinc mixtures, Collins,[47] Collins et al.,[19] and Riseberg and Schearer[48] have established that the helium metastables involved in the Penning reaction in the helium–zinc ion laser are He* 2^3S_1 metastables. Figure 5.31, in which the exponential decay rates of eight Zn-II laser lines and the metastable helium concentrations in the afterglow of a pulsed discharge are plotted as a function of neutral zinc density at constant helium pressure, illustrates this clearly.[19] The He* 2^3S_1 decay, which was determined from absorption measurements, has the same decay rate as the intensity of the Zn-II 589.4-nm and 747.8-nm lines whose upper states are the Zn-II $4s^2\,{}^2D$ doublet states. The He* 2^1S_0 metastable decay rate is seen to be much faster than that of the He* 2^3S_1 metastables and the decay of the fluorescence of the 589.4-nm and 747.8-nm lines. The decay of fluorescence of the 491.1-, 492.4-, 602.1-, 610.2-, 758.8-, and 775.7-nm lines originating on higher-lying Zn-II states of the type $3d^{10}\,nx$, differs from that of either of the two helium metastables showing that the excitation of these lines is associated with some other excited species of helium. The linear variation of the decay rate with zinc partial pressure shows that a two-body collision process involving the Zn-I ground state is responsible for the excitation of all the Zn-II lines. The only excited species of helium capable of exciting the Zn^{II} $3d^{10}\,nx$ states is the He$^+$ atomic ion.

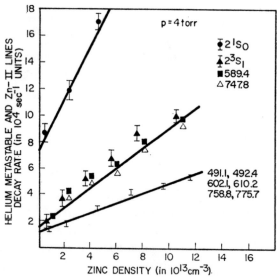

FIG. 5.31. Dependence of exponential decay rates of various Zn-II excited species and helium metastables on zinc density in the afterglow of a pulsed discharge in a He–Zn mixture. (After G. J. Collins, R. C. Jensen, and W. R. Bennett, Jr.,[19] by courtesy of the American Institute of Physics.)

The analysis by Riseberg and Schearer[48] of similar fluorescence-decay measurements to those shown in Fig. 5.31 yields a cross-section at 270°C for the Penning reaction (5.33) of $2.91 \pm 0.15 \times 10^{-15}$ cm².

Under actual helium–zinc ion laser discharge conditions, direct electron-impact excitation from the Zn-I ground state is also likely to contribute to ionization and excitation of the Penning-reaction-excited Zn-II $3d^9 4s^2$ ^2D states as these states have abnormally high cross-sections for excitation by electron impact (of the order of 5×10^{-17} cm²) at electron energies in the range 80 to 100 eV.[93]

5.3.2.2. Operating characteristics and discharge conditions

The variation of laser output power on the three transitions $3d^9 4s^2$ $^2D_{5/2}$–$3d^{10} 4p$ $^2P^\circ_{3/2}$, $3d^9 4s^2$ $^2D_{3/2}$–$3d^{10} 4p$ $^2P^\circ_{1/2}$, and $3d^{10} 5p$ $^2P_{3/2}$–$3d^{10}$ $5s$ $^2S_{1/2}$ at 747.8, 589.4 and 758.8 nm respectively, with pressure of helium, discharge current, and zinc pressure (temperature) is shown in Fig. 5.32. The laser tube used had an active length of 90-cm and an i.d. of 4 mm. The zinc vapor pressure was maintained in the active region by means of an oven surrounding the tube, zinc vapor diffused from the zinc reservoir into the active region through small holes situated along the bore of the tube. Heating of the zinc reservoirs by the discharge current was minimized by placing the zinc reservoirs far from the active bore.[47]

The optimum helium pressure for the Penning-reaction-excited 747.8-nm and 589.4-nm lines peaks sharply at approximately 4 torr, while the 758.8-nm line, which is excited indirectly in cascade processes from upper-lying charge transfer excited states, exhibits a broad maximum in output at about 6 torr (Fig. 5.32a). The optimum discharge current for the 747.8-nm and 589.4-nm lines is similar at about 100 mA, with the threshold for all lines typically 25 mA (Fig. 5.32b). The 758.8-nm line requires a higher current of between 150 to 200 mA. Figure 5.32c shows that the laser output is a sensitive function of the zinc temperature (or zinc partial pressure). The laser output of the 589.4-nm line is particularly sensitive to the zinc temperature with oscillation varying from threshold to maximum over the narrow temperature range of 325° to 335°C.

Discharge conditions for the Penning reaction excited lines at 747.8 nm and 589.4 nm (deduced from Fig. 5.32a) vary from the optimum helium-pressure and tube diameter product of 16 torr-mm for a positive-column laser discharge in a tube of 4-mm bore, to an operating range of 14 to 120 torr-mm in HCD helium–zinc lasers[94, 95] in tubes of 6- and 14-mm bore. Due to the similarity of the upper-level selective excitation mechanisms of the 747.8-nm and 589.4-nm Zn-II laser lines and the 441.6-nm Cd-II laser lines in helium–zinc and helium–cadmium discharges, it is likely that the optimum helium pressure and tube diameter product for a positive-column, helium–zinc laser will be similar at about 12 torr-mm (Table 5.8). The higher operating temperature needed, however, to give the optimum zinc pressure in the helium–zinc laser will probably introduce some difference in optimum pD products between the two laser systems. Silfvast[92] has in fact found that the optimum $p_{He}D$ product is higher at 16 torr-mm.

Likewise, as in the 441.6-nm helium–cadmium laser under optimum conditions, the gain of the 747.8-nm and 589.4-nm Zn-II laser lines probably exhibits an approximate $1/D$ relationship due the variation of the concentration of He* 2^3S_1 metastables with pressure and discharge-tube diameter. Saturation of the He* 2^3S_1 metastable concentration and (to a

235

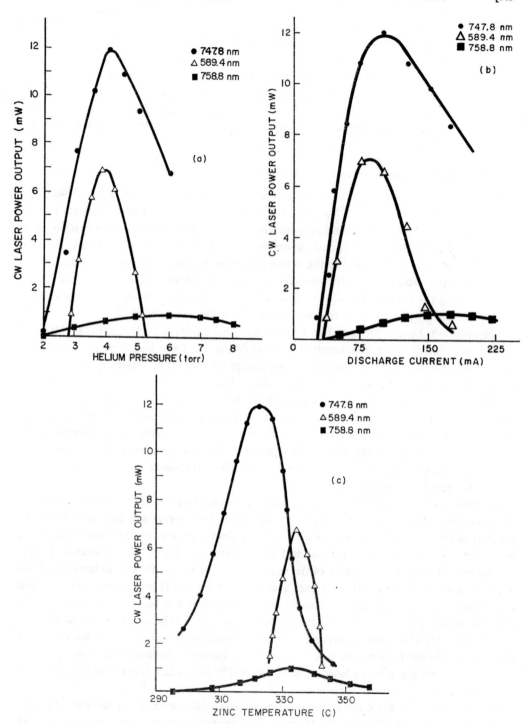

FIG. 5.32. He–Zn CW-laser output power as a function of (a) helium pressure, (b) discharge current, and (c) oven temperature, $D = 4$ mm; active region of tube heated by oven. (After R. C. Jensen, G. J. Collins, and W. R. Bennett, Jr.,[46] by courtesy of the American Institute of Physics.)

smaller extent) radiation trapping between the $4p$ $^2P^\circ_{1/2,\,3/2}$ lower laser levels and the Zn^+ $3d^{10}4s$ $^2S_{1/2}$ ion ground state on transitions at 202.6 and 206.2 nm limits the output power at high discharge currents. As well as being populated by cascade processes and direct electron-impact excitation from the Zn-I ground state, the Zn^+ $^2S_{1/2}$ ion ground state is probably also populated in the Penning reaction

$$\text{He}^* \; 2^3S_1 + \text{Zn} \; ^1S_0 \;\rightarrow\; \text{He} \; ^1S_0 + \text{Zn}^+ \; ^2S_{1/2} + e^- + 10.4 \text{ eV kinetic energy,} \qquad (5.34)$$

which produces ionization, but not simultaneous ionization and excitation of the zinc.

5.3.3. THE HELIUM–MAGNESIUM LASER

This is one metal vapor ion laser in which CW oscillation was not preceded by reports of pulsed oscillation. Lasing for the first time in magnesium, in the singly ionized spectrum, under CW conditions was reported by Hodges[96] at four wavelengths in the near-infrared. The measured wavelengths, actual wavelengths, and identification are given in Table 5.10.

TABLE 5.10. *Wavelengths and assignments for Mg-II laser transitions in the He–Mg ion laser*[a]

Measured wavelength[b] (μm)	Actual wavelength[c] (μm)	Identification[c]
0.9218	0.9218	$4p$ $^2P^\circ_{3/2}$–$4s$ $^2S_{1/2}$
0.9244	0.9244	$4p$ $^2P^\circ_{1/2}$–$4s$ $^2S_{1/2}$
	1.0915	$4p$ $^2P^\circ_{3/2}$–$3d$ $^2D_{3/2}$
1.0915		
	1.0914	$4p$ $^2P^\circ_{3/2}$–$3d$ $^2D_{5/2}$
1.0952	1.0951	$4p$ $^2P^\circ_{1/2}$–$3d$ $^2D_{3/2}$

[a] Upper laser levels are selectively excited by Penning reactions with helium metastables. Other Mg-II laser transitions have been reported in mixtures of magnesium vapor and noble gases (Appendix Table 13).
[b] Resolution ±0.00015 μm.
[c] According to ref. 97.

The basic laser tube in which oscillation was achieved was similar to that illustrated in Fig. 5.24a, in that a uniform distribution of magnesium was achieved by the cataphoretic effect from a heated magnesium reservoir situated in a sidearm near the anode. In addition, the 4-mm-bore quartz plasma tube was enclosed by a 2-cm quartz tube concentric to the bore that formed an annular air space, and provided thermal insulation for the plasma tube.

Laser oscillation was observed for helium pressures (cold-filling) of 0.5 to 5.5 torr with a magnesium-reservoir temperature of $395\pm10°C$. A minimum discharge current of 100 mA was necessary to sufficiently self-heat the plasma tube to prevent condensation of the magnesium along the bore of the tube. Contrasting with previously reported Penning-excited helium–metal vapor lasers such as the He–Cd and He–Zn ion lasers, only weak oscillation occurred for helium pressures greater than 3 torr. At helium pressures greater than 3 torr, the output was erratic, with the optimum discharge current approximately 250 mA. For pressures below 2 torr, stable oscillation at the milliwatt power-level with unoptimized mirrors was obtained, but the laser power had still not saturated at a discharge current of 800 mA, which was sufficient to heat the plasma tube to incandescence.

5.3.3.1. Upper level excitation

A partial energy-level diagram of Mg-II, also illustrating the disposition of the He^* 2^3S_1 and He^* 2^1S_0 metastable levels, the four infrared laser transitions (arrowed transitions), together with other transitions (wavelengths in μm) and lifetimes of a number of the Mg-II

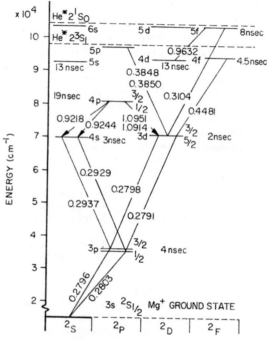

Fig. 5.33. Partial energy-level diagram of Mg-II and its coincidence with He* 2^3S_1 and He* 2^1S_0 metastable levels, including Mg II state lifetimes. Laser transitions are shown as arrow-headed transitions. Wavelengths are given in μm. (After D. T. Hodges,[96] by courtesy of the American Institute of Physics.)

states, are shown in Fig. 5.33. The $4p$ $^2P°_{3/2, 1/2}$ levels are within 17,530 cm^{-1} of the He^* 2^3S_1 state, and within 23,951 cm^{-1} of the He^* 2^1S_0 level and it is believed that the selective excitation occurs through the Penning reaction

$$He^* \; 2^3S_1, \; 2^1S_0 + Mg \; ^1S_0 \rightarrow He \; ^1S_0 + Mg^{+\prime} \; 4p \; ^2P°_{3/2, \, 1/2} + e^- + \text{kinetic energy.} \quad (5.35a)$$

If spin conservation is important here, as shown for $He^* + Cd$ and $He^* + Sr$ collisions,[68, 98] the reaction is likely to be predominantly

$$He^* \; 2^3S_1(\uparrow\uparrow) + Mg \; 3s^2 \; ^1S_0(\uparrow\downarrow) \rightarrow He \; ^1S_0(\uparrow\downarrow) + Mg^{+\prime} \; 4p \; ^2P°_{3/2, \, 1/2}(\uparrow)$$
$$+ e^-(\uparrow) + \text{kinetic energy} \quad (5.35b)$$

in which spin would be conserved. The energy discrepancy for the reaction, involved as kinetic energy of the emitted electron, is about 2 eV. This is comparable to that involved in the Penning reaction in the helium–cadmium laser and responsible for the excitation of the upper level of the strong 441.6-nm Cd-II laser line.

If the selective excitation is correctly identified as being due to a Penning reaction, it would appear that the $5s\,^2S_{1/2}$ state, within 5395 cm^{-1} of the He* 2^3S_1 level, should also be excited in the same reaction, just as is the Cd-II $6s\,^2S_{1/2}$ state in helium–cadmium discharges (Fig. 5.12). The ratios of the lifetimes of 19/2 and 19/3, given in Fig. 5.33, of the states involved in the four infrared laser transitions are seen to be larger than other allowable transitions of Mg-II in the figure, and so indicates why oscillation has been observed only on the transitions shown. If the $5p$, $4f$, $4d$, and $5s$ states are selectively excited also, a consideration of the Mg-II state lifetimes of Mg-II suggests that additional population inversion under steady-state conditions is only likely to occur between the $4f\,^2F_{7/2}$–$3d\,^2D_{5/2}$ and the $5f\,^2F_{7/2}$–$3d\,^2D_{5/2}$ states at 0.4481 and 0.3104 μm, respectively.

Although He$^+$ ground-state ions have too much potential energy to be able to ionize and excite Mg-II states, charge transfer from Ne$^+$ ground state ions to the $8s$, $7d$, $7f$, and $7g$ states of Mg-II with which they are in close energy coincidence, appears to be a good possibility for their selective excitation in a neon–magnesium discharge. Indeed, evidence for such charge-transfer reactions with large cross-sections, with a preference for f-state excitation, has already been reported by Manley and Duffendack.[99] Such evidence has been reproduced here in an earlier figure (Fig. 2.15) to illustrate the apparent variation of charge-transfer cross-sections with energy discrepancy. It is suggested here that the frequency of collisions of Ne$^+$ ions with Mg atoms will be less than that between helium metastables and Mg atoms in the helium–magnesium laser (or helium metastables and Ne atoms as in the He–Ne lasers) so that a large excitation rate from charge transfer would not be expected in a neon–magnesium ion laser, and the realizable output power would likely be low.

5.4. Electron-impact-excited Ion Lasers

Visible laser oscillation in ion species was first observed when the effect of argon as a carrier gas on the first ion laser lines in mercury was being investigated. The observation was made independently and almost simultaneously in 1964 by Bridges,[100] Convert et al.[101-2] and Bennett et al.[103] Even small traces of argon were sufficient in which to sustain oscillation in argon-ion species in a pulsed helium–mercury discharge, with simultaneous laser oscillation occurring at 488.0 (Ar-II), 615.0, and 567.7 nm (Hg-II).[100] Oscillation on ten lines in the green and blue portions of the spectrum was obtained in argon with neon or helium as a carrier gas at a few hundred millitorr, and on nine out of the ten lines in a few millitorr of argon alone. The ten lines were identified by Bridges[100] as arising from transitions in singly ionized argon. These ten lines, which include some of the strongest lines of singly ionized argon in the visible region of the spectrum, are given in Table 5.11 where their relative amplitudes are given as Strong, Moderate, or Weak. Nine of the lines arise from $4p$–$4s$ transitions as shown in Fig. 5.34 and their upper levels are approximately 35 eV above the neutral ground state. The remaining line at 501.7 nm arises from a $4p$–$3d$ transition and is not included in the figure.

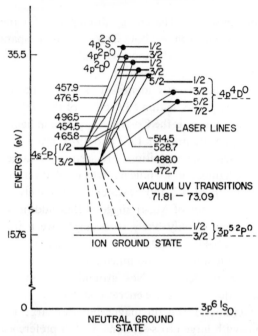

FIG. 5.34. Partial energy-level diagram of singly ionized argon showing nine of the ten laser transitions on which oscillation was first reported (wavelengths are in nm.)

TABLE 5.11. *Laser lines in singly ionized argon to the first observed*[100]

Wavelength (nm)	Identification[a]	Relative amplitude
454.5	$4p\ ^2P^\circ_{3/2}$–$4s\ ^2P_{3/2}$	Weak
457.9	$4p\ ^2S^\circ_{1/2}$–$4s\ ^2P_{1/2}$	Moderate
465.8	$4p\ ^2P^\circ_{1/2}$–$4s\ ^2P_{3/2}$	Moderate
472.7	$4p\ ^2D^\circ_{3/2}$–$4s\ ^2P_{3/2}$	Weak
476.5	$4p\ ^2P^\circ_{3/2}$–$4s\ ^2P_{1/2}$	Moderate
488.0	$4p\ ^2D^\circ_{5/2}$–$4s\ ^2P_{3/2}$	Strong
496.5	$4p\ ^2D^\circ_{3/2}$–$4s\ ^2P_{1/2}$	Moderate
501.7	$4p\ ^2F^\circ_{5/2}$–$3d\ ^2D_{3/2}$	Weak
514.5	$4p\ ^4D^\circ_{5/2}$–$4s\ ^2P_{3/2}$	Strong
528.7	$4p\ ^4D^\circ_{3/2}$–$4s\ ^2P_{1/2}$	Weak[b]

[a] As in Ionized gas lasers, by W. B. Bridges and A. N. Chester, in *Handbook of Lasers* (ed. R. J. Pressley, Chemical Rubber Co., Cleveland, Ohio, 1971), pp. 242–97.

[b] Oscillation required the presence of a buffer gas.

Unlike the first ion laser oscillation that was observed in the afterglow of a pulsed helium–mercury discharge,[7] oscillation in argon occurred during the excitation current-pulse, and it was clear that the selective excitation must be by electron impact. In the work of Bennett *et al.*[103] quasi-CW operation, longer than the time constant of any reasonable decay process, was reported for a period of about a millisecond at the (then) high power level of 10 W. By utilizing small-bore discharge tubes of a few millimeters to give high electron densities of about 10^{13} cm^{-3}, Gordon *et al.*[104] were able to achieve true CW operation of the first nine singly ionized argon laser lines listed in Table 5.11, and also on other lines in singly ionized species of krypton and xenon. The dominant CW laser lines in Ar-II are the 488.0-nm and 514.5-nm lines.

Oscillation has now been reported in more than thirty-two gaseous and nongaseous elements in singly, doubly, and triply ionized species. In most of these elements it has been established that electron-impact excitation is responsible for the selective excitation and achievement of population inversion.

Because of the similarity of the placement on an energy scale of ionic energy levels of many elements in which oscillation in ion species has been observed, a single generalized description of the excitation processes in one of the noble gases (argon) suffices to explain the majority of the salient points of most of the common ion-laser systems.

5.4.1. NOBLE GAS (ARGON) LASERS

Two types of upper laser level, selective excitation processes are involved in singly ionized ion lasers:

(1) the single-electron collision, "sudden-perturbation" (single-step) process of Bennett *et al.*,[103] shown schematically for argon in Fig. 5.35a, and

(2) the two-step process of Labuda *et al.*[105] illustrated schematically in Fig. 5.35b.

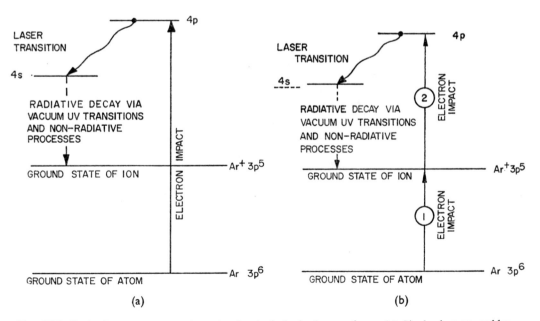

FIG. 5.35. Excitation processes pertinent to the singly ionized argon laser. (a) Single-electron, sudden-perturbation (single-step) process. (b) Two-step process.

5.4.1.1. *The single-step process*

In the first model of ion-laser excitation the upper laser level is excited by a single fast-electron collision from the $(np)^6$ neutral ground state

$$(np)^6 + e^- \rightarrow np^4(^3P)(n+1)p + 2e^- \qquad (5.36a)$$

in which $n = 2$, 3, 4, and 5 for neon, argon, krypton, and xenon, respectively. Taking the singly ionized argon laser as our example of noble gas ion-laser excitation, eqn. (5.36a) becomes

$$\text{Ar } 3p^6 + e^- \rightarrow \text{Ar}^{+\prime} \, 3p^4(^3P)4p + 2e^- , \qquad (5.36b)$$

where excitation from the Ar $3p^6$ neutral ground state occurs to the $3p^4(^3P)4p$ excited ionic Ar$^{+\prime}$ state. Although apparently a forbidden p–p transition has occurred, the parity selection rule has not been disobeyed as a change of parity has occurred through the loss of an electron in the ionization. In this sudden-perturbation collision process the ion core does not have time to change during the collision, and the main yield of excited ions is to the $3s^2 \, 3p^4(^3P)$ configuration, and excited ionic-state configurations with opposite parity to the $3p^5$ ion ground state (such as the $3s^2 \, 3p^4(^3P)4s$ state) are suppressed.[103]

The single-step, fast-electron collision process is favored in low pressure, short-pulse excited discharges, and during the breakdown period of discharges excited with long excitation pulses.[106] A very high electric field per unit pressure (E/p) is required to give the high electron temperature required to supply electrons (in argon) having an energy of over 35 eV

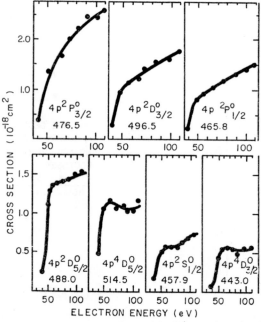

FIG. 5.36. Absolute single-step excitation cross-sections of Ar-II upper laser levels as a function of electron energy. (The wavelength used to study each state are given in nm.) (After W. R. Bennett, Jr., G. N. Mercer, P. K. Kindlmann, B. Wexler, and H. Hyman,[108] by courtesy of the American Institute of Physics.)

needed for single-step excitation of the Ar-II $4p$ levels from the neutral ground state. As discussed by Bennett[107-8] the degree to which the sudden perturbation process is satisfied depends on the energy of the incident electron in (5.36a). If the electron energies are close to the energy threshold for reaction (5.36a), resonances in the excitation functions can favor the production of excited ionic states that have the same azimuthal quantum number as the ground state. In these cases, a sharply peaked cross-section results, of the type shown in Fig. 2.20 for an electron-exchange transition. Absolute cross-section data for a number of upper laser levels in Ar-II are shown in Fig. 5.36 for electron energies a few volts above threshold to 110 eV.[109] The largest cross-sections are for the $4p\ ^2P^\circ_{3/2}$, $4p'\ ^2D^\circ_{3/2}$, $4p^2\ ^2P^\circ_{3/2}$, $4p\ ^2D^\circ_{5/2}$, $4p\ ^4D^\circ_{5/2}$ states in order of decreasing magnitude, that are predicted to have the largest cross-sections from the "sudden-perturbation" approximation of Bennett. The shape of the ionization functions of the $4p\ ^2D^\circ_{5/2}$ and $4p\ ^4D^\circ_{5/2}$ states (upper levels of the 488.0-nm and 514.5-nm laser lines) differs from those for the states predicted on the basis of the "sudden-perturbation" approximation. This has been attributed to indirect processes that contribute to the ionization functions and their magnitudes at electron energies above 50 eV.

Given selective excitation of the Ar-II $4p$ upper laser levels, as these levels have the same even parity as the ionic ground state, radiative decay cannot occur to the ground state but only to lower-lying levels having odd parity. These lower-lying levels are radiatively connected to the ion ground state by strong vacuum ultraviolet transitions at about 72 nm (720 Å), and so, in the absence of radiation trapping of these transitions, a good depopulating scheme exists that ensures that population inversion is readily achieved on the strongly allowed transitions in Ar-II of the type $4p\ ^2P^\circ\!-4s\ ^2P$.

The single-step, sudden-perturbation treatment of Bennett has been extended to higher degrees of ionization than the first, in which the excitation occurs from the ion ground-state of the preceding degree of ionization, i.e.

$$X^+ + e^- \to X^{++\prime} + 2e^- ,\tag{5.37a}$$

$$X^{++} + e^- \to X^{+++\prime} + 2e^-, \quad \text{etc.}\tag{5.37b}$$

This interpretation has been borne out by the observations of Cheo and Cooper[110] who report that doubly ionized laser lines only appear after saturation occurs of singly ionized laser lines. Saturation occurs when the appropriate ionic ground-state density becomes so large that the lower laser levels are radiation trapped and depopulation of them is prevented.

5.4.1.2. The two-step process

This process explains the behavior of the small-bore argon ion laser and other ionic systems in which laser oscillation closely follows the shape of the excitation current pulse after the cessation of the initial breakdown period of the discharge, and applies to high current density CW ion lasers. In the first step ① in Fig. 5.35b, ionization from the neutral $3p^6$ ground state produces ions in the Ar$^+$ $3p^5$ ground state by the direct electron-impact process

$$\text{Ar } 3p^6 + e^- \to \text{Ar}^+ 3p^5 + 2e^- .\tag{5.38}$$

This is followed by the second-step process ② in Fig. 5.35b,

$$Ar^+ \; 3p^5 + e^- \; \rightarrow \; Ar^{+\prime} \; 3p^4(^3P)4p + e^- , \tag{5.39}$$

that results in selective excitation of the $Ar^{+\prime}$ $4p$ excited ionic states that are nearly 20 eV above the Ar^+ $3p^5$ ion ground state. The evidence to support the two-step excitation process is that spontaneous emission intensity of lines originating from the Ar-II upper laser levels varies as the square of the discharge current, indicating that the upper laser level population N_2 varies as

$$N_2 \sim n_e n_i \sim n_e^2 \tag{5.40}$$

where n_e and n_i are the electron and positive-ion densities respectively. Since the electron density is directly proportional to the discharge current density[†]

$$N_2 \sim j^2 \tag{5.41}$$

where j is the discharge current density. This quadratic current dependence has been observed in spontaneous emission measurements over a wide range of currents, gas pressure and discharge-tube dimensions typical of CW ion laser discharge conditions.[43, 111]

The second step ② of the two-step process, the $3p^5 \; ^2P^\circ_{3/2} \rightarrow 3p^4(^3P)4p \; ^2P^\circ, \; ^2D^\circ$ excitation, is in violation of the electric-dipole selection rule and two alternative two-step processes have been proposed to account for the apparent violation:

(1) $$Ar \; 3p^6 + e^- \rightarrow Ar^* \; 3p^5 \; 4s + e^- , \tag{5.42}$$

followed by

$$Ar^* \; 3p^5 \; 4s + e^- \rightarrow Ar^{+\prime} \; 3p^4(^3P)4p + 2e^- \tag{5.43}$$

in which the Ar^* $3p^5 4s$ state is a metastable state of the neutral atom;[106] and[112, 43]

(2) $$Ar \; 3p^6 + e^- \rightarrow Ar^{+*} + 2e^- \tag{5.44}$$

where Ar^{+*} is a metastable $3p^4(^3P)3d$ ion, followed by electron-impact excitation of the Ar^{+*} atoms to the $4p$ upper laser levels

$$Ar^{+*} + e^- \rightarrow Ar^{+\prime} \; 3p^4(^3P)4p + e^- . \tag{5.45}$$

In the first alternative two-step process (5.42) and (5.43) the sudden-perturbation approximation can be used to predict the highest yield of excited-ionic states in the $Ar^{+\prime}$ $3p^4 \; 4p$ configuration in reaction (5.43). Though the density of Ar^* $3p^5 4s$ metastable atoms is expected to be less than the neutral ground state population, the larger cross-section for electron-impact excitation from a metastable state should compensate for the reduced number of atoms able to take part in the second-step process (5.43), as compared with the Ar^+ $3p^5$ ion ground-state population in the second-step process (5.39) of the first two-step process proposed by Labuda *et al.*[105]

[†] In a discharge tube of the same tube diameter this certainly is the case, though it is not clear that this is so in discharge tubes of different bore diameters.

244

Actually it has been found that a significant fraction of the population in the upper laser levels of a CW Ar-II laser is created by radiative cascade from higher-lying Ar-II states than the $3p^4(^3P)4p$ states (Rudko and Tang,[113] Bridges and Halstead,[114] and Lebedeva et al.[115]). The cascade contribution into the $4p\,^2D^\circ_{5/2}$ level (upper level of the strong 488.0-nm laser line), under CW laser conditions in small-bore discharge tubes at high current densities, amounts to as much as 50 percent of the excitation, and more than 20 percent for the $4p\,^4D^\circ_{5/2}$ level (upper level of the 514.5-nm laser line). The overall effect is that the same quadratic dependence of N_2 on n_e^2 or j^2 results, as the population of the higher-lying Ar-II states, from which the radiative cascade originates, is also proportional to n_e^2 or j^2. The involvement of the single-step process in the excitation of the Ar-II $4p$ levels under CW laser conditions is understood to be in excess of 20 percent.[116]

5.4.2. PULSED OSCILLATION BEHAVIOR

This discussion will be restricted to laser lines tabulated in Table 5.11 and first identified by Bridges.[100] Seven of these lines are now listed in Table 5.12 with their upper and lower laser levels, together with their radiative lifetimes and experimental transition probabilities. The table shows that the upper laser level lifetimes are considerably longer than those of the lower states, and that, with the exception of the 496.5-nm and 514.5-nm lines, their transition probabilities are similar. This means that under pulsed conditions (with a laser cavity that is equally reflective over the wavelength range of the lines) the order with which these laser lines that have approximately equal transition probabilities arrive in time can be taken as indicative of the rate at which their upper laser levels are excited.

TABLE 5.12. *Ar-II laser lines, upper and lower levels, lifetimes and transition probabilities*

Wavelength (nm)	Upper laser level with radiative lifetime (nsec) in parentheses	Lower laser level with radiative lifetime (nsec) in parentheses	Experimental transition probability[a] (10^7 sec^{-1})
457.9	$4p\,^2S^\circ_{1/2}$ (8.8)	$4s\,^2P_{1/2}$ (1.78)	8.42
465.7	$4p\,^2P^\circ_{1/2}$ (8.7)	$4s\,^2P_{3/2}$ (1.81)	7.55
472.7	$4p\,^2D^\circ_{3/2}$ (9.8)	$4s\,^2P_{3/2}$ (1.81)	7.27
476.5	$4p\,^2P^\circ_{3/2}$ (9.4)	$4s\,^2P_{1/2}$ (1.78)	7.15
488.0	$4p\,^2D^\circ_{5/2}$ (9.1)	$4s\,^2P_{3/2}$ (1.81)	8.96
496.5	$4p\,^2D^\circ_{3/2}$ (9.8)	$4s\,^2P_{1/2}$ (1.78)	2.63
514.5	$4p\,^4D^\circ_{5/2}$ (7.5)	$4s\,^2P_{3/2}$ (1.81)	0.71

[a] From H. Statz, F. A. Horrigan, S. H. Koozekanani, C. L. Tang, and G. F. Koster, *J. Appl. Phys.* **36**, 2278 (1965).

On the basis of the magnitudes of excitation functions of the $4p\ ^2P^\circ_{3/2}$, $4p\ ^2D^\circ_{3/2}$, $4p\ ^2P^\circ_{1/2}$, $4p\ ^2D^\circ_{5/2}$, and $4p\ ^4D^\circ_{5/2}$ states above an electron energy 50 eV, upper levels of the 476.5-, 496.5-, 465.8-, 488.0-, and 514.-5-nm laser lines respectively, the yield of these excited ionic states should occur in the order (of decreasing magnitude) in which they are listed. In the breakdown period of a pulsed discharge, electron energies are relatively higher than at other times during the discharge so it would be expected that if the electron energies are in excess of 50 eV, the 476.5-nm line would be the strongest and first to lase, and the 488.0-nm line the weakest and last to lase. (The 496.5-nm and 514.5-nm lines have transition probabilities that are much smaller than those of the 476.5-nm and 488.0-nm lines and so have a higher threshold for oscillation and cannot be intercompared time-wise as we are doing with the 476.5-nm and 488.0-nm lines. The 496.5-nm line also appears to behave differently from other laser lines.)

Following the breakdown period of the pulse, assuming that the average electron temperature is reduced to below 30 to 50 eV, one would expect that the order of oscillation would change and that the 488.0-nm laser line would in crease in intensity relative to the other laser lines. A similar behavior should result as the pressure is increased. This predicted behavior in short risetime pulsed discharges in argon at low pressures has been observed by Kobayashi et al.,[117] Smith and Dunn,[106] and Glaxunov et al.[118] It is illustrated in Fig. 5.37, in which two different modes of oscillation are displayed when a 0.02-μF capacitor charged to 6 kV is discharged through argon at 35 mtorr in a 1-mm-bore discharge tube by means of a spark

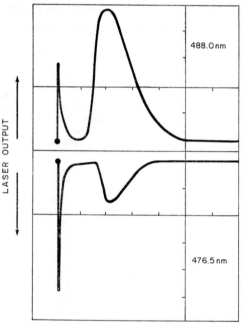

FIG. 5.37. Oscillogram of the output of a pulsed argon ion laser at 488.0 nm (upper trace) and 476.5 nm (lower trace). The single-step sudden-perturbation process is operative in the breakdown period of the discharge at the commencement of the trace, the two-step process occurs later in the excitation pulse. Discharge conditions: $p_{Ar} = 3.5 \times 10^{-2}$ torr, $D = 1$ mm, $C = 0.02\ \mu$F, charge voltage 6 kV; sweep rate 10 μsec/div. (After S. Kobayashi, T. Izawa, K. Kawamura, and M. Kamiyama,[117] by courtesy of the IEEE.)

gap (giving a high E/p of about 8600 V/cm-torr at breakdown).[117] On the initiation of the discharge, the Ar-II 476.5-nm line (lower trace) has a short risetime to peak intensity and a higher intensity than the Ar-II 488.0-nm line or the 496.5-nm line (and also the 457.9-nm and 510.6-nm lines). The second laser pulse that occurs about 4 μsec after the first shows that the 488.0-nm line oscillates over a longer period and is more intense now than the 476.5-nm line. On raising the gas pressure to about 0.01 torr the intensity of the 488.0-nm laser pulse becomes remarkably large relative to that of the first pulse as the electron temperature is lowered and two-step excitation becomes dominant.

When two distinct oscillation periods are observed in a pulsed argon discharge at high E/p values (a few thousand V/cm-torr), oscillation in the first laser pulse at breakdown of the discharge generally commences on the 476.5-nm line, then on the 496.5-nm line. This is followed by oscillation at 488.0 nm or 465.8 nm, and usually oscillation last of all at 514.5 or 501.7 nm.[106, 118] An interval of a few microseconds after the termination of the first laser pulse (as displayed in Fig. 5.37), the second laser pulse commences. The order of oscillation in the second pulse is normally 488.0, 476.5, 496.5, 514.5, 457.9, 501.7, and 465.8 nm, with the 488.0-nm line moving back if the argon pressure is increased to give a lower E/p value. As noted by Smith and Dunn[106] this order of oscillation agrees very closely with the threshold currents for a CW laser, with the 488.0-nm line requiring the smallest, and the 465.8-nm line the largest threshold current. In this second laser pulse, two-step excitation predominates.

5.4.2.1. Saturation mechanisms

Oscillation of the first pulse usually ceases at one to a few μsec after the leading edge of the current pulse, before the discharge current reaches a maximum value. And the duration of this first pulse decreases with increase in argon pressure, as indicated in Fig. 5.38. This is

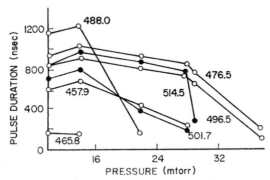

FIG. 5.38. Variation of duration of first laser pulse with argon pressure. Discharge conditions: $D = 5$ mm, 2-μsec delay line charged to 8 kV. (After A. L. S. Smith and M. H. Dunn,[106] by courtesy of the IEEE.)

believed to be due to radiation trapping of the 72.0-nm vacuum-ultraviolet transitions that connect the lower laser levels to the Ar-II ion ground state (Fig. 5.34), which causes the effective lifetime of the lower laser levels to increase.

Cheo and Cooper[119] have shown that radiation trapping affects the oscillation behavior of the Ar-II lines in the order 514.5, 457.9, 496.5, and 488.0 nm as the discharge current is

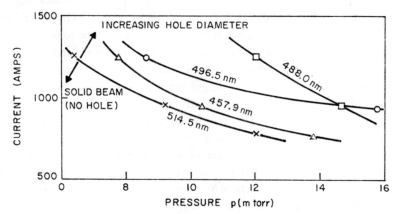

FIG. 5.39. Current and pressure conditions at onset of ring formation for pulsed singly ionized argon lines in a 5-mm-bore tube. (After P. K. Cheo and H. G. Cooper,[119] by courtesy of the American Institute of Physics.)

increased, and that the threshold current for the onset of hollow laser-beam formation decreases with increase in pressure.[†] Cheo and Cooper find that at the onset of the hollow-beam formation the product of pressure p and discharge current I is a constant for each laser line in a pulsed discharge. The behavior is illustrated in Fig. 5.39 for the 514.5-, 457.9-, 496.5-, and 488.0-nm laser lines. This order of the lines in which radiation trapping occurs goes to explain the general order of oscillation duration of laser lines during the first laser pulse (shown in Fig. 5.38). The 488.0-nm line is least affected by radiation trapping at pressures below about 16 millitorr (in a 5-mm-bore tube) and so has the longest first-pulse duration.

The recovery of gain after a period of a few microseconds after the termination of the first laser pulse is believed to be due to a decrease in the radiation trapping caused by broadening of the vacuum-ultraviolet lines as the ion temperature increases. It appears possible too that symmetric charge-transfer collisions ($Ar^{+'} + Ar \rightarrow Ar + Ar^{+'}$) play a role in the saturation and recovery behavior.

The recovery time depends on the gas pressure and the kind of gas used. It is longer for higher gas pressures, and longer in krypton and xenon than in argon.[43]

5.4.2.2. Excited state populations

Inversion densities for the 488.0-, 476.5-, and 496.5-nm Ar-II lines under pulsed conditions are of the order of 10^9 cm^{-3} at pulsed discharge currents of 120 to 160 A in a 4.5-mm-bore tube. Table 5.13 from Glaxunov et al.,[118] in which differences in level populations are tabulated at various discharge currents I, shows that the inversion density $N_2/g_2 - N_1/g_1$ of levels between which oscillation in a pulsed argon-ion laser is observed, where N_2 and N_1 are the upper and lower levels populations, and g_2 and g_1 are the usual statistical weights of the upper and lower levels respectively, is of the order of 10^9 cm^{-3} and that it decreases on the 488.0-nm and 476.5-nm lines with increase in discharge current. In the case of the

† Hollow laser-beam formation occurs when radiation trapping along the center of the laser tube is enough to reduce the gain below the threshold for oscillation along the cavity axis.

nonlaser lines at 473.6 and 484.8 nm, the lower states are more highly populated than the upper levels and make $N_1/g_1 - N_2/g_2$ positive instead of $N_2/g_2 - N_1/g_1$. With these two 473.6-nm and 484.8-nm lines, the population differences $N_1/g_1 - N_2/g_2$ increases with increase in discharge current. The inversion density $N_2/g_2 - N_1/g_1$ of approximately 10^9 cm^{-3} for upper and lower levels of the 488.0-, 476.5-, and 496.5-nm lines under pulsed conditions is approximately the same as that reported in CW argon-ion lasers.[108]

TABLE 5.13. *Differences of level populations for Ar-II lines*[118]

λ (nm)	Transition	Population difference $N_2/g_2 - N_1/g_1$ ($\times 10^9$)			
		$I = 120$ A	130 A	140 A	160 A
488.0	$4p\ ^2D^\circ_{5/2} - 4s\ ^2P_{3/2}$	1.5	0.86	0.38	no inversion
476.5	$4p\ ^2P^\circ_{3/2} - 4s\ ^2P_{1/2}$	2.3	2.1	1.9	1.3
496.5	$4p\ ^2D^\circ_{3/2} - 4s\ ^2P_{1/2}$	2.0	4.6	0.75	no inversion

λ (nm)	Transition	Population difference $N_1/g_1 - N_2/g_2$ ($\times 10^9$)			
		$I = 120$ A	130 A	140 A	160 A
473.6	$4p\ ^4P^\circ_{3/2} - 4s\ ^4P_{5/2}$	0.96	1	3.4	4.2
484.8	$4p\ ^4P^\circ_{1/2} - 4s\ ^4P_{3/2}$	0.75	2.2	2.4	2.7

Discharge conditions: Rectangular, 4.5-msec current pulses, at 5–10 kV, pulse-repetition rate 40 Hz, in an Ar–He mixture or pure argon, $D = 4.5$ mm.

5.4.3. CW-OSCILLATION BEHAVIOR

Gas filling pressures of about 0.5 torr in the CW Ar-II ion laser are normally more than an order of magnitude higher than in pulsed Ar-II lasers, and the E/p values, of the order of 10 to 100 V/cm-torr in CW lasers, are considerably less than those of 500 to 10,000 V/cm-torr of pulsed argon lasers. Under these CW conditions, the electron energies are lower than in the breakdown period of a pulsed discharge during which the first laser pulse occurs, and correspond in energy to those in the second-pulse period in the pulsed laser during which two-step electron-impact excitation predominates. Assuming that cavity losses are not wavelength dependent and that there is no preference for oscillation on any line of a particular wavelength, the 488.0-, 514.5-, and 476.5-nm lines are the strongest laser lines in a CW argon laser and require the lowest threshold currents.[43, 104, 106] Threshold currents that have been reported for CW argon lasers are given in Table 5.14.

In Table 5.14, and generally throughout the literature on ion lasers, under both pulsed and CW conditions, the value of the gas pressures given are the filling pressures when the discharge tube is cold. Under actual operating conditions the atom density in the active region of the laser will be a function of the gas temperature in that region and the ratio of the volume of the active region to the volume of the cold regions (from Charles' Law, $PV/T = $ a constant). Gordon et al.[120] have found that when one takes into account the

249

TABLE. 5.14. *Threshold currents for CW-argon laser lines*

λ (nm)	488.0	514.5	476.5	496.5	457.9	501.7	References
Threshold current (A) $\{$	0.37	1.85	1.5	1.75	2.15	2.35	104[a]
	1.45	3.6	3.8	4.0	5.2	6.0	106[b]

[a] 1 msec, quasi-CW conditions, $p_{\mathrm{Ar}} = 0.5$ torr, $D = 5$ mm.

[b] No pressure or tube-diameter values given. All but the threshold currents for the 501.7-nm line are as given by H. Statz, F. A. Horrigan, S. H. Koozekanani, C. L. Tang and G. F. Koster, *J. Appl. Phys.* **36,** 2278 (1965).

different gas temperatures of CW and pulsed argon-ion lasers, there is no difference in the optimum operating gas pressures of each. This observation should always be remembered when operating conditions or results of different experiments are being intercompared.

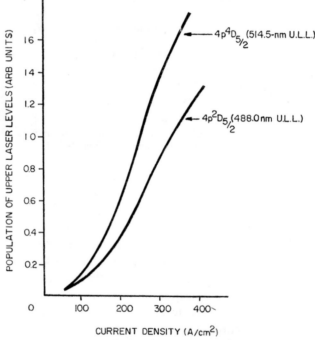

FIG. 5.40. Population of upper levels of the 488.0-nm and 514.5-nm laser lines as a function of the discharge current density at constant atom concentration in the active region. The $4p\,^2D_{5/2}$ level exhibits a (current-density)^2 behavior; the $4p\,^4D_{5/2}$ level below 200 A/cm^2 includes a (current-density)^3 component. Discharge conditions: atom concentration $= 2.2 \times 10^{15}$ cm^{-3}, $T_e = 47{,}000$ K, $D = 2.5$ mm. (Taken from V. V. Lebedeva, D. M. Mashtakov, and A. I. Odintsov,[115] by courtesy of the Optical Society of America.)

Figure 5.40 shows that when the neutral atom concentration is maintained constant during CW operation, the excited state population in the $4p\,^2D_{5/2}$ level (upper level of the 488.0-nm line) varies quadratically with discharge current density. The excitation, therefore, is clearly two-stage. The density in the $4p\,^4D_{5/2}$ level (upper level of the 514.5-nm line) appears to include a three-stage component.[115]

5.4.3.1. Saturation mechanisms

Radiation trapping, which is believed to limit population inversion in a pulsed argon discharge, is small in an argon laser operating under CW conditions.[121, 122] This is understood to be due to the increased gas and ion temperatures in a CW argon laser as compared to those in a pulsed argon laser, that causes a Doppler broadening of the 72.0-nm vacuum-ultraviolet terminal laser transitions to the ground state.[43] Contrary to this understanding, however, is the report by Merkelo *et al.*[123] that they find experimentally that the lower laser levels involved in the argon laser decay by *nonradiative* destruction processes to the ground state, and not radiatively through the strongly allowed vacuum-ultraviolet transitions. Additional support for this conclusion has been supplied by Zarowin,[124] Beigman *et al.*,[125] and Levinson *et al.*[126] It is not yet clear whether or not the nonradiative destructive process of the lower laser level can be attributed to symmetric $(Ar^{+\prime} + Ar)$ charge-transfer collisions. Increasingly there is evidence to show that these reactions are strongly resonant with ion energy (and possibly with neutral-atom gas temperature) so that it is conceivable that optimum conditions in the CW argon-ion laser are such as to optimize their reactions to the improvement of the nonradiative destructive process of the lower laser level. Kitaeva *et al.*[127] have determined that symmetric charge-transfer collisions play an important role in determining ionic motion of ground state Ar^+ ions in the CW argon-ion laser, but as yet, it is not known if they also determine the rate of decay of $Ar^{+\prime}$ ions in the lower laser levels.

5.4.3.2. Electron and ion temperatures, and electron concentrations

In the CW argon-ion laser the electron temperature increases with decrease in pressure, and unlike electron temperatures in weakly ionized plasmas, increases with increase in discharge current density. The effect, however, in a sense, is not a real one. The increase in

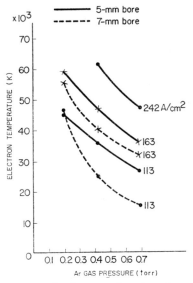

Fig. 5.41. Electron temperature in an argon-ion laser as a function of argon fill-pressure for various current densities. (After V. F. Kitaeva, Yu I. Osipov, and N. N. Sobolev,[127] by courtesy of the IEEE.)

electron temperature is due to a reduction in the "effective gas pressure" or atom concentration as the increased discharge current density heats up the gas in the active region of the discharge, and is not due to any special excitation collisions. Under discharge conditions where the atom density in the active region is maintained constant by increasing the pressure to compensate for the heating effect of the discharge,[115] the electron temperature, according to Herziger and Seelig,[128] remains constant.

Figures 5.41 and 5.42 illustrate how the electron temperature in a CW argon-ion laser varies with filling pressure of argon for various discharge current densities, and how the

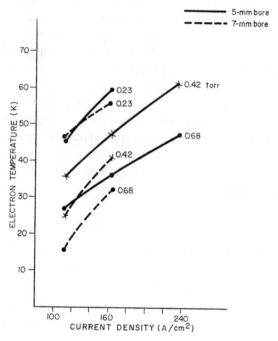

FIG. 5.42. Electron temperature in an argon-ion laser as a function of current density. (After V. F. Kitaeva, Yu. I. Osipov, and N. N. Sobolev,[127] by courtesy of the IEEE.)

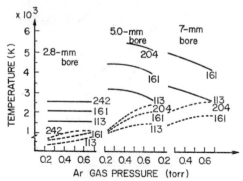

FIG. 5.43. Ionic (solid curves) and atom temperatures measured along the axis of the discharge tube versus pressure in an argon-ion laser. (After V. F. Kitaeva, Yu. I. Osipov, and N. N. Sobolev,[127] by courtesy of the IEEE.)

252

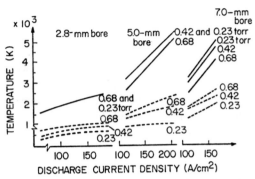

FIG. 5.44. Ionic (solid curves) and atom temperatures measured along the axis of the discharge tube versus discharge current density in an argon-ion laser. (After V. F. Kitaeva, Yu. I. Osipov, and N. N. Sobolev,[127] by courtesy of the IEEE.)

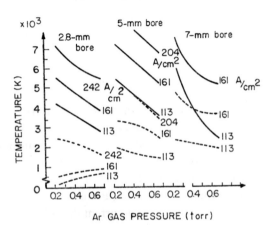

FIG. 5.45. Ionic (solid curves) and atom temperatures measured perpendicular to the axis of the discharge tube versus pressure in an argon-ion laser. (After V. F. Kitaeva, Yu. I. Osipov, and N. N. Sobolev,[127] by courtesy of the IEEE.)

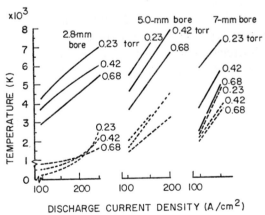

FIG. 5.46. Ionic (solid curves) and atom temperatures measured perpendicular to the axis of the discharge tube versus current density in an argon-ion laser. (After V. F. Kitaeva, Yu. I. Osipov, and N. N. Sobolev,[127] by courtesy of the IEEE.)

253

electron temperature increases for higher current densities at various (cold) argon fill-pressures for two different diameter discharge tubes.

Figures 5.43 and 5.44 illustrate how the ionic (solid curves) and atomic temperatures *measured along the axis* of a CW argon-ion laser varies with gas pressure for different discharge-tube diameters and current densities respectively. Similar curves of ionic and atomic temperatures measured perpendicular to the axis of the discharge tube are shown in Figs. 5.45 and 5.46 as a function of gas pressure and discharge current density. The higher temperatures measured perpendicular to the discharge-tube axis relative to those measured along the axis of the tube and their relative variation with discharge-tube diameter has been taken by Kitaeva *et al.*[127] as evidence that symmetric charge-transfer plays a substantial role in the argon-ion-laser plasma.

Depending on the discharge parameters covered in Figs. 5.41 to 5.46 the electron concentration in the 7-mm-bore discharge tube range from 0.7 to 2.2×10^{14} cm^{-3}. Electron densities of between 2 to 3.6×10^{14} cm^{-3} have been determined by Pleasance and George[129] to occur in a 2-mm-bore discharge tube of an Ar-II laser in the pressure range where laser oscillation occurs.

5.4.3.3. *Operating characteristics and discharge conditions*

In the development of high-efficiency, high-power ion lasers, Herziger and Seelig[122] critically examined the ion-laser literature and considered aspects of it in the light of results of their own experimental work. Their observations and conclusions are of considerable importance for the development of simple, high-power ion lasers. The majority of what follows in this section is taken from the work of Herziger and Seelig[122] and co-workers at the Physical Institute of the Technical University, Berlin.

While others have resorted to the use of high-current densities in discharge tubes of small bore (one to a few mm) and have used axial magnetic fields or segmented-bore discharge tubes to keep the highly ionized laser plasma away from the wall of the laser tube to prevent its overheating and erosion,[43, 105, 120, 130-1] Herziger and Seelig[122] went in the other direction and utilized large-bore discharge tubes, low fill-pressures, no magnetic fields, and relatively low current densities of approximately 100 A cm^{-2} compared with those of about 200 to 750 A cm^{-2} in more typical 1.8-mm-bore tubes of Ar-II lasers used by Bridges and Chester[43] and in commercial argon-ion lasers, to give output powers higher than those achieved by others. The results of Herziger and Seelig[122] nevertheless, in spite of their different approach to the design of high-power ion lasers, are applicable to the high-current density, small-bore, higher-pressure ion lasers typified by those of Bridges and Chester.[43]

The measured optimum fill-pressure p and tube diameter D of the Ar-II ion laser for the 488.0-nm line are related by the following expression:[122, 128]

$$pD = 5.0 \times 10^{-1} \text{ [torr-mm]}. \tag{5.46}$$

This relationship has been derived theoretically also,[128] and is valid for discharge tubes between 1 and 20 mm in diameter, as shown in Fig. 5.47. From this, it can be inferred that the average electron temperature in the Ar-II ion laser is the same in all Ar-II ion lasers under optimized discharge conditions.

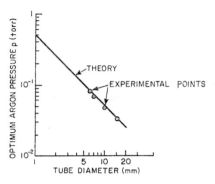

FIG. 5.47. Dependence of the optimum fill-pressure of argon on discharge-tube diameter for the Ar-II 488.0-nm laser line. (After G. Herziger and W. Seelig.[122])

In practical Ar-II lasers operating at low argon pressures, a low-impedance, gas-return path must be incorporated between the anode and cathode regions of the discharge so as to prevent gas-pumping occurring from the cathode to the anode due to a pressure differential that can exist between the cathode and anode regions of the discharge.[133–5] With the pD maintained constant at 0.5 torr-mm, the axial electric field E of the plasma column is nearly independent of the discharge current and the filling pressure, and is given by

$$E = 13/D \quad [\text{V-cm}^{-1}] \tag{5.47}$$

where D is the diameter of the discharge tube in mm.[122]

For laser tubes with diameters between 1 and 15 mm, where the pD relationship of (5.46) is maintained,

$$(I_{\min}/R^2)(L/v)^{1/2} = \text{a constant}, \tag{5.48}$$

where $I_{\min}$ is the minimum current needed for oscillation, R is the tube radius, L is the length of the plasma column, and v is the resonator loss.[132, 122]

The maximum laser output power W is given by the relationship

$$W \sim \pi R^2 L(j^2 - j_{\min}^2) \tag{5.49}$$

where j and $j_{\min}$ are the laser current density and laser threshold current density, respectively.[120, 130, 132] According to this relationship, the output power is limited only by the discharge current density j, and is independent of the optimum argon pressure. The variation of the laser power of the 488.0-nm line with argon fill-pressure for discharge currents of 37, 58, and 73 A in a 10-mm-bore discharge tube, 200-cm long, is diplayed in Fig. 5.48. It can be seen from the figure that the laser power is indeed independent of the optimum argon fill-pressure, and increases with increasing discharge current.

When the discharge current density is increased beyond the value of $jR = 50$ A cm^{-1}, where R is the tube radius, the axial electric field does not remain constant at the value of $13/D$ V-cm^{-1} given by eqn. (5.47) and the optimum pressure is no longer current independent. For discharge currents and discharge-tube radii products given by $50 < jR < 200$ A cm^{-1} the optimum pressure now has values given by

$$pR = 0.05 \ldots 0.2 \quad [\text{torr-cm}]. \tag{5.50}$$

255

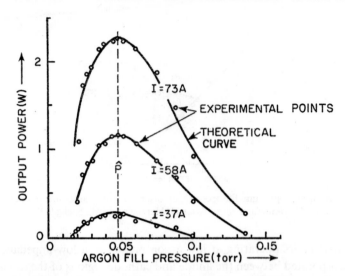

FIG. 5.48. Dependence of the 488.0-nm Ar-II laser power on the argon fill-pressure at vari ous discharge cur rents, in a 200-cm, 10-mm-bore discharge tube. (After G. Herziger and W. Seelig.[122])

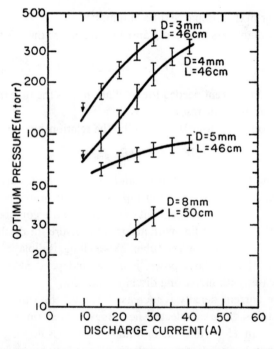

FIG. 5.49. Optimum gas pressure for maximum Ar-II 488.0-nm laser output as a function of discharge current for various bores of discharge tubes. (After W. B. Bridges and A. N. Chester,[43] reproduced from the *Handbook of Lasers* (ed. R. J. Pressley), by courtesy of the Chemical Rubber (Publishing) Co., Cleveland, Ohio, 1971.)

The values are dependent on the gas temperature, which increases with increasing current density, and the physical dimensions of the discharge container.[122] Figure 5.49 shows the typical variation with current of the optimum gas filling pressure for Ar-II ion lasers of different-bore diameters.[43] It illustrates the general trend that the optimum gas pressure increases with increasing discharge current and with decrease in tube diameter.

5.4.4. SPECTROSCOPY OF ION LASERS

5.4.4.1. Singly ionized species

The majority of singly ionized laser transitions are $p-s$, and the reverse type, $s-p$ transitions with ^3P core configurations, as shown in Fig. 5.50, where the number shown alongside each transition is the number of laser lines reported on that transition.

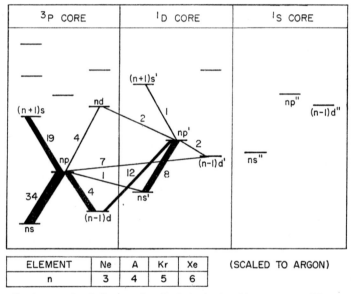

FIG. 5.50. Generalized energy-level diagram for singly ionized noble-gas atoms. The number beside each transition gives the number of laser lines that have been reported on that group transition. (Adapted from W. B. Bridges and A. N. Chester[136] and modified to include more recent data.)

The strongest laser lines are those that are strong in spontaneous emission. This indicates that where LS-coupling is applicable the strong laser lines should satisfy the preferred LS-coupling rules: $\Delta J = \Delta L$, especially $\Delta J = \Delta L = +1$ with no change in core or total spin.

Actually the only selection rules satisfied by all the laser lines are those of parity change and orbital angular momentum $\Delta l = \pm 1$.[43] Reference to the tables given in the Appendix will show that LS-coupling is violated in many cases, with changes in core configuration and total spin being relatively common. In Fig. 5.51, based on a figure from Bridges and Chester,[136] and modified to include more recently reported laser transitions, relative line strengths calculated for LS-coupling are shown for transitions in singly ionized noble gases.

$ns\,^2P$

	3/2	1/2
$np\,^2S^o_{1/2}$	1.33	0.67 ▲■
$np\,^2P^o_{1/2}$	0.67 ▲■	1.33 ●▲✳
$np\,^2P^o_{3/2}$	3.33 ●▲■	0.67 ●▲✳
$np\,^2D^o_{3/2}$	0.67 ▲	3.33 ▲■
$np\,^2D^o_{5/2}$	6.00 ▲■●	0

$ns\,^4P$

	5/2	3/2	1/2
$np\,^4S^o_{3/2}$	2.00	1.33	0.67
$np\,^4P^o_{1/2}$	0	1.67	0.33 ✳
$np\,^4P^o_{3/2}$	1.80	0.53 ▨	1.67
$np\,^4P^o_{5/2}$	4.20	1.80 ▪	0
$np\,^4D^o_{1/2}$	0	0.33	1.67
$np\,^4D^o_{3/2}$	0.20	2.13 ●✳	1.67
$np\,^4D^o_{5/2}$	1.80	4.20 ■✳	0
$np\,^4D^o_{7/2}$	8.00	0	0

$ns'\,^2D$

	5/2	3/2
$np'\,^2P^o_{1/2}$	0	2.00 ✳●
$np'\,^2P^o_{3/2}$	3.60 ●	0.40 ✳.
$np'\,^2D^o_{3/2}$	0.40	3.60 ✳
$np'\,^2D^o_{5/2}$	5.60	0.40 ✳
$np'\,^2F^o_{5/2}$	0.40	5.60 ■✳
$np'\,^2F^o_{7/2}$	8.00 ▲■	0

ELEMENT	n	SYMBOL
NEON	3	●
ARGON	4	▲
KRYPTON	5	■
XENON	6	✳

FIG. 5.51. Calculated relative line strengths of transitions in singly ionized noble-gas atoms. The observed laser lines are indicated by the appropriate symbol in each block. (Adapted from W. B. Bridges and A. N. Chester[136] and modified to include more recent data.)

The symbol in the transition boxes indicates in which noble gas the oscillation has been reported. Although the figure does not include all the singly ionized noble-gas laser lines included in the Appendix tables (or in ref. 43) it can be seen that, with only a few exceptions, oscillation occurs on the transitions that have large relative line strengths.

5.4.4.2. Further ionized species

Laser oscillation in doubly ionized species is observed when saturation in laser output occurs in singly ionized species.[110] A similar behavior is shown for oscillation in even higher ionized species and it is evident that excitation of the upper laser level is from the ground state of the ion that precedes the selectively excited ionized level. The excitation, therefore, is in the form of multiple steps of the two-step excitation process of Labuda *et al.*[105]

Figure 5.52 shows that the majority of doubly ionized laser transitions are similar to those in singly ionized species, where the number beside each transition is the number of laser lines that have that transition with the appropriate core configuration. The majority of the lines can be seen to occur on p–s, p'–s', and np' to $(n-1)d'$ transitions. Examination of the tables given in the Appendix will show that there is a prevalence for laser transitions in doubly ionized species of the noble gases that obey the LS-coupling selection rules $\Delta S = 0$, $\Delta J = \Delta L$, and particularly $\Delta S = 0$, $\Delta J = \Delta L = +1$. These are the same rules for transitions that LS-coupling predicts have the largest line strengths.

Conditions found to be conducive for the realization of population inversion in further ionized species are: high gas pressure (a few hundred millitorr to a few torr); high current density; and large-diameter discharge tubes.[110]

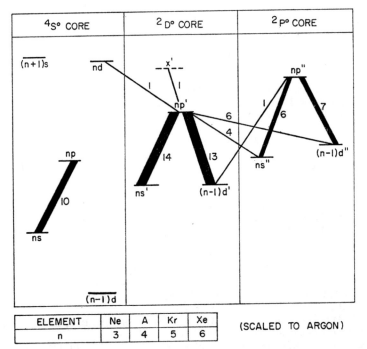

ELEMENT	Ne	A	Kr	Xe
n	3	4	5	6

(SCALED TO ARGON)

FIG. 5.52. Generalized energy-level diagram for doubly ionized noble-gas atoms. The numbers alongside each transition gives the number of laser lines observed on that transition. (Adapted from W. B. Bridges and A. N. Chester[136] and modified to include more recent data.)

References

1. LAURES, P., DANA, L. and FRAPARD, C., *Compt. Rend. Acad. Sci.*, Paris, **258**, 6363 (1964).
2. BRIDGES, W. B., *Proc. IEEE* **52**, 843 (1964).
3. DANA, L. and LAURES, P., *Proc. IEEE* **53**, 78 (1965).
4. DANA, L. and LAURES, P., Ionized gas lasers, paper presented at the IEE Conference, London (1964).
5. EGOROV, V. S., KOSTOV, YU. G. and SHUKHTIN, A. M., *Optics Spectrosc.* **15**, 458 (1963).
6. BRIDGES, W. B. and CHESTER, A. N., *IEEE J. Quant. Electr.* **QE-1**, 66 (1965).
7. BELL, W. E., *Appl. Phys. Lett.* **4**, 34 (1964).
8. BLOOM, A. L., BELL, W. E. and LOPEZ, F. O., *Phys. Rev.* **135**, A578 (1964).
9. GOLDSBOROUGH, J. P. and BLOOM, A. L., *IEEE J. Quant. Electr.* **QE-5**, 459 (1969).
10. BYER, R. L., BELL, W. E., HODGES, E. and BLOOM, A. L., *J. Opt. Soc. Am.* **55**, 1598 (1965).
11. SCHUEBEL, W. K., *IEEE J. Quant. Electr.* **QE-7**, 39 (1971).
12. GERRITSEN, H. J. and GOEDERTIER, P. V., *J. Appl. Phys.* **35**, 3060 (1964).
13. WIEDER, H., MYERS, R. A., FISHER, C. L., POWELL, C. G. and COLOMBO, J., *Rev. Sci. Instr.* **38**, 1538 (1967).
14. DYSON, D. J., *Nature* **207**, 361 (1965).
15. SCHUEBEL, W. K., *IEEE J. Quant. Electr.* **QE-6**, 574 (1970).
16. JENSEN, R. C., COLLINS, G. J. and BENNETT, W. R. Jr., *Appl. Phys. Lett.* **18**, 50 (1971).
17. SCHUEBEL, W. K., *Appl. Phys. Lett.* **16**, 470 (1970).
18. SCHUEBEL, W. K., *IEEE J. Quant. Electr.* **QE-6**, 654 (1970).
19. COLLINS, G. J., JENSEN, R. C. and BENNETT, W. R. Jr., *Appl. Phys. Lett.* **18**, 282 (1971); see also erratum **19**, 122 (1971).
20. VON ENGEL, A., *Ionized Gases*, 2nd ed., Clarendon Press, Oxford (1955), pp. 234–5.
21. WILLETT, C. S., *IEEE J. Quant. Electr.* **QE-5**, 469 (1970).
22. FOWLES, G. R. and JENSEN, R. C., *Proc. IEEE* **52**, 851 (1964).

23. SUZUKI, N., *Japan. J. Appl. Phys.* **4**, 452 (1965).
24. PASCHEN, F., *Sitzungsber. Preuss Akad. Wissen.* **32**, 536 (1928).
25. TAKAHASHI, Y., *Ann. der Physik* **3**, 49 (1929).
26. ALEINIKOV, V. S., *Optics Spectrosc.* **28**, 15 (1970).
27. BATES, D., *Atomic and Molecular Processes*, Academic Press, New York, 1962.
28. WONDERBERG, E. and MILATZ, N., *Physica* **8**, 871 (1941).
29. HYMAN, H. A., *J. Chem. Phys.* **33**, 3642 (1971).
30. BLOOM, A. L., BYER, R. L. and BELL, W. E., Emission line widths of ion lasers, in *Physics in Quantum Electronics* (eds. P. KELLY, B. LAX, and P. E. TANNEWALD), pp. 688–9, New York, McGraw-Hill, 1966.
31. TOLANSKY, S., *Proc. Phys. Soc.* **43**, 545 (1931).
32. COTMAN, R. N., JOHNSON, W. B. and LARSEN, A. B., *IEEE J. Quant. Electr.* **QE-4**, 359 (1968). In the presentation of paper 12G-10 at the Int. Quantum Electronics Conference (1968) a cross-section of 5.1×10^{-14} cm² was reported for reaction (5.8).
33. FOWLES, G. R. and SILFVAST, W. T., *IEEE J. Quant. Electr.* **QE-1**, 131 (1965).
34. SILFVAST, W. T., FOWLES, G. R. and HOPKINS, B. D., *Appl. Phys. Lett.* **8**, 318 (1966).
35. WILLETT, C. S., Population inversion mechanisms in hollow cathode, pulsed, and positive column discharges (unpublished Ph.D. thesis), University of London, 1967.
36. DUFFENDACK, O. S., HENSHAW, C. L. and GAYER, M., *Phys. Rev.* **34**, 1132 (1929).
37. TAKAHASHI, Y., *Ann. der Physik* **3**, 27 (1929).
38. FOWLES, G. R. and HOPKINS, B. D., *IEEE J. Quant. Electr.* **QE-3**, 419 (1967).
39. HOPKINS, B. D., Isotope shifts and CW oscillation in the 4416-A cadmium laser, Ph.D. thesis, University of Utah, 1968. University Microfilms, Inc., No. 68-12, 443, Ann Arbor, Michigan.
40. SILFVAST, W. T., *Appl. Phys. Letters* **13**, 169 (1968).
41. WEBB, C. E., TURNER-SMITH, A. R. and GREEN, J. M., *J. of Phys.* B, **3**, L134 (1970). Note: In this reference the order of caption of Figs. 1 and 2 is reversed.
42. COLLINS, G. J., JENSEN, R. C. and BENNETT, W. R. Jr., *Appl. Phys. Lett.* **19**, 125 (1971).
43. BRIDGES, W. B. and CHESTER, A. N., Ionized gas lasers, in *Handbook of Lasers* (ed. R. J. PRESSLEY), pp. 242–97, The Chemical Rubber Co., Cleveland, Ohio, 1971.
44. FOWLES, G. R. and JENSEN, R. C., *Appl. Optics* **3**, 1191 (1964).
45. JENSEN, R. C. and BENNETT, W. R. Jr., *IEEE J. Quant. Electr.* **QE-4**, 356 (1968).
46. JENSEN, R. C., COLLINS, G. J. and BENNETT, W. R. Jr., *Phys. Rev. Lett.* **23**, 363 (1969).
47. COLLINS, G. J., CW-oscillation and charge-exchange excitation in the zinc-ion laser (Ph.D. thesis), Yale University, 1970.
48. RISEBERG, L. A. and SCHEARER, L. D., *IEEE J. Quant. Electr.* **QE-7**, 40 (1971).
49. COLLINS, G. J., *J. Appl. Phys.* **42**, 3812 (1971).
50. BLOOM, A. L., BELL, W. E. and HARDWICK, D. L., *Bull. Am. Phys. Soc.* **9**, 143 (1964).
51. MARTIN, W. C. and CORLISS, C. H., *J. Res. Natl. Bur. Stds.* **64A**, 443 (1960).
52. JENSEN, R. C. and FOWLES, G. R., *Proc. IEEE* **52**, 1350 (1964).
53. FOWLES, G. R. and JENSEN, R. C., *Appl. Optics* **3**, 1192 (1964).
54. WILLETT, C. S. and HEAVENS, O. S., *Optica Acta* **14**, 195 (1967).
55. WILLETT, C. S., *IEEE J. Quant. Electr.* **QE-3**, 33 (1967).
56. BEREZIN, I. A. and YANOVSKAYA, G. N., *Optics Spectrosc.* **14**, 11 (1963).
57. BEREZIN, I. A., *Optics Spectrosc.* **31**, 466 (1970).
58. SILFVAST, W. T. and KLEIN, M. B., *Appl. Phys. Lett.* **17**, 400 (1970).
59. KLEIN, M. B. and SILFVAST, W. T., *Appl. Phys. Lett.* **18**, 482 (1971).
60. MARTIN, D. C., *Phys. Rev.* **48**, 938 (1935).
61. BELL, W. E., BLOOM, A. L. and GOLDSBOROUGH, J. P., *IEEE J. Quant. Electr.* **QE-1**, 400 (1965).
62. WEBB, C. E., *IEEE J. Quant. Electr.* **QE-4**, 426 (1968).
63. SILFVAST, W. T. and KLEIN, M. B., *Appl. Phys. Lett.* **20**, 501 (1972).
64. MOORE, C. E., *Atomic Energy Levels*, Vol. III, p. 99, Natl. Bur. of Stds., Circular 467, U.S. Government Printing Office, Washington, D.C. 20402 (1958).
65. SILFVAST, W. T. (private communication, June 1972).
66. NESMEGANOV, A. N., *Vapor Pressure of the Elements*, Academic Press, New York, 1963.
67. SILFVAST, W. T., *Phys. Rev. Lett.* **27**, 1489 (1971).
68. SCHEARER, L. D., *Phys. Rev. Lett.* **22**, 629 (1969).
69. HYMAN, H. A., *Chem. Phys. Lett.* **10**, 242 (1971).
70. KLEIN, M. B. and MAYDAN, D., *Appl. Phys. Lett.* **16**, 509 (1970).
71. HODGES, D. T., *Appl. Phys. Lett.* **17**, 11 (1970), and erratum **18**, 362 (1971).

72. SCHEARER, L. D. and PADOVANI, F. A., *J. Chem. Phys.* **52**, 1618 (1970).
73. SOSNOWSKI, T. P., *J. Appl. Phys.* **40**, 5138 (1969).
74. GOLDSBOROUGH, J. P., *Appl. Phys. Lett.* **15**, 159 (1969).
75. GOLDSBOROUGH, J. P. and HODGES, E. B., *IEEE J. Quant. Electr.* **QE-5**, 361 (1969).
76. SILFVAST, W. T. amd SZETO, L. H., *Appl. Optics* **9**, 1184 (1970).
77. FENDLEY, J. R., GOROG, I., HERNQUIST, K. G. and SUN, C., *R.C.A. Rev.* **30**, 422 (1969).
78. TOMPKINS, J. D., *Laser Focus* **8**, 32 (1969).
79. GIALLORENZI, T. G. and AHMED, S. A., *IEEE J. Quant. Electr.* **QE-7**, 11 (1971).
80. SOSNOWSKI, T. P. and KLEIN, M. B., *IEEE J. Quant. Electr.* **QE-7**, 425 (1971).
81. SILFVAST, W. T. and SZETO, L. H., *Appl. Phys. Lett.* **19**, 445 (1971).
82. GOTO, T., KAWAHARA, A., COLLINS, G. J. and HATTORI, S., *J. Appl. Phys.* **42**, 3816 (1971).
83. DORGELA, H. B., ALTING, H. and BOERS, J., *Physica (Haag)* **2**, 959 (1935).
84. GOLDSBOROUGH, J. P., *IEEE J. Quant. Electr.* **QE-5**, 133 (1969).
85. FENDLEY, J. R. Jr., *IEEE J. Quant. Electr.* **QE-4**, 627 (1968).
86. SILFVAST, W. T., *Appl. Phys. Lett.* **13**, 169 (1968).
87. GORDON, E. I. and WHITE, A. D., *Appl. Phys. Lett.* **3**, 199 (1963).
88. HERZIGER, G., HOLZAPFEL, W. and SEELIG, W., *Z. für Physik* **189**, 385 (1966).
89. FABRIKANT, W., *J. Exp. Theoret. Phys. (JETP)* **8**, 549 (1938).
90. YOUNG, R. T. Jr., WILLETT, C. S. and MAUPIN, R. T., *J. Appl. Phys.* **41**, 2936 (1970).
91. ZNAMENSKII, V. B., *Optics Spectrosc.* **15**, 7 (1968).
92. SILFVAST, W. T., *Appl. Phys. Lett.* **15**, 23 (1969).
93. ALEINIKOV, V. S. and USHAKOV, V. V., *Optics Spectrosc.* **29**, 111 (1970).
94. KARABUT, E. K., MIKHALEVSKII, V. S., PAPAKIN, V. F. and SEM, M. F., *Sov. Phys. Tech. Phys.* **14**, 1447 (1970).
95. SUGAWARA, Y., TOKIWA, Y. and IIJIMA, T., *Digest of Technical Papers, Quantum Electronics Conference, Kyota, Japan* (1971), pp. 320–1.
96. HODGES, D. T., *Appl. Phys. Lett.* **18**, 454 (1971).
97. STRIGANOV, A. R. and SVENTITSKII, N. S., *Tables of Spectral Lines of Neutral and Ionized Atoms,* Plenum Press, New York, 1968.
98. SCHEARER, L. D. and RISEBERG, L. A., *Phys. Rev. Lett.* **26**, 599 (1971).
99. MANLEY, J. H. and DUFFENDACK, O. S., *Phys. Rev.* **47**, 56 (1935).
100. BRIDGES, W. B., *Appl. Phys. Lett.* **4**, 128 (1964).
101. CONVERT, G., ARMAND, M. and MARTINOT-LAGARDE, P., *Compt. Rend.* **258**, 3259 (1964).
102. CONVERT, G., ARMAND, M. and MARTINOT-LAGARDE, P., *Compt. Rend.* **258**, 4467 (1964).
103. BENNETT, W. R. Jr., KNUTSON, J. W., MERCER, G. N. and DETCH, J. L., *Appl. Phys. Lett.* **4**, 180 (1964).
104. GORDON, E. I., LABUDA, E. F. and BRIDGES, W. B., *Appl. Phys. Lett.* **4**, 178 (1964).
105. LABUDA, E. F., GORDON, E. I. and MILLER, R. C., *IEEE J. Quant. Electr.* **QE-1**, 273 (1965).
106. SMITH, A. L. S. and DUNN, M. H., *IEEE J. Quant. Electr.* **QE-4**, 838 (1968).
107. BENNETT, W. R. Jr., Inversion mechanisms in gas lasers, in *Appl. Opt. Supplement on Chemical Lasers* (ed. J. N. HOWARD), pp. 3–33 (1965).
108. BENNETT, W. R. Jr., MERCER, G. N., KINDLMANN, P. J., WEXLER, B. and HYMAN, H., *Phys. Rev. Letters* **17**, 987 (1966).
109. BENNETT, W. R. Jr., Collision processes in the argon ion laser, reprinted from *The Physics of Electronic and Atomic Collisions,* Invited Papers from the Fifth International Conference, Leningrad, July 17–23 (1967) (ed. L. M. BRANSCOMB), pp. 61–85, JILA, University of Colorado, Boulder, Colorado, 1968.
110. CHEO, P. K. and COOPER, H. G., *J. Appl. Phys.* **36**, 1862 (1965).
111. BRIDGES, W. B., CHESTER, A. N., HALSTEAD, A. S. and PARKER, J. V., *Proc. IEEE* **59**, 724 (1971).
112. LABUDA, E. F., WEBB, C. E., MILLER, R. C. and GORDON, E. I., A study of capillary discharges in noble gases at high current densities, presented at the 18th Gaseous Electronics Conference, Minneapolis, Minnesota, Oct. (1965).
113. RUDKO, R. I. and TANG, C. L., *Appl. Phys. Lett.* **9**, 41 (1966).
114. BRIDGES, W. B. and HALSTEAD, A. S., Gaseous ion laser research, Tech. Rept. No. AFAL-TR-67-69, Hughes Research Laboratories, Malibu, California, May 1967.
115. LEBEDEVA, V. V., MASHTAKOV, D. M. and ODINTSOV, A. I., *Optics Spectrosc.* **28**, 187 (1970).
116. MERCER, G. N., CHEBOTAYEV, V. P. and BENNETT, W. R. Jr., *Appl. Phys. Lett.* **10**, 177 (1967).
117. KOBAYASHI, S., IZAWA, T., KAWAMURA, K. and KAMIYAMA, M., *IEEE J. Quant. Electr.* **QE-2**, 699 (1966).
118. GLAXUNOV, V. K., KITAEVA, V. F., OSTRAVSKAYA, L. YA. and SOBOLEV, N. N., *JETP-Letters* **5**, 215 (1967).
119. CHEO, P. K. and COOPER, H. G., *Appl. Phys. Lett.* **6**, 177 (1965).

120. GORDON, E. I., LABUDA, E. F., MILLER, R. C. and WEBB, C. E., Excitation mechanisms of the argon-ion laser, in *Physics of Quantum Electronics*, pp. 664–73 (ed. P. L. KELLEY, B. LAX and P. E. TANNENWALD), McGraw-Hill, 1966.
121. BOERSCH, H., HERZIGER, G., SEELIG, W. and VOLLAND, I., *Phys. Lett.* A, **24,** 695 (1967).
122. HERZIGER, G. and SEELIG, W., *Z. für Physik* **219,** 5 (1969). Available translated as FSTC-HT-23-408-70 from Defense Documentation Center, Cameron Station, Alexandria, Virginia, U.S.A., ATTN: TSR-1.
123. MERKELO, H., WRIGHT, R. H., KAPLAFKA, J. P. and BIALECKE, E. P., *Appl. Phys. Lett.* **13,** 401 (1968).
124. ZAROWIN, C. B., *Appl. Phys. Lett.* **15,** 36 (1969).
125. BEIGMAN, I. L., VAINSHTEIN, L. A., RUBIN, P. L. and SOBOLEV, N. N., *Sov. Phys. JETP* **6,** 343 (1967).
126. LEVINSON, G. R., PAPULOVSKY, V. F. and TYCHINSKY, V. P., *Rad. Engng Electr. Phys.* **13,** 578 (1968).
127. KITAEVA, V. F., OSIPOV, YU. I. and SOBOLEV, N. N., *IEEE J. Quant. Electr.* **QE-7,** 391 (1971).
128. HERZIGER, G. and SEELIG, W., *Z. für Physik* **215,** 437 (1968).
129. PLEASANCE, L. D. and GEORGE, E. V., *Appl. Phys. Lett.* **18,** 557 (1971).
130. HERNQVIST, K. G. and FENDELY, J. R. Jr., *IEEE J. Quant. Electr.* **QE-3,** 66 (1967).
131. FENDLEY, J. R. Jr., *IEEE J. Quant. Electr.* **QE-4,** 627 (1968).
132. LABUDA, E. F. and JOHNSON, A. M., *IEEE J. Quant. Electr.* **QE-2,** 700 (1966).
133. CHESTER, A. N., *Phys. Rev.* **169,** 172 (1968).
134. CHESTER, A. N., *Phys. Rev.* **169,** 184 (1968).
135. BRIDGES, W. B., CHESTER, A. N., HALSTEAD, A. S. and PARKER, J. V., *Proc. IEEE* **59,** 724 (1971).
136. BRIDGES, W. B. and CHESTER, A. N., *Appl. Optics* **4,** 573 (1965).
137. DUNN, M. H., *J. Phys. B, Atom. Molec. Phys.* **5,** 665 (1972).

CHAPTER 6

Specific Molecular Laser Systems

Introduction

This chapter deals with specific molecular laser electric discharge systems that lend themselves to analysis and that illustrate some of the selective excitation processes discussed in Chapter 2. The arrrangement follows that used in the previous two chapters for specific atomic laser systems. Omissions are made where there is no predominant inelastic collision mechanism in molecular species corresponding to that in the atomic system. Lasers covered in this chapter are selectively excited by the processes:

6.1. Resonant Excitation-energy Transfer (Molecule–Molecule);
6.2. Electron-impact Excitation;
6.3. Line Absorption;
6.4. Radiative Cascade.

The first *maser* device (in 1954) of Gordon, Zeiger, and Townes[1] was a molecular system. It operated in the microwave region at 23.8 GHz in ammonia, and the population inversion was achieved by a physical separation of molecules in the lowest doublet vibrational levels of ammonia. In 1961 Polanyi[2] pointed out the possibility that *laser* oscillation should be possible between vibrational–rotational energy levels of molecules in their electronic ground states. The first reports of actual *laser* oscillation in molecular species, however, were of oscillation on vibrational transitions between *excited* electronic energy levels. Mathias and Parker[3] obtained pulsed infrared oscillation between 0.8 and 1.2 μm on several bands of the first positive system ($B^3\pi_g - A^3\Sigma_u^+$) of molecular nitrogen, and pulsed oscillation in the red, orange, and green regions of the spectrum in the angstrom-band system ($B^1\Sigma - A^1\pi$) of carbon monoxide;[4] and Heard[5] obtained pulsed ultraviolet oscillation around 0.34 μm on bands of the second positive system ($C^3\pi_u - B^3\pi_g$) in molecular nitrogen. The first observation of CW-laser oscillation in a molecular species (carbon dioxide) in the far infrared at about 10.6 μm was reported by Patel *et al.*[6] early in 1964. Oscillation here did occur on vibrational–rotational transitions of the electronic ground state. Later, Patel[7, 8] reported output powers at the milliwatt-level, and then at the 10-watt power level in carbon dioxide–nitrogen mixtures.[9] Oscillation in molecular species has since then been extended down to the vacuum-UV region at 0.1161 μm on the Werner band ($C^1\pi_u - X^1\Sigma_g^+$) in hydrogen,[10, 11] and out beyond 773.5 μm in a molecular species that is probably HCN,[12, 13] to 1814 μm in CH_3CN (ref. 239).

263

6.1. Resonant Excitation-transfer (Molecule–Molecule) Lasers

There are at least four important high-power molecular lasers in which one of the main selective excitation processes of the upper laser levels is molecule–molecule vibrational–rotational excitation transfer. They are the nitrogen–carbon dioxide, carbon dioxide–xenon–helium, pure carbon dioxide, and the nitrogen–nitrous oxide lasers.

6.1.1. NITROGEN–CARBON DIOXIDE LASER

Molecule–molecule, vibrational resonant excitation transfer was first used as a means of selective excitation in a laser system by Patel.[8] Patel, following the proposals of Legay and Legay-Sommaire,[14] selectively excited the $00°1$ Σ_u^+ vibrational energy level of CO_2 through the transfer of vibrational energy of metastable nitrogen molecules in the $v = 1$ vibrational level of their $^1\Sigma_g^+$ ground state to CO_2 ground-state molecules, and obtained oscillation at the seven wavelengths near 10.6 µm listed in Table 6.1.

TABLE 6.1. *Laser transitions in the Patel N_2–CO_2 laser*[8]

Measured vacuum wavelength (µm)	Frequency (cm^{-1})	Rotational identification in the $00°1$–$10°0$ vibrational band
10.5322	949.47	$P(14)$
10.5519	947.70	$P(16)$
10.5716	945.93	$P(18)$
10.5915	944.15	$P(20)$[a]
10.6119	942.34	$P(22)$
10.6327	940.49	$P(24)$
10.6537	938.64	$P(26)$

[a] Strongest laser transition, giving more than 1 mW output power.

Figure 6.1 shows the vibrational energy-level coincidence in the N_2–CO_2 laser system between the first excited vibrational $N_2^*(v = 1)$ level and the $00°1$ Σ_u^+ vibrational–rotational levels of CO_2, together with pertinent transitions between various vibrational levels. Laser action at approximately 10.6 µm occurs between the $00°1$ and $10°0$ vibrational levels. Oscillation has also been reported between the $00°1$ level and other levels.[15–18]

Before we consider the N_2–CO_2 laser further, it is helpful to the understanding of it to discuss the notation that we are using. The CO_2 molecule is a linear symmetric triatomic molecule, the outer atoms being the two oxygen atoms, and it has three normal modes of vibration. Each of these modes is quantized, and can exist as a harmonic oscillator usually independent of the others. Figure 6.2 illustrates these three modes, which, with the center of gravity fixed, are:

(1) The symmetric longitudinal mode ν_1, in which the oxygen atoms vibrate in a straight line in opposition to each other. Each quantum state of this mode is labeled 100, 200, 300, etc., with a potential energy of 1388 cm^{-1} (or 0.58 eV) associated with the first ν_1 excited 100 level. A spacing of 1388 cm^{-1} exists between each successive excited 100–200, 200–300, etc., levels.

(2) A twofold degenerate bending mode ν_2 in which the oxygen atoms oscillate in two planes at right angles to lines joining the oxygen atoms and the central carbon atom. Each quantum state of the bending mode is labeled 010, 020, 030, etc., with a potential energy of 667 cm^{-1} or 0.08 eV associated with the first ν_2 excited 010 level. The potential-energy spacing between each successive level is also 667 cm^{-1}.

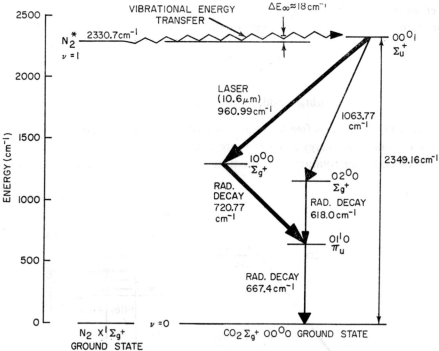

FIG. 6.1. Partial energy-level diagram of N_2 and CO_2 showing selective excitation of the CO_2 00°1 level through vibrational energy transfer from N_2 ($\nu = 1$), and CO_2 energy levels pertinent to the 10.6-μm CO_2 laser.

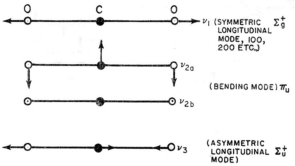

FIG. 6.2. Normal modes of oscillation of the CO_2 molecule.

(3) An asymmetric longitudinal mode v_3, in which the oxygen atoms are always moving in the same direction. Each quantum state of this mode is labeled 001, 002, 003, etc., with the excited state having $v_3 = 2349$ cm^{-1}. The spacing between each successive level is likewise equal to 2349 cm^{-1} or 0.29 eV.

The vibrational levels of the CO_2 molecule are characterized by quantum numbers specifying these three modes of vibration (n_1, n_2, n_3), where n_1 indicates excitation of the symmetric axial vibration v_1; n_2 the excitation of the twofold degenerate mode of vibration v_{2a} and v_{2b}; and n_3 the excitation of the asymmetric mode of vibration v_3. In addition since the bending mode of vibration v_2 is twofold degenerate, n_2 carries a superscript l, a quantum number for the angular momentum to indicate the degeneracy. The levels are therefore specified by three numbers n_1, n_2^l, n_3.

Transitions between vibrational levels of the type $n_1 n_2^l n_3 \rightarrow n_1' n_2''^l n_3'$ are governed by the selection rules[31]

$$|\Delta n_2| - \text{even}, \quad |\Delta l| = 0, \quad |\Delta n_3| - \text{odd} \tag{6.1}$$

or

$$|\Delta n_2| - \text{odd}, \quad |\Delta l| = 1, \quad |\Delta n_3| - \text{even}. \tag{6.2}$$

Since the CO_2 molecule is free to rotate, each vibrational level is divided into a series of sublevels, or rotational energy levels that are quantized and are designated by J, the total angular momentum or rotational quantum number, whose value is either 0 or a positive

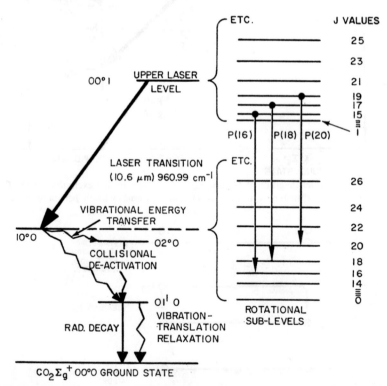

FIG. 6.3. Detailed laser-transition diagram for the $00°1-10°0$ band including rotational levels and transitions.

integer. The selection rules for vibrational–rotational transitions in the CO_2 molecule are $\Delta J = +1$, giving the so-called P-branch transitions; and $\Delta J = -1$, giving the R-branch transitions. Individual transitions are denoted by the letter P or R for either branch and include the J value of the lower level. For example, the strongest $P(20)$ transition in Patel's N_2–CO_2 laser (Table 6.1) denotes a P transition finishing on a rotational level $J = 20$ and starting on a rotational level $J = 19$ of a higher vibrational level (with $\Delta J = +1$). Rotational transitions on the P transitions around 10.6 μm are shown in Fig. 6.3. Only a few of the P transitions are shown, these are the transitions on which the strongest oscillation is observed.

6.1.1.1. Upper level excitation

From Fig. 6.1 it can be seen that the potential energy of the first vibrational level N_2^* ($v = 1$) at 2330.7 cm^{-1} above the $N_2 X^1\Sigma_g^+$ ground state is in near coincidence with the first excited vibrational level of the CO_2 molecule in the longitudinal mode v_3, the 00°1 Σ_u^+ level at 2349.16 cm^{-1}. In accordance with theoretical predictions, the smallness of the energy discrepancy ΔE_∞ of -18 cm^{-1} compared with the kT of the discharge of about 280 cm^{-1} at 400 K, makes the degree of adiabacy of the collision small (va/v in eqn. (2.13), which is equal to $(\Delta E_\infty/h)(a/v)$, where a is the distance over which the collision or interaction occurs and v is the relative velocity of the colliding atoms) and the collision is nonadiabatic. It follows that the probability of vibrational energy transfer is high in the reversible reaction

$$N_2^*(v = 1) + CO_2(00^00) \rightleftharpoons N_2(v = 0) + CO_2(00^01) - \Delta E_\infty \ (18 \ cm^{-1}). \tag{6.3}$$

Since the $N_2^*(v = 1)$ molecule is metastable to decay to the ground state, and since the $CO_2(00^01)$ molecule can decay radiatively, the reaction proceeds mainly in the forward direction.

Up to recently it has been assumed that only strong short-range repulsive forces between colliding molecules could make the collision nonadiabatic and result in effective vibrational energy transfer. This assumption lead to the derivation of the probability of vibrational energy transfer in eqn. (2.13) and was successfully used in explaining a wide variety of relaxation rates. The reasoning was this. The frequency of molecular vibrations is of the order of 10^{14} sec^{-1}, making the period of oscillation of the order of 10^{-14} sec, and at reasonable gas temperatures the time taken to traverse long-range interactions with a range of ~ 5 Å is much longer than the period of oscillation of the molecular vibration, thus making the collision adiabatic and the probability of energy transfer small. If short-range repulsive forces that operate over a much smaller distance (~ 0.2 Å) are responsible for the reaction, however, the time taken to traverse these forces can become comparable to the molecular oscillation period of 10^{-14} sec, the collision becomes nonadiabatic and the probability of vibrational energy transfer becomes high. On this basis, then, it would appear that only short-range repulsive forces between colliding systems can ensure that the probability of vibrational energy transfer is not small. An exception to this theory[19] is the vibrational-energy-transfer reaction (6.3) from $N_2^*(v = 1)$ molecules to the (00^01) excited state of CO_2. The experimental cross-section for this reaction involving $^{14}N_2$ shows a negative-temperature dependence below 1000 K in contrast with the theory that predicts an increase in cross-section or probability of energy transfer with increase in temperature. In references 19 and 20, Sharma and Brau showed that the observed negative-temperature dependence between 300 and 1000 K can be explained by taking into account only the weak long-range dipole–quadrupole forces that operate for times two orders of magnitude longer than that expected of short-range repulsive forces. Above 1000 K, the experimental cross-section data display a positive temperature dependence indicating that some other interaction (possibly short-range forces) become important.[20, 21]

The probability of vibrational-energy transfer between $N_2^*(v = 1)$ and $CO_2(00^01)$ molecules at 300 K is about 10^{-3}. This is in reasonable agreement with the value of the probability given for a $|\Delta E_\infty|$ of 18 cm^{-1} in Fig. 2.10,[283, 22, 23] a figure that illustrates the diminishing probability for vibrational energy transfer with increasing energy discrepancy (in cm^{-1}). A comparable value of the probability of energy transfer of 3.5×10^{-4} per collision has been determined for the $N_2^*(v = 1) \rightarrow CO_2(00^01)$ reaction by Offenberger and Rose.[24] This value corresponds to an effective cross-section of approximately 1.7×10^{-18} cm^2. It has been suggested that differences of gas temperatures in individual experiments could account for differences in the probabilities reported in the literature; Moore et al.[25] have reported 2×10^{-3} per collision, and Taylor et al.[26] have reported 4×10^{-4}.

The CO_2 00^01 upper laser level is optically connected to the 00^00 ground state so that radiative decay of it to the ground state is possible to the detriment of population inversion between the 00^01 and 10^00 levels. However, given a sufficient concentration of ground state CO_2 molecules, radiation trapping ensures that the main radiative decay is through the transitions indicated in Fig. 6.1.

The CO_2 Σ_g^+ 10^00 lower laser level is approximately 900 cm^{-1} below the $N_2^*(v = 1)$ highly populated level. Since 500 cm^{-1} is about the limit for efficient vibrational energy transfer (Fig. 2.10), little vibrational energy transfer occurs to it.

The 00^01 upper laser level is also excited by direct electron impact from the ground state in the overall inelastic collision process

$$CO_2(00^00) + e^- \rightarrow CO_2(00^01) + e^- \tag{6.4}$$

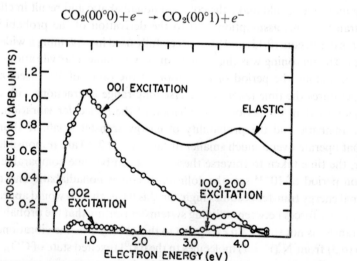

FIG. 6.4. Electron energy dependence of inelastic and elastic cross-sections in CO_2. (After M. J. W. Boness and G. J. Schulz[27] by courtesy of the American Institute of Physics.)

that results in vibrational excitation of the asymmetric stretch mode 00^01. This reaction occurs via an intermediate negatively charged CO_2^- compound state of CO_2 that has a peak in its excitation function at 0.9 eV (Boness and Schulz[27]). It has a cross-section of 3×10^{-16} cm². [27] The negatively charged CO_2^- compound state is estimated to have a lifetime of 10^{-15} sec. The excitation function for 001 and 002 vibrational excitation is shown in Fig. 6.4. Also shown in Fig. 6.4 is the excitation function of the 100 and 200 symmetric stretch modes, which would include that of the CO_2 10^00 lower laser level. The shape and position of the peak in the excitation function that is centered at 3.8 eV is associated with the formation of another negatively charged CO_2^- compound state. Both of these compound states occur as a result of resonance effects of the incident electron with the ground state CO_2 molecule. It is interesting to note that even though the first 10^00 excited state of the symmetric-stretch mode (at about 400 cm⁻¹) is lower than the first 00^01 excited state (and CO_2 upper laser level) of the asymmetric-stretch mode at 2349 cm⁻¹, its excitation function centered at 3.8 eV is higher by nearly 2.0 eV than the center of the excitation function of the asymmetric-stretch mode even though the potential energy interval of the 10^00 and 00^01 levels is only approximately 1000 cm⁻¹.

The population of the $N_2^*(v=1)$ state that is in equilibrium with the CO_2 00^01 upper laser level in CO_2 laser discharges is also established by inelastic electron-impact collisions with ground state $N_2 X^1\Sigma_g^+$ molecules. The excitation of $N_2^*(v=1)$ also involves the formation of a negatively charged compound state because of a resonance effect between the incident energetic electron and the N_2 molecular potential. The N_2 $v=2$ to $v=8$ vibrationally excited states are likewise excited. The excitation function for the $v=1$ to $v=8$ vibrationally excited states of N_2^* is shown in Fig. 6.5. It is dominated by excitation of the $v=1$ and $v=2$ states, is centered at about 2.3 eV, and has a cross-section that is larger than 2×10^{-16} cm² for electrons in the energy range 2 to 2.7 eV. [28, 29]

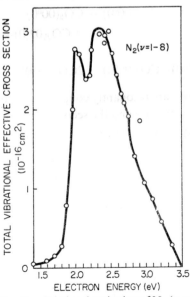

FIG. 6.5. Total effective cross-section for vibrational excitation of N_2 ($v=1$ to 8) by electron impact. (After G. J. Schulz[28] by courtesy of the American Institute of Physics.)

It is the shape, magnitude, and positions of the centers of the resonant excitation functions of the 001, 002, etc. and 100, 200, etc., CO_2 modes, and the $N_2^*(v = 1$ to 8) levels that is responsible to a large extent for specific excitation conditions that exist under optimum discharge conditions in CO_2 lasers. This will be made evident later in discussing sections: *The effect of additives* and 6.1.2 Pure Carbon Dioxide Laser.

6.1.1.2. Lifetime considerations

In the CO_2 laser, unlike CW neutral and ionic gas lasers, the radiative lifetimes in general are very large,[15] and the radiative lifetime of the upper laser level is *shorter* than that of the lower laser level. Under normal radiative-decay conditions this would of course ensure that CW laser oscillation would be impossible. Collisional mechanisms in fact determine the important effective lifetimes in the CO_2 laser, so that in spite of the adverse spontaneous radiative lifetimes CW oscillation can be realized in CO_2–noble gas–molecular gas mixtures and in pure CO_2 discharges. Radiative lifetimes vary from a few milliseconds to a few seconds at the gas pressures used in CW CO_2 lasers, whereas collisional times of molecules are of the order of 10^{-8} to 10^{-7} sec. Thus, even if only one collision in 10^4 was effective in de-exciting an excited CO_2 molecule, collisional effects would determine the effective lifetimes. It is therefore the collisional relaxation times and the relaxation mechanisms that are of primary importance in the CO_2 laser.

Relaxation of the levels involved in the laser CO_2 occurs principally by means of the following six vibration-to-translation (V–T) and vibration-to-vibration (V–V) energy transfer processes

(a) *Lower laser level*

$$CO_2(10^00)+M \rightarrow CO_2(01^10)+M+\text{kinetic energy}, \tag{6.5}$$

$$CO_2(10^00)+CO_2(00^00) \rightarrow CO_2(00^00)$$
$$+CO_2(02^00)+103 \text{ cm}^{-1}, \tag{6.6}$$

and

$$CO_2(10^00)+CO_2(00^00) \rightarrow 2CO_2(01^10) \tag{6.7}$$

in which the 10^00 and 02^00 levels are resonantly coupled by a Fermi resonance as v_1 is nearly resonant with $2v_2$ of the first symmetric and the second bending mode respectively.

$$CO_2(10^00)+M \rightarrow CO_2(00^00)+M+\text{kinetic energy} \tag{6.8}$$

and

$$CO_2(01^10)+M \rightarrow CO_2(00^00)+M+\text{kinetic energy} \tag{6.9}$$

where M throughout is CO_2, or another molecule or atom.

(b) *Upper laser level*

$$CO_2(00^01)+M \rightarrow CO_2(00^00)+M+\text{kinetic energy}. \tag{6.10}$$

Processes (6.8) and (6.10) are several orders of magnitude slower than (6.9).[30] And in all cases, the final relaxation is governed by the V–T relaxation process (6.9) of the 01^10 level, the lowest frequency mode v_2 of the bending mode.[15, 31]

6.1.1.3. Dynamics of the N_2–CO_2 laser

The dynamics of the CO_2 laser system are represented in a simplified form in Fig. 6.6. It is clear from Fig. 6.6 that the complex dynamics of the system in a gaseous discharge will be determined by a multitude of factors. These have been found to include among others: gas mixture, pressure, tube diameter, gas-flow rate, method of injection of the gas mixture, current density, electron density, electron temperature, and gas temperature.

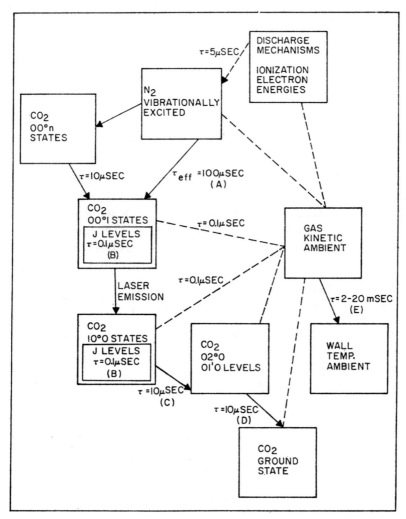

FIG. 6.6. Schematic diagram of molecular level dynamics in the CO_2 laser. (Adapted from P. A. Miles and F. A. Horrigan.[32])

Important time constants τ included in Fig. 6.6 and indicated by (A), (B), (C), (D) and (E) are associated with:

(A) The transfer of vibrational energy from $N_2^*(v = 1)$ to the CO_2 00^01 level. This time

271

constant is of the order of 100 $p(N_2)$ μsec where $p(N_2)$ is the partial pressure of nitrogen in torr.

(B) The transfer of rotational energy among the closely spaced rotational J-levels of both upper and lower laser levels ($\tau = 0.1$ μsec).

(C) The relaxation of the lower 10^00 laser level to the CO_2 02^00 level ($\tau = 10$ μsec).

(D) The relaxation of the terminal 02^00 and 01^10 levels to the CO_2 ground state ($\tau = 10$ μsec).

(E) The relaxation of the gas temperature to the discharge-wall temperature. This relaxation is dependent on the rate of diffusion of heat from the discharge region to the walls, and is affected by gas flow rate, mixing, and the tube diameter ($\tau = 2$ to 20 msec).[32]

The effect of additives. Many gases and vapors have been added to N_2–CO_2 and pure CO_2 laser mixtures in attempts to

(1) increase the excitation rate of the upper laser level;
(2) depopulate the lower laser level; and
(3) decrease the relaxation time of the terminal laser levels.

The addition of helium leads to the most important and interesting results.

Figure 6.7 shows that the addition of substantial amounts of helium to the N_2–CO_2 mixture in a CW laser increases the laser output power at 10.6 μm on the 00^01–10^00 vibrational–rotational transitions by a factor of more than 5.[33] It can increase the gain (in percent/meter) by a factor of more than 2.[15] This effect was first observed by Moeller and

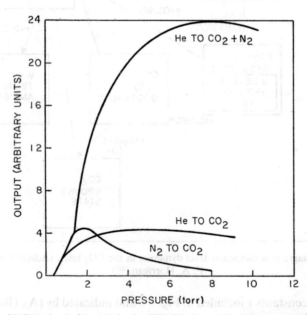

FIG. 6.7. The effect of helium (and nitrogen) in the CW CO_2 laser. The CO_2 pressure was fixed at 0.40 torr then N_2 and He were added. (After T. G. Roberts, G. J. Hutcheson, J. J. Ehrlich, W. L. Hales, and T. A. Barr, Jr.,[33] by courtesy of the IEEE.)

Rigden[34] in flowing and nonflowing CO_2 laser systems. The addition of neon to a N_2–CO_2 mixture also increases the output power but by a smaller factor of about 3, while the addition of argon, deuterium, or air decreases the output power.[33]

Effect on discharge mechanisms. The addition of helium has little effect on the discharge of a N_2–CO_2 10.6-μm CO_2 laser. The lowest $He^*2^3S_1$ energy level, at 19.8 eV above the ground state, is considerably higher than the $CO_2(00^01)$ upper laser level and the $N^*(v = 1)$ vibrational level (both of which are less than 0.3 eV above the ground level) and the ionization potentials of CO_2 or N_2. The excitation energy of 19.8 eV of the lowest excited state of helium and the ionization potential energy of 24.58 eV are also considerably higher than the typical average electron temperature of 1 to 3 eV in N_2–CO_2 CW lasers,[24, 35, 36] so that little excitation or ionization of helium is expected or observed in N_2–CO_2–He laser mixtures.

In high-pressure, pulsed CO_2 lasers, however, the addition of helium does affect the discharge. Helium addition allows use of a higher working CO_2 and N_2 gas pressure without the formation of hot arcs or filaments that seriously degrade the output power by increasing the lower laser level population. It enables more energy to be dissipated in the discharge, and appears to delay the onset of discharge filamentation after the initiation of the discharge, and enables energy to be more quickly coupled into the discharge by decreasing the discharge impedance.[37–41] The addition of hydrogen to pulsed, high pressure N_2–CO_2–He laser discharges has also been found to delay the formation of filamentary, hot-arc discharges.[42, 43]

If the additive gas has an ionization energy that is comparable to, or lower, than that of

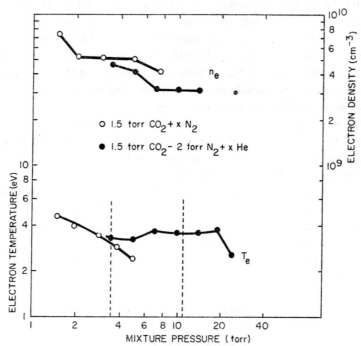

Fig. 6.8. Variation in electron temperature and density as a function of the partial gas pressures in a CW CO_2–N_2–He laser discharge, $D = 22$ mm. (After P. O. Clark and M. R. Smith[35] by courtesy of the American Institute of Physics.)

CO_2 or N_2, and has a high vapor pressure that enables it to be used at high pressures at gas-discharge temperatures, it can affect high-pressure CW gas discharges in ways that do not appear to be detrimental to laser processes. The addition of xylene has been found to give the best results.[44]

Whereas helium addition does not affect CW CO_2 laser discharges, the addition of nitrogen does. When nitrogen is added to a laser discharge in pure CO_2 at a pressure of 1.5 torr, as well as providing a means by which the CO_2 00^01 upper laser is selectively excited by vibrational energy transfer in reaction (6.3), it reduces the average electron temperature without appreciably reducing the electron concentration.[35] The effect is shown in Fig. 6.8, in which the average electron temperature and electron density are plotted as functions of the mixture pressure (nitrogen was added to 1.5 torr of CO_2). The addition of 3.5 torr of N_2, making the total mixture pressure 5 torr, reduces the average electron temperature from nearly 5 eV to about 2.5 eV, while reducing the electron density from 7×10^9 to 5×10^9 cm^{-3}. Figure 6.8 also shows that the addition of large quantities of helium (up to 20 torr) to a 1.5-torr CO_2, 2-torr N_2 laser mixture (an optimum mixture) does not affect the average electron temperature, and only slightly reduces the electron density. Similar results have been reported by Offenberger and Rose.[24]

The reduction in the average electron temperature results in more efficient excitation of vibrationally excited CO_2 and N_2 (and CO) molecules by the inelastic resonant, Schulz-type electron-impact collision-processes discussed earlier.[27-29, 45, 46] These resonant processes have maximum cross-sections at low electron energies (Figs. 6.4 and 6.5).

Though it has little effect on discharge mechanisms, the addition of helium does affect directly the equilibrium populations of the 00^01 and 10^00 upper and lower laser levels of the CW 10.6-μm CO_2 laser. It increases the population of the 00^01 upper laser level, and at the same time decreases the population of the 10^00 lower laser level. Figure 6.9, in which the spontaneous intensities of three infrared lines that originate on the upper and lower laser levels are plotted against the partial pressure of helium that is added to a discharge in pure CO_2, clearly displays its effect on these populations.

The population of the CO_2 00^01 upper laser level, represented by the amplitudes of emission from lines at 1083 cm^{-1} (00^01–02^00) and 974 cm^{-1} (00^01–10^00), increases rapidly below a partial pressure of 2 torr. Above this pressure the emission amplitudes indicate that the CO_2 00^01 population is beginning to saturate.

The population of the CO_2 10^00 lower laser level represented by the amplitude of emission of the 718-cm^{-1} line (10^00–01^10 transition) behaves quite differently. It decreases sharply on the addition of small amounts of helium at 0.5 to 1.0 torr, reaches a minimum at about 2 torr, and increases slightly above this pressure.[31]

The increase in population of the 00^01 upper laser level on the addition of helium follows from two main effects:

(1) a reduction in its thermal relaxation by causing a reduction in the gas temperature;[47, 31]
(2) a reduction in the average electron temperature to one that provides more efficient vibrational excitation of the CO_2 00^01 and $N_2^*(v = 1)$ levels (as discussed in the previous section).

274

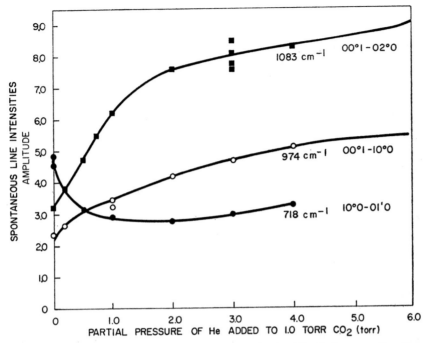

FIG. 6.9. Variation in the amplitude of infrared emission from a flowing CO_2 discharge as a function of helium addition at a constant current of 100 mA in a 3.2-cm-bore tube. (After M. J. Weber and T. F. Deutsch,[31] by courtesy of the IEEE.)

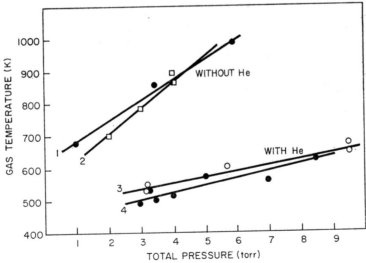

FIG. 6.10. Gas temperature versus total gas pressure in CO_2–N_2 and CO_2–N_2–He laser mixtures used to illustrate the reduction in gas temperature caused by the addition of helium. Curve 1, $I = 80$ mA, $Q(N_2) = 0.28$, $Q(CO_2) = 0.10$; curve 2, $I = 80$ mA, $Q(N_2) = 0.40$; curve 3, $I = 80$ mA, $Q(N_2) = 0.28$, $Q(CO_2) = 0.10$, $Q(He) = 2.20$; curve 4, $I = 60$ mA, $Q(N_2) = 0.40$, $Q(CO_2) = 0.13$, $Q(He = 2.40)$. $D = 25$ mm for 1 and 3, and 62 mm for 2 and 4. The flow rates Q are given in l-torr/sec. (After A. G. Sviridov, N. N. Sobolev, and G. G. Tselikov[47] by courtesy of the American Institute of Physics.)

275

Helium has a thermal conductivity that is an order of magnitude greater than either CO_2 or N_2, so that an addition of it to a discharge in these gases affects considerably the rate of dissipation of heat in the discharge. This consequently affects the gas temperature of the discharge and the rate of thermal relaxation of each laser level. Figure 6.10 shows that the addition of helium to discharge mixtures of CO_2 and N_2 at various flow rates of N_2 and CO_2, reduces the gas temperature from about 750 K to 500 K at low total gas pressures of CO_2, N_2, and He of about 2.5 torr, and lowers it from 1000 to about 600 K at a total pressure of 6 torr.[47] This reduction in gas temperature reduces the rate of thermal relaxation of the 00^01 upper laser level in CO_2–CO_2 collisions by a factor of approximately 10, under the conditions stated in the caption, from $\tau = 10^{-4}$ sec at a CO_2 pressure of 1 torr at 1000 K, to $\tau = 10^{-3}$ sec at 500 K, and increases the upper laser level population.[47] Slightly offsetting this advantage, thermal relaxation of the 10^00 lower laser level is reduced, which tends to reduce the extent of population inversion between the 00^01 and 10^00 laser levels. This disadvantage, however, is more than compensated for by increased collisional relaxation of the 10^00 and 01^11 levels by the specific reactions (6.5), (6.8) and (6.9), respectively:

$$CO_2(10^00) + He \;\Big<\!\!\begin{array}{l} \longrightarrow CO_2(01^10) + He + kinetic\ energy \\[2ex] \longrightarrow CO_2(00^00) + He + kinetic\ energy \end{array}$$

and $CO_2(01^10) + He \rightarrow CO_2(00^00) + He + kinetic\ energy.$

These three reactions depend little on gas temperature so that they are not adversely affected by the reduction in gas temperature when helium is added. Water cooling the walls of the discharge is found to reduce the gas temperature by about 50 K.[47] It too, by decreasing the thermal relaxation of the 00^01 upper laser level, increases the laser output.[48]

The gas-flow rate in a nonsealed-off N_2–CO_2–He laser also affects considerably the gas temperature and thereby affects the relaxation of the upper laser level. In a nonsupersonic

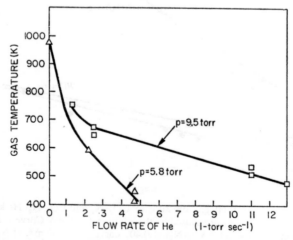

FIG. 6.11. Effect of helium flow rate on the gas temperature of a N_2–CO_2–He discharge mixture. Discharge conditions: $I = 80$ mA, $Q(N_2) = 0.28$, $Q(CO_2) = 0.10$ l-torr/sec, $D = 25$ mm. (After A. G. Sviridov, N. N. Sobolev, and G. G. Tselikov[47] by courtesy of the American Institute of Physics.)

flowing discharge system, flowing the gas decreases the gas temperature rapidly at low flow rates, with a greater reduction occurring at lower total pressures (Fig. 6.11). The gain in a N_2–CO_2–He laser can be increased by a factor of nearly two by the use of even low flow rates ($\lesssim 20\ cm^3/min$).[15, 49] This effect is also related to the removal of dissociation products that are produced in the discharge.

As well as affecting the upper and lower laser levels in the ways described, helium produces two further beneficial results. These are related to the distribution of population among the rotational sublevels of the CO_2 vibrational levels.

(1) It cools the rotational levels without affecting the vibrational levels and enables the closely spaced rotational levels to come quickly into thermal equilibrium with the gas temperature. This leads to a shift of the strongest oscillation on the P transitions to a lower value, as predicted by Patel.[50]

(2) Once oscillation commences on a single P transition, which is usually the strongest $P(18)$ transition in a CO_2–N_2–He laser mixture, it helps feed in population from neighboring J levels in an attempt to rethermalize the distortion in the Boltzmann distribution in the rotational levels of the 00^01 upper laser level produced by the laser oscillation. In this way most of the population of the upper laser vibrational state is directed through one laser transition. These two effects are illustrated diagrammatically in Fig. 6.12. In the absence of helium, strongest oscillation is observed on the $P(20)$ transition; with helium present, strongest oscillation occurs on the cooler $P(18)$ transition with additional oscillation at a reduced magnitude now present only on the two neighboring $P(16)$ and $P(20)$ transitions.

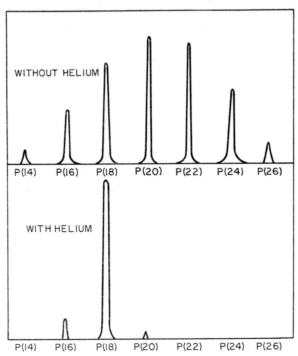

Fig. 6.12. Illustrating the effect of helium on laser oscillation on vibrational–rotational transitions in the N_2–CO_2 laser.[32] (By courtesy of P. A. Miles and F. A. Horrigan, Raytheon Corp.)

19*

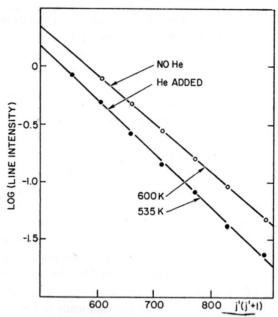

FIG. 6.13. Rotational temperature of the second positive system of N_2 with and without helium in N_2–CO_2 laser mixtures. (After N. Djeu, T. Kan, and G. J. Wolga[51] by courtesy of the IEEE.)

Quantitative data, which clearly shows that helium "cools" the rotational energy levels, has been obtained by Djeu et al.,[51] who measured the intensity profile of spontaneous emission in the R branch of the 0–0 band of the second positive system $(C^3\pi_u$–$B^3\pi_g)$ of electronically excited molecular nitrogen in a N_2–CO_2 laser mixture with and without helium. This data is shown plotted as log (line intensity) versus $j'(j'+1)$ in Fig. 6.13. Since the log (line intensity) = a constant:

$$\frac{B_{v'} J'(J'+1)}{kT},$$

where $B_{v'}$ is a rotational constant for a given vibrational level v' and J' is the upper quantum number of the center component,[52] the straight lines indicate that local thermodynamic equilibrium exists in the 0–0 band of the second positive system of N_2, with corresponding temperatures of 600 K without helium, and a lower temperature of 535 K with helium. As noted by Djeu et al.[51] this is much higher than the 340-K rotational temperature of CO_2 levels that are in equilibrium with the translational motion of the gas.

Depopulation of the 10^00 lower laser level by additives via V–T collisions can also occur to give increased gain on the 00^01–10^00 laser transition. The final relaxation rate of the 10^00 level, because of inverse excitation processes, is determined by the relaxation rate of the 01^10 level in the specific V–T reaction (6.9) where

$$CO_2(01^10) + He \rightarrow CO_2(00^00) + He + \text{kinetic energy}.$$

In order to maintain population inversion between the 00^01 and 10^00 levels the relaxation time of the 01^10 level must not be longer than that of the 10^00 CO_2 lower laser level. This maximum allowable relaxation time has been estimated to be approximately 10^{-4} sec.[53]

278

The relaxation time τ of the 01^10 level of CO_2 can be calculated from the relationship given by Sobolev and Sokovikov[53]

$$1/\tau = \Sigma_i(p_i/\tau_{CO_2,i}) \tag{6.11}$$

where p_i are the partial pressures of the components and $\tau_{CO_2,i}$ is the relaxation time of the CO_2 molecules in the gas i at a pressure of 1 torr. Values of the relaxation times of the 01^10 first bending mode of CO_2 in pure CO_2 and in mixtures with other gases for 300 K at 1 torr are given in Table 6.2.

TABLE 6.2. *Relaxation times of the 01^10 level of CO_2 in various gases (at a pressure of 1 torr and a temperature of 300 K)* [53]

Gas	Relaxation time τ (sec)[a]	Average number of gas-kinetic collisions leading to the loss of a vibrational quantum
CO_2	4.4×10^{-3}	51,300
N_2	1.04×10^{-4}	1,200
CO	2.0×10^{-5}	230
NO	2.26×10^{-5}	260
H_2	2.6×10^{-5}	300
H_2O	5.2×10^{-6}	60
He	2.26×10^{-4}	2600
Ar	4.1×10^{-3}	47,000

[a] The number of gas-kinetic collisions is taken to be the same for all gases, 1.15×10^7 sec.$^{-1}$

The table shows that the addition of nitrogen at 1 torr, as well as providing a pump for exciting the upper laser level by resonant vibrational energy transfer, leads to a relaxation time of approximately 10^{-4} sec for the 01^10 level. The relaxation time of the 01^10 level is also reduced by $CO_2(01^10)$–CO collisions since considerable dissociation of CO_2 to CO and O_2 occurs in the discharge.

Table 6.2 also indicates why the addition of water vapor to a N_2–CO_2–He laser discharge can give a large increase in output power (Witteman,[54, 55] and Carbone and Witteman[56]). The table shows that water vapor causes the largest decrease in $\tau(01^10)$ of all the additions listed. An addition of only 0.1 torr of water vapor leads to a $\tau(01^10)$ of about 50 μsec. Water vapor is also an effective relaxant of excited CO_2 molecules in 10^00 lower laser levels because the lowest frequency of vibration of the angle between two O–H bonds in H_2O at 1596 cm^{-1} is within kT of the CO_2 10^00 level, so that vibrational energy transfer occurs to H_2O in only a few collisions.[55] In turn, the vibrationally excited H_2O molecules are rapidly de-excited in V–T collisions with other H_2O molecules. Hydrogen is also seen to be effective in reducing $\tau(01^10)$, and can possibly account for the effect that small quantities of hydrogen can have on the gain in CO_2 laser discharges.[57] Both hydrogen and water vapor, however, reduce $\tau(00^01)$ slightly and so can adversely affect population inversion between the 00^01 and 10^00 levels.

6.1.1.4. Electron energy distributions and vibrational excitation rates in CO_2 lasers

In considering the effects of additives on the CO_2 laser discharges we have loosely spoken about the average electron temperature. This presupposes that the electron energy distribution can be characterized by a distribution that can be described mathematically. So far in talking about the average electron temperature in CO_2 laser discharges, it has been assumed that the electron energy distributions are Maxwellian (for instance ref. 35 and Fig. 6.8). Nighan and Bennett[45] and Novgorodov, Sviridov and Sobolev[58] have shown theoretically and experimentally, respectively, that this assumption does not hold in the weakly ionized molecular plasmas of CO_2 laser discharges. This is because the electron energy distribution is determined almost entirely by collisions of electrons with molecules in elastic collisions in the production of vibrational and electrical excitation and ionization and not by electron–electron or electron–ion interactions. The result is that the electron energy distributions are

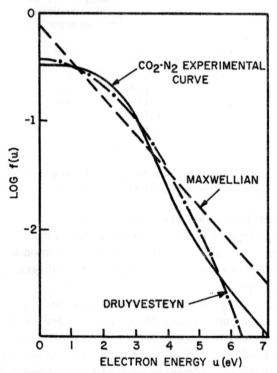

FIG. 6.14. Electron energy distribution in a CO_2-N_2 (1 : 1) mixture, $p = 2.3$ torr, $D = 20$ mm; $\bar{\ } = 1.9$ eV; compared with Maxwellian and Druyvesteyn distributions. (After M. Z. Novgorodov, A. G. Sviridov, and N. N. Sobolev[58] by courtesy of the American Institute of Physics.)

neither Maxwellian nor Druyvesteyn and cannot be characterized by a single electron temperature. This is illustrated in Fig. 6.14 in which the measured electron energy distribution (on a semilogarithmic scale) in a nonflowing CO_2-N_2 (1:1) mixture is compared with Maxwellian or Druyvesteyn distributions with the same mean energy $\bar{\varepsilon}$ of 1.9 eV. The measured distribution is clearly neither Maxwellian nor Druyvesteyn, having two

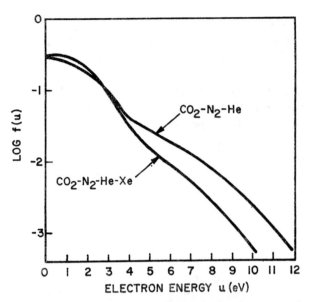

FIG. 6.15. Electron energy distribution in a CO_2–N_2–He $(1:1:6)$ mixture, $p = 4$ torr; $\bar{u} = 3.1$ eV; in a CO_2–N_2–He–Xe $(1:1:6:1)$ mixture, $p = 4.5$ torr, $\bar{u} = 2.5$ eV; $D = 20$ mm. (After M. Z. Novgorodov, A. G. Sviridov, and N. N. Sobolev,[58] by courtesy of the American Institute of Physics.)

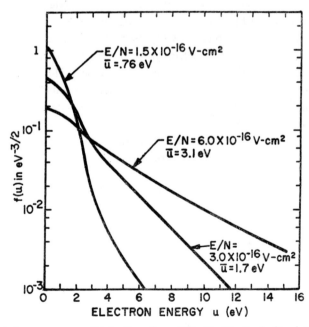

FIG. 6.16. Calculated electron energy distributions for a CO_2–N_2–He $(1:1:8)$ mixture. The distribution function is defined such that $\int_0^\infty u^{1/2} f(u)\, du = 1$, and the average energy such that $\bar{u} = \frac{2}{3} \int_0^\infty u^{3/2} f(u)\, du$, where $\bar{u}$ is in eV; for a Maxwellian distribution $\bar{u}$ is the electron temperature. (After W. L. Nighan and J. H. Bennett,[45] by courtesy of the American Institute of Physics.)

distinctly different portions. One portion above approximately 4.5 eV could be approximated by a straight line and given a Maxwellian average electron temperature; below 4.5 eV the distribution is more Druyvesteyn-like. Similar deviations from Maxwellian or Druyvesteyn distributions in discharges in CO_2–N_2–He and CO_2–N_2–He–Xe laser mixtures have been observed by Novgorodov et al.[58] (Fig. 6.15). Calculated electron energy distributions for CO_2–N_2–He mixtures for E/N values typical of CO_2 laser discharges (2 to 3×10^{-16} V-cm²) exhibit the two distinctly different portions that have been determined experimentally to exist. (The E/N value is the ratio of the electric field in volts per centimeter to the total neutral-particle density in number per cubic centimeter; it is a parameter to which the average electron temperature in the positive column of a glow discharge can be related, just as can the product of the pressure p and tube-diameter D.) Shown in Fig. 6.16 are electron energy distributions $f(u)$ for various E/N values as calculated for a $1:1:8$ CO_2–N_2–He mixture.[45] The change-over from a more Druyvesteyn-like distribution to a distribution that could be attributed a Maxwellian average electron temperature is seen to occur in the calculated distribution at slightly lower electron energies than the point at which it has been determined experimentally to occur and exhibited in Figs. 6.14 and 6.15.

With a knowledge of the electron energy distribution function it is possible to perform an energy balance by integrating the electron kinetic equation over all electron energies.[59] This procedure leads to an electron energy conservation equation that equates the rate of energy input to electrons from the electric field to the rate of energy transfer from the electrons to various molecular states that are excited in inelastic electron-impact collisions.[45] Having determined the electron energy distribution $f(u)$, Nighan and Bennett[45] have calculated the various contributions to the inelastic energy-loss processes as a function of E/N for CO_2–N_2–He mixtures typical of CO_2 lasers. By dividing each contribution by the power input to electrons from the electron field, the fractional rate of energy transfer or power transfer to each inelastic collision process is obtained. The fractional power transfer is closely related to laser efficiency. Nighan and Bennett's results[45] are shown in Fig. 6.17 where the fractional power transferred from electrons to excited states of CO_2 and N_2 is plotted as a function of E/N and the average electron energy $\bar{u}$. The range of 10^{-16} to 10^{-15} V-cm² for E/N in the figure covers the operating range of CO_2 lasers. Although the calculations were made for a CO_2–N_2–He mixture with particle densities in the ratio $1:1:8$, small variations in the mixture are not considered to influence the fractional power transfer significantly.[45] These calculations of the fractional power transfer indeed vividly illustrate the sensitivity of electron energy exchange processes to variations in E/N. Because of this fact, Nighan and Bennett[45] have suggested that the parameter E/N be measured routinely in experimental studies of CO_2 laser operation.

Figure 6.17 shows that for an E/N of approximately 10^{-16} V-cm², nearly 65 percent of the electron energy goes directly into the CO_2 00^01 upper laser level and less than 8 percent into the CO_2 01^00 lower laser level. Effectively all the electron energy goes into vibrational excitation of either CO_2 or N_2, as it can be seen that the fractional power transfer into electronic excitation and ionization of CO_2 and N_2 is negligible. The very large low-energy cross-section for electron excitation of the CO_2 00^01 level is responsible for the dominance of upper laser level excitation for values of $\bar{u}$ near 0.5 eV.[60] At an E/N value of about 10^{-15} V-cm², more than 80 percent of the total electron energy goes into the electronic excitation

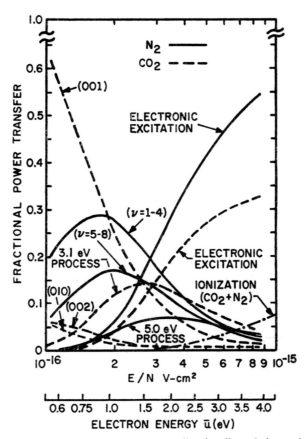

Fig. 6.17. Fractional power transferred from electrons to vibrationally and electronically excited states of CO_2 and N_2 as a function of E/N and $\bar{u}$ for a CO_2-N_2-He (1 : 1 : 8) mixture. After W. L. Nighan and J. H. Bennett.[45] The figure also includes a curve for ionization of CO_2 and N_2 taken from R. H. Bullis, W. L. Nighan, M. C. Fowler, and W. J. Wiegand, United Aircraft Research Lab. Rpt. J-291 (1970).

of N_2 and CO_2, about 8 percent into ionization of N_2 and CO_2, and minor amounts into vibrational excitation. It is clear, therefore, that there are definite advantages to be had if a CO_2 laser discharge could be sustained in the presence of low ionization at the lower E/N values covered by the figure at reasonable electron densities. As ionization is, of course, necessary for the maintenance of the discharge, and as it probably proceeds by means of electron impact dissociation of CO_2 to CO and O and ionization of O_2 to O_2^+ as well as by direct electron impact of CO_2 from the ground state, practically, one has to operate between the two extremes of 10^{-16} and 10^{-15} V-cm² such that the electron power transfer is shared between vibrational and electronic excitation and ionization. The calculations of Nighan and Bennett[45] enable good agreement to be obtained between the estimated output power and experimentally determined output power of CO_2 lasers. For a typical CO_2 laser-discharge E/N-value of 3×10^{-16} V-cm², approximately 42 percent of the total power goes into the first eight vibrational levels of N_2 (from which it is transferred to the CO_2 00⁰1 upper laser level) and into the 001 and 002 levels of CO_2, and is expected to contribute to the CO_2 00⁰1 upper laser level excitation. This 42-percent power transfer, combined with the quantum

283

efficiency of 41 percent of the CO_2 laser transition, indicates that approximately 17 percent of the power coupled into the N_2 and CO_2 is available for conversion to optical power. This is in agreement with typical experimental values.[16, 45]

6.1.1.5. Upper and lower level populations

Calculations of the 00^01 upper, and 10^00 lower laser level populations have been made by Djeu et al.;[51] and Gordiets et al.[61] have analysed and computed the population inversion in various CO_2 laser mixtures under a number of CW discharge conditions.

The calculations by Gordiets et al.[61] are based on the following assumptions: The electrons excite the vibrational levels of N_2 directly but do not excite CO_2 molecules. This is followed by resonant exchange of vibrational energy from N_2 to the $00n$ levels of CO_2, transfer of vibrational energy of CO_2 molecules in the $00n$ levels to the $0n0$ bending modes by collisional deactivation, and V–T energy transfer from the $0n0$ levels in collisions with CO_2, N_2, and He atoms and molecules. Direct V–T energy transfer at the 10^00 and 00^01 levels and radiative processes can be neglected.

The calculations were carried out assuming an average electron temperature (kT_e) of 3 eV in a discharge tube of 25-mm bore. As the computations of Gordiets et al.[61] of the population inversion are in good agreement with experimental measurements in spite of the simplifying assumptions made in the computation, a number of their results are reproduced in

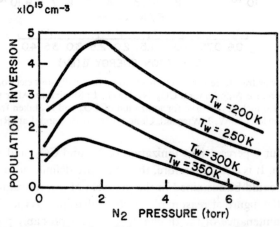

FIG. 6.18. Population inversion as a function of N_2 pressure in a 2-torr CO_2, x-torr N_2 mixture at various discharge-tube wall temperatures, with $n_e = 7.5 \times 10^9$ cm^{-3} (in a 25-mm-bore discharge tube). (After B. F. Gordiets, N. N. Sobolev, and L. A. Shelepin,[61] by courtesy of the American Institute of Physics.)

Figs. 6.18 to 6.22. These results can be directly related to the excitation and de-excitation processes that we have discussed in the previous sections. The analysis of these results follows closely that of Gordiets et al.[61]

Figure 6.18 shows the population inversion as a function of nitrogen pressure in a CO_2–N_2 mixture with CO_2 at 2 torr, at various temperatures T_w of the wall of the discharge tube. As nitrogen is first added, energy transfer from $N_2^*(v = 1)$ to the CO_2 00^01 upper laser level increases, and the population inversion increases. Further addition of N_2 causes the gas temperature to rise. This results in an increased relaxation rate and reduced population of the 00^01 level, and an increased population of the 10^00 lower laser level. These factors combine to produce an optimum N_2 pressure of 1 to 2 torr.

Figure 6.19 shows the population inversion as a function of CO_2 pressure at a fixed N_2 pressure of 2 torr. The optimum CO_2 pressure of 1 to 2 torr is due to competition of two factors: an increased number of CO_2 molecules, and an increased decay rate of the 00^01 upper laser level due to increased gas temperature and increased collisional $CO_2(00^01)$–CO_2 collisions. It can be seen from Fig. 6.19 that the population inversion is sensitive to the

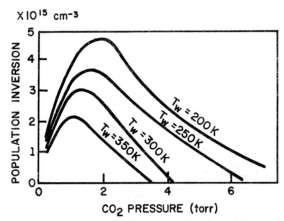

FIG. 6.19. Population inversion as a function of CO_2 pressure in a 2-torr N_2, x-torr CO_2 mixture at various discharge-tube wall temperatures with $n_e = 7.5 \times 10^9$ cm^{-3} (in a 25-mm-bore discharge tube). (After B. F. Gordiets, N. N. Sobolev, and L. A. Shelepin,[61] by courtesy of the American Institute of Physics.)

partial pressure of CO_2. This is particularly true at higher discharge-tube wall temperatures. The maximum inversion in the curves between 1 and 2 torr of CO_2 corresponds approximately to that pressure at which the total decay rate of the 00^01 upper laser level due to volume quenching and wall collisions has a minimum,[62] and maximum gain is observed in pure CO_2 and CO_2–N_2 laser tubes of the same bore size (~ 20 mm).[15]

The effect of helium on the population inversion in a CO_2–N_2 (1 : 1) mixture at 4 torr is illustrated in Fig. 6.20. The inversion (solid curve) is seen to increase rapidly at all discharge-tube wall temperatures up to a helium pressure 3 to 5 torr, after which it flattens out but continues to increase up to the maximum pressure of 16 torr considered in the calculations. When collisional deactivation of the 00^01 upper laser level by helium is considered (dashed curves) a broad inversion maximum is observed to occur between 4 and 8 torr of helium. This corresponds to the optimum helium pressure that is observed experimentally in CO_2–N_2–He lasers.

The population inversion as a function of the average electron density $\bar{n}_e$ in an optimum (1 : 1) mixture of CO_2 and N_2 at 4 torr, and a (1 : 1 : 3) mixture of CO_2–N_2–He at a total pressure of 10 torr is illustrated in Fig. 6.21. The optimum $\bar{n}_e$ in the CO_2–N_2–He mixture is shifted towards higher electron densities due to the cooling effect that helium has on the gas temperature that enables the population of the 00^01 level to increase and the population of the 10^00 lower laser level to decrease.

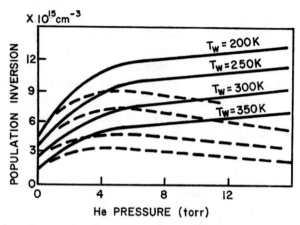

FIG. 6.20. Population inversion as a function of He pressure in a 2-torr N_2, 2-torr CO_2, x-torr He mixture at various discharge-tube wall temperatures with $n_e = 7.5 \times 10^9$ cm^{-7} (in a 25-mm-bore discharge tube). The dashed lines denote a rough estimate of the effect of the helium addition on the relaxation rate of the $00^0 1$ level. (After B. F. Gordiets, N. N. Sobolev, and L. A. Shelepin,[61] by courtesy of the American Institute of Physics.)

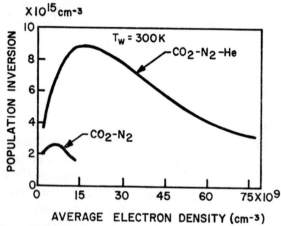

FIG. 6.21. Population inversion as a function of average electron density in a 2 torr N_2, 2-torr CO_2 laser mixture, and in a 2-torr N_2, 2-torr CO_2, 6-torr He mixture in a 25-mm-bore discharge tube. (After B. F. Gordiets, N. N. Sobolev, and L. A. Shelepin,[61] by courtesy of the American Institute of Physics.)

Lastly, the effect of the discharge-tube diameter on the calculated population inversion between the $00^0 1$ and the $10^0 0$ levels of the CO_2 laser is shown in Fig. 6.22 as a function of the CO_2 pressure for various tube diameters at a constant current density $\bar{n}_e D^2/4 = 1.17 \times 10^{10}$ cm^{-1}, where $\bar{n}_e = 7.5 \times 10^9$ cm^{-3} would correspond to a discharge current of about 20 mA. The figure shows that the maximum obtainable inversion and the corresponding optimum pressure of CO_2 both increase with decrease in diameter of the discharge tube. The population inversion (and hence the gain) at the optimum CO_2 pressure is shown to be approximately inversely proportional to the tube diameter D, in agreement once more with experimental observations.[49, 63, 15] Under optimum conditions for population inversion,

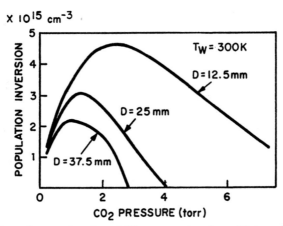

FIG. 6.22. Population inversion as a function of CO_2 pressure in a 2-torr N_2, x-torr CO_2 mixture at various discharge-tube diameters, with $n_eD^2/4$ maintained constant at 1.17×10^{10} cm^{-1}. (After B. F. Gordiets, N. N. Sobolev and L. A. Shelepin,[61] by courtesy of the American Institute of Physics.)

the figure demonstrates that for a CO_2–N_2 laser mixture

$$p(CO_2)D = 33 \text{ torr-mm}. \tag{6.12}$$

This result is also in good agreement with experimental results[49, 63, 15] on CO_2 laser discharges in which the constancy of the pD-product has been observed. Here again, this optimum $p(CO_2)D$ of about 30 torr-mm for maximum population inversion, corresponds to that at which the total decay rate of the 00^01 upper laser level due to volume quenching and wall collisions is a minimum.[62] This would indicate that the population inversion is dominated by upper-laser-level excitation processes just as it is in neutral and ion gas lasers covered in Chapters 4 and 5.

Values for the maximum population inversions $N(00^01)$–$N(10^00)$ calculated by Gordiets et al.[61] are

(1) 9×10^{15} cm^{-3} for a CO_2–N_2–He (1 : 1 : 3) mixture, $D = 25$ mm.
(2) $\sim3\times10^{15}$ cm^{-3} for a CO_2–N_2(1 : 1) mixture, $D = 25$ mm; both at a discharge-tube wall temperature of 300 K.

Values of absolute population densities in a flowing 0.65 torr CO_2, 1.4 torr N_2, 2.9 torr He laser–amplifier mixture in a tube of 25-mm bore have been determined by Djeu et al.[51] to be

$$N(00^01) = 3.3\times10^{15} \text{ cm}^{-3},$$
$$N(10^00) = 1.4\times10^{15} \text{ cm}^{-3},$$

giving a population inversion of nearly 2×10^{15} cm^{-3}. These experimentally determined values are possibly low compared with those calculated by Gordiets et al.[61] because the gas pressures used are lower than the optimum ones for a tube of 25-mm bore. The CO_2 pressure is considerably lower than the optimum one of 1.3 to 1.5 torr for a 25-mm-bore tube according to the $p(CO_2)D$ relationship (6.12).

287

6.1.1.6. Operating characteristics and discharge conditions

CW operation. Throughout this consideration of the 10.6-μm CO_2 laser, emphasis has been on CW operation at low gas pressures up to approximately 10 torr of total pressure, and on the processes that are operative at low pressure and that bring about population inversion. The important operating characteristics of CW low pressure CO_2 lasers have been essentially covered in Section 6.1.1.5 that precedes this section and will not be repeated here. All that remains to cover the operating characteristics and discharge conditions of this basic type of CW CO_2 laser is to mention that large increases in output power can be realized from CO_2 laser mixtures if the gas mixture is flowed rapidly through the active region of the discharge.

Prior to a significant advance in CW CO_2 laser technology made by Tiffany, Targ, and Foster, it appeared that the maximum CW 10.6 μm output power that could be obtained from 1 meter of discharge was limited by heating of the gas to about 200 W.[61] Tiffany *et al.*,[65] by using electrical excitation transverse to the laser axis and rapidly flowing the CO_2–N_2–He laser mixture across the discharge region were able to demonstrate that more than 1 kW could be obtained from 1 meter of discharge. In this work, the gas flow was such that molecules crossed the discharge region within the lifetime of the upper laser level, and the gas was cooled by forced convection instead of relying on diffusion processes and water cooling of the walls of the discharge tube to keep the gas temperature down. The gas used by Tiffany *et al.*[65] consisted of a 2-torr CO_2, 11-torr N_2, and 5-torr He mixture (the gas flow was above 30 m/sec) and 1 kW of output power was obtained for about 13 kW of electrical discharge input power. A high CW output of 140 W has also been obtained from 10-cm long, relatively high pressure (10 to 120 torr), high-speed flowing CO_2 lasers in which the gas flow at several hundred meters per second is along the discharge and optical axis.[66, 15]

Pulsed operation. Output power considerably higher than those obtained from CW CO_2 lasers can be obtained if pulsed electrical excitation is used, powers in excess of megawatts being readily obtainable. The higher output power of pulsed CO_2 lasers results from operating at pressures that are higher than those in CW CO_2 lasers as first indicated by work of a number of researchers.[67–71]

Under proper excitation conditions, the realizable gain is determined principally by the pressure at which a laser can be operated (as shown in Chapter 1). The relationship, gain is proportional to pressure, breaks down at high pressures if the pressure becomes the dominant line-broadening mechanism and the line width $\Delta\nu$ becomes proportional to the pressure. Under these conditions, the population inversion (N_2–N_1) divided by $\Delta\nu$ tends to become a constant. While the gain can reach a maximum at high pressures, the laser energy depends only on the population inversion (N_2–N_1), and so can continue to increase linearly with pressure as the pressure is increased as long as the optimum E/N, or E/p where p is the pressure, is maintained.

Essentially, the obtaining of high pulse energies at high powers is a discharge problem.

To operate at high gas pressures it is necessary to use high charge voltages to maintain the optimum E/N or E/p, where p is the gas pressure, required for efficient excitation. In axially-excited pulsed CO_2 lasers in which the excitation is along the axis of the discharge tube, operation at high pressures in discharge tubes of a meter or more in length requires that the potential applied between the electrodes be in the 100 kV to MV range. For example, Hill[71, 72] used electrical pulses in the range of 200 kV to 1 MV to obtain 10-joule, 1-MW pulses in a 2.5-m-long axially excited CO_2 laser operating at a total pressure of approximately 60 torr. And Dezenberg et al.[73] have used a 360-kV Marx generator to obtain 24-joule, 3.6-MW pulses in a 3.5-m-long CO_2 laser operating at approximately one-tenth of an atmosphere.

Hill[70, 72] found that uniform discharges could be maintained as long as the excitation-current-pulse rise times were short compared with arc-formation times, that the gain fell off more slowly with increasing tube diameters than in a CW CO_2 lasers, and that some pre-ionization of the gas was beneficial in ensuring a uniform discharge. Using the experimental arrangement shown in Fig. 6.23, Dezenberg et al.[73, 74] found the pulsed output energy

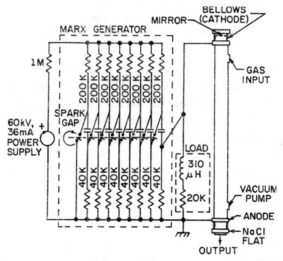

FIG. 6.23. Experimental arrangement: an eight-stage Marx generater giving 480-kV pulses from a 60-kV supply used to excite a CO_2 laser axially. (After G. J. Dezenberg, E. L. Roy, and W. B. McKnight,[74] by courtesy of the IEEE.)

increased as the applied voltage was mixed until an arc discharge occurred. An arc discharge was characterized by an order of magnitude increase in tube current and a change in the appearance of the discharge from a reddish glow that filled the entire tube to a bright white arc. When an arc occurred the peak power was reduced, or the power rapidly decreased to zero at the onset of the arc due to gas heating and/or a reduction in the electron temperature in the discharge due to a drop in the electric field. The dependence of the output energy per unit length of discharge on the applied electric field with a 7-torr CO_2, 26-torr N_2, 26-torr He mixture, optimum capacitance for each tube, is shown in Fig. 6.24 for 5-, 10-, and 15-cm-bore laser tubes. The dashed continuations of the curves corresponds to the drop in laser output as arc discharges are formed at the high E/p values. The onset of the arc always occurs after the initial current pulse at breakdown of the gas. The delay time from the beginning of the current pulse and the onset of the arc is variable and depends mainly

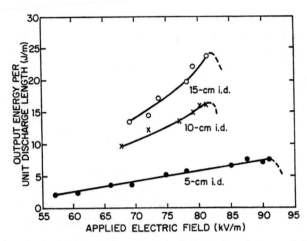

FIG. 6.24. Dependence of pulsed output energy per unit discharge length on applied axial electric field of CO_2, N_2, He laser mixtures in tubes of differing internal diameters. (After G J. Dezenberg, E. L. Roy, and W. B. McKnight,[74] by courtesy of the IEEE.)

on the value of the applied field. At a constant gas mixture, the delay time decreases with increase in electric field to a minimum value. In addition to the output-pulse-energy dependence on the electric field, it was observed that when the electric field increased, the peak output power of the laser pulse increased, the laser pulsewidth decreased, and the delay time from the beginning of the current pulse on breakdown and the onset of laser action decreased.[74] This behavior is illustrated in Fig. 6.25 for the 15-cm-bore laser at two values of applied electric field of 65 and 80 kV/m. At 65 kV/m the peak power was 1.5 MW, the pulse half-width (FWHI) was 40 μsec, and delay time was 10 μsec. At 80 kV/m the peak power was 4 MW, the pulse half-width was 20 μsec, and the delay time was reduced to 3 μsec.

Figure 6.25 also shows the shape of a typical output laser pulse in an axially excited N_2–CO_2–He laser. The output pulse exhibits an initial spike followed by a second maximum in the tail. The initial spike can have a larger amplitude than the second maximum. The spike results from direct electron-impact excitation of both the upper and lower laser levels

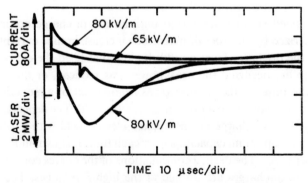

FIG. 6.25. Laser and current pulses in a 15-cm-bore CO_2 laser for applied axial electric fields of 65 and 80 kV/m in a 7-torr CO_2, 26-torr N_2, and 26-torr He mixture, and a discharge capacitance of 0.032 μF. (After G. J. Dezenberg, E. L. Roy, and W. B. McKnight,[74] by courtesy of the IEEE.)

of CO_2. The nitrogen concentration has little effect on this initial spike though it has a considerable effect on the second maximum. The risetime of the initial output spike is of the order of 100 nsec, which is comparable to that of a Q-switched laser pulse.

The second maximum, about 2-μsec wide, can be delayed and varied in amplitude by changing the CO_2–N_2 mixture ratio. This maximum is not observed when nitrogen is omitted. Transfer of vibrational energy from N_2 to CO_2 is responsible for this second maximum[75] in which it appears that excitation transfer occurs directly to the 00^01 upper laser level via the reaction

$$CO_2(00^00) + N_2'(v = n) \rightarrow CO_2(00^01) + N_2'(v = n-1), \qquad (6.13)$$

in which a highly excited N_2 molecule with $v = n$ takes part in a number of excitation-transfer collisions with CO_2 molecules. The high percentage of CO_2 molecules that contribute quanta to the large output pulse makes it appear that each CO_2 molecule is selectively excited more than once in the deactivation cascade of $N_2'(v = n)$[74] in eqn. (6.13).

The E/p ratios in this work at which glow discharge to arc discharge transitions occur were approximately 14 to 16 V/cm-torr. This E/p is comparable to optimum E/p ratios in CW CO_2 lasers. The similarity, however, should not be taken further. The E/p ratio of 14 V/cm-torr in the axially pulsed CO_2 laser is not an optimum ratio related to any require-ment for a specific E/p or E/N as per the analysis of Nighan and Bennett.[45] Here it is simply the maximum value of E/p after which arc discharges occur. Similarly, optimum E/p values given in the literature for other laser systems are not *optimum* values, but values above which discharge instabilities set in to limit or reduce the pulse output power.

The dangerously high voltages (MV) needed to axially excite a long column of gas at high pressure, and the difficulty of exciting uniformly large volumes of gas at high gas pressures, are two reasons why the transverse-excitation technique has been applied to CO_2 lasers, and transversely excited atmospheric pressure (TEA) lasers have been developed.

Transverse excitation was first used as a means of exciting gas lasers by D. A. Leonard at the Avco Everett Research Laboratory.[76] In 1965 Leonard reported a pulsed transversely excited nitrogen laser that achieved E/p ratios of 200 V/cm-torr at the then relatively high nitrogen pressure of about 40 torr. Numerous short discharges transverse to the laser axis generated the high field strengths necessary for direct electron-impact of the molecular nitrogen to the C $^3\pi_u$ state about 9 eV above the ground state. The discharge voltage, a reasonable one of 25 kV, was independent of the length of the lasing region. This compares to a voltage of approximately 1.5 MV that would be required to generate a similar field strength axially in a laser one meter long at 40 torr. In this first transversely excited laser the peak output energy was found to be directly proportional to the applied electric field, just as it is in axially excited CO_2 lasers and in the more recent transversely excited CO_2 lasers.

Parallel solutions to the problem of exciting a gas uniformly at high pressure under desired E/p conditions at reasonable applied voltages was provided by a Canadian group at the Defense Research Establishment Valcartier (DREV), and a French group at the Compagnie Générale d'Électricité (CGE).

In early 1970 Beaulieu,[77, 78] and later colleagues,[79–82] reported the development of CO_2 TEA lasers. In the first pulsed CO_2 TEA laser[77, 78] many separate transverse discharges

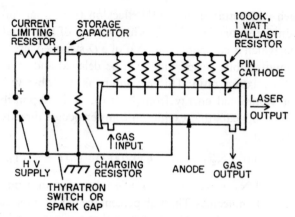

FIG. 6.26. Transversely excited, resistively loaded cathode laser that operates at high gas pressures.

occur between separate pin cathodes and a common bar-anode as shown in Fig. 6.26. To prevent arc discharges forming at one or two pins and draining off the available charge on the capacitor, and to limit the individual discharge currents, each pin cathode is loaded separately with a carbon resistor. A second technique is to use a distributed energy-storage capacitance in the form of a small capacitor attached directly to each pin.[82] Individually uniform discharges without resistive or capacitive pin loading can be obtained also by using an

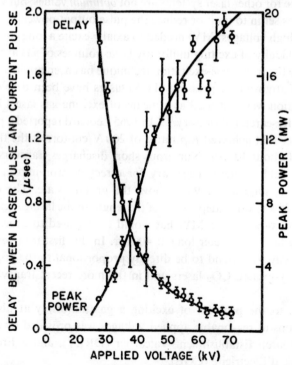

FIG. 6.27. Typical output peak power and laser pulse delay versus applied voltage in a nonresistively loaded, multiple pin-rod CO_2 TEA laser using a flowing 9.1-percent CO_2 : 4.8-percent N_2 : 86.1-percent He laser mixture. (After K. A. Laurie and M. M. Hale,[83] by courtesy of the IEEE.)

electrical pulse at higher voltages than usual with a risetime so short that all electrodes are sufficiently overvolted to establish a low-energy discharge before any one of them can convert from a uniform glow discharge to an arc (Laurie and Hale,[83] extension of prior work[79]). In the latter technique, the use of applied voltages up to 70 kV enables less helium but more CO_2 to be used than is usual in loaded TEA lasers, to give peak powers of up to 20 MW at efficiencies of up to 15 percent. The behavior of a nonresistor loaded, pin-rod electrode system when such a voltage is applied is shown in Fig. 6.27, in which the variation of peak laser output power with applied voltage is shown. Also displayed in Fig. 6.27 is quantitative data of the relationship of the delay between the start of the initiating current pulse and the peak of the laser pulse, and the charge voltage. A minimum delay of slightly more than 100 nsec appears to be inherent in the system. A similar qualitative relationship between laser-pulse delay and applied voltage in axially excited CO_2 lasers was observed by Dezenberg et al.[74] (shown earlier in Fig. 6.25). Pearson and Lamberton[84] have related this effect in another type of CO_2 TEA laser to a relationship between peak power and time delay, as being one between the gain (which is proportional to the peak power) and the pulse build-up time (which should be inversely proportional to the gain), and show that the curve of peak power P and delay time t approximates to a rectangular hyperbola obeying the relationship

$$P = 0.04/(t-0.02)^{1.6} \text{ MW} \qquad (6.14)$$

where t is in microseconds with an asymptote $t = 200$ nsec.

As stated earlier, the main problem with all discharges at high pressure is insuring uniformity of the discharge over the excited volume. While the Canadians initially used individually loaded pin electrodes, the CGE researchers of Dumanchin et al.[85, 86] built a segmented cathode with secondary trigger electrodes to produce preionization and create a zone of high and nearly uniform electron density close to the cathode, resulting in quasi-uniform initiation of the main discharge between the cathode and anode. Figure 6.28 is a schematic showing the cathode and trigger-electrode configuration and the discharge circuitry used by Dumanchin et al.[86] The cathode consists of parallel-mounted metal fins with glass-insulated trigger electrodes between the strips. The trigger wires are connected together and to the anode, a flat Duraluminum plate, by means of a small coupling capacitor.

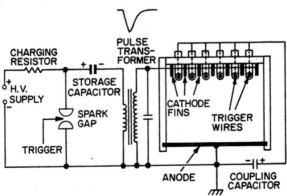

FIG. 6.28. Cathode and trigger-electrode configuration and discharge circuitry developed to give uniform volume excitation of high pressure gas laser mixtures. (After R. Dumanchin, M. Michon, J. C. Farcy, G. Boudinet, and J. Rocca-Serra,[86] by courtesy of the IEEE.)

When a negative-going discharge pulse is applied to the cathode fins, the applied voltage appears over the small 2-mm spacing between the cathode fins and the trigger wires. The ensuing field of about 10^5 V/cm causes electron emission to occur from the cathode fins and provides the sheet of electrons that ensures a uniform discharge between the cathode fins and the anode. With the preionization technique, a CO_2 laser 10 by 10 cm in cross-section and 3 meters long has produced 130 joule pulses having a 2-μsec duration. LaFlamme[80] at DREV has developed a further preionization technique giving similar results, and Lamberton and Pearson at the Services Electronics Research Laboratory (SERL) have developed another more rugged technique[87, 88] giving output energy densities of 10 mJ per cm^3 of active discharge in a high percentage of CO_2–N_2 in CO_2–N_2–He mixtures. (Photo-ionization is a newly developed preionization technique for high pressure gas lasers.[278])

Figure 6.29 shows how the gain in a CO_2 TEA laser that incorporates distributed energy storage capacitance in the form of a separate capacitor attached to an electrode in an array

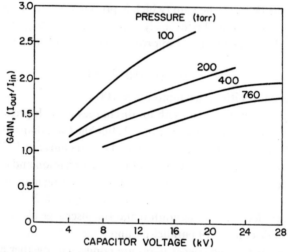

FIG. 6.29. Peak small-signal gain versus capacitor voltage in a multi-capacitor-loaded, transversely excited CO_2 laser. (After D. C. Smith and A. J. DeMaria,[89] by courtesy of the American Institute of Physics.)

of separate discharges varies with capacitor charge voltage for a range of total pressures of 100 torr to atmospheric pressure.[89] It shows that the maximum gain decreases with increase in pressure at a fixed capacitor voltage, and it can be deduced from the figure that at a constant voltage-to-pressure ratio of 60 V/torr (corresponding to a constant E/p of 20 V/cm-torr with the 3-cm electrode spacing used) the maximum or small-signal gain remains approximately constant (at 1.7) as might be expected at these high pressures where the line-width is determined by pressure broadening.

Figure 6.30 shows that the output pulse energy from a CO_2 TEA laser increases with increase in total pressure at a fixed charge voltage, and that the pulsewidth decreases with increase in pressure. These two effects combine to cause the peak power to increase approximately as the square of the pressure. It has been observed also in CO_2 TEA lasers that the output pulse energy is directly proportional to the input pulse energy.[89]

A later development has been volume preionization with an electron beam to initiate a

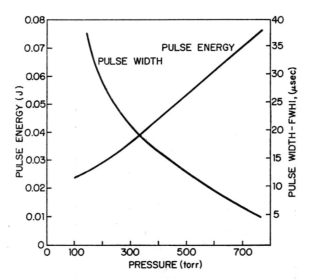

FIG. 6.30. Output pulse energy and pulse width (full width at half intensity, FWHI) versus total gas pressure in a multi-capacitor-loaded, transversely excited CO_2 laser. (After D. C. Smith and A. J. DeMaria,[89] by courtesy of the American Institute of Physics.)

large volume electrical discharge between planar electrodes. A hot-cathode, pulsed electron gun injects a 20-A electron beam with an energy of 200 keV through the cathode to provide the initial ionization of a CO_2–N_2–He mixture at atmospheric pressure and above.[90–92, 279]

The main advantage with the preionization technique is that it enables the main discharge conditions to be optimized for vibrational excitation of CO_2 and N_2 (low E/N) in isolation from the different conditions required for producing ionization (high E/N) as per Nigham and Bennett[45] (Fig. 6.17).

Also, the output pulse energy has been found to be directly proportional to the initiation pulse energy input. This implies that the output pulse energy is limited by the number of electrons that one can produce in ionization of the gas mixture by ion-pair production, a possible fact that suggests that there are apparently unlimited possibilities for the realization of extremely high output pulse energies from CO_2 gas mixtures at pressures considerably higher than atmospheric pressure.[†]

6.1.2. PURE CARBON DIOXIDE LASER

Oscillation in pure CO_2 under CW conditions was first reported in 1964 by Patel et al.[93] Later, Patel[94, 7] reported CW and pulsed oscillation on 14 lines in the 00^01–10^00 band of CO_2 in the wavelength range 10.5135–10.7880 μm, and on seven lines in the 00^01–02^00 band in the range 9.5691–9.6762 μm (vacuum wavelengths). Strongest oscillation occurred on the $P(24)$ rotational transition in the 00^01–02^00 band with a total CW output power of 1 mW. Slightly increased output power was realized under pulsed conditions.

[†] It has been found that gas breakdown, caused by ionization of the gas by the laser radiation, sets an upper limit on the operating pressure of pulsed CO_2 lasers (P. J. Smith and D. C. Smith, *Appl. Phys. Letters* **21**, 167 (1972)).

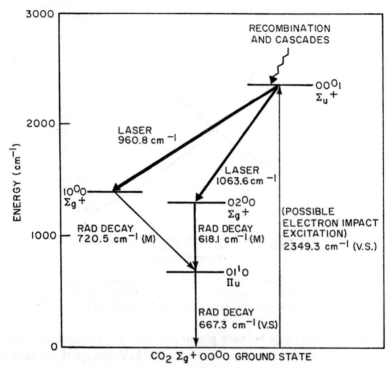

FIG. 6.31. Partial vibrational energy-level diagram of CO_2 showing laser transitions and paths for excitation of the 00^01 upper laser level in pure CO_2 or CO_2-He mixtures. (After C. K. N. Patel,[94] by courtesy of the American Institute of Physics.)

Figure 6.31, a partial energy-level diagram of CO_2, shows energy levels pertinent to the CO_2 laser, together with various radiative decay paths, line strengths (V.S. = very strong, M = medium), and possible excitation paths for the 00^01 Σ_u^+ upper laser level. The frequencies given in the figure are frequencies in air.

6.1.2.1. Upper level excitation

Patel suggested that selective excitation of 00^01 upper laser level occurred by direct electron impact from the 00^00 ground state, recombination of CO and O or cascading from higher CO_2 levels.[94] Studies under pulsed excitation conditions by Patel indicated that recombination of CO and oxygen atoms or cascade processes were the most probable excitation processes, as long time delays of the order of 300 μsec occurred between the current pulse and laser oscillation. These time delays were not consistent with direct electron impact excitation of the upper and lower laser levels, and a differential decay rate between the lower level and the upper level, since this would have meant that the lower laser level was being populated via a forbidden transition. Whereas recombination and cascade processes (and also direct electron impact excitation of both upper and lower laser levels) have been found to be important excitation processes in pulsed, pure CO_2 laser discharges[75, 95, 96] under various conditions of CO_2 pressure, discharge current, and flow velocities, further studies by Clark and Wada[97] and Sobolev and Sokovnikov[98–100] on CW CO_2 laser systems have shown that the dominant excitation process of the upper laser level is due to the production of dissociated CO and excitation of excited CO molecules by electron impact and then vibrational energy transfer from the vibrationally excited CO molecules to the $CO_2\,00^01$ upper laser level in the reaction

$$CO(v = 1) + CO_2(00^00) \rightleftarrows CO(v = 0) + CO_2(00^01) - \Delta E_\infty \; (206 \text{ cm}^{-1}). \qquad (6.15)$$

In a pure CO_2 discharge, considerable dissociation of CO_2 to CO and O occurs so that excited CO molecules are available in a discharge to take part in reaction (6.15). They are produced in the resonant reaction

$$CO(v = 0) + e^- \rightarrow CO^- \rightarrow CO(v = 1) + e^-. \qquad (6.16)$$

This reaction is similar to that involved in the production of $N_2^*(v = 1-8)$ molecules in which the reaction proceeds via a negatively charged compound state. Both reactions have large cross-sections in the region of 10^{-16} cm^2 for electrons of about 2 eV energy. The resonant nature of the excitation function of vibrationally excited CO molecules is illustrated

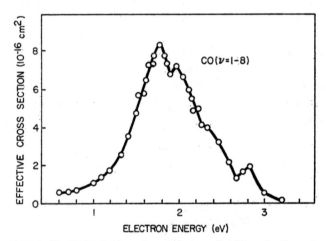

FIG. 6.32. Excitation function for inelastic scattering of electrons by CO molecules at low electron energies (After G. J. Schulz,[28] by courtesy of the American Institute of Physics.)

in Fig. 6.32; it is broader than that of $N_2 + e^- \rightarrow N_2(v = 1 \text{ to } 8) + e^-$, and has a maximum about 0.5 eV lower in energy.[28] Because of the large cross-section at low electron energies of reaction (6.16), large numbers of $CO(v = 1)$ molecules are formed in the discharge, and are available for transferring their vibrational energy to CO_2 molecules.

The laser transitions and energy levels involved in decay processes in the pure CO_2 laser are the same as those involved in the N_2–CO_2–He laser, and the treatment of the nitrogen–carbon dioxide–helium laser applies to the pure CO_2 laser. In a similar manner, population inversion only occurs in the pure CO_2 laser or in CO_2–He lasers because of collisional de-excitation of the lower 10^00 and 02^00 and terminal (01^10) laser levels. In pure CO_2 and CO_2–He laser discharges, CO molecules also play an important role in depleting the lower laser level population[96, 101] as well as strongly selectively exciting the 00^01 upper laser level.

The effect of additives. The addition of additives to the pure CO_2 laser has similar effects, as might be expected, as the addition of additives to the N_2–CO_2 laser. Helium is one additive that improves the gain considerably, and xenon is another. The effect of xenon we have not considered previously, and will consider it separately in this section.

The CW CO_2–He–Xe laser has been found to be nearly as efficient as the N_2–CO_2–He laser in which molecule–molecule vibrational excitation transfer between $N_2^*(v = 1)$ molecules and CO_2 is an important selective excitation process. The addition of xenon to a N_2–CO_2–He laser improves the output power by a factor of approximately 35 percent, and the efficiency by a factor of about 30 percent under optimized conditions.[97, 102] Xenon has no energy levels in coincidence with CO_2 (or CO) vibrational levels that could give selective excitation of the upper laser level, and, being a heavy atom, is not effective (as is helium) in thermalization of rotational levels, relaxation of lower or terminal laser levels, or in keeping the gas cool by conducting heat from the discharge to the walls of the discharge tube. Nevertheless, the addition of fairly large quantities of xenon to a CO_2–He laser discharge in a 1-cm-bore discharge tube leads to a system giving an increase of approximately 25 percent of the output power of an optimized CO_2–He laser mixture, a slightly increased efficiency of about 15 percent, and a vast improvement in sealed-off life.[97] According to Clark and Wada,[97] observations about the beneficial effects of xenon as an additive were first made by R. Rempel and J. P. Goldsborough.

The practical improvements that accrue from xenon addition are believed to follow entirely from the effects xenon has on the discharge plasma. It increases the optimum discharge current and hence the electron density and apparently alters the shape of the electron energy distribution to one that is more favorable for electron-impact excitation of vibrational levels.

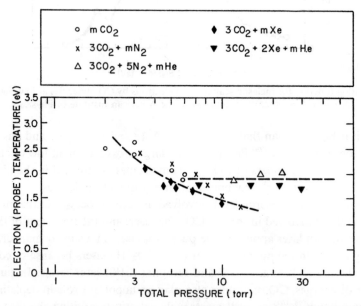

FIG. 6.33. Pressure dependence of electron temperature in sealed-off CO_2–Xe–He and CO_2–N_2–He laser discharges; $D = 22$ mm. (After P. O. Clark and J. Y. Wada,[97] by courtesy of the IEEE.)

Figure 6.33 graphically illustrates results of average electron temperature measurements for CO_2, CO_2–Xe, CO_2–Xe–He, CO_2–N_2, and CO_2–N_2–He laser mixtures in a 1-cm-bore discharge tube under nonflowing (sealed-off) conditions. The general variation of electron temperature as a function of gas pressure is the same as either xenon or nitrogen is added to CO_2. The electron temperature drops rapidly as either the xenon or nitrogen pressure is increased; however, the addition of fairly large quantities of helium to a fixed mixture of CO_2–Xe produces no further change just as it does when added to a CO_2–N_2 laser discharge (in this figure, and Fig. 6.8). The average electron temperature corresponding to the optimum[†] mixture of CO_2–N_2–He (3 : 5 : 14 torr) is approximately 2 eV. The discharge that contains 3 torr CO_2, 2 torr Xe (instead of 5 torr N_2) and helium has a slightly lower average electron temperature than the (3 : 5 : m torr) CO_2–N_2–He mixture. With a mixture, however, that contains only 1 torr Xe, 3 torr CO_2, and 14 torr He, which is an optimum mixture for a CO_2–Xe–He laser in a 1-cm-bore discharge tube, the electron temperature is the same as it is in the CO_2–N_2–He mixture.[(97)]

Whereas the electron temperature in the CO_2–Xe–He laser discharge is approximately the same as that for the CO_2–N_2–He mixture (under operating conditions for both), the optimum discharge current for the CO_2–Xe–He mixture is shifted to higher values. The optimum current for a CO_2–Xe–He mixture is also higher than that for a CO_2–He mixture. This shift to a higher optimum operating current in the CO_2–Xe–He mixture increases the electron density by about 45 percent over that in an optimized CO_2–N_2–He mixture.

Clark and Wada[(97)] postulate that the improvements follow from increased excitation of vibrationally excited CO molecules, and subsequently an increased rate of excitation transfer from them to the CO_2 00^01 upper laser level. Even though the cross-section for selectively exciting the 00^01 level of CO_2 from CO by reaction (6.15),

$$CO(\nu = 1) + CO_2(00^00) \rightleftarrows CO(\nu = 0) + CO_2(00^01) - \Delta E_\infty \ (206 \ cm^{-1}),$$

is smaller than the cross-section for vibrational energy transfer from $N_2^*(\nu = 1)$ to CO_2 in

$$N_2^*(\nu = 1) + CO_2(00^00) \rightleftarrows N_2(\nu = 0) + CO_2(00^01) - \Delta E_\infty \ (18 \ cm^{-1}),$$

the larger cross-section of 6 to $8 \times 10^{-16} \ cm^2$ for excitation of low-lying vibrational levels of CO by electrons of 1.5 to 2.2 eV energy compared with that of 2 to $3 \times 10^{-16} \ cm^2$ for excitation of $N_2(\nu = 1$ to 8) by electrons in the higher energy range of 2 to 2.7 eV (Figs. 6.32 and 6.5 respectively), combined with a 45 percent increase in electron concentration in the CO_2–Xe–He laser discharge compared with that in a CO_2–N_2–He discharge, make up for the difference between the vibrational excitation cross-sections for $N_2^*(\nu = 1)$ to CO_2 and $CO(\nu = 1)$ to CO_2.

Throughout, we have again loosely spoken about the average electron temperatures in the laser mixtures that have been compared. There are clear indications that we are dealing with electron energy distributions in these lasers that cannot be characterized by simple distributions. For instance, although the average electron temperatures in CO_2–Xe–He and CO_2–N_2–He laser discharges are approximately the same, as Figs. 6.33 and 6.8 illustrate, Deutsch and Horrigan[(103)] report that whereas CO angstrom bands (with an upper level at

---

† "Optimum" in this section is taken to be that which gives the maximum output power.

10.8 eV) are observed in a CO_2–N_2–He laser discharge, they are absent in the CO_2–Xe–He mixture. Instead of the CO angstrom bands, only the CO triplet bands, with an upper level 7.8 eV above the ground state, can be observed. This indicates that the electron energy distribution is being altered in the high-energy tail of the distribution, and that it is an effect that is not being reflected in a straightforward determination of the average electron temperature using standard double-probe techniques. Further support for this deduction is provided in an earlier figure (Fig. 6.15). Figure 6.15 clearly shows that the electron energy distribution in CO_2–N_2–He–Xe and CO_2–N_2–He laser mixtures is neither Maxwellian or Druyvesteyn. It also shows that the addition of xenon to a CO_2–N_2–He discharge mixture does reduce the number of energetic electrons above an energy of 2.5 eV, but at the same time the number of slower electrons with an energy less than 2.5 eV is increased noticeably.

The improved efficiency obtained on the substitution of xenon for nitrogen in CO_2–He lasers follows directly from a reduction of about 20 to 25 percent of the discharge voltage.[97] On the other hand, the reason for the vast improvement in sealed-off life of CO_2 lasers that contain xenon is not entirely clear. Clark and Wada[97] have suggested that it is probably associated with clean-up processes of CO_2 in the discharge that involve positive ions. Since xenon has the lowest ionization potential and largest ionization cross-section of all the major constituents of the discharge, it probably substitutes for ionization of the CO_2 in the discharge and reduces subsequent clean-up of CO_2 by ion bombardment of the cathode. Xenon also reduces the number of high energy electrons in the discharge, so that electron impact ionization of CO_2 and O_2, and dissociation of CO_2 is reduced. Cascade ionization of xenon from metastable $6s_5$ and $6s_3$ states (Paschen notation) at 8.3 and 9.4 eV respectively, that are in resonance with the $B^3\pi_u$ state of molecular nitrogen, could possibly be playing a role in the ionization processes[104] and so reduce the number of high energy electrons that are required in discharges of CO_2–N_2–He that include xenon as an additive.

6.1.2.2. Operating conditions

The output power of CW sealed-off pure CO_2 lasers and CO_2–He lasers has been investigated by Whitney[105] as a function of the initial fill-pressures of CO_2 and He, the discharge current, and the discharge-tube diameter. He finds that for producing maximum output power and efficiency, the optimum fill-pressure of pure CO_2 is $5/D$ (torr), where D is the internal diameter of the discharge tube in centimeters; and the optimum current is $7D$ (mA). With the addition of helium, the optimum current is increased to $18D$ (mA), and the output power is increased by a factor of 5. The optimum CO_2 pressure, however, remains the same as in the pure CO_2 laser.

The optimum CO_2 pressure and tube-diameter product of 5 torr-cm for pure CO_2 and CO_2–He sealed-off lasers, is larger than the optimum 3.3 torr-cm of (6.12) for CO_2–N_2 laser mixtures derived from Fig. 6.22. It is also larger than the optimum of $p(CO_2)D$ of about 3 torr-cm at which the total decay rate of the 00^01 upper laser level due to volume quenching and wall collisions is a minimum.[62] The difference in optimum pD-products probably arises because more dissociation of CO_2 occurs in closed-off pure CO_2 and CO_2–He laser discharges than in CO_2 laser discharges that can contain N_2 and other gaseous additives, thus causing the fill pressure of CO_2 to need to be increased to compensate for the increased CO_2 loss by dissociation.

The constancy of the $p(CO_2)D$ product for pure CO_2 and CO_2–He laser discharges implies again that there is some optimum mean electron energy for producing maximum laser output. The optimum E/N value for the discharges is about 5×10^{-16} V-cm^2.

6.1.3. NITROUS OXIDE–NITROGEN LASER

Patel[106] was responsible again for the discovery of this continuous-flow laser system in which CW laser oscillation, on the P-branch rotational transitions of the 001–10^{0}0 vibrational band in the ground electronic state of the N_2O molecule, was due to transfer of vibrational energy of $N_2^*(\nu = 1)$ to N_2O. Laser oscillation occurred on 26 lines between 10.77 and 11.04 μm with an output power of approximately 1 mW on the strongest $P(19)$ transition at 10.8416 μm (vacuum wavelength). Some of the lines on which CW oscillation was reported, correspond to those reported earlier under pulsed conditions in pure N_2O by Mathias et al.[107] and tentatively identified by them as occurring on the $P(20)$ to $P(28)$ lines of the 001–10^{0}0 transition, or on the $P(18)$ to $P(26)$ of the 002–101 transitions in N_2O.

Although the N_2O molecule resembles the CO_2 molecule in that they are both linear triatomic molecules and have roughly the same energy levels, and although the N_2O laser is a laser that is capable of high power operation, it has not received the attention that the CO_2 laser has had. This is probably because lower ultimate power is expected from N_2O compared with CO_2 by a factor of approximately 4,[108] and because dissociation of N_2O (to N_2 and O_2) is much more extensive in electrically excited N_2O lasers than in CO_2 lasers, so that closed-off, nonflowing laser systems are unlikely to be realizable.

6.1.3.1. Upper level excitation

Figure 6.34, a partial energy-level diagram of N_2 and N_2O drawn with respect to the N_2 $(\nu = 0)$ and N_2O (00^{0}0) ground states, shows that there is a close vibrational energy-level coincidence between the $N_2^*(\nu = 1)$ level at 2330.7 cm^{-1} and the N_2O Σ^+ 001 level at a lower potential energy of 2223.7 cm^{-1}. Because of the smallness of the energy discrepancy ΔE_∞ of 107 cm^{-1} between the two levels compared with kT of about 280 cm^{-1} at a gas temperature of 400 K, the reaction

$$N_2^*(\nu = 1) + N_2O(00^00) \rightleftharpoons N_2(\nu = 0) + N_2O(001) + \Delta E_\infty \ (107 \ cm^{-1}) \qquad (6.17)$$

is nonadiabatic and the probability of vibrational energy transfer is high.

The N_2O 100 lower laser level at 1284.96 cm^{-1} is more than 1000 cm^{-1} from the $N_2^*(\nu = 1)$ level so that it is not selectively excited by the $N_2^*(\nu = 1)$ level so that it is not selectively excited by the $N_2^*(\nu = 1)$ or other higher-lying nitrogen levels. The N_2O 100 level is optically connected to the ground state, unlike the situation in the 10.6-μm CO_2 laser, and to the 01^{1}0 level at 588.78 cm^{-1} so that it can decay radiatively, and because of the effectiveness of the vibrational energy-transfer reaction (6.17), population inversion is readily achieved between the 001 and the 100 levels.

Unlike the pure CO_2 CW laser in which CO from dissociated CO_2 molecules selectively excites the CO_2 00^{0}1 upper laser level, CW operation in pure N_2O is not observed. Though, of course, a good energy coincidence exists between $N_2^*(\nu = 1)$ and the $N_2O(001)$ level

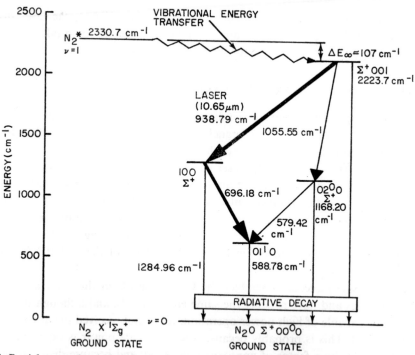

FIG. 6.34. Partial energy-level diagram for N_2 and N_2O showing laser transitions around 10.65 μm in N_2O, and various radiative transitions.

with $\Delta E_\infty = 107$ cm^{-1} in eqn. (6.17), and sufficient N_2 molecules would appear to be produced in dissociation of N_2O in the discharge to enable enough $N_2^*(v = 1)$ molecules to be produced in electron-impact excitation and so enable eqn. (6.17) to proceed effectively, surprisingly there have been no reports (to the author's knowledge) of CW operation of a pure N_2O, or $N_2O + He$ laser. As CW laser oscillation occurs in $N_2O + He$ mixtures to which N_2 is added, nonobservation of CW oscillation in N_2O or $N_2O + He$ mixtures where dissociation can be almost complete at low gas flow-rates,[108] must apparently be attributed to the "more than sufficient" dissociation of N_2O that reduces the working population of N_2O molecules.

Population inversion in pure N_2O under axial pulsed excitation[107, 109] and under pulsed TEA laser conditions in $N_2O + He$[110] laser mixtures can be attributed to direct electron impact excitation of both upper and lower laser levels and subsequent differing decay rates of the two levels, and possibly cascade processes.

The effect of additives. The addition of CO to optimized N_2O-N_2-He laser mixtures under both CW and pulsed excitation conditions has been found to enhance the laser output.[108, 109] The output power of a N_2O-N_2-He laser can be nearly doubled by the addition of CO in quantities comparable to, and more than, the N_2O fill-pressure under flowing conditions (Fig. 6.35).

The effect is due to better matching of the energy levels of the 001 upper level of N_2O and vibrationally excited CO molecules (as compared with N_2) in the reaction

$$CO(v = n) + N_2O(00^0 0) \rightarrow CO(v = n-1) + N_2O(001) + \Delta E_\infty \qquad (6.18)$$

302

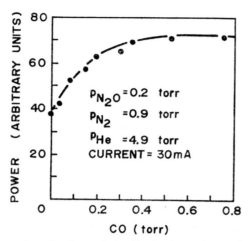

FIG. 6.35. Power enhancement in a flowing optimized N_2O–N_2–He laser mixture by the addition of CO. (After N. Djeu, T. Kan, and G. Wolga,[108] by courtesy of the IEEE.)

where n = number of vibrational quanta, and ΔE_∞ is equal to -81 cm^{-1} for $n = 1$. The effect is not believed to be due to CO_2 formation from CO and then subsequent pumping of N_2O from CO_2 as the substitution of CO_2 for CO gives significantly less N_2O laser power enhancement than the CO addition. In fact the formation of CO_2 from CO in repetitively pumped N_2O–CO–He discharges can actually reduce the gain and output power.[109]

TABLE 6.3. *Comparison of N_2O and CO_2 laser parameters*[108-9]

Parameter	N₂O	CO₂
Upper-level spontaneous-emission lifetime (sec)	3.6	9.4
Maximum gain on single-line transition (m⁻¹):		
Low pressure (5 torr)	0.24	1
High pressure (11 torr)	0.13	0.5
Inversion ratio at 400 K	2.12	2.13
Upper level/ground-state population ratio	3	17
Relaxation of upper laser level (in torr⁻¹ sec⁻¹)	700	385
Saturation-output power parameter at high pressure (W/cm²)	18[b]	22[a]

[a] From D. F. Hotz and J. W. Austin, *Appl. Phys. Letters* **11**, 60 (1967).

[b] Peak powers of 250 kW under mode-locked conditions using TEA-laser excitation have been achieved from N_2O–N_2–He mixtures.[111] MW output powers are anticipated with the addition of CO.

6.1.3.2. Comparison of N_2O and CO_2 laser parameters

Important parameters for N_2O and CO_2 lasers that can be related to processes already discussed with respect to the CO_2 are given in Table 6.3 presented on the previous page.

6.2. Electron-impact-excited Lasers

A large number of laser transitions in molecular species have been observed in which the population inversion is established predominantly by electron-impact excitation from the ground state of the molecule. In most cases the excitation has been by means of various types of pulsed, high-voltage electrical discharges in pure molecular gases, or mixtures of molecular gases, vapors, and noble gases. Electron-beam pumping at very high electron energies has also been used as an excitation method. In a number of molecular laser systems, particularly those that operate under pulsed conditions, cascade processes from higher energy levels into the upper laser levels are involved, and in only a few systems have the main excitation paths been established with any certainty. However, even when the main excitation paths have not been determined, it is informative to discuss the possible types of excitation that are involved and that lead to the subsequent establishment of population inversion.

Typical axial, pulsed-excitation conditions that have been used in electron-impact-excited molecular lasers follow those first used by Mathias and Parker[3, 4] to obtain oscillation in molecular nitrogen and carbon monoxide in the near infrared and visible respectively: peak voltage 20 kV; peak current a few tens of amperes; pulse lengths of approximately 1 μsec; pulse repetition rates up to a kilohertz; and gas pressures of a few torr in discharge tubes of a few millimeter bore. Transverse-excitation conditions range from a few torr to a few atmospheres, at discharge voltages that can approach 100 kV, and peak discharge currents of a few thousands of amperes. In both pulsed axially excited and transversely excited systems the risetime of the current pulse (and sometimes the actual duration of the excitation pulse) is of prime importance in determining the output power obtainable from the system. Invariably the shorter the risetime of the excitation current pulse, the more the laser output power. This is particularly true in the electron-impact excitation of self-terminating (or transient) gas lasers.[112, 113] In these laser systems the pulse risetime must be comparable to and if possible less than the radiative lifetime of the upper laser level to produce population inversion (Chapter 4, Section 4.3.2). Excitation conditions in these pulsed discharges are dependent on the energy dissipated per unit time; the charge or applied voltage and the gas pressure (E/p); the anode–cathode geometry, and shape of the electrodes. Because of the finite breakdown velocity of a longitudinal column of gas ($\sim 10^9$ cm/sec) and the development of high discharge currents, transverse-discharge excitation has a definite advantage over longitudinal excitation as a means of short risetime pulsed excitation.

The molecular lasers that are known to be selectively excited in inelastic collisions with electrons have outputs that range from the vacuum ultraviolet around 0.1100 μm out to the far infrared to at least 200 μm. We shall consider the excitation mechanisms and operating characteristics of five molecular lasers. These five systems illustrate specific aspects of inelastic electron-impact excitation and operate from the vacuum ultraviolet at about 0.1100 μm to 220 μm in the submillimeter region of the spectrum. The laser systems to be covered operate in: H_2 (and HD and D_2); CO; N_2; Xe_2 (and other (molecular) noble gases); and H_2O.

6.2.1. H_2 (AND HD AND D_2) LASERS

Laser oscillation in molecular hydrogen has been observed in the two wavelength regions of the near infrared and vacuum ultraviolet on three separate band systems. Oscillation has been observed in the corresponding band systems in deuterated hydrogen (HD) and deuterium.

6.2.1.1. Near-infrared oscillation

Bazhulin et al.[114-15] were the first to report in 1964 the observation of pulsed laser oscillation in molecular hydrogen in the near infrared region of the spectrum. Following later work, Bockasten et al.[116] and Fromm[117] also reported similar laser oscillation under pulsed conditions affirming that the excitation was by inelastic electron-impact collisions on ground-state hydrogen molecules.

TABLE 6.4. Near-infrared laser lines in molecular H_2, HD, and D_2

Molecule	Wavelength (air) (μm)[a]	Vibrational transition bands	Rotational transition $P(J)$	References
H_2	0.834961[b]	(2,1)	2	114, 116
	0.887625[b]	(1,0)	4	114, 116
	0.889884[b]	(1,0)	2	114, 116
H_2	1.116214[b]	(0,0)	4	114–116
	1.122200[b]	(0,0)	2	114–116
	1.30578[b]	(0,1)	4	115
	1.31623	(0,1)	2	116
HD	0.916021	(1,0)	Unidentified	114–116
D_2	0.827753	(2,0)	3	114–116
	0.952999	(1,0)	3	114–116

[a] Measured by K. Bockasten, T. Lundholm, and D. Andrade, J. Opt. Soc. Amer. **56**, 1260 (1966); ref. 116.
[b] First reported oscillation (ref. 114).
Excitation conditions: Refs. 114 and 115: 35-kV pulses, in a 1.45-m-long, 15-mm-i.d. discharge tube, optimum pressure, $H_2 = 2.5$ torr, $D_2 = 3.5$ torr, also in H_2, D_2, HD mixtures. Ref. 116; capacitor discharge, 3 to 6 nF charged to 10 to 20 kV, in a 51 to 102-cm-long discharge tube, i.d. 7 mm; with H_2 at a few torr.

Oscillation in molecular hydrogen in the near infrared occurs on vibrational–rotational P-branch transitions between the $(E)2s\sigma^1\Sigma_g^+$ and $(B)2p\sigma^1\Sigma_u^+$ bands. The wavelengths and transitions on which oscillation occurs in H_2, HD, and D_2 in the near infrared are tabulated in Table 6.4.

The near-infrared laser transitions and potential-energy curves for the levels of immediate interest are illustrated in Fig. 6.36. Potential-energy curves for the isotopic molecules HD

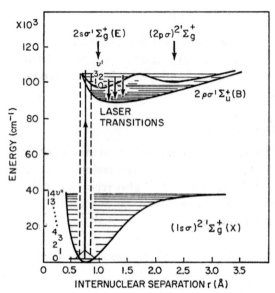

FIG. 6.36. Potential-energy curves for the H_2 molecule involved in near-infrared laser transitions showing laser transitions and excitation path of the $2s\sigma\ ^1\Sigma_g^+(E)$ state. (After P. A. Bazhulin, I. N. Knyazev, and G. G. Petrash,[115] by courtesy of the American Institute of Physics.)

and D_2 depend little on the masses of the nuclei and are thus similar to those illustrated for H_2. The excitation processes will apply similarly to all three molecular systems. Omitted from the figure is the curve for the C state, and the $B \leftrightarrow X$ Lyman-band and the $C \leftrightarrow X$ Werner-band transitions in the vacuum ultraviolet.

Laser oscillation occurs only during the excitation current pulse following a short delay of a few hundred nsec from the commencement of the current pulse, and it has only a limited duration. The half-power duration (FWHM) of about 200 nsec does not depend on the length of the excitation pulse, pulse voltage, or the gas pressure, and is very similar to the duration of spontaneous and stimulated emission at the laser wavelengths. The excitation of the $(E)2s\sigma^1\Sigma_g^+$ upper laser level, therefore, can be attributed with some degree of certainty to direct electron-impact excitation by fast electrons from the $H_2(v=0)$ ground state in the reaction

$$H_2(X^1\Sigma_g^+ v'' = 0) + e^- \rightarrow H_2(E^1\Sigma_g^+ v' = 3, 2, 1, 0) + e^- . \tag{6.19}$$

The $H_2(B)^1\Sigma_u^+$ lower laser level is similarly excited by fast electrons. It can be seen from Fig. 6.36, however, that whereas vertical transitions from the $H_2(v=0)$ ground state can occur to the $0, 1, 2, 3$, etc. vibrational levels of the $(E)^1\Sigma_g^+$ state in accordance with the Franck–

Condon principle, vertical transitions to the $(B)^1\Sigma_u^+$ lower laser state can only occur to those vibrational levels with $v' \geqslant 5$.[118–19] This is due to the fact that the equilibrium nuclear separations of the (E) and (B) states are larger than that of the ground state (X) and are displaced, therefore, to the right of it; and because the equilibrium nuclear separation of the lower laser (B) state differs from that of the upper-laser-level (E) state. The lines that show strongest oscillation are those having $v' = 1$ for an upper vibrational level. It is not known if this vibrational level is preferentially excited or if the strongest oscillation is caused by transitions from the $v' = 1$ level having high transition probabilities due to perturbation of the $2s\sigma$ and $(2p\sigma)^2$ states.

It is not clear why laser oscillation and the spontaneous emission is observed for only a limited duration of a fraction of a microsecond. If the laser oscillation does indeed follow closely the intensity of the spontaneous emission, the limited duration must be due to mechanisms that affect the upper-laser-level excitation rate. Mechanisms that came to mind are: excitation transfer between the $2s\sigma$ and the $(2p\sigma)^2$ configurations of the first excited $^1\Sigma_g^+$ state; a large reduction in the number of high-energy electrons following the initiation of the current pulse, or dissociation of molecular hydrogen to atomic hydrogen to give a lower molecular hydrogen gas pressure. The results of Bazhulin et al.[114–15] and Bockasten et al.,[116] differ from those of Fromm.[117] In the work of Fromm flowing gases were used, and oscillation was observed concurrent with the excitation current pulse for a period that could be as long as 10 μsec. Such an increase in duration of the laser oscillation over that of the other workers, one must assume, in the absence of differing electrical excitation conditions, would indicate that the transient nature of the population inversion is not due to excitation transfer between the $2s\sigma$ and $(2p\sigma)^2$ states (or cascading into the $^1\Sigma_u^+(B)$ lower laser level).

6.2.1.2. Vacuum-ultraviolet oscillation

Laser oscillation in molecular hydrogen in the vacuum ultraviolet has been reported in the Lyman band (0.1400–0.1650 μm) by Hodgson,[120] and Waynant et al.,[121] and in the Werner band (0.1000–0.1200 μm) by Waynant,[11] and Hodgson and Dreyfus.[10] Incidentally, the oscillation in the Lyman band was the first to be reported in the vacuum ultraviolet. In both cases of oscillation in the Lyman and Werner bands, laser oscillation is produced by inverting the population of low vibrational states of excited electronic states of molecular hydrogen with respect to high vibrational-rotational levels of the $X^1\Sigma_g^+$ ground state, following the general scheme for producing inversion suggested by Bazhulin et al.[122] Figure 6.37 shows a partial electronic-energy-level diagram for molecular hydrogen, with energy levels, electron-impact-excitation path (solid vertical line), and radiative-decay paths (dashed vertical lines) via the Lyman and Werner bands relevant to the vacuum-ultraviolet hydrogen lasers. It can be seen that both laser systems are two-level systems, and thus should be efficient as far as quantum efficiency is concerned.

Oscillation in the Lyman band (0.1400–0.1650 μm). Oscillation in the Lyman band occurs between vibrational levels of the $(B)2p\sigma^1\Sigma_u^+$ state (the lower laser level of the near infrared laser lines in H$_2$ discussed in Section 6.2.1) and higher vibrational levels of the $(X)1s\sigma^1\Sigma_g^+$ ground state of molecular hydrogen. The wavelengths and vibrational transitions of ten

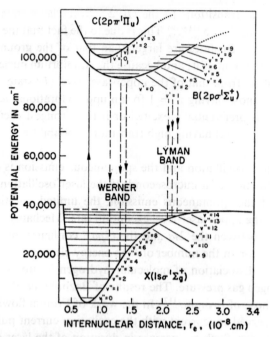

FIG. 6.37. Partial energy-level diagram of the hydrogen molecule showing the levels involved in vacuum-UV-laser oscillation on the Lyman and Werner bands together with electron-impact excitation path (solid vertical line). (After R. W. Waynant,[125] by courtesy of the Industrial and Scientific Conference Management, Inc.)

TABLE 6.5. *Wavelength and vibration transitions of observed laser lines in the Lyman band*[121]

Transition $B(v')-X(v'')$	Wavelength$_{(vac)}$ (μm)[a]		
	P (1)	P (3)	R (1)
2– 9	0.15718 (1)	—	—
3–10[b]	0.15915 (4)	0.15961 (8)	—
4–11[b]	0.16047 (6)	0.16088 (10)	—
5–12[b]	—	0.16133 (10)	—
6–13[b]	0.16074 (8)	—	—
7–13[b]	—	0.15805 (6)	0.15771 (6)
8–14	0.15673 (1)	—	—

[a] The numbers in parenthesis indicate the relative film densities on a scale of 1 to 10 produced by exposures of each line.

[b] Observed in ref. 120, using transverse-discharge excitation by means of a Blumlein-circuit parallel-plate transmission-line.

In ref. 121, a transverse-discharge, travelling-wave-excitation method was used, shown later as Fig. 6.39.

lines between 0.1567 and 0.1613 μm on which oscillation has been observed in the Lyman band are given in Table 6.5. Transitions between $(J', J'') = (0, 1)$, $(2, 3)$ and $(2, 1)$ are designated $P(1)$, $P(3)$, and $R(1)$ respectively.

Because of the large vibrational and rotational constants of H_2, only a few of the lower rotational levels of the $X^1\Sigma_g^+(v = 0)$ state are occupied at room temperature. In this state, the molecules are distributed among the rotational levels $J'' = 0, 1, 2$, and 3 in the percentages 13, 66, 12 and 8 percent respectively. In electronically induced transitions to the excited electronic B and C states, therefore, $J' = 2$ and $J' = 1$ levels would be expected to be preferentially excited. Using the total cross-section for excitation of the B-state from the X-state, taking into account the vibrational structure of the B-state, and using Franck–Condon factors $q_{v'v''}$, the excitation rate can be determined for each vibrational level. It is found that the $v' = 2$ to 11 levels of the B-state will be preferentially populated with $v' = 6$ and 8 the most highly populated.[123, 121] Since the upper vibrational levels of the $X^1\Sigma_g^+$ ground state can be considered empty, population inversion and oscillation can occur between the $B^1\Sigma_u^+(v' = 2$ to $11; J' = 0, 2)$ and $X^1\Sigma_g^+(v'' = 9$ to $14; J'' = 1, 3)$ states, with intensities proportional to the Franck–Condon factors for the transition. Table 6.6 gives the largest Franck–Condon factors[123] calculated for the Lyman band $B(v') \to X(v'')$ transitions.

TABLE 6.6. *Franck–Condon factors for the $H_2(B^1\Sigma_u^+ - X^1\Sigma_g^+)$ Lyman band system*

Upper state v'	Lower State v''						
	0...	9	10	11	12	13	14
2	0.0273...	0.2552	0.1251				
3	0.0427...		0.2952	0.1693			
4	0.0563...			0.3442*	0.1600		
5	0.0663...				0.4347*		
6	0.0723...					0.4366*	
7	0.0740...					0.2600	
8	0.0725...						0.2781
9	0.0688...						
10	0.0634...						
11	0.0575...						0.1546
12	0.0513...						
13	0.0452...						
14	0.0394...						

* Indicates strong transitions having large Franck–Condon factors.

According to Table 6.6 the strongest lines should be on the $v' = 4$ to $v'' = 11; v' = 5$ to $v'' = 12$, and $v' = 6$ to $v'' = 13$ transitions marked by an asterisk, as these transitions have the largest Franck–Condon factors. Referring to Table 6.5 it can be seen that strongest oscillation is observed on these transitions. A spectrogram showing more intensive laser emission in the Lyman band than that given in Table 6.5 is shown in Fig. 6.38. These lines occur in a 1-nsec pulse, with a peak power of more than 1 MW. The wavelengths of the lines are given in Å.

Hodgson[120] and Waynant,[125] and Waynant et al.[121] used Blumlein circuitry,[126] developed by Shipman,[127] to provide the fast risetime, high-voltage, high-current excitation needed to provide fast excitation and to prevent rapid spontaneous emission through strong ultraviolet transitions depleting upper-laser-level populations to the detriment of population

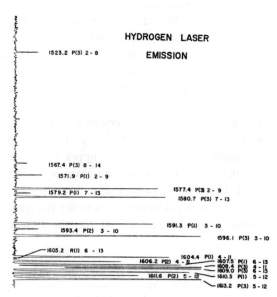

FIG. 6.38. Spectrogram of the laser emission in the $B^1\Sigma_u^+ X^1\Sigma_g^+$ Lyman band of H_2. (After R. W. Waynant,[125] by courtesy of the Industrial and Scientific Conference Management, Inc.)

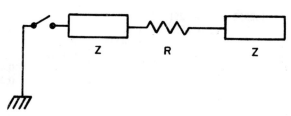

FIG. 6.39. Simplified version of a Blumlein circuit. (After **R. W. Waynant**,[125] by courtesy of the Industrial and Scientific Conference Management, Inc.)

inversion. A simplified version of Blumlein's circuit is shown in Fig. 6.39, where two coaxial cables of impedance Z are separated by the load R, and a switch is connected to one end of the center conductor and to ground. The cable is externally charged to a voltage $+V$ and the switch is closed. The resulting short circuit causes a $-V$ voltage wave to travel toward the load and, when $R \gg Z$, doubles the voltage V across the load. Starting with this circuit, Shipman constructed the device sketched in Fig. 6.40.[125] Flat-plate transmission lines replace the coaxial cables of the original Blumlein circuit, the laser gas is the load, and seven solid dielectric switches are spaced equally across one end of the flat-(aluminum) plate transmission line. The seven switches are fired by a single solid dielectric switch in an oil capacitor. They can be fired sequentially by a suitable choice of the length of connecting cable between each switch and the initiating switch. By making the connecting cable progressively longer for each successive switch equally spaced across the flat-plate transmission line, the excitation wave is inclined to the laser discharge channel that is situated between the electrodes. When the correct inclination of the excitation to the laser channel is achieved, the excitation and the light emission travel down the discharge channel at a velocity (once

310

thought to be of the velocity of light, but which is dependent on the gain) of approximately 0.8 the velocity of light[128] (in agreement with the prediction of Casperson and Yariv[129]). The novelty of this type of excitation is that excited molecules in the upper laser level do not have a chance to radiate spontaneously and populate the upper vibrational levels of the ground state as the stimulated emission wave arrives simultaneously with the excitation wave that excites them by electron impact. In effect the system works as a travelling-wave amplifier at optical frequencies, and almost all of the stimulated emission comes out of one end of the device into a vacuum path (for vacuum-ultraviolet wavelengths). The lifetime of the upper laser level is so short that stimulated emission can only occur for a period of a few nanoseconds before the inversion is suppressed. The shortness of the stimulated-emission pulse means that optical feedback (from cavity mirrors) plays no role in this system.

More than 1 MW of total power on transitions in the Lyman band in 1-nsec pulses have been obtained using the apparatus described.[125] The charge voltage needed to produce this output power was 80 kV, and the discharge current was several hundred kiloamperes, with a rise-time of about 2.5 nsec. The optimum H_2 pressure for oscillation in the Lyman band is between 20 and 40 torr.

Deuterium has also lased in the system illustrated in Fig. 6.40. The output power obtainable is less than in H_2, but has the same pressure dependence, and it is extremely sensitive to the angle between the excitation wave and the discharge channel. That is to say, it is sensitive to the relative velocity of the excitation wave along the discharge channel and the propagation velocity of the stimulated emission along the channel.

The excitation technique used by Hodgson[120] to obtain stimulated emission in the Lyman band is simpler than that used by Waynant[125] and Waynant et al.[121] It does not utilize multidielectric switches but only a single switch. The excitation wave is then transverse to the discharge channel and optical axis, and cannot be made to incline to the discharge-channel axis to give the longitudinal travelling-wave effect. Stimulated emission under these conditions is emitted equally from both ends of the discharge channel. In more recent experiments Hodgson and Dreyfus[10] have used a high-energy electron-beam-excitation technique.

Oscillation in the Werner band (0.1000–0.1200 μm). Transient stimulated emission in the Lyman bands has been extended to the H_2 Werner band ($C^1\pi_u$–$X^1\Sigma_g^+$) with emission from 0.1161 to 0.1240 μm by Waynant[11] using the travelling-wave excitation technique previously described, and by Hodgson and Dreyfus[10] using a field-emission, electron-beam generator.

As in excitation of the upper vibrational levels in the Lyman-band vacuum-ultraviolet laser, inelastic electron-impact collisions of fast electrons with H_2 molecules in the $X^1\Sigma^+$ ($v'' = 0$) ground state cause vertical transitions to occur to low vibrational levels ($v' = 1, 0$) of the excited $C^1\pi_u$ state of molecular hydrogen (Fig. 6.37). These low vibrational levels of the $C^1\pi_u$ state are excited more efficiently than high vibrational levels of the $X^1\Sigma_g^+$ ground state. Again, because of the displacement of the equilibrium nuclear separation of the $C^1\pi_u$ state with respect to the $X^1\Sigma_g^+$ ground state, population inversion can be achieved between the low vibrational levels of the upper $C^1\pi_u$ state and the high vibrational levels of the ground state, and laser oscillation can be obtained before spontaneous emission populates the upper vibrational levels of the ground state.

TABLE 6.7. *Franck–Condon factors*[a] *for the most probable excitation from the* $H_2\ ^1\Sigma_g^+(v'' = 0)$ *ground state, and for possible laser transitions in the* $C^1\pi_u-X^1\Sigma_g^+$ *Werner band*[124]

Upper state v'	Lower state v''								
	0	1	2	3	4	5	6	7	8
0	0.1248	0.3254	0.3318	0.1679	—	—	—	—	—
1	0.1968	0.1397	—	0.2138	0.2900	0.1306	—	—	—
2	0.1936	—	0.1157	—	—	0.2824	0.2255	—	—
3	0.1539	—	0.1044	—	0.1319	—	0.1934	0.3	—
4	0.1093	—	—	—	—	—	—	—	0.3479

[a] Transitions with Franck–Condon factors below 0.1 are neglected.

Table 6.8, which includes the $q_{0v'}q_{v'v''}$ products for the $v' = 1$ to $v'' = 4$, and $v' = 2$ to $v'' = 5$ transitions, gives corresponding products in order of the expected intensity (relative probability of emission) for nine of the most likely laser transitions. The table shows that the 1–4 and 2–5 transitions are followed by the 3–7 and 2–6 transitions in order of their probability of emission.

According to the Franck–Condon factors given in Table 6.7, in the $X^1\Sigma_g^+(v'' = 0)-$ $C^1\pi_u(v' = 0, 1, 2, \ldots, n)$ excitation transition, the $v' = 0, 1, 2, 3$, and 4 vibrational levels of the C-state will be preferentially excited in inelastic electron-impact collisions from the $X^1\Sigma_g^+(v'' = 0)$ ground state, with the $v' = 1, 2$, and 3 levels excited the most. The factors given in Table 6.7 also show that, given that the $v' = 1$ vibrational level of the $C^1\pi_u$ is excited the most, then the most intense radiative transition in spontaneous emission from the $v' = 1$ level should be the $C^1\pi_u(v' = 1)-X^1\Sigma_g^+(v'' = 4)$ transition. According to Table

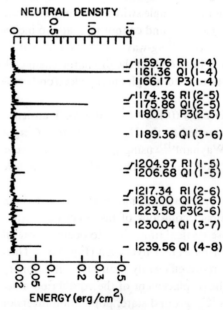

FIG. 6.41. Microdensitometer trace of the Werner-band stimulated-emission spectrum. The measured neutral density is plotted on the left as a function of the wavelength (Å) and the corresponding film exposure is indicated on the right. (After R. T. Hodgson and R. W. Dreyfus,[10] by courtesy of the American Institute of Physics.)

6.7, the next highest product of the Franck–Condon factors $(q_{0v'}q_{v'v''})$ that gives the relative intensity of line emission within the Werner band, is the product $(v'' = 0$ to $v' = 2)$ of 0.1936 and $(v' = 2$ to $v'' = 5)$ of 0.2824 which is equal to 0.0546. This transition should give the line with the next highest intensity.

Figure 6.41, a densitometer trace of the Werner-band stimulated-emission spectrum observed by Hodgson and Dreyfus,[10] using electron-beam excitation of H_2, shows that the strongest lines are on the 1–4, 2–5, 3–7, and 2–6 transitions as predicted from the $q_{0v'}q_{v'v''}$ products of the Franck–Condon factors of Table 6.7 given in Table 6.8.

Hodgson and Dreyfus[10] have calculated that the inversion number density in their system is 6×10^{11}, and the gain is 0.07 per cm for the strongest transition.

TABLE 6.8. *Relative probability of emission in the Werner band according to the Franck–Condon factors in Table 6.7, and the resulting wavelengths of the $Q-1$ transitions*[11]

$v'-v''$	$q_{0v'}q_{v'v''}$	$Q-1$ wavelength (μm)
1–4	0.0571	0.116132
2–5	0.0546	0.117587
3–7	0.0462	0.122998
2–6	0.0436	0.121894
0–2	0.0414	0.109945
1–3	0.0413	0.111634
0–1	0.0406	0.105418
4–8	0.0380	0.123953
3–6	0.0297	0.118938

6.2.2. CARBON MONOXIDE LASERS

Laser oscillation in carbon monoxide is observed in three widely separated regions of the spectrum ranging from the infrared at approximately 5 μm, through the visible at 0.6 μm, to the vacuum ultraviolet at approximately 0.18 μm. The oscillation occurs in vibrational-rotational transitions belonging to the $X^1\Sigma^+$ ground electronic state at 4.87 to 6.7 μm;[130, 131] on the angström band on vibrational–rotational transitions between the $B^1\Sigma^+ - A^1\Sigma^+$ electronic states at 0.52 to 0.66 μm;[4] and on vibrational–rotational cascade transitions between the $A^1\Sigma^+ - X^1\Sigma^+$ electronic states at 0.18 to 0.20 μm[132] (see Appendix Tables). Oscillation on the infrared vibrational–rotational transitions in the $X^1\Sigma^+$ ground electronic state has been obtained under both pulsed and CW discharge conditions, but the visible and vacuum-uv transitions have been obtained only under pulsed discharge conditions. Chemical reactions too have been successfully used to give oscillation on the $X^1\Sigma^+$ ground state vibrational–rotational transitions.[133–4]

A partial energy-level diagram of CO is given in Fig. 6.42. This figure shows the electronic laser transitions (dashed lines) belonging to the angström band, those belonging to the

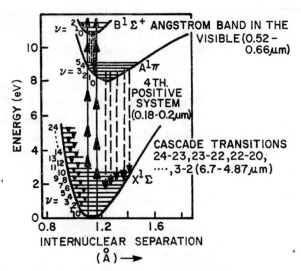

FIG. 6.42. Potential energy curves of the $B^1\Sigma^+$, $A^1\pi$, and $X^1\Sigma^+$ states of the CO molecule showing electronic laser transitions belonging to the angström band, the fourth positive system, and vibrational–rotational cascade laser transitions in the $X^1\Sigma^+$ ground electronic state, together with excitation paths. (Adapted from a figure given by G. G. Petrash.[135])

fourth positive system; and the vibrational–rotational cascade laser transitions 24 to 23, 23 to 22, 22 to 21, ..., 3 to 2 in the $X^1\Sigma^+$ electronic ground state. The solid arrowed lines show the excitation path of the $B^1\Sigma^+(v = 2, 1, 0)$, the $A^1\pi(v = 5, 4, 3, 2, 1, 0)$, and the vibrational levels of the $X^1\Sigma$ ground state by electron impact from the $v = 0$ level of the $X^1\Sigma$ ground state. Rotational transitions have been omitted to avoid any further complexity of the figure.

The CO molecule has only one vibrational mode of oscillation (in contrast to CO_2, which has three different modes of vibration) and so it has a single sequence of almost equally spaced vibrational energy levels, associated with each of which are rotational energy levels. Vibrational–rotational transitions involve a change in J and v within a given electronic state, with $\Delta v = \pm 1$, $\Delta J = 0, \pm 1 (0 \nrightarrow 0)$. Transitions with $\Delta J = 0$ form the Q-branch transitions; $J-1$ to J transitions with $\Delta J + 1$ give rise to P-branch transitions; and $J+1$ to J transitions with $\Delta J = -1$ form R-branch transitions.

6.2.2.1. Oscillation in the infrared (4.87 to 6.7 μm)

Following the reports of pulsed laser oscillation[136] in CO and CW-oscillation in a flowing mixture of CO and N_2 by Patel[137] and Legay-Sommaire et al.,[138] oscillation has been obtained on many vibrational–rotational transitions of the $X^1\Sigma^+$ CO ground state. High power CW operation at 95 W has been realized,[139-40] as well as high peak powers in excess of the 100-W level under pulsed, transverse-discharge-excitation conditions.[140-2]

Pulsed laser oscillation in vibrational–rotational transitions of the $X^1\Sigma^+$ ground state of CO exhibits a cascading behavior from one vibrational level to another. This causes the time-behavior after an excitation-current pulse of that illustrated in Fig. 6.43. The energy-level diagram at the left-hand side (energy scale not given) shows pertinent vibrational–

rotational levels and the laser transitions for which the laser output pulses are shown on the
right.[136] The discharge current pulse starts each trace. The first transition to start oscillating
is the $P_{6-5}(12)$ in (a) after a delay of 60 μsec after the current pulse. After a total delay of
70 μsec the $P_{7-6}(11)$ transition starts oscillating in (b) and can be seen to increase the output
on the $P_{6-5}(12)$ transition. After a total delay of 100 μsec the $P_{8-7}(10)$ transition starts
oscillating (c) and increases the population of the upper laser level for the $P_{7-6}(11)$ transition

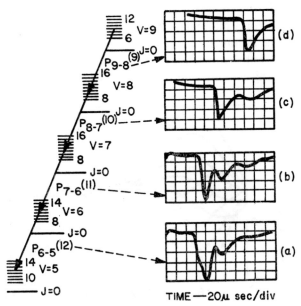

FIG. 6.43. Typical cascade laser-transition pulsed power outputs for vibrational–rotational transitions of the
CO $X^1\Sigma^+$ electronic ground state. The energy-level diagram at the left shows pertinent vibrational–rotational
levels and the laser transitions. (After C. K. N. Patel and R. J. Kerl,[136] by courtesy of the American Institute of Physics.)

and subsequently the population of the upper level of the $P_{6-5}(12)$ laser transition. In the
topmost trace, after a delay of 150 μsec, the $P_{9-8}(9)$ transition starts oscillating causing
additional population inversion and laser output down through the $P_{8-7}(10) \rightarrow P_{7-6}(11) \rightarrow$
$P_{6-5}(12)$ transition train. This cascading is observed for all the chains in the vibrational–
rotational transitions. The important thing to note from Fig. 6.43 is that the cascade behavior on each transition is caused by laser oscillation on the transition beneath it in an
upward sequence. Thus oscillation on the $P_{6-5}(11)$ transition depletes the population of
the lower laser $v = 6$, $J = 11$ level of the $P_{7-6}(11)$ transition to produce partial vibrational
population inversion and oscillation on the $P_{7-6}(11)$ transition. Oscillation in turn on the
$P_{7-6}(11)$ transition depletes the population of the $v = 7$, $J = 10$ level and in so doing
produces inversion between the $v = 8$, $J = 9$ and the $v = 7$, $J = 10$ levels and oscillation
occurs on the $P_{8-7}(10)$ transition. This happens stepwise up the transition train to produce
laser oscillation in the downward cascade sequence that is exhibited by the time behavior
of the pulsed output power illustrated in Fig. 6.43. The whole pulsed behavior, therefore,
appears to be triggered by oscillation on the lowest transition in the cascade sequence.

315

An analysis by Patel *et al.*[6] and Patel[7] of laser oscillation on vibrational–rotational transitions of linear polyatomic molecules, that is applicable to laser action on vibrational–rotational transitions of diatomic molecules, predicts that for nearly equal populations in two adjacent vibrational levels, the strongest rotational laser transitions at a molecular gas temperature of 400 K should be $P(11)$. Apart from oscillation in the 9–8 vibrational band at 5.274 μm that occurs on the $P(10)$ transition, Patel observed that all the other strongest laser transitions were $P(11)$ transitions in agreement with his prediction. The absence of oscillation on R-branch transitions for all the vibrational bands is also in agreement with theoretical predictions. For a molecular-gas temperature of 300 K, Patel[7] predicted that gain is possible on some P-branch transitions even where there is no complete population inversion $(N_v/N_{v'} < 1)$, but that positive gain for such partial inversion is not possible for R-branch transitions. This is illustrated in Fig. 6.44 where the normalized gain is plotted

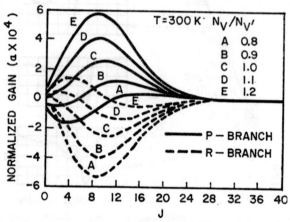

FIG. 6.44. Normalized gain for P-branch $(J-1$ to $J)$ transitions and R-branch $(J+1$ to $J)$ transitions for the 7–6 vibrational–rotational band of CO as a function of upper level J, for $T = 300$ K and $N_v/N_{v'} = 0.8, 0.9, 1.0, 1.1,$ and 1.2. (After C. K. N. Patel,[7] by courtesy of the American Institute of Physics.)

as a function of upper level J for transitions that are inhomogeneously (Doppler) broadened. The upper- and lower-level populations N_v and $N_{v'}$ are described by a Boltzmann distribution at the same temperature T. Clearly for P-branch transitions, gain is possible for a partial population inversion of 0.9 with J-levels above 6 at $T = 300$ K, and gain is possible for $N_v/N_{v'} = 1, 1.1,$ and 1.2 for all J-values. Only for low values of J is gain possible on R-branch transitions for $N_v/N_{v'}$ ratios above 1.1. For both P-branch and R-branch transitions, Patel[7] showed that as T is lowered the gain increases and the J-value of the highest-gain transition is lowered. This predicted increase in gain with reduction in temperature is utilized in the operation at low temperatures of CW CO lasers operating on vibrational–rotational transitions of the $X^1\Sigma^+$ ground state.[139, 143]

The experimentally observed shift of the gain of laser transitions towards lower J-values at lower molecular temperatures is now known to be due to the rapid near-resonant V–V anharmonic relaxation-pumping process of CO[143–50] in which a single quantum of vibrational energy is transferred in the reaction

$$CO(v = n_2) + CO(v = 0) \rightarrow CO(v = n_2 - 1) + CO(v = 1) - \Delta E_\infty \qquad (6.20)$$

in which the energy defect $\Delta E_\infty = 2\omega_e x_e(v = 1)^{(151)}$ is supplied by the translational energy of the gas.

The reverse of (6.20), the pumping reaction

$$\text{CO}(v = n_2 - 1) + \text{CO}(v = 1) \rightarrow \text{CO}(v = n_2) + \text{CO}(v = 0) + \Delta E_\infty \qquad (6.21)$$

can also occur, but because the higher levels are less populated than the lower levels has a lower probability than (6.20), in spite of it being an exothermic reaction.

The primary process of exciting vibrational levels of the CO $X^1\Sigma^+$ ground state under pulsed excitation conditions is by resonant inelastic electron-impact collisions reaction (6.16) with CO molecules in the ground state to form negatively charged CO compound states of CO, which then decay into excited vibrational states.[28] Nighan[146-7] has calculated that between 30 and 90 percent of the total electrical power (typical CO laser mixtures and discharge conditions) can be transferred to CO vibrational levels in this way. It has been proposed too, that the production of excited vibrational states of CO might proceed by the production of metastable $a^3\pi$ metastable CO molecules by electron impact followed by electronic to vibrational energy transfer. This latter quenching process would be expected to have the inverse pressure dependence that has been observed, and could account for the delay between the current pulse and the laser oscillation.[141]

Pulsed excitation of vibrationally excited CO molecules in the $X^1\Sigma^+$ state by electron-impact is not expected to produce any selective excitation of a particular vibrational energy level, but merely a population distribution among approximately the first ten vibrational levels[28, 144] that are not capable of supporting laser oscillation. Relaxation of this non-inverted population distribution then occurs on a time-scale that is longer than the time it takes the current pulse to produce the initial excited-states populations by anharmonic V–V relaxation of each of the $v = 1$ to 10 states. This produces population inversion between the vibrational states in a sequential way.[144, 147-50] This relaxation process due to molecular collisions is much faster than spontaneous radiative-decay and V–T relaxation processes, and is the dominant relaxation process in the CO laser. A value of the equilibration time τ of vibrationally excited states in CO has been obtained by Yardley;[149] it is given by the relationship

$$p_{\text{CO}}\tau = 208 \pm 30 \text{ } \mu\text{sec-torr}. \qquad (6.22)$$

This equilibration time is considerably less than decay times by spontaneous emission of 33 msec[152] and the collisional relaxation time of approximately 2 sec-atmospheres for collisional V–T relaxation[153-4] with helium at 77 K. The inverse dependence of the equilibration time on CO pressure exhibited by eqn. (6.22) can also account for the inverse dependence on CO pressure of the delay between the excitation current pulse and the onset of laser oscillation observed in ref. 141.

The initial population distribution in vibrationally excited states of the CO $X^1\Sigma^+$ ground state in a CO laser following an excitation pulse has been calculated by Jeffers and Wiswall.[148] They assumed that the excitation is by direct electron impact, and using cross-sections for vibrational excitation of Schulz[28] for $v = 1$ through 8 and an electron-energy distribution that was close to Maxwellian for the $E/N \approx 2 \times 10^{-14}$ V-cm^2 appropriate to a pulsed pure CO laser[146] as predicted from the calculations of Nighan,[146] calculate that the rate of vibrational excitation $r(0, v, kT_e)$ of $v = 0$ molecules to level v at a particular average

317

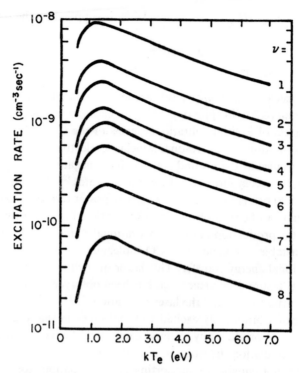

FIG. 6.45. Calculated values of the rate of excitation of vibrational states of the CO ground-state molecule from $v = 0$ to $v = 8$ as a function of the electron temperature (eV) using a Maxwellian electron-energy distribution and vibrational excitation cross-sections of ref. 28. (After W. Q. Jeffers and C. E. Wiswall,[148] by courtesy of the American Institute of Physics.)

electron temperature T_e for values of kT_e in the range 0.5 to 7 eV, have values given by Fig. 6.45. The peaks of the rate-of-vibrational-excitation curves shift from $kT_e = 1.10$ eV for $v = 1$, to $kT_e = 1.60$ eV for $v = 8$. In addition the figure shows that the $v = 1$ state is excited more than the $v = 8$ state by over two orders of magnitude. The calculated vibrationally excited-state populations in the $X^1\Sigma^+$ ground state in a CO laser, after the occurrence of a rectangular current pulse of 0.6-μsec duration, are shown in Fig. 6.46. The excited populations shown (curve A) are closely described by two approximate vibrational temperatures; for $v = 1$ to 6, $T_v = 7000$ K, while for $v = 6$, 7 and 8, a T_v of 2200 K matches the calculated data. A linear extrapolation of the excitation rates at $v = 6$, 7, and 8 were used to get the excitation rate $r(0, v, 1.6 \text{ eV})$ for $9 \leqslant v \leqslant 15$. The rapid decrease in the excitation rates at $v = 6$, 7, and 8 exhibited by Fig. 6.45 follow from the use of the experimental data of Schulz[28] and result in the sharp break in the population curve in curve A of Fig. 6.46 between $v = 6$ and $v = 7$. Curve B was obtained by assuming that the Schulz cross-sections are low at $v = 7$ and 8 and that a Boltzmann distribution among the excited states holds for all v, and that $T_v \approx 7000$ K as for $v = 1$ to 6 of curve A, in which[147]

$$N_v(t = 0) = (fN_0/Q_v) \exp (-E_v/kT_v), \quad v \geqslant 1, \qquad (6.23)$$

where f is the fractional excited-state population (equal to 0.01), E_v is the vibrational energy, Q_v is the vibrational partition function, and T_v is the effective vibrational temperature.

318

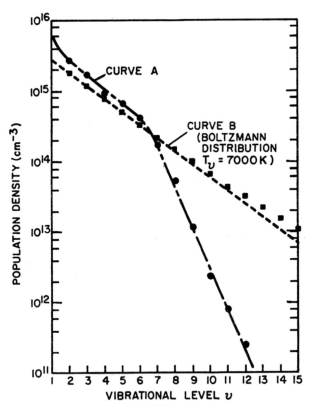

FIG. 6.46. Calculated vibrational population distributions in the CO $^1\Sigma^+$ ground state. Curve A obtained using cross-section data from ref. 28, $N_0 = 1.24 \times 10^{18}$ molecules cm^{-3}, $n_e = 1.0 \times 10^{12}$ electrons cm^{-3}, and a pulse duration of 600 nsec. Curve B obtained assuming a Boltzmann distribution throughout, $T_v = 7000$ K and $f = 0.01$. (After W. Q. Jeffers and C. E. Wiswall,[148] by courtesy of the American Institute of Physics.)

Given the initial excited-state populations of the vibrational levels of the $X^1\Sigma^+$ CO molecular ground state, there is general agreement that the rapid anharmonic selective relaxation process (6.20)

$$CO(v = n_2) + CO(v = 0) \rightarrow CO(v = n_2 - 1) + CO(v = 1) - \Delta E_\infty$$

is responsible for the subsequent excited states populations at any time following the cessation of the excitation current pulse. This relaxation process, a quasi-resonant/near-resonant V–V energy-transfer process, results from anharmonicity in the vibrational structures of diatomic molecules. This occurs because the diatomic molecule is not a simple harmonic oscillator, and has a potential-energy curve that is not symmetrical about the internuclear equilibrium position. As a result the energy levels are not equidistant as would be the case for a simple harmonic oscillator. The anharmonicity causes the spacing between successive vibrational levels to progressively decrease as the vibrational quantum number v of the levels increases.

Term values $G(v)$ are given by[151]

$$G(v) = \omega_e\left(v + \tfrac{1}{2}\right) - \omega_e x_e\left(v + \tfrac{1}{2}\right)^2 \cdots, \tag{6.24}$$

319

where v is again the vibrational quantum number, ω_e is the vibration frequency at the equilibrium position measured in cm^{-1}, and $\omega_e x_e \ll \omega_e$. The spacing ΔG between vibrational levels v and $v-1$, differs from that of $v = 1$ and 0 by the relation

$$\Delta G_{1,0} - \Delta G_{v,\,v-1} = 2\omega_e x_e(v-1). \qquad (6.25)$$

In the harmonic case, $\omega_e x_e = 0$, and the relaxation process (6.21) is resonant for all levels. Rapid relaxation then occurs to the $v = 1$ state with subsequent collisions merely interchanging molecules in the $v = 1$ and $v = 0$ states.

In the anharmonic case, $\omega_e x_e \neq 0$, reaction (6.20) is nonresonant and the energy discrepancy ΔE_∞ must be made up by the translational energy of the colliding molecules. When

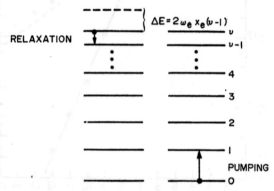

FIG. 6.47. Qualitative description of anharmonic relaxation. If $\Delta E > kT_g$ the process $CO(v = 5)+CO$ $(v = 0) \rightarrow CO(v = 4)+CO(v = 1) - \Delta E$ is inefficient and the population of $v = 5$ is "trapped".

the energy discrepancy ΔE_∞, which is equal to $\Delta G_{1,\,0} - \Delta G_{v,\,v-1}$, becomes comparable to or greater than kT_g, molecules in quantum levels above $v-1$ cannot relax by collisions with $CO(v = 0)$, and they are collisionally trapped (Fig. 6.47). The value of this vibrational quantum number v is given by the relationship

$$\Delta E = 2\omega_e x_e(v-1) \approx kT_g \qquad (6.26)$$

and can be calculated from the gas temperature T_g. For CO,[151] $\omega_e x_e = 13.461\ cm^{-1}$, and with $kT_g\ cm^{-1} = 0.695T_g$ where T_g is in degrees K, $v = 3$ at 77 K, 6.0 at 195 K, and 8.7 at 300 K. Corresponding values of v for the first overtone ($\Delta v = 2$) are 3.0, 4.5, and 5.85 respectively.[144] Based on this simple analysis, at 77 K, given an initial distribution of vibrational population, only levels $v = 3$ and 2 can relax by the anharmonic process (6.20) and levels above $v = 3$ remain populated. This can, therefore, rapidly create a population inversion between levels $v = 4$ and $v = 3$. Stimulated emission on the 4–3 transition empties level 4, and creates an inversion between the 5 and 4 levels. As a result of the initial collisional trapping of the $v = 4$ level, population inversion and laser oscillation progresses upwards through the vibrational energy levels at progressively later times.

At a higher gas temperature of 195 K, collisional trapping occurs at levels above $v = 6$, so that initial lasing would be predicted on the 7–6 transition, with a minimum delay between the onset of oscillation and the termination of the current pulse occurring for this

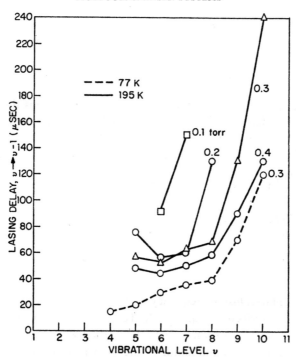

FIG. 6.48. Lasing delay of $v \to v-1$ transitions as a function of v for several partial pressures of CO, with 5 torr of helium added in each case. (After W. J. Graham, J. Kershenstein, J. T. Jensen, Jr., and K. Kershenstein,[144] by courtesy of the American Institute of Physics.)

transition. At 300 K, initial laser oscillation would be expected above $v = 9$. Figure 6.48 shows experimental results of the delay in the onset of laser oscillation relative to the start of the excitation current pulse at a number of gas pressures at two temperatures, given by Graham et al.[144] The delay plotted is the minimum value observed within each vibrational–rotational band. The broken curve is for the onset of oscillation at 77 K; it shows that the minimum delay occurs on the $v = 4$ (or lower) vibrational level as predicted. At 195 K the minimum delay is seen to occur for lasing on the 6–5 transition, which is close to the 7–6 transition on which initial oscillation would be expected. Laser action at 195 K also occurs on the lower 5–4 transition at a later time than the onset of oscillation on the 6–5 transition. This occurs as a direct result of cascade pumping into the $v = 5$ level by laser oscillation on the 6–5 transition.

According to the simple qualitative treatment of anharmonic relaxation of CO $X^1\Sigma^+$ vibrational levels given here and applied to the case where $T_g = 195$ K, at a gas temperature of 300 K, initial laser oscillation would be predicted from vibrational levels above $v = 9$. This, however is not in agreement with what is observed experimentally as Osgood et al.[141] have reported that oscillation in a room-temperature, transversely excited CO laser occurs first of all on the $v = 6 \to v = 5$ transition. In addition, this is in agreement with the observation by Patel and Kerl[136] that oscillation in their work on a room-temperature axially excited, low-pressure CO laser commenced on the 6–5 transition.

Jeffers and Wiswall[148] and Dawson and Tam[145] have treated the question of the time evolution of an initial distribution of CO excited-state populations in a different way to the

321

simple treatment of anharmonic relaxation considered above. Given an initial distribution of excited-states populations, they calculate the probability of V–V transfer, assuming that the endothermic anharmonic relaxation process [of eqn. 6.20)]

$$CO(v) + CO(0) \rightarrow CO(v-1) + CO(1) - \Delta E_\infty \qquad (6.27)$$

is the dominant relaxation process, and solve relaxation equations that result from eqn. (6.27). In eqn. (6.27) we have changed from the way the vibrational quantum numbers are specified in eqn. (6.20) so as to be able to follow the form of the treatment of Jeffers and Wiswall[148] and to be able to more easily display the relaxation equations given by them as

$$\left(\frac{dN_0}{dt}\right)_{relax.} = -\sum_{v>1} v p_{v,\,v-1} N_{v'},$$

$$\left(\frac{dN_1}{dt}\right)_{relax.} = 2v p_{2,\,1} N_2 + \sum_{v>2} v p_{v,\,v-1} N_v, \qquad (6.28)$$

$$\left(\frac{dN_v}{dt}\right)_{relax.} = - v p_{v,\,v-1} N_{v'} + v p_{v+1,\,v} N_{v+1} \quad \text{for} \quad v > 1$$

where v is the molecular-collision frequency and $p_{v,\,v-1}$ is the probability of V–V energy transfer per collision described by eqn. (6.27). Terms describing V–T relaxation have been omitted from (6.28) as V–T relaxation is much slower than V–V anharmonic relaxation.

Data for $p_{v,\,v-1}$ from Hancock and Smith[155] are shown plotted in Fig. 6.49 for values of v between 4 and 12 together with calculated values that extend the $p_{v,\,v-1}$ data from $v = 4$ down to $v = 1$ using the Sharma–Brau theory for V–V energy transfer,[150] while a linear

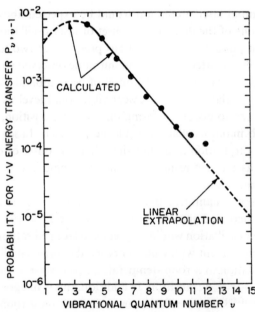

FIG. 6.49. Probabilities for V–V energy transfer in the CO $^1\Sigma^+$ ground state per collision at 300 K. The data points are from Hancock and Smith.[155] The solid curve and the broken curve below $v = 4$ are calculated. The broken curve above $v = 12$ is extrapolated from the calculated curve. (After W. Q. Jeffers and C. E. Wiswall,[148] by courtesy of the American Institute of Physics.)

extrapolation (dashed curve) was used to give values for $v > 12$. The data show that the probability for V–V energy transfer *decrease* monotonically for $v > 3$, so that the relaxation rates *decrease* with increasing vibrational quantum number. The effect of this decreasing relaxation rate with increasing vibrational number v on the evolution of the initial excited states populations is shown in Fig. 6.50 which is a solution of the relaxation equations (6.28) using the Boltzmann excited-state-population distribution (curve B) in Fig. 6.46. The figure clearly shows that population inversion occurs first of all between the $v = 6$ and $v = 5$ levels at a time $t = 2.0$ μsec. At $t = 4.0$ μsec, the 6–5 inversion has increased further, and inversion has developed up the vibrational chain between the $v = 7$ and $v = 6$ levels. At a later time of 8.0 μsec, the 6–5 inversion has increased still more, likewise that between the 7–6 levels; while inversion has developed on the 8–7 transition and also between the $v = 5$ and $v = 4$ levels. At still later times, inversion has increased between all levels greater than $v = 4$, while the inversion between the $v = 4$ and $v = 3$ levels has remained approximately constant.

Figure 6.50, therefore, enables one to predict three things:

(1) on which transition oscillation should first occur;
(2) the progression of inversion to *higher* vibrational quantum numbers;

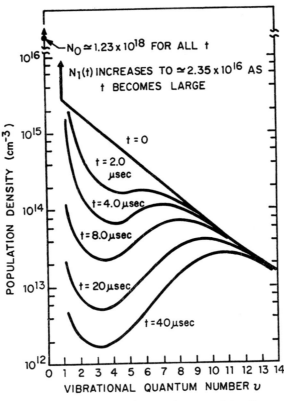

Fig. 6.50. Population distributions in the CO $^1\Sigma^+$ ground state evolving from a Boltzmann distribution at time $t = 0$. $N_0 = 1.24 \times 10^{18}$ cm^{-3}, $f = 0.01$, and $T_v = 7000$ K. (After W. Q. Jeffers and C. E. Wiswall,[148] by courtesy of the American Institute of Physics.)

22

(3) the progression of population inversion to *lower* quantum numbers, e.g. on the 5–4 transition, at a time when inversion has just developed up the vibrational chain on the 8–7 transition.

In Table 6.9 a comparison is made between the measured time delay between the current-excitation pulse and the time when a particular transition reaches the threshold for oscillation (called τ_{delay} (α peak) in the table), and the calculated time-delay, assuming an initial Boltzmann distribution (6.23) of excited-states populations with $T_v = 7000$ K, $f = 0.01$, and probabilities for V–V energy transfer as shown by the broken and solid curves of Fig. 6.49.[148] Included also are the predicted peak gain coefficients and the J-values for which they occur.

TABLE 6.9. *Measured and calculated time-delays prior to oscillation on CO $X^1\Sigma^+$ vibrational transitions*[148]

| Band | τ_{delay} (α peak) | | α_{peak} (cm^{-1}) | Strongest J-transition |
	Measured (μsec)	Calculated[a] (μsec)		
5 → 4	—	6.8	8.10×10^{-4}	13
6 → 5	3.5	3.9	6.26×10^{-3}	10
7 → 6	4.0	4.2	1.23×10^{-2}	10
8 → 7	5.0	6.6	1.23×10^{-2}	9
9 → 8	12.5	11.2	1.01×10^{-2}	9
10 → 9	20.0	20.4	6.91×10^{-3}	10
11 → 10	31.0	36.0	6.27×10^{-3}	10
12 → 11	—	66.0	4.89×10^{-3}	10

[a] Calculated assuming the CO ground-state population density N_0 (at 35 torr and 300 K) is equal to 1.24×10^{18} cm^{-3}.

It can be seen from Table 6.8 that there is good agreement between the experimentally determined time-delays and the calculated time-delays using the data and assumptions stated. The shortest time delay occurs on the 6–5 transition in agreement with the Patel prediction,[136] and the Patel and Kerl[130] observation that oscillation at 300 K occurs first of all on the 6–5 transition. This is an observation that cannot be predicted from the simple theory of anharmonic relaxation given earlier. On the simple theory, the first oscillation at 300 K would have been predicted to occur from around the $v = 9$ level, and not as observed on the 6–5 transition.

Summarizing, the treatment here by Jeffers and Wiswall[148] enables one to predict completely the major characteristics of the pulsed, infrared CO laser operating on vibrational transitions of the CO $X^1\Sigma^+$ ground-state *at room temperature*. More recent calculations by Jeffers and Kelley[156] of the probability of V–V transfer in CO–CO anharmonic relaxation have shown that the probability depends not only on the energy defect ΔE_∞ but also on the gas temperature. Their calculations, which are in reasonable agreement with experimental data, show that long-range interactions dominate in determining the V–V transfer probability at low temperatures, while at 300 K short-range interactions between the colliding

CO molecules dominate the probability for energy defects ΔE_∞ greater than 210 cm^{-1}. Using the newer calculations of probability for anharmonic relaxation, and electron-impact collision cross-sections of CO (suitably modified if they are also found to depend on gas temperature as might occur because of the resonant nature of the excitation process), it would appear that the treatment of Jeffers and Wiswall[158] should enable all the observed major characteristics of the pulsed CO laser at any gas temperature to be satisfactorily explained.

CW oscillation. The CO laser is the most powerful and efficient CW laser source in the 5-μm region of the spectrum. Output powers of nearly 100 W, at efficiencies exceeding 20 percent from 3.2 liter of active medium have been reported.[140] As in the pulsed CO laser, the CW laser output can be extremely broadband if wavelength selection is not used. The broadband oscillation follows from the cascading nature of the laser transition in the CO $X^1\Sigma^+$ ground state. It is this cascading of the laser transitions (Fig. 6.43) down through the single set of ladder-like vibrational levels of the $X^1\Sigma^+$ ground state in which the lower level of one laser transition becomes the upper level of the succeeding laser transition, that is mainly responsible for efficiencies approaching 50 percent that have been reported.[143] Just as under pulsed-excitation conditions, the CW 5-μm CO laser operates most efficiently at low discharge-wall temperatures, the use of cryogenic cooling of the discharge tube increases the gain under conditions of only partial population inversion, as predicted by Patel,[7] and causes a shift in oscillation to transitions having lower vibrational quantum numbers as illustrated in a later figure, Fig. 6.54.

In the preceding analysis of the pulsed CO laser we concentrated our attention on the inversion mechanisms in pure CO. In the case of the CW CO laser, optimum lasing mixtures can consist of a number of gases: CO, N_2, O_2, He, air, Xe, and Hg with the addition of other gases giving beneficial effects. The usual mixtures include CO, N_2, He, and Xe or O_2 as their primary constituents,[143, 157] with the expensive Xe often omitted.[140, 143] The effect of additives on processes operating in the CO laser operating under CW conditions will be discussed as was done for the CO_2 laser (see page 331).

In the electrically excited CW CO laser there are a number of processes by which vibrational excitation of CO $X^1\Sigma^+$ ground-state molecules can occur. These processes include: direct electron impact; V–V energy transfer from excited N_2 molecules in CO laser mixtures that contain N_2; radiative cascade into the CO ground state from electronic levels that have been excited by inelastic electron-impact collisions; dissociation of CO followed by recombination into excited vibrational states; energy transfer from electronic states of CO that have been excited by electron impact; and chemical reactions such as $C + O_3 \rightarrow CO + O_2$, where the carbon and ozone have been produced in the discharge.

The excitation processes that are most consistent with experimental results from high-output-power, high-efficiency CO lasers involve (1) direct excitation of vibrational states of the CO $X^1\Sigma^+$ ground-state molecule by the the electron-impact resonant process

$$CO(v = 0) + e^- \rightarrow (CO)^- \rightarrow CO(v = n) + e^- \tag{6.29}$$

that proceeds via the formation of the negatively charged $(CO)^-$ complex and has a large cross-section for excitation of vibrational states $v = 1$ to 8, and (2) the near-resonant, V–V anharmonic relaxation energy-transfer process (6.20).

Vibrational–vibrational (V–V) and electronic energy transfer from excited N_2 molecules to CO is a process that can be responsible for some additional selective excitation of vibrational states of the CO ground states in CO laser mixtures that contain N_2[158] because of the similarity and closeness of some of the energy levels of CO and N_2. It of course cannot be a selective excitation in a CO laser discharge that does not contain N_2. Anharmonic V–V energy transfer is also believed to occur between N_2 and CO molecules;[159] but the extent to which it occurs in an electrically excited CO laser discharge is not known. Any beneficial effect that N_2 does have on upper level excitation of vibrationally excited CO molecules is somewhat offset by the formation of CN $X^2\Sigma$ molecules (from CO and nascent N atoms) that detrimentally depopulate the vibrational levels of CO and N_2 in the reactions[158]:

$$ \text{CN}\,(X^2\Sigma) + \begin{cases} N_2(v) \\ \text{CO}\,(v) \end{cases} \rightarrow \text{CN} \begin{bmatrix} A\,^2\pi_i \\ B\,^2\Sigma^+ \end{bmatrix} + \begin{cases} N_2\,(0) \\ \text{CO}\,(0). \end{cases} \tag{6.30} $$

The importance of this reaction in CO lasers is an open question.[143]

Radiative cascading from higher excited electronic levels as a dominant excitation mechanism in CO lasers can be ruled out for the following reasons. The electronic levels that would have to be excited by electron impact and from which cascading to the vibrationally excited states of the ground state could occur are the $A^1\Pi$ and the $B^1\Sigma^+$ levels at approximately 8 and 10.5 eV, respectively, above the $X\,^1\Sigma^+$ ground state (Fig. 6.42). Although the $B\,^1\Sigma^+$ state has a large cross-section for excitation by electron impact, in CO laser discharges that are characterized by low average electron temperatures centered about 2 to 3 eV, excitation of the B-state (and also the A-state) must be small and cannot account for the high output powers and high efficiencies that obtain from CW CO laser discharges, that as well as having low electron temperatures, have low electron concentrations[143, 146, 160] in the region of 10^9 to 10^{10} cm^{-3}. In addition, an argument against the cascade-excitation mechanism is the observed slow recovery time of CO lasers that is four orders of magnitude slower than the lifetimes of 25 nsec of the $A\,^1\pi$ and $B\,^1\Sigma^+$ electronically excited states, respectively.[140]

Dissociation of CO followed by recombination into excited vibrational states can be ruled out as an important excitation mechanism in the CW CO laser because of the requirement for electrons that have energies in excess of 14 eV necessary to produce dissociation of CO molecules. In typical low electron-temperature CO laser discharges there are insufficient numbers of high-energy electrons present that could give the high powers and efficiencies that are observed. As pointed out by Osgood et al.[140] the highest quantum efficiency of this process in producing cascade laser transitions should be 26 percent, and, requiring 14-eV high energy electrons, the total efficiency of excitation should be considerably less than the experimentally observed 20- to 47-percent values.

A similar argument and the requirement for high energy electrons based on efficiency rules out chemical reactions such as $C+O_3 \rightarrow CO+O_2$ as dominant processes in the CW CO laser.

The only upper-laser-level excitation process that satisfactorily explains the characteristics of CW CO lasers operating on vibrational–rotational transitions of the $X^1\Sigma^+$ ground state is electron-impact excitation of vibrational levels via CO$^-$ formation (6.29) followed by

V–V anharmonic relaxation and cascading from higher to lower vibrational quantum states (as discussed in the analysis of the pulsed infrared CO laser). In some cases, oscillation on certain vibrational transitions is only possible when operating free-running with multi-wavelength oscillation, and is not possible using intra-cavity wavelength selection.[140] This is clear evidence that cascading is of importance in the CW CO laser.

Electron energy distributions and vibrational excitation rates in CW CO lasers. As well as determining the electron-energy distributions in CO_2 laser discharges, Nighan[146] has evaluated the electron-energy distributions for plasma conditions typical of low-pressure, CW, electrically excited CO lasers on the assumptions that the E/N of the plasma is low and the average electron energy is only a few eV, and that electron-molecule and electron-atom ionization-cross-section data are known. The range of E/N values that he chose is representative of CO-laser-discharge conditions and was found to yield mean electron energies in the range 0.5 to 2.0 eV, which is a range typical of electric discharges in molecular gases. E/N is defined as the ratio of the electric field intensity to the total neutral particle number density. Calculated distribution functions are shown in Fig. 6.51 for a CO–N_2–Xe–He gas mixture having the number density proportions $1:1:1:17$. The distribution function is defined such that $\int_0^\infty u^{1/2} f(u)\, du = 1$ and the reduced average electron energy such that

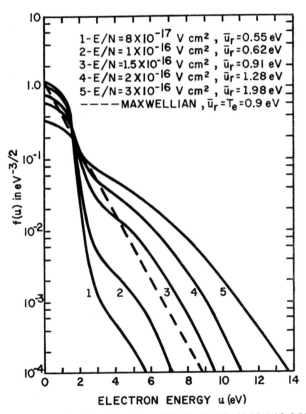

FIG. 6.51. Electron-energy-distribution functions for a CO–N_2–Xe–He (0.05–0.05–0.05–0.85) laser mixture. (After W. L. Nighan,[146] by courtesy of the American Institute of Physics.)

$\bar{u}_r = 2/3 \int_0^\infty u^{3/2} f(u)\, du$. For comparison purposes a Maxwellian distribution with an average electron energy of 0.9 eV is included in Fig. 6.51.

The distribution functions are clearly neither Maxwellian nor Druyvesteyn, having distinct varying curvatures at all electron energies, with the exception that curve 5 becomes Druyvesteyn-like at electron energies above 4 eV. The pronounced dips in the distribution functions between 1.5 to 3 eV that are emphasized at low electron temperatures occur as a result of the very large cross-sections for vibrational excitation of CO and N_2 in this electron-energy range in the Schulz-type resonant reactions. The distribution functions shown here are very similar to those calculated for CO_2 laser plasmas and shown earlier in Chapter 6 as Figs. 6.14 to 6.16.

With a knowledge of the electron-energy-distribution function the power transferred in electron-molecule collision processes can be calculated as was done for the CO_2 laser.[45, 147] Figure 6.52 presents the fractional power transfer from electrons for the range of E/N values

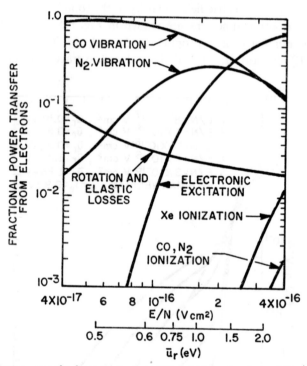

FIG. 6.52. Fractional power transfer from electrons to CO and N_2 molecules and Xe atoms as a function of E/N and reduced average electron energy $\bar{u}_r$ appropriate to CO–N_2–Xe–He laser-mixture conditions of Fig. 6.51. (After W. L. Nighan,[146] by courtesy of the American Institute of Physics.)

covered by Fig. 6.51. For average electron energies $\bar{u}_r$ below 1.0 eV, vibrational and electronic excitation dominates in energy transfer from the electrons, with ionization of both CO, N_2, or Xe negligible in comparison with these excitation processes. It can be seen from Fig. 6.52 that in the average electron-energy region of 0.75 to 1.0 eV the fractional power transferred in various excitation processes is an extremely sensitive function of the actual average electron energy. This fact is more clearly illustrated in Fig. 6.53, in which the vibrational

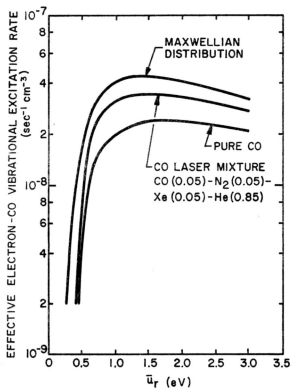

FIG. 6.53. Effective electron–CO vibrational excitation rate as a function of reduced average electron energy (temperature) in CO laser discharges. (After W. L. Nighan,[146] by courtesy of the American Institute of Physics.)

rate per particle of CO is plotted as a function of reduced average electron energy $\bar{u}_r$. The computed rate for pure CO is also representative for mixtures not containing N_2. Below a $\bar{u}_r$ of 0.8 eV the vibrational excitation falls off drastically, while in the region of 1.0 to 3.0 eV covered by the figure, the vibrational excitation rate does not vary appreciably. The difference in excitation rates between the curve for pure CO and the CO laser mixture amounts to approximately 50 percent at the optimum reduced average electron energy of about 1.4 eV. This is due to the presence of N_2 which, it is suggested by Nighan,[146] evens out the electron-energy-distribution function and causes it to increase in the 1.5 to 3-eV range within which vibrational excitation cross-sections for CO are largest. It also couples in vibrational energy to CO molecules by V–V energy transfer. Nighan[146] has also shown that the presence of vibrationally excited CO molecules in the $X^1\Sigma^+$ ground state is likely to have a minor effect on the calculated electron-energy-distribution functions shown in Fig. 6.51 because the majority of molecules are in the $v = 0\text{–}2$ levels, which have approximately equal cross-sections for electron-impact excitation to higher levels.[161] This result might not have been expected in view of the resonant nature of the electron-impact-collision process that is responsible for the excitation of vibrational levels in the CO $X^1\Sigma^+$ ground state.

Effects of temperature. The temperature of the wall of the discharge tube of an axially excited CW infrared CO laser under optimized gas-mixture and discharge-current condi-

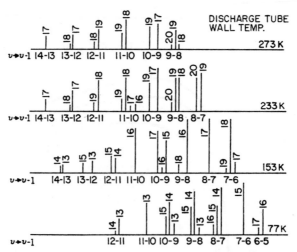

FIG. 6.54. CO laser-output spectra at various discharge-tube wall temperatures (under unoptimized conditions). The numbers along the axes denote vibrational bands. Numbers on top of the spectral lines give values of T for $P(T)$ transitions. (After M. L. Blaumik, W. B. Lacina, and M. M. Mann,[143] by courtesy of the IEEE.)

tions determines to a large extent the laser output spectra, the maximum laser efficiency, and the realizable gain. It is one of the most important operating parameters of such a laser.

The effect that the discharge-tube wall temperature has on the laser-output spectra of an axially excited CW CO laser is illustrated in Fig. 6.54. The gas mixture consisted of CO, N_2, He and O_2 with partial pressures of 0.35, 2.1, 12.0, and 0.01 torr respectively. The mixture was not optimized at each temperature. The output spectrum clearly shifts towards lower vibrational transitions as the wall temperature is decreased. There is also a general tendency for oscillation in a particular vibrational band to shift to lower rotational transitions as the wall temperature, and hence the molecular kinetic temperature, is decreased.

The shift of the laser output spectrum is attributed to the endothermic near-resonant anharmonic relaxation process (6.27)

$$CO(v) + CO(0) \rightarrow CO(v-1) + CO(1) - \Delta E_\infty,$$

in which vibrationally excited CO molecules are de-excited progressively by the transfer of individual quanta of vibrational energy to ground state CO molecules. A reduction in translational temperature of CO ground state molecules apparently causes a redistribution of population among the vibrational levels by deactivating molecules in the lower vibrational levels more quickly than those in higher vibrational levels. The overall result is that the magnitude of the population inversion increases between all vibrational levels (Figs. 6.55 and 6.56). A related effect has been observed to occur in a CO chemical laser where the addition of ($v = 0$) vibrationally cold CO molecules and other gases enhances the population inversion.[162, 145]

Figures 6.55 and 6.56 show the results of small-signal gain measurements on a CO laser amplifier containing CO–N_2–He–O_2 mixtures at two discharge-tube wall temperatures of 293 and 77 K, respectively. The gas mixture for operation at room temperature (293 K) was 1.1 torr CO, 6.5 torr N_2, 15 torr He, and 0.02 torr O_2; while that for 77 K was 0.18 torr

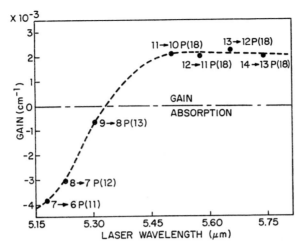

FIG. 6.55. The CW 5-μm CO-laser peak small-signal gains at a 293 K discharge-tube wall temperature. Flowing-gas mixture: 1.1-torr CO, 6.5-torr N_2, 15-torr He, and 0.02-torr O_2; $D = 2.1$ cm. (After M. L. Blaumik, W. B. Lacina, and M. M. Mann,[143] by courtesy of the IEEE.)

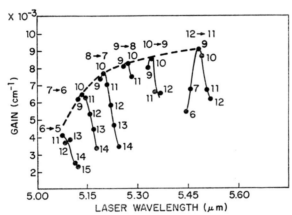

FIG. 6.56. The CW 5-μm CO-laser small-signal gains at a 77 K discharge-tube wall temperature, with $T_g = 143$ K. Flowing-gas mixture: 0.18-torr CO, 0.27-torr N_2, 12-torr He, and 0.01-torr O_2; $D = 2.1$ cm. (After M. L. Blaumik, W. B. Lacina, and M. M. Mann,[143] by courtesy of the IEEE.)

CO, 0.27 torr N_2, 12 torr He, and 0.01 torr O_2. The gain in any particular vibrational band varies as a function of the rotational transition. In Fig. 6.55 only the peak gains are plotted, while in Fig. 6.56 gains on individual rotational transitions of each vibrational band are included together with a broken curve of the peak measured gains. At room temperature, the only transitions to exhibit gain are the 11 → 10, 12 → 11, 13 → 12, and 14 → 13 transitions with a gain of slightly more than 0.002 cm⁻¹. At 77 K, gain occurs on all the transitions indicated (6 → 5, 7 → 6, 8 → 7, 9 → 8, 10 → 9, and 12 → 11). It has a value of about 0.009 cm⁻¹ on the 12 → 11 transition.

Effects of additives. Though infrared oscillation under both CW and pulsed conditions can be obtained in pure CO, beneficial results are realized by adding various gases in judicious quantities. Gas mixtures that are most frequently used are CO, N_2, He, and Xe; or CO,

331

N_2, He, and O_2 with or without Xe. Detailed information on specific partial pressures of these laser mixtures is given in *Operating characteristics and discharge conditions*, p. 336. Qualitative observations on the effects of various gas mixtures on laser output power and their effects on discharge mechanisms and excitation processes are presented here.

The addition of helium to a CW CO laser discharge has little effect on the mechanisms of the discharge as it produces little change in the discharge-sustaining voltage or the current. Its main role, as in the CO_2 laser, appears to be due to its increasing the thermal conductivity of the gas mixture and thus lowering the translational and rotational temperatures near the axis of the discharge tube.[140] As a result of this, it is found that the optimum discharge current is raised, and presumably because this increases the CO vibrational-level electron-impact excitation rate, the laser output power is considerably increased. The effect of the addition of helium on the output power of a CO laser containing CO and air is shown in Fig. 6.57. The addition of large amounts of helium has been shown by Champagne[280] to increase the output pulse energy by a factor of 70.

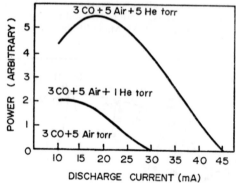

FIG. 6.57. The CW 5-μm CO laser output power versus discharge current for three different partial pressures of He; $D = 4$ cm. (After R. M. Osgood, W. C. Eppers, Jr., and E. R. Nichols,[140] by courtesy of the IEEE.)

It is not clear what function N_2 has in the CW 5-μm CO laser. Results and calculations indicate that the most important effect its presence has is to even out the electron-energy distribution and cause it to increase in the electron-energy range of 1.5 to 3 eV, at which energies vibrational cross-sections of CO by electron impact are largest. The addition of N_2, however, does not lead only to beneficial effects. Its presence causes nascent nitrogen atoms to be produced that react with CO molecules to form CN molecules, which de-excite vibrationally excited $v = 12$ or 13 CO $X^1\Sigma^+$ molecules in V–V energy transfer to the detriment of population inversion on the $12 \to 11$ transition.[158]

The addition of small quantities of O_2 to a CO–N_2–He laser discharge can increase the laser output power by factors of two or more.[140, 163] It is only at low discharge-wall temperatures (77 K), however, that it improves laser performance significantly.[164] It has been suggested by Osgood *et al.*[140] and Legay *et al.*[158] that the beneficial effect of adding O_2 to electrically excited CO laser discharges containing N_2 is due to removal of CN radicals by the oxidation reaction

$$O + CN(X^2\Sigma^+) \to CO(v \geqslant 14) + N. \qquad (6.31)$$

In this reaction, CN molecules, which can deleteriously de-excite vibrationally excited CO molecules, are removed from the discharge, but at the same time additional vibrationally excited $CO(\nu \geqslant 14)$ molecules are proeuced to the benefit of population inversion and cascading oscillation on lover vibrational transitions of theCO $X^1\Sigma^+$ ground state.

At low temperatures, the addition of O_2 prevents the build up of carbon on the walls of the discharge tub and prevents dissociation of CO that can occur rapidly in the absence of O_2 addition.[140, 163] It is not clear if the observed prevention of CO dissociation is causied by the O_2 forcing the reversible reaction $2\,CO \rightarrow 2\,C + O_2$ to shift to the CO side, or if it is caused by a quenching of the high-energy tail of the electron-energy distribution (as xenon does in the CO_2 laser) that reduces dissociation produced by electron impact.

Hartwick and Walder[163] have observed that when the O_2 concentration added to a CO–He laser mixture is increased from zero, the CO laser output power rises to a well-defined maximum and then declines. This variation in output power is furthermore correlated to a decrease to a minimum in gas discharge kinetic temperature in the absence of laser oscillation followed by an increase for the same variation in O_2 concentration. Their results are shown in Fig. 6.58, in which the CO laser output power and the independently measured discharge temperature are plotted as a function of the flow rate of O_2. As the flow rate of O_2 is increased the output power rises smoothly to a maximum (at a partial pressure of about 0.1 torr) and then declines. At the O_2 flow rate that gives the maximum laser output power, the discharge temperature measured along the axis of the laser tube with one of the cavity mirrors removed is a minimum. Similar results were observed for different input powers to the laser discharge. The addition of N_2 only produced a monotonic increase in temperature in contrast to the variation exhibited in Fig. 6.58 for the O_2 addition. *Note:* The tempera-

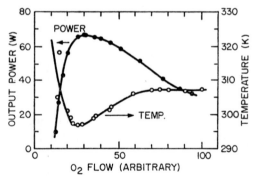

FIG. 6.58. The CW 5-μm CO laser output power and gas kinetic temperature of the discharge for an input power of 395 W versus various O_2 flow rates. The temperature measurements were made independently in the absence of laser action. (After T. S. Hartwick and J. Walder,[163] by courtesy of the IEEE.)

ture measurements were taken in the absence of laser oscillation so that the cooling effect is not produced by the extraction of laser power that could be expected to reduce the gas temperature by a few degrees. If the independently measured curves of Fig. 6.58 are used to plot the output power versus the measured discharge temperature, it is found that the output power decreases linearly with increase in discharge temperature. The variation of output power with change in discharge temperature, however, is too large to be associated with the explicit variation of power with discharge temperature, and it is postulated by Hartwick and Walder[163] that the observed correlation could be associated with a modification of the electron energy distribution or electron density produced by the O_2 addition.

The addition of xenon produces both an increase in efficiency[165-6] to as much as 40 percent, and also a change in the spectral intensity distribution of the laser output.[165] The increased efficiency follows from a substantial decrease in the operating voltage when xenon is added to a CO–He–N_2–O_2 laser discharge, and, it is believed, from a lowering of the average electron energy of the discharge to one that brings the peak of the electron energy distribution closer to coincidence with the peaks of the electron excitation cross-sections for N_2 and CO that occur around 2 eV. This latter effect has been observed in studies of the influence of xenon on CO_2 lasers.[97, 167] As mentioned earlier, and illustrated by Fig. 6.53, the fractional power transferred from electrons to CO molecules in various excitation processes is an extremely sensitive function of the average electron energy, and probably of quite small changes in the actual shape of the electron energy distribution. Because of its high ionization efficiency xenon is able to take over ion production in the discharge, thus reducing the ionization of CO, O_2, and N_2 that was necessary to maintain the discharge. In doing this, the E/p of the discharge is reduced and the average electron temperature is lowered. The optimum operating discharge current remains approximately the same or increases slightly (unlike the CO_2 laser discharge where it increases), but a lowering of the average electron temperature has the effect of increasing the excitation rate as more electrons are available with the required lower electron energy. Associated with the reduction in ionization of the molecular constituents, a reduction in average electron energy can lead to a decrease in the dissociation of CO in the discharge by a factor of approximately 4, depending on the discharge current, from between 19 to 13 percent to 5 to 3 percent, when xenon is added to a CO–He laser discharge.[168]

Together with increase in efficiency, xenon addition to the 5-μm CO laser causes a shift in the intensity distribution of the laser output spectra toward shorter wavelengths (i.e. lower vibrational transitions).[165] This probably follows from an increase in the excitation rate of vibrational excitation of the CO molecules in the $X^1\Sigma^+$ ground state.

The addition of mercury to 5-μm CO laser discharges produces results similar to those produced by xenon addition, and enables high power CW operation to be achieved at room temperature.[169, 166] The results are similarly postulated to accrue from a reduction in the average electron temperature at the same or increased electron concentration with a subsequent increase in vibrational excitation rate of the CO. It has also been suggested[169] that interactions between electronically excited metastable xenon[170] and mercury atoms[153, 171] and molecular constituents in CO laser discharges that contain xenon or mercury might be playing a role in producing preferential excitation of particular vibrational levels.

Various efficiency enhancements and discharge conditions with xenon and mercury addition to 5-μm CO laser discharges are given later in Table 6.10, *Operating characteristics and discharge conditions*, on page 337.

Vibrational population densities. Relative vibrational populations in the CO $X^1\Sigma^+$ ground state in CO laser discharges under different excitation conditions have been determined by Rich and Thompson[172] and by Legay *et al.*[173] They have observed that the relative vibrational population distribution is nonBoltzmann, and that the distribution is not inverted. Figure 6.59 after Rich and Thompson[172] shows this clearly. The relative vibrational population N_v/N_1 is seen to be significantly nonBoltzmann with the upper vibrational states rela tively highly populated. Vibrational temperatures based on upper-level population ratios can exceed 15,000 K, contrasting with the vibrational temperature of the $v = 4$ to 3 band of only approximately 3000 K. Self-absorption occurs on the $v = 2$ to 1 and $v = 1$ to 0 bands to distort the relative populations in Fig. 6.59; accordingly the vibrational population

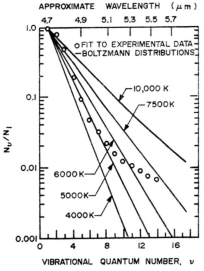

FIG. 6.59. Vibrational population distributions in a CW liquid-nitrogen-cooled 5-μm CO–He–O$_2$ laser. Discharge conditions: $p(CO) = 0.7$ torr, $p(O_2) = 0.03$ torr, $p(He) = 7.3$ torr, flowing at 3 m/sec; $I = 22$ mA. (After J. W. Rich and H. M. Thompson,[172] by courtesy of the American Institute of Physics.)

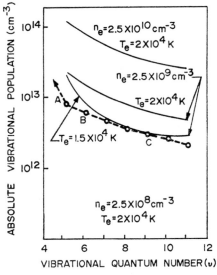

FIG. 6.60. Absolute vibrational populations in the CO $X^1\Sigma^+$ ground state in a CW liquid-nitrogen-cooled CO–He–O$_2$ laser discharge: experimental points for a 2-cm-bore tube, $I = 15$ mA; 0.2-torr CO, 0.03-torr O$_2$, 6.0-torr He mixture; (solid line) theory curve of ref. 157. In region A ($v < 5$), $T_v \approx 2500$ K, in B, $T_v \approx 9000$ K; and in C, $T_v \approx 13,500$ K. (After E. N. Lotkova, G. N. Mercer, and N. N. Sobolev,[160] by courtesy of the American Institute of Physics.)

for states $v = 1$ and 2 are underestimates of the actual populations, and the apparent rise in vibrational temperature on these bands is not real.[172]

Absolute vibrational population densities in CO laser discharges under various conditions have been experimentally determined by means of gain measurements by Lotkova *et al.*,[160] Blaumik *et al.*[143] and Corcoran *et al.*[174] In Fig. 6.60 are shown the absolute vibrational populations for the levels $v = 4$ to 12 for a 0.2-torr CO, 6.0-torr He, 0.03-torr O_2 laser discharge in a 2-cm-bore, liquid-nitrogen-cooled discharge tube. The gas mixture had a flow velocity of 2 m/sec. Again the experimental population distribution can be seen to be distinctly nonBoltzmann having three separate slopes in the regions marked A, B, and C to which can be ascribed three differing vibrational temperatures. The population ratio that can be deduced from the curve in region A below $v = 5$ indicates a vibrational temperature of 2500 K; in B between $v = 8$ and $v = 5$ the temperature is approximately 9000 K; and in region C between $v = 11$ and $v = 8$ the vibrational temperature is higher at about 13,500 K. Shown also in the figure are theoretical vibrational population distributions of Rich[157] for various values of electron temperature and electron concentration, and for a gas temperature and electron concentration, and for a gas temperature of 175 K. Best agreement between theory and experiment occurs for a theoretical electron temperature of 15,000 K and an electron concentration of 2.5×10^9 cm^{-3}; the theoretical concentration is believed to be close to the actual concentration under the experimental conditions.[157] The population densities given by Fig. 6.60 are a factor of about 10 lower than of those of Corcoran *et al.*,[174] and more than 10 lower than those of Blaumik *et al.*[143]

Operating characteristics and discharge conditions. Some of the details of the operating characteristics of the CW CO laser have already been covered in the earlier discussion "Effects of temperature" and the accompanying figures: 6.54—Effect of discharge-tube-wall temperature on the laser-output spectra; and 6.55 and 6.56—small signal gain at discharge-wall temperatures of 293 K and 77 K, respectively. And the variation of output power with discharge current is illustrated in Fig. 6.57.

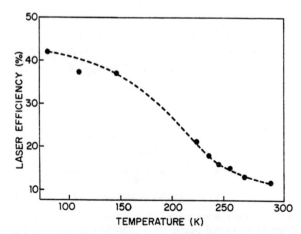

FIG. 6.61. Maximum efficiency of a CW 5-μm CO laser versus discharge-tube wall temperature, obtained by optimizing the gas mixture and discharge current, and using the cavity configuration C of Table 6.10. (After M. L. Blaumik, W. B. Lacina, and M. M. Mann,[143] by courtesy of the IEEE.)

Figure 6.61 illustrates the general effect the discharge-tube wall temperature has on the efficiency of a CO laser with optimized gas mixtures of He, CO, H_2, O_2, and Xe, at optimized discharge currents. The efficiency clearly increases with decreasing wall temperature. An approximate threefold increase in efficiency can be seen to be achieved by operating at 77 K instead of at room temperature.

A list of CO laser operating characteristics and discharge conditions is given in Table 6.10. The table includes characteristics of CO laser discharges that contain mixtures of He CO, N_2, and O_2; and He, CO, N_2, O_2, and Xe; and He, CO, N_2, O_2, and Hg at two discharge wall temperatures of 77 and 300 K.

TABLE 6.10. *CO laser operating characteristics and discharge conditions*

Configuration temperature (K)		Pressure (torr)						Discharge voltage (kV)	Discharge current (mA)	Output (W)	Efficiency (%)	Reference
		He	CO	N_2	O_2	Xe	Hg					
D	77	16	0.5	1.5	0.01	0.3	—	12.4	12	70	46.9	143
A	77	15	0.4	0.4	0.01	0.15	—	12.8	10	51	39.8	143
A	77	15.25	0.25	0.6	0.01	—	—	13.6	18	60	29.2	143
A	77	15.25	0.25	0.6	0.01	0.15	—	12.8	18	70	36.7	143
B	77	14.9	1.6	1.8	0.01	—	—	12.0	27	91	28	143
B	150	14.9	1.5	1.9	0.02	—	—	12.0	10	25	20.8	143
C	220	12.5	0.45	3.0	0.02	0.3	—	9.0	22.5	42.5	21	143
C	300	12.5	0.45	3.0	0.02	—	—	14	13	10	5.5	143
C	300	12.5	0.45	3.0	0.02	0.3	—	9.2	22.5	25	12.1	143
E	300	12.5	0.45	3.0	0.02	0.4	—	14	12	12.5	7.0	143
F	77	30.5	0.5	1.2	< 0.05	—	—	13.6	18	60	29.2	165
F	77	30.5	0.5	1.2	< 0.05	0.3	—	12.8	18	70	36.7	165
F	77	7.5	0.6	0.9	< 0.05	—	—	7.5	10	7.5	10.0	165
F	77	7.5	0.6	0.9	< 0.05	0.5	—	6.1	10	12.5	20.5	165
G	300	50	1.8	12	~ 0.2	—	—	14	13	10	5.5	166
G	300	50	1.8	12	~ 0.2	2.5	—	11	13	15	10.5	166
G	300	50	1.8	12	~ 0.2	2.5	—	9.2	22.5	25	12.1	166
H	300	62.6	2.4	15	~ 0.2	—	0.001[a]	7	25	14	8	166
H	300	62.6	2.4	15	~ 0.2	2.5	0.001[a]	5.9	25	25	17	166

Configuration: A–E, discharge length 126 cm, tube diameter 2.1 cm, cooling jacket 115 cm, a 10-m total reflector and a plane output mirror of reflectivity R in each case. Cavity length for A, B, and C – 215 cm with two Brewster windows; cavity length for B and F – 165 cm with internal mirrors. Values of reflectivity: A – 95 percent, B – 90 percent, C – 85 percent, D – 80 percent, E – 85 percent. F, discharge length 126 cm, tube diameter 2.5 cm, cooling jacket 116 cm, a 10-m total reflector and 95 percent reflecting flat output mirror between 4.6 and 5.5 μm, cavity length 215 cm, with Brewster windows.

G, as in ref. 165, except for 85 percent output mirror.

H, as in ref. 169, discharge length 124 cm, tube diameter 2 cm; cooled over 110 cm, total reflector as in F with 95 percent reflectivity.

[a] The electrode wells were filled with mercury to within 1 cm of the bore; pressure assumed here to be that at 300 K.

6.2.2.2. Oscillation in the visible (0.52 to 0.66 μm)

Limited attention has been paid to pulsed oscillation in the visible at numerous wavelengths in the red, orange, and green regions of the spectrum that was first observed by Mathias and Parker.[4] The optical transitions belong to the angström band system ($B^1\Sigma^+$

to $A^1\pi$) of the CO molecule, with oscillation occurring on at least 24 lines. Twenty of these lines have been identified as being $P(J)$, $Q(J)$, $R(J)$ lines of the 0–5, 0–4, and 0–3 bands[4] and the shortest wavelength lines at around 0.52 μm have been identified tentatively as lines of the 0–2 band.[4, 135]

The excitation conditions used by Mathias and Parker[4] consisted of longitudinal excitation by means of high voltage d.c. pulses having an approximately triangular shape, provided by a line-type pulse modulator and a pulse transformer. The risetime of the current pulse was 500 nsec, and the overall length 2 μsec. Oscillation was achieved at CO pressures above 0.5 torr and peak voltages above 16 kV; the peak discharge current in the 10-mm-bore discharge tube was 80 A. The discharge length was 117 cm.

Oscillation commenced 200 nsec after the start of the current pulse and ceased just after the peak current was reached. The time between half-power points of the laser pulse was 180 nsec, and the overall pulse duration was approximately 400 nsec. Cheo and Cooper[175] later used somewhat similar axial high-voltage excitation conditions in spectral and time-resolved studies of the spontaneous and stimulated radiation of this laser system. Figure 6.62 shows the time relationship observed by Cheo and Cooper[175] between

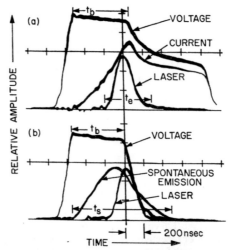

FIG. 6.62. Time relationship in a pulsed CO laser between: (a) voltage, current, and laser pulses for an excitation pulse of duration $> t_b$; (b) voltage, spontaneous emission from upper laser level, and laser amplitude ($E = E_{opt}$, $\lambda = 0.6080$ μm, 1-torr CO, in 4-mm-bore tube. (After P. K. Cheo and H. G. Cooper,[175] by courtesy of the American Institute of Physics.)

voltage V, discharge current I, and the laser pulse L (Fig. 6.62a), and voltage V, spontaneous emission S, and the laser pulse L (Fig. 6.62b) under optimum electric field conditions, at $\lambda = 0.6080$ μm, with CO at a pressure of 1 torr in a 4-mm-bore tube. In figures (a) and (b), t_b is the time for gas breakdown, t_s the duration of the spontaneous-emission, and t_l the duration of the laser pulse. In both cases the current begins to rise when the voltage pulse has reached its peak, and the current pulse risetime is equal to t_b. Laser action occurred over a wide range of t_b (of 200 nsec to 2 μsec). At optimum conditions, $t_b = t_l = 300$ nsec, and $t_s = 600$ nsec for all pressures. At the end of the breakdown period t_b, the electric field decreases to a lower value followed closely by the discharge current, and pumping of the upper laser level, as indicated by the spontaneous emission S, decreases rapidly (Fig. 6.62b). Cheo and Cooper also found that a linear relationship existed between the gas pressure and the optimum electric field E_{opt} for maximum laser intensity at the pressures of a few torr that they used. The linear relationship is relatively independent of the discharge-tube diameter, the Q of the laser cavity, and the discharge length. At 1 torr the E_{opt}/p ratio is approximately 190 V/cm-torr; at 2 torr it is 140 V/cm-torr; and at 3 torr it is only approximately 120 V/cm-torr. Spontaneous intensity measurements, of the laser line in the $B^1\Sigma^+ \rightarrow A^1\pi$ band, of transitions from the lower laser band $A^1\pi$ to the $X^1\Sigma^+$ ground state, and transitions from the higher-lying $C^1\Sigma^+$ levels to the $A^1\pi$ lower-laser levels versus electric field, and the corresponding variation of discharge current with electric field, are given in Fig. 6.63.

It was also observed that the fourth positive emission, emission from the lower laser level $A^1\pi$ to the $X^1\Sigma^+$ ground state occurs at higher values of electric field, as the trend of the $A^1\pi$–$X^1\Sigma^+$ curve in Fig. 6.63 would indicate. Although it could be that the $A^1\Sigma^+$ lower laser level is excited more effectively at higher E/p values by electrons with higher energies, it is believed that the $A^1\Sigma^+$ level is populated primarily by transitions from higher-lying levels such as the $C^1\Sigma^+$ level.

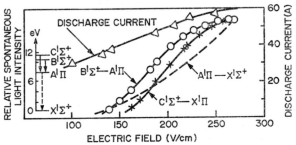

FIG. 6.63. Pulsed CO laser variation of spontaneous light intensity with electric field, and peak discharge current versus electric field; $p = 1$ torr CO, $D = 4$ mm. (After P. K. Cheo and H. G. Cooper,[175] by courtesy of the American Institute of Physics.)

Upper level excitation. Referring back to the potential-energy curves of the CO molecule shown in Fig. 6.42, it can be seen that the equilibrium internuclear separations for the $X^1\Sigma^+$ ground state and the $B^1\Sigma^+$ upper state of the visible CO laser lines are practically coincident. It follows because of this coincidence that vertical transitions caused by inelastic collisions of CO ground state molecules in the $v'' = 0$ vibrational level with high energy electrons only occur effectively to one single level, the $v = 0$ vibrational level of the $B^1\Sigma^+$ upper laser state. Since the equilibrium internuclear separation of the $A^1\pi$ lower laser state is more than that of the $A^1\Sigma^+$ ground state, according to the Franck–Condon principle, inelastic electron-impact collisions cause vertical transitions to many vibrational levels with relatively little excitation of the $v = 0$ vibrational level of the $A^1\pi$ lower laser state.

The fact that the spontaneous emission from the $B^1\Sigma^+$ upper laser level is emitted strongly during the risetime of the current pulse (Fig. 6.62), and is of short duration, suggests that excitation of the upper laser level is mainly by electron impact with high energy electrons that are only of high energy during the breakdown period of the discharge.[175] If this is the case, the vibrational level of the $B^1\Sigma^+$ state that would be expected to be predominantly excited would be the $v = 0$ level. This is precisely the level, the only vibrational level, from which oscillation has been observed (from $v' = 0$ to $v'' = 5, 4, 3,$ and 2).

Radiative decay of the $A^1\pi$ lower laser level with a lifetime of less than 200 nsec[176] can occur rapidly by means of near vacuum-ultraviolet transitions in the 4th positive system to the $X^1\Sigma^+$ ground state. Again, due to the relationship of the potential-energy curves of the $A^1\pi$ and the $X^1\Sigma^+$ states, transitions from the $A^1\pi$ state occur mainly to highly vibrationally excited states of the $X^1\Sigma^+$ ground state. Initially these vibrationally excited states are empty but are rapidly filled by the vacuum-UV transitions from the $A^1\pi$ state. These vibrationally excited states decay radiatively to lower-lying vibrationally excited levels through the rotational–vibrational laser transitions at 5 μm, and via the anharmonic relaxation process discussed in Section 6.2.2.1.

Up to now it is not known if it is the limited time during which high energy electrons are available in a high-voltage pulsed discharge for providing effective excitation of the upper laser level; rapid cascade transitions into the lower laser level; or the decay rate of the vibrationally excited ground state molecules and radiation trapping that occurs on them in the vacuum-ultraviolet transitions to the lower laser level, that determines the duration of the laser output pulse. However, in view of the relationship between the intensity of spontaneous emission from the upper laser level, the laser output pulse, and the breakdown period of the discharge illustrated in Fig. 6.62, in which the spontaneous-emission intensity decreases rapidly following the breakdown period t_b of the discharge, it would appear that excitation of the upper laser level is the determining factor.

6.2.2.3. Oscillation in the vacuum-UV (0.18 to 0.2 μm)

Stimulated emission has been reported on transitions of the CO fourth positive system ($A^1\pi–X^1\Sigma^+$) that are similar to the molecular hydrogen, Lyman-band transitions discussed in Section 6.2.1.2.[132] The upper laser levels are the lower laser levels of the visible angström

TABLE 6.11. *Stimulated emission spectrum of the 4th positive system of CO $A^1\pi–X^1E^+$* [132]

Band λ_{head} (μm)	$v'–v''$	Lines	Remarks
0.181085	2–6	Q5–13	Not all resolved – Q7 missing
		R2–9	All R lines would be blended
0.187831	2–7	Q5–13	Not resolved – Q7 probably missing
		R2–9	Blended
0.195006	2–8	Q5–11	All resolved – Q7 missing
		R2–9	Blended
0.189784	3–8	Q5–12	All resolved
		R2–9	Blended
0.197013	3–9	Q5–11	All resolved, R lines missing

Excitation conditions: Blumlein circuit, parallel-plate transmission line similar to that described in ref. 120; discharge channel $0.05 \times 1.2 \times 120$ cm³, with flowing CO at 60 torr.

band ($B^1\Sigma^+ - A^1\pi$) laser transitions in CO, and the lower laser levels are highly vibrationally excited levels of the $X^1\Sigma^+$ ground state of CO that form upper and lower levels of the vibrational–rotational cascade laser transitions of the 5-μm CO laser. Figure 6.42 again shows the disposition of the $A^1\pi$ and the $X^1\Sigma^+$ levels, the $v'–v''$ 2–6, 2–7, 2–8, 3–8, and 3–9 laser transitions reported by Hodgson,[132] and the electron-impact excitation path of $A^1\pi$ upper laser levels. The wavelengths and vibrational transitions and remarks about the laser lines are given in Table 6.11.

High-resolution, stimulated-emission spectra show that the Q-branch lines are most intense; the R-branch lines are much less intense, or, in one band are missing; and the P-branch lines are missing. The $Q7$ line is missing in the 2–8 band and appears to be weak or missing in the 2–6 and 2–7 bands as well.[132]

Upper level excitation. The stimulated emission can be qualitatively explained by considering that the upper level excitation is by inelastic electron-impact collisions with high energy electrons well above threshold for excitation, and by considering the stimulated emission coefficients of the P-, Q-, and R-branch transitions from the excited states, and the effect of perturbations and molecular collisions on the level populations.[132]

The $v' = 2$ and 3 vibrational levels of the $A^1\pi$ state from which laser emission is observed are those levels which, according to the Franck–Condon principle, are the most likely to be excited by electron impact in vertical transitions from the $v'' = 0$ vibrational level of the CO $X^1\Sigma^+$ ground state, as Fig. 6.42 shows. Whereas excitation of the $A^1\pi$ levels from the ground state is favored in electron-impact collisions with high energy electrons, excitation of vibrational levels of the ground state that comprise the lower laser levels ($v'' = 6, 7, 8$, and 9) is favored in collisions with low energy electrons via the formation of a quasi-resonant CO^- negative ion state. During the breakdown period of an electrical discharge, therefore, when electron energies are relatively higher than at other times in the current pulse, conditions are favorable for producing population inversion between the $A^1\pi$ state and the vibrationally excited levels of the $X^1\Sigma^+$ ground state of the CO molecule. Because of the very short lifetime of the $A^1\pi$ states, spontaneous emission can rapidly populate the vibrational levels of the ground state and reduce the inversion, and excitation pulses that have a very short current risetime are necessary to produce amplified stimulated radiation and the short superradiant pulses produced in this type of vacuum-ultraviolet gas laser. The pulse width

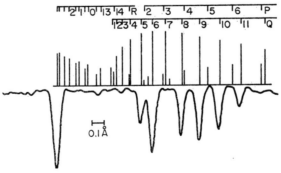

FIG. 6.64. Microdensitometer tracing of the 2–8 band stimulated emission spectrum in the pulsed CO vacuum-UV laser. (After R. T. Hodgson,[132] by courtesy of the American Institute of Physics.)

of the superradiant pulses at half-maximum was 1.5 nsec, and the peak output power was 6 W in the work reported by Hodgson.[132]

Figure 6.64 shows a microdensitometer tracing of the $v' = 2$ to $v'' = 8$ band stimulated emission, high-resolution spectra. The P-, Q-, and R-branch line positions and spontaneous emission intensities are given by the positions and lengths of the lines drawn above the micro-densitometer tracing. As shown by the figure, of those lines that are resolved, the Q-branch 6, 8, 9, and 10 lines are the strongest, with the $Q7$ line missing; and no P-branch transitions are observed. The gain for the Q-branch transitions is expected to be higher than either the R- or P-branch transitions because of the spontaneous emission coefficients that are proportional to $2J'+1$, $J'+1$, and J' respectively, given equal populations in the upper and lower Λ states of each rotational level. Gain is also reduced on the R- and P-branch transitions by Q-branch transitions that populate the lower energy states and reduce the inversion density. Absence of superradiance on the $Q7$ line in the $v'-v''$ 2–6, 2–7, and 2–8 transitions is attributed by Hodgson[132] to perturbation of the lower Λ level of the $J' = 7$, $v' = 2$ level of the $A^1\pi$ state by a rotational level of a $^1\Sigma^-$ state.[177]

6.2.3. NITROGEN LASERS

Laser oscillation in molecular nitrogen has been observed on three identified groups of transitions: the second positive system ($C^3\pi_u$–$B^3\pi_g$) in the ultraviolet at 0.337 μm, 0.358 μm and 0.316 μm; the first positive system ($B^3\pi_g$–$A^3\Sigma_u^+$) from about 0.745 μm to 1.235 μm; the $a^1\pi_g$–$a'^1\Sigma_u^-$ system between 3.29 μm and 3.47 μm and at approximately 8.15 μm to 8.21 μm; and the $w^1\Delta_u$–$a^1\pi_g$ system at approximately 3.7 μm.[13] Oscillation has also been observed on about 80 lines on unidentified transitions at wavelengths between 5.4 μm and 8.07 μm.[178] In none of these systems is it known with certainty what the exact mechanisms are that produce the population inversions. It is clear, however, that electrons are primarily responsible for the excitation, and calculations based on models that assume that the excitation of both upper and lower laser levels is by direct electron impact from the ground state give good agreement with experimental observations.

Just two N_2 laser systems are considered, the second positive system ($C^3\pi_u$–$B^3\pi_g$) in the ultraviolet in which oscillation was first observed by Heard,[5] and the first positive system ($B^3\pi_g$–$A^3\Sigma_u^+$) in the near-infrared in which oscillation was first achieved by Mathias and Parker.[3] In both systems the lower-laser-level lifetimes are much longer than the upper-state lifetimes and both are inherently transient laser systems and have been operated only under various pulsed excitation conditions. These excitation conditions range from pulsed high-voltage, low-pressure axial excitation,[3, 175, 179–183] higher-pressure transmission-line and transverse excitation under static and high-speed gas-flow conditions,[183–189] excitation at atmospheric pressure,[190] and excitation by high-energy electron beams.[191, 192]

It will be noted that the $B^3\pi_g$ lower level for the ultraviolet laser transitions is the upper level for the infrared laser transitions.

Figure 6.65 shows a partial energy-level diagram relevant to the $N_2 C^3\pi_u - B^3\pi_g$ laser system and the $B^3\pi_g$–$A^3\Sigma_u^+$ near-infrared laser system. (The energy scale is incorrectly shown, and and it should be multiplied by a factor of 1.25.) A number of vibrational energy levels of the $C^3\pi_u$, $B^3\pi_g$, and $A^3\Sigma_u^+$ electronic states are included in the figure between which oscillation has been reported. In the second positive system, laser oscillation has only been observed

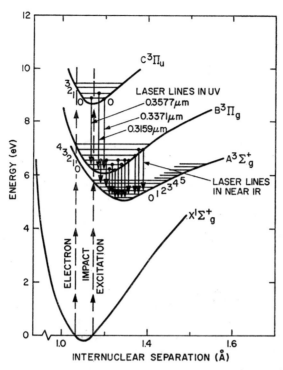

FIG. 6.65. Partial potential energy level diagram of N_2 relevant to the $C^3\pi_u \rightarrow B^3\pi_g$, and the $B^3\pi_g$–$A^3\Sigma_u^+$ laser systems, showing observed laser lines and excitation paths. (The spacings between the vibrational levels are not to scale.)

from two vibrationally excited levels of the $C^3\pi_u$ upper state ($v' = 0, 1$) giving rise to laser lines as indicated at 0.3371 μm ($v' = 0$ to $v'' = 0$), 0.3577 μm ($v' = 0$ to $v'' = 1$),[179, 181, 193] and 0.3159 μm ($v' = 1$ to $v'' = 0$).[193] In the first positive system, oscillation has been reported from the four lowest vibrational levels of the $B^3\pi_g$ state to the four lowest vibrational levels of the $A^3\Sigma_u^+$ state.[179, 180] A low-temperature environment of the discharge tube is needed to obtain oscillation from the 4, 3, and 2 vibrational levels of the B-state. Oscillation can also be obtained using intra-cavity-prism wavelength selection to suppress oscillation from other vibrational levels.[281] As already mentioned, both the UV and IR laser systems are essentially transient in operation. This follows as the C-state has a short radiative lifetime of about 40 nsec compared with that of about 10 μsec (at a few torr of N_2 pressure[195]) of the B-state,[184] which in turn is short compared with the lifetime of the metastable A-state, the lower laser level of the IR-line, that cannot decay radiatively and has an effective lifetime in excess of a millisecond. The short lifetimes of the respective upper levels of the UV and IR transitions means that inversion can only be obtained by utilizing short-risetime, high-voltage pulsed excitation in which the current-pulse risetime is comparable or less than that of the effective lifetimes of the upper laser levels. Typical laser output-pulse durations, depending on the operating gas pressure, range from a few nanoseconds to about 20 nanoseconds for the UV-laser transitions and a few tens of nanoseconds to a few hundred nanoseconds for the IR-laser transitions. The UV system will operate superradiantly without cavity mirrors, but for obtaining oscillation on the lower

343

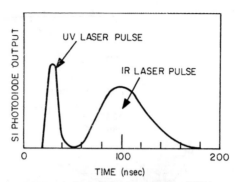

FIG. 6.66. Time relationship between emission of the superradiant UV laser pulse and the IR laser pulse using transverse electrical, pin-rod excitation, $p(N_2) = 35$ torr, electrode separation 1 cm. (Unpublished work, J. S. Kruger and R. M. Curnutt, Harry Diamond Laboratories.)

gain IR transitions, feedback by means of cavity mirrors has so far been found to be necessary. It appears that the difference in gain between the two systems gives rise to the temporal differences in UV and IR laser output illustrated in Fig. 6.66, where the start of the oscillogram-trace is triggered by the initiation of the discharge. Whereas the UV-pulse is superradiant and does not require any cavity-build-up time, the IR-pulse requires a number of round trips in the cavity for it to reach threshold, and it appears to be this that produces the time delay of a few tens of nanoseconds between the commencement of the UV laser pulse and the IR-pulse. The spontaneous emission on both transitions begins on the commencement of the rise of the current pulse,[176] so that the delay cannot be caused by (laser) cascade processes between the C-state and the B-state. Also, as the A-state, the lower laser level, is metastable the delay cannot be caused by decay processes involving the A-state.

The optimum excitation conditions for the UV and IR laser transitions are not the same. Using axial excitation of low-pressure N_2 discharges in small-bore tubes of a 3 to 10 mm, the optimum nitrogen pressure is lower at 0.4 to 1.1 torr for the second positive system than that of 1.2 to 2.5 torr for the first positive system at room temperature. At lower temperatures the optimum pressures still differ.[179] If transverse excitation is used, we[212] have found that is little difference in the optimum pressures for both systems.

The UV laser transitions also require higher discharge currents than the IR transitions.[182, 194] Presumably this arises because of the need for a higher excitation rate of the $C^3\pi_u$ state than the $B^3\pi_g$ state in the presence of spontaneous and stimulated emission that depopulates the C-state faster than the B-state.

6.2.3.1. Oscillation in the UV in the second positive system ($C^3\pi_u - B^3\pi_g$)

Oscillation in the second positive system has only been observed on the three electronic-vibrational transitions $v' = 0$ to $v'' = 0$, $v' = 0$ to $v' = 1$, and $v' = 1$ to $v'' = 0$ that give rise to the three UV laser bands shown in Fig. 6.65 at 0.3371, 0.3577, and 0.3159 μm respectively. The presence of rotational structure of the vibrational levels gives rise to the emission of laser band emission at numerous wavelengths close to the stated wavelengths. At liquid-air temperatures, the maximum output shifts to rotational quantum numbers that are lower than those of the maximum output under room-temperature excitation conditions,[179, 193] as might be expected from a reduction in the translational temperature.

344

Upper level excitation. The emission of spontaneous emission from the $C^3\pi_u$ upper laser level immediately on the commencement of the excitation current pulse, and superradiant emission during the risetime of the excitation pulse, is strong supporting evidence for excitation of the $C^3\pi_u$ state being that of electron impact by high-energy electrons.

The time-scale of the excitation is such that it precludes a number of excitation processes that could conceivably populate the $C^3\pi_u$ state. Figure 6.67 shows a time-history chart of

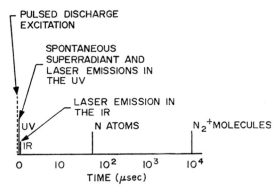

FIG. 6.67. Time-history chart of N_2 excitation and observation of spontaneous and stimulated UV- and IR-emission, together with the formation of N atoms and N_2^+ molecules. (After C. A. Massone, M. Garavaglia, M. Gallardo, J. A. E. Calatroni and A. A. Tagliaferri,[179] by courtesy of the Optical Society of America.)

nitrogen gas excitation and observation of spontaneous emission and stimulated UV (and IR) emission, together with the formation of N atoms and N_2^+ molecules. It is clear from Fig. 6.67 that recombination of dissociated nitrogen atoms that can only occur towards 100 μsec after the initiation of the discharge cannot account for excitation of the C-state. Likewise, recombination of N_2^+ molecules with electrons and cascading into the C-state that can only be of significance 10 msec after the initiation of the current pulse cannot be the excitation process of the $C^3\pi_u$ state.

The only known selective excitation processes that could account for the observed spontaneous emission on the commencement of the excitation current pulse are: direct electron-impact excitation, or indirect production (in a resonant process) of a short-lived temporary N_2-ion state that changes to the $C^3\pi_u$ state. The production of such a short-lived negatively charged molecule has been observed by Schulz[196] at an electron excitation energy of 11.2 eV which he attributed to excitation of the $C^3\pi_u$ state (as noted by Massone *et al*.[179]). Later work, however, attributes this resonance to excitation, not of the C-state, but of the E-state.[197] Whatever the detail of the excitation might be, electrons are clearly involved intimately in the selective excitation process of the C-state.

The emission of spontaneous emission from the $B^3\pi_g$ lower laser level of the N_2 UV laser transitions immediately on the commencement of the excitation current pulse[175] can likewise be taken to indicate strongly that the selective excitation of the B-state in a pulsed discharge in nitrogen occurs in inelastic collisions of N_2 $X^1\Sigma_g$ ground-state molecules with high-energy electrons that are present in sufficient numbers only during the risetime of the current pulse.

Assuming that both the $C^3\pi_u$ upper laser level and the $B^3\pi_g$ lower laser level are excited in inelastic electron-impact collisions with fast electrons that cause vertical transitions to both states (as per the Franck–Condon principle) and referring to the N_2 molecular potential diagram of Fig. 6.65, it can be deduced qualitatively that because of the equilibrium internuclear separations (r_e) of the C- and B-states that are larger than that of the $X^1\Sigma_g^+$ N_2 ground state $\left(r_e(B^3\pi_g) > r_e(C^3\pi_u) > r_e(X^1\Sigma_g^+)\right)$, excitation of the C-state will be predominantly to low vibrational quantum numbers $v = 0$, 1, and 2 while excitation of the B-state will be to a larger number of vibrational levels.

Quantitatively, the excitation probability of each vibrational level of the C- and B-states from the $X^1\Sigma_g^+$ $(v'' = 0)$ ground state is nearly proportional to the Franck–Condon factor $Q_{0v'}$. Since the electronic transition moment is reasonably constant with internuclear separation, the Franck–Condon factors are a good measure of excitation cross-sections and line intensities for N_2.[198, 199]

Calculations of excitation of vibrational levels of the N_2 $C^3\pi_u$ state from the $X^1\Sigma_g^+$ state have been made by Bates,[200] Pillow,[199] and Nichols.[205] Cross-sections $Q_{0v'}$ for excitation from the $v'' = 0$ vibrational level to the $C^3\pi_u$ state for the first four vibrational levels are given in Table 6.12. The table indicates that as predicted qualitatively, the cross-sections for excitation of vibrational levels of the C-state decrease rather quickly as v' increases, from Q_{01} of 6.2×10^{-18} cm² to Q_{02} of approximately 2×10^{-18} cm², to between 0.1 and 0.4×10^{-18} for Q_{04}. Values of excitation cross-sections for excitation from vibrational levels other than $v'' = 0$ to $v' = 0$, 1, 2, and 3 are practically the same as from $v'' = 0$.[201]

TABLE 6.12. *Cross-sections $Q_{0v'}$ for excitation of vibrational levels of the $C^3\pi_u$ state of N_2 from the $X^1\Sigma_g^+$ $(v'' = 0)$ ground state at an electron energy of 35 eV*

Cross-section $Q_{0v'}$ (10^{-18} cm²)	$C^3\pi_u$ upper state v'				
	0	1	2	3	4
Cross-section	6.2[a]	4.6[a]	1.9[a]	1.7[a]	< 0.2[a]
		3.5[b]	2.0[b]	0.3[b]	< 0.1[b]
		3.8[c]	1.6[c]	0.7[c]	0.3[c]

[a] From ref. 201, or deduced from the same reference.
[b] Deduced from data[198] tabulated in ref. 202 of predicted populations of the $v' = 0$, 1, 2, 3, and 4 states using the cross-section of 6.2×10^{-18} cm² for $Q_{0,0}$ at 35 eV from ref. 201 as a reference cross-section, and assuming $Q_{0v'}$ is directly proportional to the relative population rates and populations of the states.
[c] Deduced from data[199] given in ref. 203.

Summing the cross-sections Q_{00}, Q_{01}, Q_{02}, Q_{03}, and Q_{04} given in Table 6.12 gives a total cross-section for excitation of the $C^3\pi$ state from the $X^1\Sigma_g^+(v'' = 0)$ ground state of approximately 15×10^{-18} cm². Extrapolating up to v' values of 8, the total cross-section is not likely to exceed a value of 16×10^{-18} cm². This is in reasonable agreement with a value of 11×10^{-18} cm² at an electron energy of 35 eV for excitation of the $C^3\pi_u$ state determined by Zapesochnyi and Skubenich.[204] As they are relevant to this discussion of excitation of the $C^3\pi_u$ state and inversion on the $C^3\pi_u$–$B^3\pi_g$ transition, excitation functions for the $C^3\pi_u$ and $B^3\pi_g$ states are presented in Fig. 6.68.

The rapid rise of the curves to sharp maxima (that are slightly displaced from each other) followed by a rapid decrease with increasing electron energy is characteristic of a transition involving a spin change where electron exchange is the predominant mechanism of excitation (see Fig. 2.20). The small, second maximum on the excitation function of the $B^3\pi_g$ state coincides within 0.1 eV of the maximum in the excitation curve of the $C^3\pi_u$ state, and is due to cascading from the C-state. The excitation cross-section for the $C^3\pi_u$ state was obtained by adding the cross-sections of the first five vibrational levels; that of the $B^3\pi_g$ state by adding the cross-sections of the first thirteen vibrational levels. Cross-sections for the higher vibrational levels are relatively small and were neglected.[204]

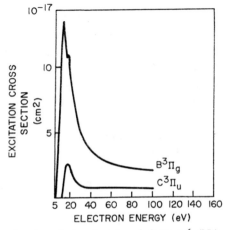

FIG. 6.68. Excitation functions of various electronic states of nitrogen ($p(N_2) = 0.004$ to 0.005 torr, electron flux density 0.0001 to 0.005 A/cm²). (After I. P. Zapesochnyi and V. V. Skubenich,[204] by courtesy of the Optical Society of America.)

Figure 6.68 shows that the total maximum cross-section of the $B^3\pi_g$ state is about 13×10^{-17} cm², and approximately five times larger than that of the $C^3\pi_u$ state (about 2.5×10^{-17} cm²). Based on these values of the total cross-sections, the similarity of the shape of the excitation functions, and the occurrence of their maximum cross-sections within a few eV of each other, it would appear that it would be impossible to produce population inversion between the C- and B-states by electron impact excitation from the ground state under any discharge conditions. What is important, however, in determining whether or not population inversion is realizable, is the magnitude of the cross-sections to *individual* vibrational levels of each of the electronic $C^3\pi_u$ and $B^3\pi_g$ states. Whereas the total cross-section of the $B^3\pi_g$ state is five times larger than that of the $C^3\pi_u$ state, a large number of vibrational levels contribute to the total cross-section of the B-state, while only a few levels ($\nu' = 0$, 1, and 2) contribute significantly to the cross-section of the C-state.

Unfortunately cross-section data for excitation of individual vibrational levels of the $B^3\pi_g$ state does not appear to have been published. Relative populations of vibrational levels of the B-state have, however, been calculated by Broadfoot and Hunten[203] based on Franck–

347

Condon factors calculated by Pillow.[199] The calculations assume that the B-state is popu-lated by electron impact excitation of N_2 molecules in the zero vibrational level of the ground state. The relative populations of the $B^3\pi_g$ state are given in Table 6.13.

TABLE 6.13

B-state vibrational quantum number	v'										
	0	1	2	3	4	5	6	7	8	9	10
Relative population	25	64	92	100	100	68	50	28	17	12	9

It can be seen that the $v' = 2, 3$, and 4 vibrational levels are the most highly populated and must have the largest cross-sections $Q_{0v'}$ for excitation from $v'' = 0$ of the N_2 ground state.

From these calculated relative populations of vibrational levels of the B-state, and know-ing the total cross-section for its excitation from $v'' = 0$ of the ground state (from Fig. 6.68), it is possible to calculate the individual cross-sections for excitation of each of the vibrational levels of the B-state. The individual cross-sections for each vibrational state are directly proportional to the ratio of relative population of the state divided by the sum of the relative populations, multiplied by the total cross-section for excitation of the B-state from the ground state. As we wish to compare the individual vibrational level cross-sections of the B-state with those of the C-state that are given in Table 6.12 for an electron energy of 35 eV it is appropriate to use the total cross-section of the B-state at an electron energy of 35 eV also. From Fig. 6.68 the total cross-section $Q_{0v'}$ for the B-state is approximately 4×10^{-17} cm^2 at an electron energy of 35 eV. Based on the value of this cross-section, individual cross-sections of vibrational levels of the B-state of N_2 are as given in Table 6.14.

TABLE 6.14

B-state vibrational quantum number	v'										
	0	1	2	3	4	5	6	7	8	9	10
Cross-section $Q_{0v'}$ $(10^{-18} cm^2)$[a]	1.8	4.5	6.5	7.1	7.1	4.8	3.6	2.0	1.2	0.9	0.6

[a] Where $\Sigma Q_{0v'} = 4 \times 10^{-17} cm^2$.

The cross-section Q_{00} of 1.8×10^{-18} cm^2 is shown to be smaller than that of any other vibrational level up to $v' = 8$. It is approximately a factor of four smaller than either Q_{03} or Q_{04}, which are the maximum vibrational-state cross-sections of the $B^3\pi_g$ electronic state for excitation from the ground state.

It can now be seen that although the total cross-section for excitation of the B-state is larger than that of the C-state at all electron energies (Fig. 6.68), Q_{00} of the B-state (of 1.8×10^{-18} cm^2) is smaller than Q_{00} of the C-state (of 6.2×10^{-18} cm^2 from Table 6.12) by a factor of about 3.5. This is in qualitative agreement with unpublished calculations by

Spindler (mentioned by Gerry[184]) that the cross-section of the $C^3\pi_u$ state is a factor of 10 larger than that of the $B^3\pi_g$ state. The assumption being made here is that the calculations of Spindler did not refer to total cross-sections of each state but to excitation of the $v = 0$ level of each state. If it was otherwise, it would be in disagreement with the cross-sectional data of Zapesochnyi and Skubenich[204] presented in Fig. 6.68.

The cross-section data presented above and in Table 6.12, indicate that the only cross-sections of the C-states (low vibrational-state quantum numbers) larger than those for excitation of vibrational levels of the B-state are:

$$Q_{00}(C^3\pi_u) = 6.2\times10^{-18} \text{ cm}^2 > Q_{01}(B^3\pi_g) = 4.5\times10^{-18} \text{ cm}^2;$$
$$> Q_{00}(B^3\pi_g) = 1.8\times10^{-18} \text{ cm}^2;$$

and

$$Q_{01}(C^3\pi_u) = 3.5\times10^{-18} \text{ cm}^2 > Q_{00}(B^3\pi_g) = 1.8\times10^{-18} \text{ cm}^2.$$

Based on the relationships of these cross-sections, population inversion between vibrational levels of the $C^3\pi_u$ and $B^3\pi_g$ states of N_2 that could be produced by electron-impact excitation from the N_2 ground state during the risetime of the discharge current pulse should only be possible between

$$C^3\pi_u(v' = 0) \quad \text{and} \quad B^3\pi_g(v'' = 0 \text{ and } 1),$$
$$C^3\pi_u(v' = 1) \quad \text{and} \quad B^3\pi_g(v'' = 0).$$

The fact that oscillation has been observed on only the $v' = 0$ to $v'' = 0$ and 1, and $v' = 1$ to $v'' = 0$ transitions on which population would be predicted, in spite of a large amount of experimental work, would indicate that the main assumption that excitation of the C- and B-states is principally by direct electron impact excitation from the $N_2 X^1\Sigma_g^+$ ($v = 0$) ground state is a valid assumption.

The relative probability of emission in the $C^3\pi_u$–$B^3\pi_g$ band system is directly proportional to the product of the excitation cross-section Q_{0v}, and the Franck–Condon factors for emission $q_{v'v''}$. Values of $q_{v'v''}$ for the $C^3\pi_u$–$B^3\pi_g$ transition with $v' = 0, 1, 2, 3$, and 4 and $v'' = 0$, 1, 2, 3, 4, and 5 are given in Table 6.15.[205] They are also displayed in a block diagram in Fig. 6.69[206] where the largest values of the transition probabilities are more readily seen than in the table.

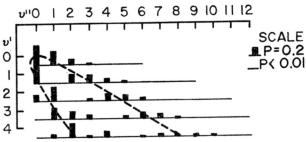

FIG. 6.69. Block diagram of the vibrational transition array, for the second positive band system $(C^3\pi_u$–$B^3\pi_g)$ of N_2, together with a Condon parabola showing the strongest transitions. (After W. R. Jarmain and R. W. Nicholls. Reproduced by permission of the National Research Council of Canada from the *Canadian Journal of Physics* 32, 201–4 (1954).)

TABLE 6.15. *Franck–Condon factors for low vibrational quantum numbers of the second positive* $(C^3\pi_u - B^3\pi_g)$ *band system of* N_2[205]

Upper state v'	Lower state v''					
	0	1	2	3	4	5
0	0.449*	0.329*	0.147	0.052	0.016	0.005
1	0.390*	0.019	0.204	0.200	0.112	0.048
2	0.135	0.322*	0.033	0.060	0.161	0.143
3	0.024	0.252	0.163	0.118	0.002	0.089
4	0.002	0.070	0.303*	0.048	0.157	0.014

* Indicates relatively large Franck–Condon factor.

The transitions with the largest Franck–Condon factors (or transition probabilities) are: $v' = 0$ to $v'' = 0$, $v' = 1$ to $v'' = 0$, $v' = 0$ to $v'' = 1$, $v' = 2$ to $v'' = 1$, and $v' = 4$ to $v'' = 2$. Combining some of these $qv'v''$ factors with the excitation cross-sections $Q_{0v'}$ given in Table 6.12 appropriate to each upper vibrational level to give the $Q_{0v'} q_{v'v''}$ products in order to find the strongest lines, the relative probabilities of emission are proportional numerically to: 2.8 for $v' = 0$ to $v'' = 0$; 2.0 for $v' = 0$ to $v'' = 1$; 1.6 for $v' = 1$ to $v'' = 0$; and 0.9 for $v' = 0$ to $v'' = 2$. The rather rapid decrease in the $Q_{0v'} q_{v'v''}$ products is due primarily to the rapid decrease in the excitation cross-sections $Q_{0v'}$ with increasing vibrational quantum number. The excitation rates $q_{0v'}$, of course, vary in the same way as Q_{0v} with vibrational quantum number.

The first three transitions that have the largest $Q_{0v'} q_{v'v''}$ products are just those transitions, the only three transitions, on which oscillation has been reported, and on which oscillation would have been first predicted assuming that electron impact excitation was the primary excitation process of the upper $C^3\pi_u$ state and the lower $B^3\pi_g$ state of the ultraviolet N_2 laser transition.

Additional evidence to support the belief that direct electron-impact excitation from the N_2 ground state is the excitation process of the $C^3\pi_u$ and $B^3\pi_g$ states, and not "collisions of the second kind" between molecules in the metastable a' $^1\pi_g$ state,[3, 207] is the behavior of N_2 ultraviolet (and IR) lasers with temperature.[179] It has been found that cooling axially excited N_2 lasers to liquid-air temperatures gives substantial increases in laser output over that obtainable at room temperature. The effect is not due to any increased molecular ground state concentration N_0 at the lower temperature at a fixed pressure that follows from the relation $N_0 = p/kT$, as the output maximizes at both temperatures at one value of N_0 (illustrated in a later figure, Fig. 6.81). If "collisions of the second kind" were responsible for selective excitation of the C- and B-states, it would be expected that the laser intensity would decrease as the collision frequency decreased with decrease in gas temperature (as in the He–Ne laser). This is contrary to what is observed in the N_2 UV and IR lasers.

Electron energy distributions and excitation rates in the N_2 $C^3\pi_u - B^3\pi_g$ *laser.* Ali[208] has determined the laser power density and its time history and the effect of various discharge-circuitry parameters on the electron temperature and electron density in the N_2 UV laser.

It is important to note that the laser power density P is essentially proportional to the excitation rate of the upper laser level, i.e. $P \propto X_{0c} = n_e\langle\sigma v\rangle$, where X_{0c} is the excitation rate of the $C^3\pi_u$ state from the ground state, and $n_e\langle\sigma v\rangle$ is the product of the electron concentration n_e and the velocity-averaged excitation cross-section. The velocity-averaged cross-section is a function of the electron temperature.

The electron density, at a particular time following the initiation of the excitation current pulse in a discharge is determined by the available energy from the source (e.g. the charged capacitor, delay-line, or transmission-line), the charge voltage, and the circuit inductance that limits the rate of rise of the discharge current and hence the rate of rise of the electron density. The electron temperature too is determined mainly by the circuit inductance.

It does not follow *a priori* that a high electron temperature is required in the nitrogen laser. It only follows that a high electron temperature is required in the ultraviolet nitrogen laser because of the electron energy loss rates in N_2 as a function of electron temperature in excitation of upper and lower laser levels and vibrational excitation of the ground state, and because of electron energy loss rates in ionization that is required to provide electrons that are used subsequently for excitation processes.

Figure 6.70 shows electron energy loss rates in molecular nitrogen as a function of electron temperature. For electron temperatures below 4 eV, the rate of energy loss by electrons into excitation of vibrational levels of the ground state predominates over the other (desirable) loss processes of excitation of the $C^3\pi_u$ upper laser level and ionization of N_2 molecules. Further, because of the similar shape of the curves of electron energy loss rate of excitation of the $C^3\pi_u$ and $B^3\pi_g$ states, the shape of ionization curve as a function of electron temperature, and their relation to each other with respect to the electron energies covered in Fig. 6.70, as high an electron temperature above 4 eV as possible is desirable for high efficiency. As both cross-sections for excitation of the C- and B-states decrease rapidly for electron ener-

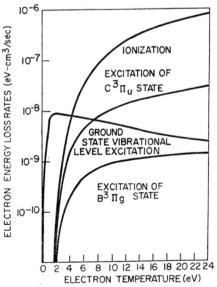

FIG. 6.70. Electron energy loss rates in N_2 as a **function of** electron temperature. (After A. W. Ali,[208] by courtesy of the **Optical** Society of America.)

gies (not electron temperatures) above about 18 eV, while the cross-section for ionization of $A^2\pi_g \, N_2^+$ continues to increase up to a broad maximum at approximately 60 eV, it is to be expected that this laser system will have some optimum electron temperature for operation under combined simultaneous discharge conditions of excitation and ionization.

The time history of the electron temperature early in a pulsed discharge in N_2 for different discharge-circuit inductances is shown in Fig. 6.71. To produce electron energy distributions with electron temperatures much above 4 eV, the figure indicates that circuit inductances should preferably be less than 10 nH. In practical, transient-laser-discharge circuits that incorporate simple spark-gaps (Fig. 4.51), this is not easy to achieve as a single-channel spark gap can have a typical inductance of 20 nH. To reduce the circuit inductance

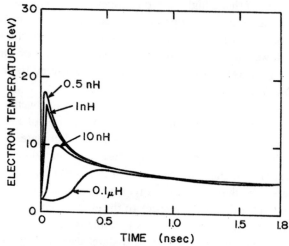

FIG. 6.71. Time history of the electron temperature in a pulsed nitrogen discharge for different values of circuit inductance. (After A. W. Ali,[208] by courtesy of the Optical Society of America.)

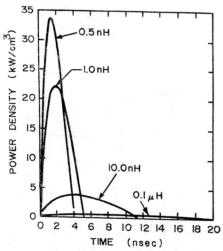

FIG. 6.72. N_2 laser UV power density as a function of time for several different values of circuit inductance. The peak power density (and upper level excitation rate) rises with decreasing inductance values (or decreasing current pulse risetimes). (After A. W. Ali,[208] by courtesy of the Optical Society of America.)

it becomes necessary to utilize multi-channel, triggered, pressurized spark-gaps or more complicated discharge circuitry such as that illustrated in Fig. 6.40.

The effect that circuit inductance has on the laser power density (and the rate of excitation of the upper laser level) as a function of time is indicated in Fig. 6.72. The way the power density (and excitation rate) is affected by the circuit inductance is a result of the combined effect the inductance has on the electron density (Fig. 6.73) and electron temperature (Fig. 6.71) as a function of time. Once again it is made clear that the power density and excitation rate is strongly affected by the magnitude of the circuit inductance.

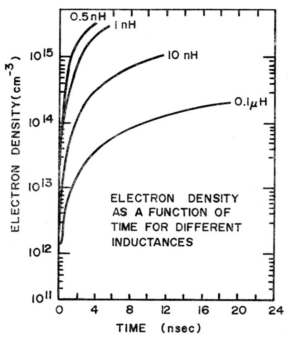

FIG. 6.73. The effect of the current pulse risetime on the production of electrons is indicated for different values of circuit inductance. (After A. W. Ali,[208] by courtesy of the Optical Society of America.)

The behavior of the power density as a function of time for different initial capacitor charge voltages is shown in Fig. 6.74. The laser power density is seen to increase with increasing charge voltage. This can be attributed (at a fixed nitrogen pressure) to an increased E/p that raises the electron temperature, and to a faster production rate of electrons at higher charge voltages (fig. 8 of ref. 208). The power density and excitation rate of the upper laser level is affected only slightly by the capacitance of the circuit for at least the range 0.01 to 5 μF (fig. 9 of ref. 208).

Finally, the dependence of the laser power density (and upper level excitation rate) and its duration on the N_2 fill-pressure is shown in Fig. 6.75. The peak power density is higher and occurs earlier, while the duration of the laser pulse is shorter with increasing N_2 fill-pressure. The energy density (power density integrated over the laser-pulse duration) has an optimum value at a fill-pressure of 28 torr. This is in agreement with experimentally observed optimum values of 20–30 torr.[183, 186]

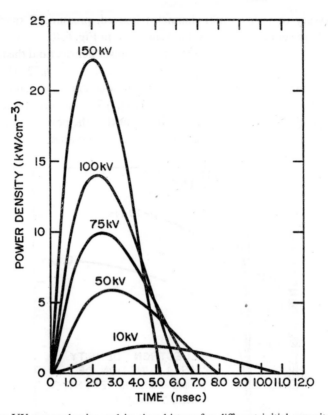

FIG. 6.74. N₂ laser UV power density and its time history for different initial capacitor charge voltages. (After A. W. Ali,[208] by courtesy of the Optical Society of America.)

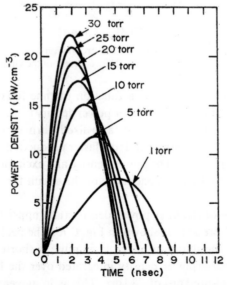

FIG. 6.75. N₂ laser UV power density as a function of time for different N₂ fill-pressures. (After A. W. Ali,[208] by courtesy of the Optical Society of America.)

Operating characteristics and discharge conditions. Some of the operating characteristics have been covered already in the preceding section in Figs. 6.72, 6.74 and 6.75 where the calculated laser power density as a function of time is shown for various circuit-inductance values, charge voltage, and N_2 fill-pressure, respectively.

Axial and transverse electric field excitation has been employed in investigations of the ultraviolet N_2 laser system. The investigations have generally taken two forms: (1) Analytical and spectroscopic studies using axial excitation in conventional, narrow-bore discharge tubes at low nitrogen pressures of up to a few torr,[180-2] and (2) studies based on realizing high output power using transverse excitation at relatively high nitrogen pressures,[183, 184, 186-9] and at pressures up to atmospheric pressure.[190-1]

At this time it is not yet clear if the excitation conditions in discharges using both forms of excitation (axial or transverse) are similar. It is clear, however, that high electric field values are required for producing laser oscillation on the $C^3\pi_u-B^3\pi_g$ transition with both forms of excitation.[180, 183] The magnitude of the electric field depends on the nitrogen gas pressure in such a way that the E/p does not remain constant but decreases as the gas pres-

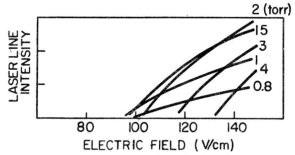

FIG. 6.76. Variation of relative intensity of molecular $C^3\pi_u-B^3\pi_g$, $0 \to 0$ nitrogen laser lines as a function of electric field for various fill-pressures. (After T. Kasuya and D. R. Lide, Jr.,[180] by courtesy of the Optical Society of America.)

sure increases. The variation of the intensity of the 0–0 band of the ultraviolet laser as a function of the applied electric field at various gas pressures using axial excitation is shown in Fig. 6.76. Whereas the laser intensity at a particular gas pressure increases with increasing electric field, to produce the same intensity the E/p is 55 V/cm-torr at 2 torr but is more than 110 V/cm-torr at 1 torr. Likewise, the threshold electric field for oscillation decreases as the gas pressure is increased. It decreases approximately linearly from about 75 V/cm-torr at 1.2 torr to about 34 V/cm-torr at 4 torr.[180]

All indications are that the E/p in ultraviolet N_2 lasers for giving maximum output power should be as high as possible, subject to uniform, non arc-like discharges being formed. This would suggest that output power is limited by the ionization that is produced. If this is the case, some limit would be expected to be set on the optimum E/p as ionization efficiency falls off at high values of E/p. This limiting value would not occur in nitrogen at E/p values of less than 10^4 V/cm-torr.[209]

In 1965 Leonard[183] reported the development of an ultraviolet transversely excited N_2 laser that achieved E/p values of 200 V/cm-torr at pressures from 1 torr to several tens of torr of nitrogen. Over this range the output power was found to be approximately proportional to E so that high electric fields were clearly desirable. The optimum nitrogen pres-

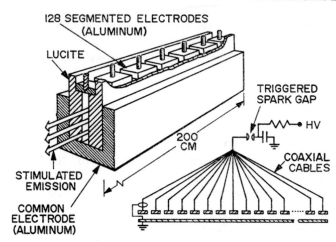

FIG. 6.77. Schematic of crossed-field laser structure in which the electric field is transverse to the stimulated emission or laser axis. Circuit parameters typically: $L = 100$ nH; $C = 0.03$ μF, initial capacitor voltage 15 to 25 kV. The discharge channel is 2.5 cm by 0.3 cm in cross-section. (After D. A. Leonard,[183] by courtesy of the American Institute of Physics.)

sure was approximately 20 torr, and the system has an "optimum" E/p of about 200 V/cm-torr. This optimum was in fact the maximum that could be achieved before problems of discharge instability degraded the laser output. In the transverse excitation technique developed by Leonard, high-voltage excitation pulses applied across the discharge channel provided the high electric fields and produced numerous short discharges transverse to the laser axis. By applying the electric field perpendicular to the laser axis, it becomes possible to construct discharge channels of arbitrary length with high electric fields without the need for increasing the charge voltage as would be the case for an axially excited laser system. A schematic of Leonard's original "crossed-field" laser structure is shown in Fig. 6.77. The segmented electrodes were introduced in order to ensure uniform current density along the entire length of the discharge channel and to permit the active length to be varied. The segmented electrodes are connected by coaxial cables through a triggered spark gap to a capacitor. Later it was found that the current distribution was essentially inductance controlled, and extremely uniform discharges could be produced along the entire length of the channel with an unsegmented upper electrode but with it still fed through numerous low-inductance transmission lines (coaxial cables).[210] With the arrangement shown in Fig. 6.77, laser pulses of 200 kW and 20-nsec duration have been produced.[184] If this laser was axially excited, it would require voltages in the MV-range to produce the same electric field as is produced by a crossed-field voltage of 20 kV. Above an active discharge length of about 1 meter, Leonard[183] found that the laser output power varied linearly with active discharge length up to the 2-m maximum discharge length he used.

Whereas the optimum pressure was 20 torr in Leonard's laser structure, which had a 2.5×0.3-cm² discharge-channel cross-section, using a similar structure with a 1×0.3-cm² discharge channel, Kobayashi et al.[211] found that the optimum nitrogen pressure was nearly 100 torr. The maximum output power of 20 kW that they obtained at this pressure from a discharge length of 1 meter is directly comparable to that obtained by Leonard from a similar discharge length. By adding helium to the discharge, Kobayashi et al.[211] found

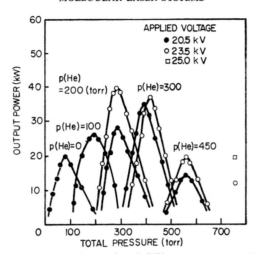

FIG. 6.78. N_2 and N_2–He pressure dependence of peak UV output power at 0.3371 μm of a crossed-field transversely excited N_2 laser similar to that shown in Fig. 6.77. (After T. Kobayashi, M. Takemura, H. Shinizu, and H. Inaba,[211] by courtesy of the IEEE.)

that the output power could be increased by more than 25 percent at an optimized total pressure of 300 torr for the same charge voltage. With a slight increase in charge voltage the output power could be doubled. Furthermore, it was found that the laser could be operated at atmospheric pressure (with an optimum N_2–He mixture ratio of 0.13) at an output power of 20 kW. This is equal to the maximum output power obtained in the absence of helium. These results of output power for different pressures of N_2 and He against total pressure from 20 to 760 torr are shown plotted in Fig. 6.78. It is postulated[211] that the beneficial effect of the helium addition follows from improving the uniformity of the discharge. It is suggested here that it could also follow from an increase in the electron density and a decrease in the current risetime both of which produce an increase in the upper-laser-level excitation rate.

6.2.3.2. Oscillation in the IR in the first positive system ($B^3\pi_g$–$A^3\Sigma_u^+$)

Oscillation in the first positive system has been observed from the four lowest vibrational levels of the $B^3\pi_g$ state on transitions to the metastable $A^3\Sigma_u^+$ state at wavelengths that range from 0.7504 μm to 1.4983 μm[3, 179] (Fig. 6.65). Low discharge-tube temperatures are needed to observe oscillation from the higher vibrational levels ($v = 4$ and 3) of the $B^3\pi_g$ state.[179] Whereas laser output powers in excess of 100 kW have been reported on the $C^3\pi_u$–$B^3\pi_g$ UV-transition, the highest output power reported for the IR laser system is the relatively low one of 500 W.[212] (*Added in proof*: kilowatts now reported.[282])

A lot of the previous treatment of the second positive system laser, in which the excitation processes and excitation rates of the $B^3\pi_g$ are discussed is relevant to the material presented here. Particular figures that relate almost as much to the IR laser as to the UV laser and should be referred to are Figs. 6.66 to 6.68 and 6.70. For sake of completeness in this discussion, some of the material given in Section 6.2.3.1 is repeated again in this section.

Upper level and lower level excitation. Emission of spontaneous emission from the $B^3\pi_g$ state immediately on the commencement of the current pulse during the breakdown period of the gas when the current is rising (similar to that shown in Fig. 6.62b for the spontaneous emission from the upper level involved in the pulsed visible CO laser[175]), indicates that the excitation is by direct electron impact from the ground state with possibly some contribution by fast radiative decay from the $C^3\pi_u$ state. The delay that occurs between the superradiant UV laser pulse and the occurrence of the nonsuperradiant but reasonably high gain IR laser pulse (illustrated in Fig. 6.66) later in the current pulse is most likely due to the time it takes the cavity to build up. It is not related to any delayed excitation processes of the upper laser levels nor to any depopulating processes that involve the lower laser level.

Referring to the N_2 molecular potential diagram of Fig. 6.65, it is clear that due to the displacement of the equilibrium internuclear potential of the $A^3\Sigma_u^+$ lower laser level and the $X^1\Sigma_g^+$ N_2 ground state, vertical transitions from the ground state caused by electron-impact collisions will occur predominantly to vibrational levels above $v = 4$ of the $A^3\Sigma_u^+$ state. On the other hand, we have already seen in the preceding section that electron-impact excitation of the $B^3\pi_g$ state from the N_2 ground state leads to excitation of a number of its lowest vibrational quantum states. The only way that low vibrational quantum states of the $A^3\Sigma_u^+$ lower laser level can be populated is by radiative processes from the upper laser level and by collisional processes from other states. Both processes would be much slower than the electron-impact excitation process of the $B^3\pi_g$ state. During the breakdown period of the discharge, when it is anticipated that any transient electron energy distribution is biased to high electron energies, and excitation of the $A^3\Sigma_u^+$ lower laser level (that could proceed by electron-impact excitation of molecules in high vibrational quantum states of the $X^1\Sigma_g^+$ ground state in collisions with slow electron) is insignificant, it follows that population inversion between the $B^3\pi_g$ and the $A^3\Sigma_u^+$ levels should be easily achieved.

In the case of the $A^3\Sigma_u^+$ state, the relative excitation rate of its vibrational levels can be estimated from the Franck–Condon factors $q_{v''v'}$ for the Vegard–Kaplan system ($A^3\pi_u^+ \rightarrow X^1\Sigma_g^+$) calculated by Nicholls.[205] Table 6.16 lists Franck–Condon factors for emission from the first sixteen vibrational levels v' of the $A^3\Sigma_u^+$ state to the $v'' = 0$ and 1 of the N_2 $X^1\Sigma_g^+$ ground state.

TABLE 6.16. *Franck–Condon factors for the Vegard–Kaplan* ($A^3\pi_u^+ \rightarrow X^1\Sigma_g^+$) *system of* N_2[205]

Upper state v'	Lower state v''		Upper state v'	Lower state v''	
	0	1		0	1
0	5.90 (10⁻⁴)	5.34 (10⁻³)	9	8.73 (10⁻²)	6.43 (10⁻³)
1	3.32–3	2.28–2	10	8.43–2	2.05–4
2	9.98–3	5.08–2	11	7.81–2	2.01–3
3	2.13–2	8.85–2	12	6.99–2	9.01–3
4	3.64–2	9.33–2	13	6.07–2	1.82–2
5	5.29–2	8.96–2	14	5.15–2	2.71–2
6	6.80–2	7.08–2	15	4.28–2	3.42–2
7	7.94–2	4.54–2	16	3.50–2	3.86–2
8	8.59–2	2.21–2			

Note: In the table, the minus values indicate powers to the base 10; e.g. 3.32–3 is 3.32×10^{-3}.

Table 6.16 shows that the Franck–Condon factors are small for transitions between low vibrational quantum numbers $v' = 0$ and 1 to $v'' = 0$ and 1, but that they rise to a maximum of 0.09 for $v' = 9$ to $v'' = 0$, and 0.09 for $v' = 4$ to $v'' = 1$, and decrease for higher v' values. The more extensive table given in ref. 205 shows that the largest factors occur for $v' = 0$ to $v'' = 5, 6$, and 7. This is in accord with a qualitative interpretation from Fig. 6.65 that only levels above $v = 4$ show overlap with $v = 0$ of the $X^1\Sigma_g^+$ ground state and can be excited in vertical transitions from the $X^1\Sigma_g^+$ ($v = 0$) state.

Based on the Franck–Condon factors for the Vegard–Kaplan $A^3\pi_u^+ \to X^1\Sigma_g^+$ system, excitation of the first few (0, 1, 2, 3) vibrational levels of the $A^3\pi_u$ state from $v = 0$ and 1 of the $X^1\Sigma_g^+$ ground state appears to be negligible. These first few vibrational levels of the $A^3\pi_u$ state are only likely to be increasingly populated as vibrational levels above $v = 4$ of the $X^1\Sigma_g^+$ N_2 ground state become progressively populated during the duration of the current pulse. As the electron temperature decreases, the number of vertical transitions to the $A^3\pi_u$ ($v = 0$ and 1) states can then occur with reasonable probability.

Relative population rates $q_{0v'}$ of the B-states by electron impact with molecules of N_2 ($v = 0$) of the ground state, and relative populations $N_{v'}$ that are essentially the same as the population rates, have been calculated[199]. For the first ten vibrational levels of the $B^3\pi_g$ state, the values of $q_{0v'}$ (or $N_{0v'}$) are:

$v' =$	0	1	2	3	4	5	6	7	8	9	10
$q_{0v'} =$	25	64	92	100	100	68	50	28	17	12	9

$\Sigma q_{0v'} = 565$.

The data confirm our earlier qualitative estimate that vibrational levels of the B-state above $v' = 1$ will be preferentially excited from the $X^1\Sigma_g^+$ ($v = 0$) ground state in electron-impact collisions because of the displacement of the internuclear equilibrium separations of the B- and X-states.

Making the assumption that the total cross-section of the B-state is directly proportional to the sum of the excitation rates $\Sigma q_{0v'}$ and knowing the total cross-section for excitation of the B-state from the ground state, the individual cross-sections $Q_{0v'}$ (B) for excitation of the B-state can be calculated. At an electron energy of 35 eV, the total cross-section for excitation of the B-state from $v'' = 0$ of the $X^1\Sigma_u^+$ ground state is approximately 4×10^{-17} cm^2 (from Fig. 6.68), and it follows that individual cross-sections $Q_{0v'}$ (B) are:

v'	0	1	2	3	4	5	6	7	8	9	10
$Q_{0-v'}$ (10^{-18} cm^2)	1.8	4.5	6.5	7.1	7.1	4.8	3.6	2.0	1.2	0.9	0.6

The value of Q_{00} (B) of 1.8×10^{-18} cm^2 is therefore seen to be smaller than that of Q_{00} (C) of 6.2×10^{-18} cm^2 by a factor of 3.5, and though smaller than the factor of 10 calculated by Spindler for Gerry[184] for the ratio of the cross-sections, confirms that Q_{00} (C) is larger than Q_{00} (B).

The expected intensity and occurrence of IR laser lines on the $B^3\pi_g - A^3\Sigma_u^+$ transition should follow the product of the excitation rates $q_{0v'}$ and the probability of emission $q_{v'v''}$ of the Franck–Condon factors for the transition. A partial listing of these factors for low

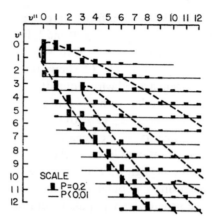

FIG. 6.79. Block diagram of the vibrational transition array together with Condon parabolae showing the strongest transitions for the first positive band system ($B^3\pi_g$–$A^3\Sigma_u^+$) of N_2. (After W. R. Jarmain and R. W. Nicholls. Reproduced by permission of the National Research Council of Canada from the *Canadian Journal of Physics* **32**, 201–4 (1954).)

vibrational quantum numbers is given in Table 6.17. They are displayed also in a block diagram in Fig. 6.79 that covers $v' = 0$ to 12 to $v'' = 0$ to 12 transitions.[206]

TABLE 6.17. *Franck–Condon factors for low vibrational quantum number of the first positive* ($B^3\pi_g$–$A^3\Sigma_u^+$) *system of* N_2[205]

Upper state v'	Lower state v''						
	0	1	2	3	4	5	6
0	0.338	0.325	0.190	0.089	0.036	0.014	0.005
1	0.406	0.002	0.103	0.178	0.145	0.086	0.044
2	0.197	0.212	0.113	0.001	0.077	0.128	0.113
3	0.050	0.299	0.039	0.162	0.032	0.001	0.069
4	0.007	0.132	0.274	0.002	0.114	0.088	0.005
5	0.001	0.027	0.211	0.181	0.048	0.043	0.106

The $q_{0v'}\, q_{v'v''}$ products (in arbitrary units) in order of decreasing magnitude for each transition, together with wavelength of the laser emission band, are given in Table 6.18.

Oscillation has been reported on all but five of the sixteen transitions listed in Table 6.18. Somewhat surprisingly, some of the transitions that have relatively high $q_{0v'}$, $q_{v'v''}$ products are transitions on which oscillation has not been observed, while the 0–0 and 0–1 transitions that have relatively small products are transitions on which oscillation is achieved readily. Unfortunately relative power measurements of the laser output on each transition under optimized conditions have not been published that would enable one to check whether or not the laser relative photon emission rates are in agreement with the $q_{0v'}\, q_{v'v''}$ products in Table 6.18.

It appears likely that the cascade superradiant ultraviolet $C^3\pi_u$ ($v' = 0$)–$B^3\pi_g$ ($v'' = 0, 1$) laser transitions have a considerable effect on the excitation rate of vibrational levels of the B-state, and subsequently affect, through interaction of the IR transitions,[182] the degree

TABLE 6.18. *Products of excitation rates and emission probabilities for the first positive* $B^3\pi_g{-}A^3\Sigma_u^+$ *system of* N_2

$q_{0v'}\,q_{v'v''}$ (arb. units)	$v'{-}v''$	Wavelength of laser bands$_{\text{vac}}$ (μm)[179, 180]
30	3–1	0.7574–0.7627
27	4–2	0.7485–0.7504
26	1–0	0.8836–0.8911
19	2–1	0.8656–0.8725
18	2–0	0.7714–0.7755
16	3–3	0.9599
13	4–1	—
11	1–3	—
11	4–4	—
10	2–2	—
9	1–4	—
8	0–0	1.0439–1.0538
8	0–1	1.2306–1.2350
7	2–4	1.3646
7	1–2	1.1933
5	0–2	1.4983

of inversion that is possible on the various transitions in the first positive system. This arises because of the inability of the metastable $A^3\Sigma_u^+$ lower laser level to decay on a time scale comparable to the lifetime of the $B^3\pi_g$ upper laser level.

Operating characteristics and discharge conditions. Information on the N_2 IR laser system is much more limited than on the UV laser system though the IR system has been operated using both axial[3, 179–80] and transverse excitation.[212–13] Just as with the N_2 laser operating in the second positive system, high electric fields are necessary for producing population inversion in the $B^3\pi_g{-}A^3\Sigma_u^+$ first positive system. This is illustrated by Fig. 6.80 in which the dependence of the optimum electric field (for producing maximum laser intensity using axial excitation) on N_2 pressure for a number of bands of the $B^3\pi_g{-}A^3\Sigma_u^+$ transition is

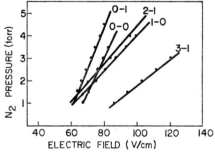

FIG. 6.80. Dependence of the optimum applied electric field on nitrogen pressure for producing maximum laser intensity of $B^3\pi_g{-}A^3\Sigma_u^+$ bands. (After T. Kasuya and D. R. Lide, Jr.,[180] by courtesy of the Optical Society of America.)

plotted. The optimum electric fields of the order of 70 to 120 V/cm (depending on the laser band) are seen to be directly proportional to the gas pressure. It has also been shown that the laser intensity increases with increasing electric field under axial excitation conditions in discharge tubes of 2 to 8-mm bore,[180] and that larger-bore tubes favor the IR laser.[179] The optimum electric fields, however, for the IR laser system are less than they are for the UV system.[179] This is also true of the current density. Allen et al.[182] found that using axial excitation at a pressure of 2.5 torr in a 4-mm-bore discharge tube, only the first positive system lased for currents between 3 and 50 A; in the range 50–100 A both first and second positive systems lased; and in the range 100–160 A the second positive system continued to oscillate and grow in intensity but many first positive system lines were either quenched or weakened. Allen et al.[182] also observed interaction between certain bands of the IR with the UV system. This they explained in terms of the strong transitions of the UV system reducing the inversion of certain bands of the IR system, and they predicted an intensification of the 0–0 band of the same system. Later work by Garavaglia et al.[194] has shown, however, that all the bands of the IR system behave in the same way in disagreement with this prediction.

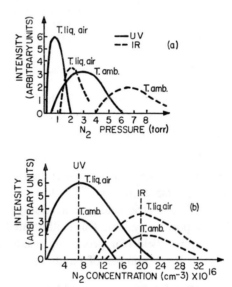

FIG. 6.81. N_2 laser IR and UV intensities as a function of nitrogen (a) pressure, and (b) N_2 molecular concentration, at two temperatures. (After C. A. Massone, M. Garavaglia, M. Gallardo, J. A. E. Calstroni, and A. A. Tagliaferri,[179] by courtesy of the Optical Society of America.)

The variation of laser intensity on gas pressure and discharge-tube temperature for the IR laser (and also the UV laser) is shown in Fig. 6.81 (a) and (b). The optimum pressure at room temperature for the IR laser can be seen to be higher (at approximately 6.5 torr) than that of the UV laser (at about 3 torr). At the temperature of liquid air, the optimum pressures of both are reduced with the result that a constant optimum N_2 molecule concentration for each system is maintained. At room temperature with transverse excitation, oscillation can be achieved on the 0–0 band of the $B^3\pi_u - A^3\Sigma_u^+$ transition in pure nitrogen over the pressure range of 20 to 90 torr (with an optimum pressure of 45 torr), and anode–cathode separation of 2 cm, and an applied voltage of between 10 to 25 kV.[212] The E/p values used here in transverse excitation to obtain oscillation on the 0–0 band (of about 200 V/cm-torr) are considerably higher than the optimum E/p values of between 70 V/cm-torr at 1 torr and 90 V/cm-torr at 4 torr for producing oscillation on the same band using axial excitation.[180]

Again, as with the $C^3\pi_u$–$B^3\pi_u$ laser system, low inductance circuitry and short risetime excitation-current pulses are needed to give high output powers in the $B^3\pi_u$–$A_3\Sigma_u^+$ infrared laser system.[282]

6.2.4. NOBLE GAS LASERS

Stimulated emission in the range 0.1600 to 0.1700 μm has been reported by Basov *et al.*[214–15] following excitation of liquid xenon with energetic electrons from an electron gun source.[284] The stimulated emission occurs in a broad band as a result of transitions from a stable excited state of Xe_2 to the ground state of Xe_2, which is an unstable repulsive state. The attraction in this molecular system and in other noble-gas systems lies in the possibility of high efficiency of operation with small active volumes, and operation in the vacuum-UV region of the spectrum. In the case of Xe_2 the maximum conversion efficiency, which is limited by the maximum energy that can be delivered to Xe_2, is 65 percent.[216]

As the molecular ground states of Xe_2 is repulsive except for a small shallow potential well at large internuclear separations that does not correspond with the equilibrium separation of the stable state of Xe_2 from which vertical transitions to the ground state can occur, the upper laser level cannot be populated directly by electron-impact excitation from the ground state. It is populated by collisions of metastable and ground-state Xe atoms in the molecular formation reaction:

$$Xe^* + Xe \rightarrow Xe_2(^1\Sigma_u^+, {}^3\Sigma_u^+). \tag{6.32}$$

This reaction is similar to the molecular-helium formation reaction

$$He^* \, 2^1S_0 + He \, {}^1S_0 \rightarrow He_2 \, {}^1\Sigma_u^+$$

(mentioned in Chapter 4 in the discussion of the He–Ne laser) that has a small potential barrier at an internuclear separation at about 2 Å (Fig. 4.19).

The metastable Xe^* atoms are produced as a result of recombination of Xe^+ ions, and radiative-cascade processes from excited states of xenon produced by impact of ground-state Xe atoms with high energy and secondary electrons. It is the formation of the metastable Xe^* atoms and the subsequent formation of Xe_2 that limits the maximum conversion efficiency at high atom (liquid) densities of xenon to 65 percent.

A generalized, partial molecular potential diagram of the noble gases showing the band of radiative transitions (broken lines) on which stimulated emission is observed is given in Fig. 6.82. It can be seen that although the inversion is produced between the stable state and the ground state, the ground state is repulsive at the position of the internuclear equilibrium separation of the stable molecular state A_2' and thus has practically zero population that could limit the population inversion.

In theory, even though inversion is being produced between an excited state and the ground state, CW operation should be possible *if* molecules in the stable Xe_2 state can be

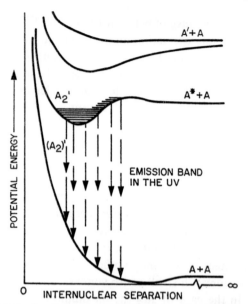

FIG. 6.82. Generalized diagram of molecular potentials of the noble gases showing the UV emission band between the stable A_2' excited st ate and the repulsive $(A_2)'$ ground state. The equilibrium internuclear separation of the A_2' state occurs between 0.1 to a few tenths of a nanometer. The potential well in the lower curve is much shallower than as drawn

produced quickly enough to supply the fast radiative decay in the presence of stimulated emission of the ultraviolet laser transitions.

Molecular band systems arising from transitions to the ground state of the type shown in Fig. 6.82 occur in: xenon at 0.1470 to 0.2250 μm; krypton at 0.1250 to 0.1850 μm; argon at 0.1070 to 0.1650 μm; neon at 0.0745 to 0.1000 μm; and in helium the Hopfield continuum at 0.0600 to 0.1100 μm.[217-18]

In normal discharges in the noble gases at pressures as low as a hundred torr, molecular emission in the vacuum-UV in these bands can be the dominant emission.[217-20] At pressures of 800 torr (of xenon), using 250 keV electrons to produce excitation, nearly 50 percent of the electron energy is emitted as molecular radiation.[216]

The excitation and emission model described here is applicable to other van der Waals molecular systems such as Hg, Cd, Zn and the alkaline earths.[216] With Hg, the possible laser-band emission from Hg_2 would be in the visible at 0.4800 μm. Spectra in I_2 and Br_2 apparently also arise from transitions from stable excited states to unstable states arising from normal atoms,[221-2] so that these two are possible candidates for excitation in the manner described.

6.2.5. FAR-INFRARED H_2O LASER

Following the first report of stimulated emission in the far-infrared from pulsed discharges in water vapor and deuterium oxide by Mathias and Crocker[223] in 1964, oscillation in water vapor has been observed on more than 100 pulsed lines and over a dozen CW lines that range from 2.28 μm to 220 μm.[13, 224] The lines are tabulated in the index.

364

The general time-behavior of laser pulses from H_2O is that the pulses occur during the excitation current pulse and cease shortly before, or shortly after, the end of the current pulse (at pressures of approximately 0.2 torr)[225–6] and lengthen at lower pressures where CW operation of some lines can be achieved. The addition of helium is found to improve the power output under both pulsed and CW operating conditions.[227]

6.2.5.1. Upper level excitation

Of the four far-infrared lasers, H_2O, HCN, SO_2 and H_2S, which will lase under both pulsed and CW-discharge conditions, details of the excitation mechanisms involved in the upper level excitation have only been investigated with H_2O. Even though it has not been established with certainty by which mechanism or mechanisms the H_2O laser is excited,[228] the most probable mechanism that fits the pulsed output behavior appears to be that of selective excitation of the upper laser level by direct electron impact (Pichamuthu et al.[229]). The studies by Pichamuthu et al.[229] were restricted to the observation of the time behavior and pressure dependence of the H_2O laser output at 28 μm, one wavelength of the many H_2O laser lines, with respect to the current pulse. Apart from where cascade transitions are clearly involved in upper-laser-level selective excitation of a number of the many laser transitions, the general characteristic time behavior of the H_2O laser lines indicates that direct electron impact excitation is also the excitation mechanism involved in other transitions.

The majority of the identified H_2O laser lines involve vibrational–rotational transitions in the $\nu_1(100)$ to $2\nu_2(020)$, $\nu_3(001)$ to $2\nu_2(020)$ vibrational transitions of the H_2O molecule.

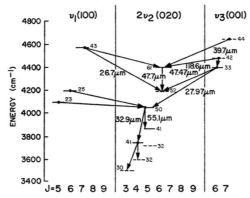

FIG. 6.83. Partial energy-level diagram for H_2O^{16} showing in particular the 27.97-μm transition and a number of other observed H_2O laser transitions. Each level is identified by the quantum numbers K_a and K_c. The vibrational state and J value are indicated by the column in which the level is found. (Adapted from ref. 13.)

The 27.97-μm line is a vibrational–rotational transition of the $\nu_3(001)$ to $2\nu_2(020)$ transition, with the upper laser level about 4400 cm^{-1} (about 0.55 eV) above the (000) ground state (Fig. 6.83).

The observed laser-output wave-shapes are shown in Figs. 6.84(a) and 6.84(b) for typical low- and high-pressure operation. The discharge was excited by a 10-kV pulser, that could provide 0 to 100-μsec, rectangular 2- to 3-A current pulses with a risetime of less than 1 μsec. The laser output shows a sharp rise followed by an exponential decay whose time constant

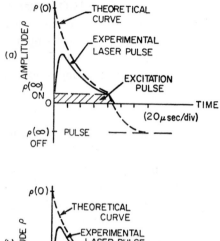

FIG. 6.84. Observed 28-μm H_2O laser output wave-shapes (solid curves) for typical low-pressure (a) and high-pressure operation (b), compared with theoretical curves (broken curves). (a) $p = 0.1$ torr, $\tau_D = 20$ μsec, current turned off at t_e. (b) $p = 0.2$ torr, $\tau_D = 10.5$ μsec. (After J. P. Pichamuthu, J. C. Hassler, and P. D. Coleman,[229] by courtesy of the American Institute of Physics.)

is τ_D. The value of $p\tau_D$ was estimated to be 1.9 ± 0.4 μsec-torr.[229] As the H_2O pressure is increased, the output pulse terminates as in Fig. 6.84(b), but $p\tau_D$ remains constant. If the current excitation pulse is shortened so that it ends during the laser pulse, the laser pulse stops decaying exponentially and follows the current pulse. The laser pulse then follows the solid experimental curve in Fig. 6.84 a. If excitation occurred through cascading from higher-lying vibrational levels, a different pulse shape and a slower decay after the current pulse was terminated would be expected.[229]

A comparison is made in Figs. 6.84 a and 6.84 b of theoretical calculations of the laser output behavior, based on the basic model that the upper laser level is (1) excited by direct electron collisions with ground state H_2O molecules, (2) can decay radiatively to the ground state as well as through the laser transition, and (3) that electron impact excitation of the lower laser level is negligible. A solution of the rate equations involved leads to the relationship[229]

$$\varrho(t) = [\varrho(0) - (\infty)] \exp(-t/\tau_D) + \varrho(\infty) \tag{6.33}$$

where ϱ is the laser amplitude. This relationship describes the behavior of the laser output. Representative solutions of it are plotted as theoretical curves in Fig. 6.84. It is seen that there is reasonable agreement between the decaying portions of the experimental laser-output curves and relationship (6.33). It is concluded[229] that the simple, selective, direct-electron-impact-excitation model adequately explains the behavior of the pulsed H_2O laser, and (by considering the case in eqn. (6.33) where $\varrho(\infty) > 0$) explains the behavior of CW operation.

6.2.5.2. *Operating characteristics and discharge conditions*

Pulsed operation. The water vapor laser has been operated at peak output power levels of 5 kW (in all lines) at an average power of 0.25 W.[231] The 28-μm line, which has been operated at peak powers well in excess of 1 kW,[224, 231-2] is the strongest line in the H_2O laser spectrum. Pulsed output power in excess of 1 kW has also been reported on the 26.7- and 33.03-μm lines.[231] In the realization of high-power operation at the 5-kW peak power level, McFarlane and Fretz[231] used a 75-mm-diameter discharge tube with an active discharge length of 3.9 meters that was axially excited with an excitation current pulse that had a width of 2 μsec and a peak value of 1600 A. Water vapor was continuously pumped through the tube at a system pressure of approximately 1 torr. The 28-μm laser pulse had a pulse length of 1.3 μsec and a pulse energy of 1.9 mJ, and the 26.7-μm line a pulse duration of 0.7 μsec and a pulse energy of 1.12 mJ[231] under these discharge conditions.

The variation of peak pulsed gain of the 28- and 118.6-μm H_2O lines with vapor pressure, and the peak gain of the 27.97-μm line versus peak discharge current at an H_2O pressure

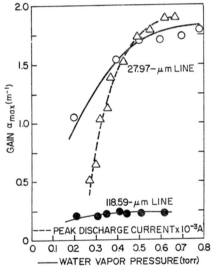

FIG. 6.85. Peak pulsed gain of the 27.97- and 118.59-μm H_2O laser lines as a function of H_2O pressure, and peak gain of the 27.97-μm line versus peak discharge current at an H_2O pressure of 0.75 torr. (After W. Q. Jeffers and P. D. Coleman,[233] by courtesy of the American Institute of Physics.)

of 0.75 torr, is given in Fig. 6.85. The peak gains were measured with the discharge current adjusted to give maximum gain. The optimum current increases with increasing pressure, and at low pressure the gain saturates and decreases with increasing current. The results of the peak gain as a function of peak current at the highest pressure indicate that to obtain maximum pulsed gain the highest pressure should be used together with the (high) optimum discharge current.[233]

CW operation. Information on CW operation of the H_2O laser is very limited although there have been reports of the operation of CW systems (Flesher and Muller,[234-5] Frenkel *et al.*,[236] and Dyubko *et al.*[237]).

In the Flesher and Muller work,[234-5] continuous oscillation at the microwatt power level of a number of submillimeter laser lines was achieved using gaseous compounds of C, N, and H or D, or NH_3 with and without the addition of water vapor or helium. In the region between 107.71 and 336.83 μm, ten laser lines were obtained with d.c. operation. Six of these lines have been identified subsequently as HCN or DCN lines: two as D_2O lines (107.7 and 171.7 μm), and two as the 118.6 and 220.2-μm lines of H_2O.[13] Oscillation of the 118.6-μm H_2O line occurred simultaneously with the 337 and 331-μm HCN lines in continuously flowing mixtures of $(CN)_2$ and H_2O with $(CN)_2$ at 0.07 torr, and H_2O at 0.15 torr. The discharge tube had a 75-mm bore and a length of 2.15 meters; discharge currents of 0.6 to 1.8 A were used. It was not reported what the optimum discharge current was.

Dyubko et al.[237] improved considerably on the μW-power outputs obtained in the previously described work in which the emphasis was on HCN and DCN (and not H_2O) laser oscillation at wavelengths at about 100 μm, and obtained 3 mW of power at 27.9 μm, 0.15 mW at 79 μm, and 0.2 mW at 118 μm. Their laser had a 70-mm bore and a length of 3 meters and continuous gas flow was utilized. The optimum discharge current for the three H_2O laser lines was 500 mA. The optimum water-vapor pressure was not reported, but it was likely to be about 0.4 torr. This pressure was used in a similar-sized discharge tube that had a bore of 75 mm and gave a CW output of 0.1 mW of laser power at 118.6 μm (Frenkel et al.[236]). The 27.9-μm line had a higher gain than the 79-μm or 118-μm laser lines that needed longitudinal resonator-mirror scanning to make them oscillate, as did the other longer wavelength lines.[237]

6.3. Line Absorption (or Optically Pumped) Lasers

The advent of extremely high-power lasers of both the gas and crystalline-host type, and high-power parametric oscillators (tunable lasers) has led to extensive use of optical pumping or line absorption as an upper laser level selective excitation mechanism of molecular species.

High power is of course not the only requirement for effective use of the optical pumping technique; just as important is the requirement for a close coincidence between the incident (pumping) radiation and that of the resonance transition (Chapter 2, Section 2.8). Fortunately, a number of high-power molecular lasers that operate at the megawatt power level under pulsed conditions (HBr, CO_2, N_2O, HF), and can operate over a wide range of wavelengths at discrete wavelengths using simple frequency-selection techniques (such as the use of piezoelectrically scanned cavity mirrors, and diffraction gratings) have outputs that match the resonance radiation of a number of molecules and so can be useful optical pump sources. They can also operate at photon conversion efficiencies in excess of 1 percent.

Actually, the extremely high megawatt pulsed power levels of the molecular lasers are not essential for producing laser oscillation at photon conversion efficiencies of about 1 percent. For example, oscillation has been achieved using input pulse energies of only a few mJ at a power level of about 6 W, and peak output powers of 1 W have been obtained on a single line (0.618 μm) at 0.5 per cent conversion efficiency from molecular iodine vapor pumped with a Q-switched Nd : YAG laser operating at various second-harmonic wavelengths.[238] In the case of optically pumped molecular iodine, oscillation has been observed on more than 150 individual lines that span the spectral region 0.544 to 1.355 μm, and it is anticipated that approximately 10^6 individual laser transitions may be excited.[238]

As well as covering the visible to near-infrared spectral region, optically pumped laser systems now provide coherent sources of far-infrared radiation at wavelengths between 50 to more than 1220 μm,[239-40] and it appears likely that they will replace the H_2O and the HCN lasers that have been the main sources of coherent radiation in this far-infrared and sub-millimeter wavelength region.

A consideration of three laser systems serves to illustrate upper level selective excitation by optical pumping in molecular species. The three systems described are: the CO_2 laser that is optically pumped with the 4.23-μm line of an HBr laser; the NH_3 laser, pumped with the 10.78-μm line of an N_2O laser; and the I_2 laser that is pumped with the second harmonic output of three Nd : YAG-laser lines at 530.6, 532.0, and 536.7 nm.

6.3.1. 10.6-μm CO_2, HBr-LASER-PUMPED LASER

The 10.6-μm CO_2 laser has been successfully optically pumped in two ways:

(1) by incoherent radiation from a CO–air flame-source that emitted 4.23-μm radiation to produce ~ 1 mW of CW output power (Wieder[241]);
(2) by the 4.23-μm line of a high-power transversely excited atmospheric pressure (TEA) HBr laser to give more than 80 W of pulsed output power (Chang and Wood[242]).

Figure 6.86 shows a greatly simplified vibrational energy-level diagram for CO_2 that includes pertinent energy levels and excitation pumping paths. As we have seen in Section 6.1.1, the 00^01 upper level of the 10.6-μm CO_2 line is normally excited in a gas discharge by direct electron-impact excitation from the 00^00 ground state of CO_2, or by V–V energy transfer from excited N_2 molecules. An alternative method of excitation is by optical pumping on the 00^01 to 00^00 transition near 4.3 μm, shown as a vertical transition R(20) in Fig. 6.86.

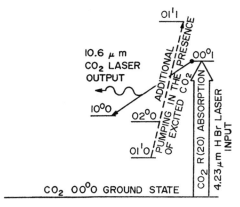

FIG. 6.86. Energy levels relevant to the optically pumped 10.6-μm CO_2 laser, together with the main 4.23-μm optical pump path from the 00^00 ground state to the 00^01 upper laser level, and an additional pump path from vibrationally excited CO_2 molecules to the 01^11 state. (Adapted from a figure given in ref. 242.)

369

The HBr laser, which oscillates on eighteen wavelengths between 4.0 and 4.6 μm with a peak power of 32 kW,[243-4] provides the optical pump, though it has been found that only two HBr lines are absorbed by low-pressure CO_2 gas. The $P(6)$ line of the 2–1 band of HBr (at 4.23 μm) is strongly absorbed, with the most absorption occurring on the $R(20)$ line of the $00^00 \rightarrow 00^01$ vibrational band of CO_2.

Figure 6.87 shows an oscilloscope trace of the observed output pulses from an optically pumped CO_2 laser.[242] The first pulse is the 4.23-μm pump pulse from the HBr TEA laser,

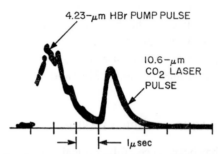

4.23-μm HBr PUMP PULSE

10.6-μm
CO_2 LASER
PULSE

1μ sec

FIG. 6.87. Observed output pulses from a 10.6-μm CO_2, 4.23-μm HBr-pumped laser. The first pulse is the pump pulse, the second is the 10.6-μm CO_2 pulse. (After T. Y. Chang and O. R. Wood II,[242] by courtesy of the American Institute of Physics.)

and the second pulse is the 10.6-μm CO_2 laser pulse delayed by about 26 μsec from the peak of the 4.23-μm pump pulse. The amplitude of the CO_2 laser pulse decreases monotonically with increasing CO_2 pressure in the CO_2 absorption/laser cell. For the CO_2 laser pulse there is an optimum pressure for which the pulse amplitude is a maximum and the pulse delay is a minimum. As the absorption/laser-cell pressure is varied away from the optimum value, the pulse amplitude decreases and the pulse delay increases, and finally oscillation ceases at both high- and low-pressure values. Nevertheless, oscillation can be obtained from CO_2 at atmospheric pressure.[242]

The optimum pressure for producing maximum output depends on the length of the absorbing cell, and the tightness of the focusing of the HBr pump beam into the absorption/laser cell.

Chang and Wood[242] report nearly unity quantum efficiency for the 10.6-μm CO_2, HBr-pumped laser system. They observed that nearly 40 percent of the input power at 4.23 μm (measured to be 260 W in their experimental arrangement) appeared in the 10.6-μm laser output. With conditions optimized for maximizing the conversion efficiency, there is a negligible delay between the pump pulse and the 10.6-μm CO_2 laser pulse showing that the delay is caused by the time it takes for the cavity to build up and overcome system losses prior to oscillation. Unlike normal electrically excited CO_2 lasers, relaxation of lower laser levels is not involved before inversion can occur so that the CO_2 laser pulse can occur promptly.

Optical pumping of vibrationally excited CO_2 molecules has also been observed to lead to inversion on the $00^01–10^00$ transition and to oscillation occurring at 10.6 μm.[242] With excited CO_2 molecules present in the absorption cell (produced by an electric discharge), any one (or all) of the following six lines of an HBr laser produced oscillation: $P_{2-1}(4)$; $P_{2-1}(6)$; $P_{2-10}(7)$; $P_{2-1}(8)$; $P_{3-2}(5)$; and $P_{3-2}(6)$, contrasting with the case when excited CO_2 molecules are absent and only the $P(6)$ line of the 2–1 band of HBr is absorbed. In this case it is believed that the optical pumping is occurring on the $01^{1}0 \rightarrow 01^{1}1$ absorption band of CO_2 (shown as "Additional pumping in the presence of excited CO_2" in Fig. 6.86), with the possible exception that the $P_{2-1}(7)$ and the $P_{2-1}(4)$ lines of HBr are optically pumping on other absorption bands.

6.3.2. 81.48-μm NH_3, N_2O-LASER PUMPED, LASER

CW laser action at 81.5 μm and 263.4 μm has been produced in gaseous NH_3, optically pumped by an N_2O laser operating at 10.78 μm.[245] Figure 6.88 shows some energy levels of NH_3 relevant to the laser transitions and the optical pumping path. The 81.5-μm line arises

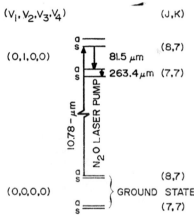

FIG. 6.88. Energy levels relevant to the optically pumped 81.5- and 263.4-μm NH_3 laser, showing the $P(13)$ line (at 10.78 μm) pumping the $\nu_2 s(8, 7)$ state on the $a-s\,Q(8, 7)$ line of NH_3. (After T. Y. Chang, T. J. Bridges, and E. G. Burkhardt,[245] by courtesy of the American Institute of Physics.)

from a $J = 8 \rightarrow 7$, $s \rightarrow a$ rotational transition in the ν_2 axial stretching state of NH_3, and the 263.4-μm line arises from an $a \rightarrow s$ inversion transition in cascade from the 81.5-μm line. The rotational levels of NH_3 are specified by quantum numbers J and K. Each (J, K) level is split into s and a states (lower and upper respectively) due to tunneling of the N atom through the H_3 triangle of the NH_3 molecule. This "inversion" splitting is approximately 0.8 cm^{-1} in the ground vibrational state, and about 36 cm^{-1} in the ν_2 vibrational state and corresponds to the $\Delta\nu$ of the 263.4-μm laser line. The selection rules $a \leftrightarrow s$ and $\Delta k = 0$ apply to the transitions. The 263.4-μm cascade laser line will also oscillate independently of the 81.5-μm line.

The optical pumping occurs on the $a \rightarrow s$ transition between the $J=8, K=7$ of the ground state to $J = 8, K = 7$ of the ν_2 state of NH_3 by means of the $P(13)$ line at 10.78 μm of the CW N_2O laser.[106] Chang et al.[245] used an optical pump power of 1.5 W on the $P(13)$ N_2O line and obtained submillimeter powers of the order of 0.1 mW. The NH_3 pressure was approximately 20 mtorr.

6.3.3. IODINE VAPOR (FREQUENCY-DOUBLED, YAG-LASER PUMPED) LASER

Upper level selective excitation in the molecular iodine laser is through optical pumping of various vibrational–rotational levels of the electronic $B^3\pi_{0+u}$ state from the $X^1\Sigma_g^+$ ground state by means of a Q-switched Nd : YAG laser operating at various second-harmonic wavelengths. The selective pumping process leads to population inversion between the excited B-state and higher vibrational levels of the $X^1\Sigma^+$ ground state, and oscillation occurs on a series of molecular transitions from 0.544 to 1.335 μm. Table 6.19 lists a number of I_2 laser wavelengths that have been observed with the appropriate frequency-doubled Nd : YAG laser pump wavelength.

TABLE 6.19. *Partial list of observed optically pumped*
I_2 laser wavelengths[238]

Pump wave-length (nm)	Iodine laser wavelengths (nm)					
530.6	555.0	574.4	588.5	617.5	649.0	LMR[a]
	568.0	581.5	602.5	633.0	664.5	
531.9	544.3	651.1	906.0	1005.3	1107.3	1329.1
	556.7	659.2	928.8	1022.5	1125.5	1331.0
	569.7	676.3	929.5	1024.5	1135.0	1332.4
	576.4	693.6	630.5	1025.5	1287.0	1333.3
	590.5	711.4	954.5	1053.4	1292.5	1334.9
	604.9	LMR[a]	955.5	1077.5	1315.3	
	619.8	881.3	996.3	1078.8	1319.2	
	635.2	904.7	997.3	1106.6	1328.2	

[a] Limited by mirror reflectance.

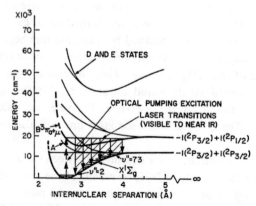

FIG. 6.89. Simplified partial energy-level diagram of the iodine molecule showing the disposition of the $B^3\pi_{0+u}$ and $X^1\Sigma_g^+$ states of the optically pumped visible to near-infrared iodine vapor laser. The dashed arrowed line indicates the optical pumping path between $v'' = 0$ of the $X^1\Sigma_g^+$ ground tate of I_2 and the $B^3\pi_{0+}$ upper laser level. Laser transitions are shown as the vertical transitions between the $v'' = 2$ and $v'' = 73$ levels of the ground state. (Adapted from a figure given by R. Mulliken.[246])

Figure 6.89, a simplified partial energy-level diagram of I_2, shows the relative disposition of the $B^3\pi_{0^+u}$ upper laser level and the $X^1\Sigma_g^+$ ground state, the optical pumping path from $v'' = 0$, and the laser band emission from the $B^3\pi_{0^+u}$ state to vibrational levels between $v'' = 2$ to $v'' = 73$ of the ground state. Laser action occurs to approximately every second vibrational level of the $X^1\Sigma_g^+$ ground state from $v'' = 2$ to $v'' = 73$. The missing oscillations on alternate vibrational transitions are due to unfavorable Franck–Condon factors for emission. The laser wavelengths given in Table 6.19 are average wavelengths as each of the main lines of molecular iodine consist of doublets, caused by pumping of more than one transition by the 0.02-nm width of the pump radiation overlapping a number of the numerous iodine lines in the region of 0.53 µm. By reducing the bandwidth of the pump radiation a factor of 8, pumping could be achieved on each individual transition.

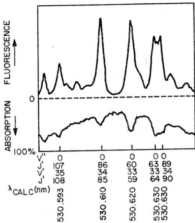

FIG. 6.90. High-resolution absorption and fluorescence spectra of I_2 in the region of 530.6 nm. (After R. L. Byer, R. L. Herbst, H. Kildal, and M. D. Levenson,[238] by courtesy of the American Institute of Physics.)

Figure 6.90 shows the recorded high-resolution absorption and fluorescence spectrum of molecular iodine in the region of 530.6 nm. Laser oscillation was achieved when pumping at each of the five indicated absorption lines.

The absorption cell was operated at 25°C, giving an iodine vapor pressure of 0.32 torr. The maximum output energy per pulse is limited by the total number of iodine molecules that are in the rotational–vibrational ground state, and by upper-state-population losses through collisional quenching and fluorescence to the ground state, and to losses to a repulsive state of I_2 during the build-up time of the laser cavity.

Byer et al.[238] estimate the gain, which is expected to be approximately the same for all the strong transitions, to be 19 percent for the 3-cm-long absorption cell. The inversion density which is limited by the number of iodine molecules in the rotational–vibrational ground state that can be pumped during the pump pulse is about one-half of the ground-state population (or about 3.65×10^{13} cm^{-3}).

6.4. Radiative Cascade Pumped Lasers

A number of molecular gas laser systems exist in which radiative cascade pumping occurs. Often it occurs in conjunction with other selective excitation processes such as

25*

electron-impact excitation or line absorption. Systems that have been covered in the text in which radiative cascade pumping occurs with other processes that have been discussed, include the infrared (4.87 to 6.7-μm) CO laser, which operates on vibrational–rotational transitions of the $X^1\Sigma^+$ CO ground state (Fig. 6.42; covered in Section 6.2.2.1); the near-infrared N_2 laser operating in the $B^3\pi_g$–$A^3\Sigma_u^+$ first positive system that is partially cascade-pumped by strong UV laser transitions in the $C^3\pi_u$–$B^3\pi_g$ second positive system (Fig. 6.65; covered in Section 6.2.3); the far-infrared H_2O laser, in which numerous cascade laser transitions occur, e.g. the 55.1- and 32.9-μm lines in series cascade from the strongest H_2O laser line at 28.0 μm, and the 47.7-μm line that is in series cascade with the 118.6-μm line (Fig. 6.83; covered in Section 6.2.5); and the NH_3 laser operating at 81.5 and 263.4 μm, in which the 263.4-μm inversion-transition line operates in cascade with the optically pumped 81.5-μm line (Fig. 6.88, covered in Section 6.3.2). No further discussion of these systems will be made.

Three other molecular laser systems in which radiative cascade pumping occurs are particularly interesting for a number of reasons. The systems occur in HCN, DCN, and SO_2. Although the mechanism of the primary selective excitation in them is not known (and in some cases the assignments of the transitions involved has not been established), the interactions involved between their energy levels and the cascade transitions together with their operation at far infrared to submillimeter wavelengths warrants their consideration as separate systems.

6.4.1. THE HCN (AND DCN) SUBMILLIMETER LASERS

Oscillation in the submillimeter and far-infrared regions of the spectrum that was first observed in volatile vapors of various cyanides,[247] and later in related compounds[248-9]

TABLE 6.20. *Strongest laser lines of the HCN laser between 12.85 and 774 μm*

Measured wavelength λ_{vac} (μm)	Vibrational transition	Rotational transition	Power Pulsed (W)	Power CW (mW)	References
98.693	—		0.8	—	251
116.132	—		0.5	—	251
126.164	(12^20)–$(05^10)^a$	$R(26)$	3	—	250, 251
128.629	(12^20)–$(05^10)^a$	$R(25)$	9	0.2	235,[c, d] 251,[e] 267
130.834	(12^00)–$(05^10)^a$	$R(25)$	4	—	250, 251
134.932	(12^00)–$(05^10)^a$	$R(24)$	0.8	—	250, 251
309.731	(11^10)–$(11^10)^b$	$R(10)$	0.4	—	250, 251
310.908	(11^10)–$(04^00)^b$	$R(10)$	140	3	235,[e] 237,[c, d] 248, 250, 251, 264,[e] 265, 268, 234,[e] 252
336.579	(11^10)–$(04^00)^b$	$R(9)$	1000	100	234,[e] 235,[e] 247, 250, 251, 252, 261,[c, d] 262,[c] 263,[c] 264,[e] 265,[c] 266, 267,[c] 268[e]
372.547	(04^00)–$(04^00)^b$	$R(8)$	0.6	—	250, 251

[a] Identification as in ref. 259.   [b] Identification as in ref. 257.   [c] Reference to CW operation.   [d] Reference to highest power CW operation.   [e] Reference to highest pulsed power.

and gas mixtures of compounds that contained C, N, and H or D,[250-3] and ascribed to oscillation in CN,[248, 254-6] is now known to be due to transitions in HCN (and DCN).[257-60]

Of the thirty-eight laser lines, which cover the wavelength range 12.85 μm to 773.5 μm,[13] that have been reported and allocated to HCN, DCN, and HCN[15], twenty-eight are believed to be HCN laser lines; six DCN lines; and four HCN[15]. The strongest lines with their assignments (when known), peak pulsed output power, and CW power (if operated CW), are given in Table 6.20.

The energy levels identified[257, 13] with the laser transitions in HCN between 311 and 373 μm are shown in Fig. 6.91 together with the five lines observed in this wavelength range. The vibrational levels involved are the 11^10 and 04^00 levels of HCN which lie about 2800 cm^{-1}

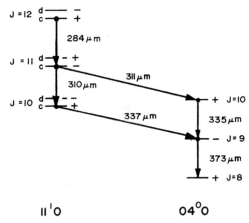

FIG. 6.91. Partial energy-level diagram of HCN showing far-infrared 300-μm laser transitions in series and cross radiative cascade. Adapted from a figure given in ref. 13.

above the ground state and differ in vibrational energy by 2.64 cm^{-1} and have rotational levels of common J that cross near $J = 10$. The 337-μm and the 311-μm laser transitions are allowed by dipole selection rules but would normally have small transition probabilities. However, a strong Coriolis perturbation[269] occurs at the close energy level coincidence of the levels at the crossing at $J = 10$ (and to a lesser extent at the $J = 9$ and $J = 11$ levels) and causes the wave functions of the 311- and 337-μm line vibrational–rotational transitions to be sufficiently mixed to give a large enough transition moment for laser action.[257]

In an electrically excited gas discharge it appears that the HCN, which is formed in some chemical reaction from H, C, and N that are present in the discharge, is produced preferentially in the ν_1 normal mode and gives a vibrational population inversion of 11^10 relative to the 04^00 vibrational level. Only where the mixing of the 11^10 and 04^00 wave functions occurs, however, are the transition moments large enough for the gain to be sufficient to allow oscillation to occur. The 337-μm line is the strongest line and oscillation commences here. This causes a rotational population inversion to occur between $J = 9$ and $J = 8$ of the 04^00 state, and laser oscillation occurs at 373 μm in cascade with the 337-μm line. The 311-μm line also produces cascade oscillation of the 335-μm pure rotational line by producing inversion between the $J = 10$ and $J = 9$ levels of the 04^00 vibrational level. As the 337-μm

line is stronger than the 311-μm, it suppresses oscillation of the 335-μm line in CW operation, and only the 337- and 311-μm lines oscillate.[257, 263] The high gain of the 337- and
311-μm lines is due chiefly to the sharp peaking of the mixing coefficient, and hence the
vibrational transition moment, at $J = 10$; also the rotational factors that affect the gain are
optimized at about a J-value of 10.[267] The gain of the 337-μm line is about 2.5 times that
of the 311-μm line, which is consistent with the observed intensities of these laser lines.[257]

Figure 6.92 shows the assignments for the HCN lines near 130 μm.[259] The solid lines
indicate transitions on which oscillation has been observed, and the dashed vertical lines

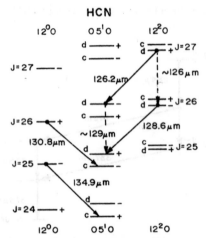

FIG. 6.92. Partial energy-level diagram of HCN showing far-infrared laser transitions at approximately
130 μm. The dashed lines show predicted laser transitions. (After A. N. Maki,[259] by courtesy of the American Institute of Physics.)

indicate rotational transitions on which oscillation is predicted. The assignments are based
on the prediction that the $12^{2d}0$ and $05^{1d}0$ levels will cross at $J = 26$, and $12^0 0$ and $05^{1c}0$
levels will cross at' $J = 25$.[259] At the crossing points, the levels are perturbed by the same
Coriolis resonance that perturbs the $J = 10$ levels of the $04^0 0$ and $11^{1c}0$ states.

At present the limited accuracy with which term values of HCN are known, and the
limited accuracy of measurement of the HCN lines has not enabled any further HCD laser
transitions to be assigned.

A similar situation of pumping and cascade laser transitions is observed around 180 to
205 μm in the vibrational levels of the ground electronic state of DCN (Fig. 6.93). The
190.01-μm line will only oscillate when the 189.94-μm line is oscillating strongly, and the
194.74-μm line will only oscillate when the 194.71-μm line is oscillating strongly. The 204.4-
and 189.94-μm transitions oscillate simultaneously with the 194.71-μm line; and the 194.74-
μm line oscillates with the 189.94-μm line.[267]

The way in which the four laser lines at 190.01 μm and 189.94 μm originate on the $22^0 0$,
$J = 22$ state, and the 194.71- and 194.74-μm lines terminate together at the $09^1 0$, $J = 20$
state and form a parallelogram of laser transitions, enables an accurate check to be made on
the accuracy of the wavelength (frequency) measurements. The sum of the frequencies of the
189.94- and 194.74-μm lines in the right-hand side of the parallelogram should be exactly

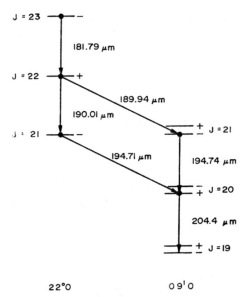

FIG. 6.93. Partial energy-level diagram of DCN showing far-infrared laser transitions at approximately 200 μm. Adapted from a figure given in ref. 266.

equal to the sum of the frequencies of the 190.01-μm and the 194.71-μm lines in the left-hand side of the parallelogram. By accurately measuring the frequencies of the lines in the parallelogram, Hocker and Javan[257] were able to establish the assignment of the transitions shown in Fig. 6.93, and confirm the predicted assignments by Maki[259] of six lines of DCN.

6.4.1.1. Operating characteristics and discharge conditions

Oscillation in HCN has been obtained under both CW and pulsed discharge conditions in mixtures of a whole multitude of chemical compounds. Such mixtures include: $CH_4 + NH_3$; $CH_4 + N_2$; CH_3CN; $(CH_3)_2NH$; HCN or ICN (plus impurities); and CD_3CN; CH_2H_5CN; $(C_2H_5)_2O + N_2$; $CH_3OH + N_2$; C_2H_5OH and $(C_3H_7)_2O + N_2$. A lot of these mixtures have a serious drawback in that tar-like deposits caused by polymerization of some of the reaction products of the discharge quickly degrade the discharge tube.[235, 237, 252, 262]

The best of these mixtures is the $CH_4 + N_2$ mixture. Its use leads to the cleanest discharge products[252] and has been used in experiments that have realized high output power (of the 337-μm line) under pulsed and CW conditions.[264]

The majority of studies of the HCN laser have been made on the strongest laser transition at 337 μm. It is on this transition that the highest pulsed power and highest CW output power have been reported, with more than 1 kW of pulsed output power,[264] and in excess of 60 mW of CW power[252, 261, 264] using axial excitation. Experiments have indicated that the maximum observed pulsed output power of 1 kW can be increased by as much as an order of magnitude using axial excitation,[264] and it is suggested here that considerably more power should be realizable by operating at higher gas pressures using transverse electrical excitation techniques.

Pulsed operation. Extensive experimental studies of the 337-μm HCN laser operating under pulsed conditions have been made by Sharp and Wetherell,[264] Kon *et al.*,[265] and Turner and Poehler.[267]

Operating characteristics using pulsed excitation have been found to be dependent on the discharge conditions: gas pressure and composition of the gas; flow rate; pulsed energy input; length and amplitude of the current excitation pulse; and wavelength of the laser emission. The realizable power is of course dependent on the resonator geometry and output coupling as well as the discharge conditions that affect the basic operating characteristics.

The variation of peak laser output power at 337 μm with total gas pressure of CH_4 and N_2 for different flow rates is given in Fig. 6.94. The output power is seen to be dependent on either the gas pressure and the flow rate. Oscillation occurs over the pressure range 0.2 to 1.2 torr with an optimum between 0.6 and 0.8 torr, and there is an optimum flow rate between 1.1 and 2.0 l./min. (of 1.4 l./min) in a discharge tube of 15-cm bore.[264]

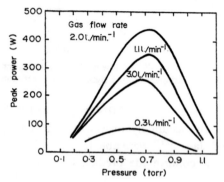

Fig. 6.94. Peak pulsed 337-μm laser output power as a function of pressure for different gas flow rates of a 1 : 1 CH_4–N_2 mixture in a 15-mm-bore discharge tube. (At an optimum charge voltage of 16 kV, input energy of 90 J, I of 560 A, and pulse repetition rate of 3 Hz.) (After L. E. Sharp and A. T. Wetherell,[264] by courtesy of the Optical Society of America.)

The output power according to Sharp and Wetherell[264] is a sensitive function of the gas composition for at least a CH_4 and N_2 mixture, with a 1 : 1 mixture the nominal mixture ratio. Nevertheless, an 87 per cent DH_4 and 13 percent N_2 mixture appears also to be a good mixture (Turner and Poehler[267]). Though, as mentioned earlier, use of CH_4 and N_2 leads to the cleanest operating discharge, $[(CH_3)_2NH]$ (diamethylamine)[250-1] and C_2H_5CN (ethyl cyanide)[265] have been used extensively in pulsed far-infrared HCN lasers. Using these latter compounds, the optimum operating gas pressure appears to be in the range of 0.1 to 0.3 torr for discharge tubes of about 8 to 10-mm bore for both static[265] and flow-ing[251] gas fills. The sensitivity of the output power to mixture composition[264] at low flow rates (0.3 to 2 l./min) is possibly due to air leaks into the discharge tube. A similar sensitivity to the gas composition is observed in certain types of transversely excited CO_2 lasers operating below atmospheric pressure at low gas-flow rates.

The output power of the pulsed 337-μm HCN laser is affected by the discharge voltage[264] and charge capacity.[265] At high input energies ($\sim$90 J), for a charge voltage of 16 kV applied across a discharge length of 3.4 meters at a $CH_4 + N_2$ total pressure of 0.7 torr, the output power starts to decrease after increasing slowly to the maximum at a constant capacity (by a factor of about 4) from a charge voltage of 10 kV. The output also increases at a fixed voltage with increase in capacity[265] so that the output depends only on the input energy. The decrease in power is believed to be due to a saturation effect and also to discharge instabilities and radial contraction of the discharge column that became more pronounced as the input energy is increased.

The maximum input energy for a C_2H_5CN gas fill (in a similar-sized discharge tube to that used in ref. 264) is about 10 J,[265] which is a factor of 6 less than the maximum input energy of a $CH_4 + N_2$ gas fill.[264] It is not known if the various reported maximum input energies are related more to the length of excitation current pulse used than to any inherent input-power limitation per unit volume of the discharge.

The type of output emitted from the pulsed HCN is a strong function of the duration and amplitude of the current pulse. The output can occur as a train of pulses emitted during the discharge period for long, relatively low-current excitation pulses,[265] or in the afterglow of the discharge for short, high-current excitation pulses,[267] or as a single pulse in the afterglow for short, low-current excitation pulses.[265] This behavior of the pulse output of the 337- and 311-μm lines as a function of current is shown in Fig. 6.95.

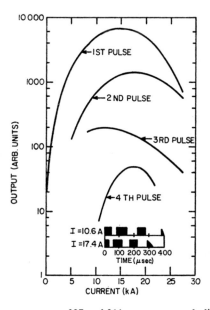

FIG. 6.95. HCN laser pulsed output power at 337 and 311 μm versus peak discharge current from a 220-cm-long, 15-cm-bore laser tube using a 1 : 1 CH_4-N_2 mixture at 0.75 torr. The current pulse was 1 μsec wide. The insert shows the time sequences of the train of pulses at the two peak currents. (After R. Turner and T. O. Poehler,[267] by courtesy of the American Institute of Physics.)

The initial output pulse occurs several μsec[267] (to several tens of μsec[265]) after the current pulse. As the discharge current is progressively increased, the amplitude of the initial pulse increases and additional output pulses appear. Four pulses are shown in Fig. 6.95 for the output from a $CH_4 + N_2$ laser mixture in a tube of 15-cm bore. The spacing of the pulses in the pulse train is a function of the discharge current. The spacing is shorter at the (high) optimum discharge current of 17.4 kA than at the lower discharge current of 10.6 kA, as illustrated in the insert of Fig. 6.95.

The pulsed nature of the output illustrated in Fig. 6.95 has been shown by Turner and Poehler[267] to be caused by radial acoustic oscillations that are produced in the gas by the high current and which modulate the electron density in the cavity. In the experiments of Turner and Poehler the electron density had an annular distribution and was highest close to the wall of the discharge tube. The acoustic oscillations, that had periods as short as 95 μsec, produce radial density and temperature gradients in the gas, that modulate the laser output. When the gas density is a maximum along the axis of the discharge tube, the laser output is a minimum and vice versa.[267] The effect described here should be, and was observed to be, independent of wavelength and cavity length.

In other pulsed HCN laser experiments[251] at much lower discharge currents than those used by Turner and Poehler,[267] while the oscillation occurred in the afterglow of the discharge, the output pulses at different wavelengths commenced at different times after the end of the current pulse. The shorter-wavelength, higher-frequency laser pulses always appeared earlier than pulses at longer wavelengths. In this case, at lower discharge currents when it is to be expected that the electron density would be higher along the axis of the tube than close to the wall of the tube, McCaul and Schawlow[270] have suggested that the multiple-output pulses could be due to cavity-mode sweeping of the gain profile as the apparent cavity length increases with electron recombination. They calculated a restricted set of mirror curvatures that should support laser oscillation for $M \leqslant 1$, where

$$M = (r_0/L)^2 (n_p/n_e), \tag{6.34}$$

in which r_0 and L are the laser tube radius and length, respectively, and n_p and n_e are the plasma resonance electron density and the electron density of the tube center respectively. As M increases through unity as n_e rapidly decreases through electron–ion recombination, the predicted stable-mirror curvatures approach their empty-cavity values and become stable, thus effectively Q-switching the laser cavity. Since M varies as ω^2, the effect is frequency dependent, and high-frequency laser lines are predicted to occur earlier following the current pulse. This is in agreement with observation.[251]

The validity of this model hinges on the magnitude, and the decay rate of the electron density following the current pulse. According to Turner and Poehler,[267] even with high-excitation currents, the electron density does not greatly exceed 2×10^{12} cm^{-3} and drops to about 10^{11} cm^{-3} through ion–electron recombination during the first 15 μsec in the afterglow (as evidenced by the emission of H_β radiation in the afterglow), so that after 15 μsec the change in refractive index of the gas is to too small to produce cavity-mode scanning and bring about the emission of the train of output pulses. The multiple pulses, of course, commence at times much more than 15 μsec into the afterglow of the current pulse.

The delay between the current pulse and the emission of the first laser pulse in HCN (and also H_2O) lasers appears to be due to a thermal-diffusion effect. It can be attributed to the time it takes heat to be conducted to the walls to cool the hot gas in the discharge tube in order to depopulate the lower laser level. The reduction in delay time between the first laser pulse and the discharge-current pulse that occurs when small amounts of helium are added to the discharge is consistent with this interpretation.[267] The increase in output power (by a factor of 3) and longer laser emission that is observed when hydrogen is added to a CH_3CN laser discharge is possibly also related to a decreased thermal-diffusion time in the presence of hydrogen.[270-1]

CW operation. The CW operating characteristics of the HCN laser are necessarily less involved than those observed in pulsed operation. The output power and the gain have been found to be dependent on the composition of the excited gas, the gas pressure, and the discharge current. Some of the organic compounds, and combinations of organic and inorganic compounds that have been utilized in the continuous wave laser are listed in Table 6.21.

TABLE 6.21. *HCN laser materials*[a]

Material	Remarks
$CH_3CN + NH_3$	1 : 1 mixture has given oscillation at 337, 311, 211 μm and highest reported output at 129 μm[235]
$(C_2H_5)_2O + N_2$	Gives usual HCN lines at 130 and 300 μm, plus lines at 281 and 291 μm, and highest reported output at 311 μm[237]
$CH_3OH + N_2$	Relatively clean discharge
C_2H_5OH	Relatively clean discharge
$(C_3H_7)_2O + N_2$	Very dirty discharge
$CH_4 + N_2$	Cleanest discharge products, very high power, best mixture ratio 1 : 1

[a] With the exception of the first two inclusions, the material presented in the table is from ref. 252.

The best mixture for the realization of high gain and high output power, coupled with cleanliness of operation is the $CH_4 + N_2$ mixture. Figure 6.96 shows the relative CW power output at 337 μm for this mixture as a function of CH_4 pressure for various N_2 pressures

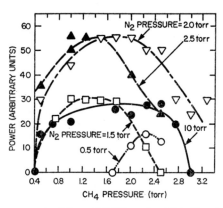

FIG. 6.96. Relative CW power output of a 337-μm HCN laser for CH_4–N_2 mixtures as a function of pressure with a constant discharge current of 500 mA; $D = 7.5$ cm. (After O. M. Stafsudd, F. A. Haak, and K. Radisavljevic,[252] by courtesy of the IEEE.)

in a discharge tube of 7.5-cm bore. The best mixture can be seen to be a 1 : 1 CH_4 : N_2 mixture at a total pressure of approximately 4 torr.

The typical dependence of the gain at 337-μm in a 1 : 1 CH_4 : N_2 laser mixture on pressure at different discharge currents in a 5-mm-bore discharge tube is shown in Fig. 6.97. If the operating pressure and current are kept constant, the fractional gain/meter has been found to be directly proportional to the flow rate over (at least) the flow-rate range of 0 to 28 cm³ (STP)/min.[268]

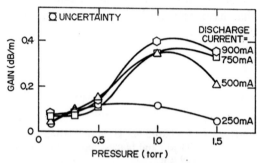

FIG. 6.97. Gain of a CW 337-μm HCN laser as a function of pressure for various discharge currents for a 1 : 1 CH_4 : N_2 mixture in a 5-cm-bore discharge tube. (After O. M. Stafsudd and Y. C. Yeh,[268] by courtesy of the IEEE.)

6.4.2. FAR-INFRARED SO_2 LASER

Laser oscillation in SO_2 in the far-infrared between 140 to 215 μm occurs on ten lines, nine of which are in a double, series-parallel cascade sequence. The observed and calculated laser transitions are presented in Table 6.22.

All the lines in Table 6.22 except the one at 215 μm form a closed-parallelogram, double, series-parallel, radiative-cascade sequence that is characteristic of laser transitions involving perturbed energy levels. This is illustrated in Fig. 6.98.

TABLE 6.22. *Observed and calculated laser transitions in SO_2*

λ(approx) (μm)	Frequencies		Assignment		References
	Obs. (cm⁻¹)	Calc. (cm⁻¹)	Upper level	Lower level	
215	46.44	—	—	—	273, 275, 276
206	48.44	48.420	$\nu_1 26_{15}$	$\nu_1 26_{14}$	273
193	51.89	51.868	$\nu_1 28_{16}$	$\nu_1 28_{15}$	273, 274, 276
152	—	65.603	$\nu_1 27_{15}$	$2\nu_2 25_{15}$	273
151'	66.09	66.074	$\nu_1 27_{15}$	$2\nu_2 26_{16}$	273, 275, 276
151	66.14	66.192	$\nu_1 28_{15}$	$2\nu_2 27_{16}$	273
150	66.67	66.693	$\nu_1 28_{15}$	$\nu_1 27_{14}$	273
142	70.43	70.421	$2\nu_2 26_{16}$	$2\nu_2 25_{15}$	273
141'	70.98	70.892	$\nu_1 26_{14}$	$2\nu_2 25_{15}$	273
141	71.03	71.013	$\nu_1 27_{14}$	$2\nu_2 26_{15}$	273, 276
140	71.53	71.514	$2\nu_2 27_{16}$	$2\nu_2 26_{15}$	273

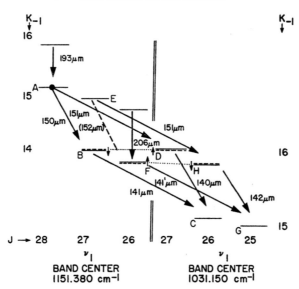

FIG. 6.98. Partial energy-level diagram for SO_2 with assigned laser transitions. The perturbed levels B and D; and F and H are connected by dotted lines. (Adapted from a figure given by G. Hubner, J. C. Hassler, P. D. Coleman, and G. Steenbeckeliers.[273])

The series-parallel cascade sequences are similar to those that occur in the far-infrared and submillimeter HCN and DCN lasers (Figs. 6.90–6.92) and also in the far-infrared H_2O laser. They form the parallelograms between the energy levels marked ABCDA and EFGHE. The first parallelogram involving the 150-, 141-, 140-, and 151-μm lines is complete, but the 152-μm line marked (152 μm) in the EFGHE parallelogram, involving the lines at (152), 141', 142, and 151' μm, is a predicted line. The position of its lowest laser level marked F has been established through the prediction and observation of another laser line at 206 μm.

The levels that are irregularly perturbed (by a Fermi resonance[277]) are the levels marked B, F, D, and H. Levels marked B and D are shifted down and up respectively from their unperturbed positions, and F and H are shifted up and down respectively. The perturbation occurs between the vibrational energy levels v_1, $K_{-1} = 14$ and $2v_2$, $K_{-1} = 16$, which are nearly resonant in energy between $J = 26$ and $J = 27$. The observed energy difference between the perturbed levels marked B and D is only 0.5 cm^{-1}, and only 0.47 cm^{-1} between the levels marked F and H.

The mechanism by which the levels involved in the laser oscillation are excited has not been determined. Though the 193-μm line feeds into the level marked A from which the 150- and 151-μm lines are in cascade, it is not known if it is responsible for the selective excitation of that level.

References

1. GORDON, J. P., ZEIGER, H. J. and TOWNES, C. H., *Phys. Rev.* **95**, 282 (1954).
2. POLANYI, J. C., *J. Phys. Chem.* **34**, 347 (1961).
3. MATHIAS, L. E. S. and PARKER, J. T., *Appl. Phys. Lett.* **3**, 16 (1963).
4. MATHIAS, L. E. S. and PARKER, J. T., *Phys. Lett.* **7**, 194 (1963).

5. HEARD, H. G., *Nature* **200**, 667 (1963).
6. PATEL, C. K. N., FAUST, W. L. and McFARLANE, R. A., *Bull. Am. Phys. Soc.* **9**, 500 (1964).
7. PATEL, C. K. N., *Phys. Rev. Lett.* **12**, 588 (1965).
8. PATEL, C. K. N., *Phys. Rev. Lett.* **13**, 617 (1965).
9. PATEL, C. K. N., *Appl. Phys. Lett.* **7**, 15 (1965).
10. HODGSON, R. T. and DREYFUS, R. W., *Phys. Rev. Lett.* **28**, 536 (1972).
11. WAYNANT, R. W., *Phys. Rev. Lett.* **28**, 533 (1972).
12. MATHIAS, L. E. S., CROCKER, A. and WILLS, M. S., *IEEE J. Quant. Electr.* **QE-4**, 205 (1968).
13. POLLACK, M. A., Molecular gas lasers, in *Handbook of Lasers*, pp. 298–349, PRESSLEY, R. J. (Ed.), Chemical Rubber Co., Cleveland, Ohio, 1971.
14. LEGAY, F. and LEGAY-SOMMAIRE, N., *Compt. Rend.* **260**, 3339 (1964).
15. CHEO, P. K., CO_2 Lasers, in *Lasers: a Series of Advances*, vol. III, pp. 111–267, LEVINE, A. K. and DeMARIA, A. J. (Eds.), Dekker, New York, 1971.
16. SOBOLEV, N. N. and SOKOVIKOV, V. V., *Sov. Phys. Uspekhi* **10**, 153 (1967).
17. TYCHINSKII, V. P., *Sov. Phys. Uspekhi* **10**, 131 (and 152) (1967).
18. PATEL, C. K. N., Gas lasers, in *Lasers: a Series of Advances*, vol. II, LEVINE, A. K. (Ed.), Dekker, New York, 1968.
19. SHARMA, R. D. and BRAU, C. A., *Phys. Rev. Lett.* **19**, 1273 (1967).
20. SHARMA, R. D. and BRAU, C. A., *J. Chem. Phys.* **50**, 924 (1969).
21. TAYLOR, R. L. and BITTERMAN, S., *Revs. Mod. Phys.* **41**, 26 (1969).
22. CALLEAR, A. B., *Discussions of the Faraday Society*, no. 33, 28 (1962).
23. CALLEAR, A. B., *Appl. Optics*, Supplement 2, *Chemical Lasers*, pp. 145–70, HOWARD, J. N. (Ed.), 1965.
24. OFFENBERGER, A. A. and ROSE, D. J., *J. Appl. Phys.* **41**, 3908 (1970).
25. MOORE, C. B., WOOD, R. E., HU, B. and YARDLEY, J. T., *J. Chem. Phys.* **46**, 4222 (1967).
26. TAYLOR, R. L., CAMAC, M. and FEINBERG, R. M., *Proc. Int. Symp. Combustion*, pp. 49–65, Pittsburgh, 1966.
27. BONESS, M. J. W. and SCHULZ, G. J., *Phys. Rev. Lett.* **21**, 1031 (1968).
28. SCHULZ, G. J., *Phys. Rev.* **135A**, 988 (1964).
29. SCHULZ, G. J., *Phys. Rev.* **125**, 229 (1962).
30. MARRIOTT, R., *Proc. Phys. Soc.* **84**, 877 (1964).
31. WEBER, M. J. and DEUTSCH, T. F., *IEEE J. Quant. Electr.* **QE-2**, 369 (1966).
32. MILES, P. A. and HORRIGAN, F. A., Raytheon Co., Semiannual Technical Summary Report, *Research Study of the CO_2 Laser Radar Transmitter*, June (1967).
33. ROBERTS, T. G., HUTCHESON, G. J., EHRLICH, J. J., HALES, W. L. and BARR, T. A. Jr., *IEEE J. Quant. Electr.* **QE-3**, 605 (1967).
34. MOELLER, G. and RIGDEN, J. D., *Appl. Phys. Lett.* **7**, 274 (1965).
35. CLARK, P. O. and SMITH, M. R., *Appl. Phys. Lett.* **9**, 367 (1966).
36. TYTE, D. C. and SAGE, R. W., *S.E.R.L. (England) Technical Journal*, **19**, Paper no. 12 (1969).
37. JOHNSON, A. M., *IEEE J. Quant. Electr.* **QE-7**, 199 (1971).
38. JOHNS, T. W. and NATION, J. A., *Appl. Phys. Lett.* **20**, 495 (1972).
39. LAURIE, K. A. and HALE, M. M., *IEEE J. Quant. Electr.* **QE-7**, 530 (1971).
40. JOHNSON, D. C., *IEEE J. Quant. Electr.* **QE-7**, 185 (1971).
41. FORTIN, R., *Can. J. Phys.* **49**, 257 (1971).
42. DEUTSCH, T. F., *Appl. Phys. Lett.* **20**, 315 (1972).
43. DEUTSCH, T. F. and RUDKO, R. I., *Appl. Phys. Lett.* **20**, 423 (1972).
44. SCHRIEVER, R. L., *Appl. Phys. Lett.* **20**, 354 (1972).
45. NIGHAN, W. L. and BENNETT, J. H., *Appl. Phys. Lett.* **14**, 240 (1969).
46. NIGHAN, W. L., *Appl. Phys. Lett.* **15**, 355 (1969).
47. SVIRIDOV, A. G., SOBOLEV, N. N. and TSELIKOV, G. G., *JETP Lett.* **6**, 62 (1967).
48. BRIDGES, T. J. and PATEL, C. K. N., *Appl. Phys. Lett.* **7**, 244 (1965).
49. CHEO, P. K., *IEEE J. Quant. Electr.* **QE-3**, 683 (1967).
50. PATEL, C. K. N., *Phys. Rev.* **136**, A1187 (1964).
51. DJEU, N., KAN, T. and WOLGA, G. J., *IEEE J. Quant. Electr.* **QE-4**, 246 (1968).
52. ROBINSON, D., *Quantitative Spectroscopy and Radiation Transfer*, vol. 4, pp. 335–42, Pergamon Press, Oxford.
53. SOBOLEV, N. N. and SOKOVIKOV, V. V., *JETP Lett.* **5**, 99 (1967).
54. WITTEMAN, W. J., *Phys. Lett.* **18**, 125 (1965).
55. WITTEMAN, W. J., *IEEE J. Quant. Electr.* **QE-2**, 375 (1966).
56. CARBONE, R. J. and WITTEMAN, W. J., *IEEE J. Quant. Electr.* **QE-5**, 442 (1969).

384

57. CHEO, P. K., *Appl. Phys. Lett.* **11**, 38 (1967).
58. NOVGORODOV, M. Z., SVIRIDOV, A. G. and SOBOLEV, N. N., *IEEE J. Quant. Electr.* **QE-7**, 508 (1971).
59. FROST, L. S. and PHELPS, A. V., *Phys. Rev.* **127**, 1621 (1962).
60. BJORKHOLM, J. E., *Appl. Phys. Lett.* **13**, 36 (1968).
61. GORDIETS, B. F., SOBOLEV, N. N. and SHELEPIN, L. A., *Sov. Phys-JETP* **26**, 1039 (1968).
62. KOVACS, M., RAO, D. R. and JAVAN, A., *J. Chem. Phys.* **48**, 3339 (1968).
63. CHEO, P. K., *IEEE J. Quant. Electr.* **QE-3**, 683 (1967).
64. CROCKER, A. and WILLS, M. S., *Electr. Lett.* **5**, 63 (1969).
65. TIFFANY, W. B., TARG, R. and FOSTER, J. D., *Appl. Phys. Lett.* **15**, 91 (1969).
66. DEUTSCH, T. F., HORRIGAN, F. A. and RUDKO, R. I., *Appl. Phys. Lett.* **15**, 88 (1969).
67. FRAPARD, C., ROULOT, M. and ZIEGLER, X., *Phys. Lett.* **20**, 384 (1966).
68. CLARK, P. O. and SMITH, M. R., *Appl. Phys. Lett.* **9**, 369 (1966).
69. McKNIGHT, W. B., *J. Appl. Phys.* **40**, 2810 (1969).
70. HILL, A. E., *Appl. Phys. Lett.* **12**, 324 (1968).
71. McKNIGHT, W. B., *IEEE J. Quant. Electr.* **QE-5**, 420 (1969).
72. HILL, A. E., *Appl. Phys. Lett.* **16**, 423 (1970).
73. DEZENBERG, G. J., McKNIGHT, W. B., McCLUSKY, L. N. and ROY, E. L., *IEEE J. Quant. Electr.* **QE-6**, 652 (1970).
74. DEZENBERG, G. J., ROY, E. L. and McKNIGHT, W. B., *IEEE J. Quant. Electr.* **QE-8**, 58 (1972).
75. McKNIGHT, W. B., *J. Appl. Phys.* **40**, 2810 (1969).
76. LEONARD, D. A., *Appl. Phys. Lett.* **7**, 4 (1965).
77. BEAULIEU, A. J., *A New Generation of High Power Lasers*, DREV Report M-2005/70, Jan. 1970.
78. BEAULIEU, A. J., *Appl. Phys. Lett.* **16**, 504 (1970).
79. LAURIE, K. A. and HALE, M. M., *IEEE J. Quant. Electr.* **QE-6**, 530 (1970).
80. LaFLAMME, A. K., *Rev. Sci. Instr.* **41**, 1578 (1970).
81. FORTIN, R., *Can. J. Phys.* **49**, 257 (1970).
82. JOHNSON, D. C., *IEEE J. Quant. Electr.* **QE-7**, 185 (1971).
83. LAURIE, K. A. and HALE, M. M., *IEEE J. Quant. Electr.* **QE-7**, 530 (1971).
84. PEARSON, P. R. and LAMBERTON, H. M., Atmospheric pressure CO_2 lasers giving high output energy per unit volume, *S.E.R.L. Tech. J.* **21**, no. 3 (Oct. 1971).
85. DUMANCHIN, R. and ROCCA-SERRA, J., High power density pulsed molecular laser, Paper 18-6 presented at the *Sixth Q.E.C., Kyoto, Japan*, Sept. 1970.
86. DUMANCHIN, R., MICHON, M., FARCY, J. C., BOUDINET, G. and ROCCA-SERRA, J., *IEEE J. Quant. Electr.* **QE-8**, 163 (1972).
87. LAMBERTON, H. M. and PEARSON, P. R., *Electr. Lett.* **7**, 141 (1971).
88. PEARSON, P. R. and LAMBERTON, H. M., *IEEE J. Quant. Electr.* **QE-8**, 145 (1972).
89. SMITH, D. C. and DeMARIA, A. J., *J. Appl. Phys.* **41**, 5212 (1970).
90. FENSTERMACHER, C. A., NUTTER, M. J., LELAND, W. T. and BOYER, K., *Bull. Am. Phys. Soc.* **17**, 399 (1972).
91. FENSTERMACHER, C. A., NUTTER, M. J., RINK, J. P. and BOYER, K., *Bull. Am. Phys.. Soc.* **16**, 42 (1971).
92. FENSTERMACHER, C. A., NUTTER, M. J., LELAND, W. T. and BOYER, K., *Appl. Phys Lett.* **20**, 56 (1972).
93. PATEL, C. K. N., FAUST, W. L. and McFARLANE, R. A., *Bull. Am. Phys. Soc.* **9**, 500 (1964).
94. PATEL, C. K. N., *Phys. Rev.* **136**, A1187 (1964).
95. CHEN, C. J., *J. Appl. Phys.* **42**, 1016 (1971).
96. McQUILLAN, A. K., YEH, G. and CARSWELL, A. I., *Can. J. Phys.* **49**, 1611 (1971).
97. CLARK, P. O. and WADA, J. Y., *IEEE J. Quant. Electr.* **QE-4**, 263 (1968).
98. SOBOLEV, N. N. and SOKOVIKOV, V. V., *Sov. Phys.-JETP* **4**, 303 (1966).
99. SOBOLEV, N. N. and SOKOVIKOV, V. V., *Sov. Phys.-JETP* **5**, 122 (1967).
100. SOBOLEV, N. N. and SOKOVIKOV, V. V., *Sov. Phys.-Usp.* **10**, 153 (1967).
101. BRINKSCHULTE, H. W., *IEEE J. Quant. Electr.* **QE-4**, 948 (1968).
102. PAANANEN, R. A., *Proc. IEEE* **55**, 2035 (1967).
103. DEUTSCH, T. F. and HORRIGAN, F. A., *IEEE J. Quant. Electr.* **QE-4**, 972 (1968).
104. CHIRKOV, V. N. and YAKOVLEVA, A. V., *Opt. Spectry.* **28**, 441 (1970).
105. WHITNEY, W. T., *Bull. Am. Phys. Soc.* **17**, 398 (1972).
106. PATEL, C. K. N., *Appl. Phys. Lett.* **6**, 12 (1965).
107. MATHIAS, L. E. S., CROCKER, A. and WILLS, M. S., *Phys. Lett.* **13**, 303 (1964).
108. DJEU, N., KAN, T. and WOLGA, G., *IEEE J. Quant. Electr.* **QE-4**, 783 (1968).
109. MULLONEY, G. J., AHLSTROM, H. G. and CHRISTIANSEN, W. H., *IEEE J. Quant. Electr.* **QE-7**, 551 (1971).

110. WOOD, O. R., BURKHARDT, E. G., POLLACK, M. A. and BRIDGES, T. J., *Appl. Phys. Lett.* **18**, 261 (1971).
111. GILBERT, J., LACHAMBRE, J. L. and RHEAULT, F., *IEEE J. Quant. Electr.* **QE-7**, 462 (1971).
112. WALTER, W. T., SOLIMENE, N., PILTCH, M. and GOULD, G., *IEEE J. Quant. Electr.* **QE-2**, 474 (1966).
113. BENNETT, W. R. Jr., *Appl. Optics*, Supplement 2: *Chemical Lasers* (1965), pp. 3–33.
114. BAZHULIN, P. A., KNYAZEV, I. N. and PETRASH, G. G., *Sov. Phys. JETP* **20**, 1068 (1965).
115. BAZHULIN, P. A., KNYAZEV, I. N. and PETRASH, G. G., *Sov. Phys. JETP* **22**, 11 (1966).
116. BOCKASTEN, K., LUNDHOLM, T. and ANDRADE, D., *J. Opt. Soc. Am.* **56**, 1260 (1966).
117. FROMM, D., *Molecular Hydrogen and Deuterium Laser*, Report no. 3, Night Vision Laboratory, U.S.A.E.C., Fort Belvior, Va., April 1967.
118. TOBIAS, J. and VANDESLICE, J. T., *J. Chem. Phys.* **35**, 1852 (1961).
119. KOLOS, W. and ROOTHAM, C. C. J., *Revs. Mod. Phys.* **32**, 219 (1960).
120. HODGSON, R. T., *Phys. Rev. Lett.* **25**, 494 (1970).
121. WAYNANT, R. W., SHIPMAN, J. D., ELTON, R. C. and ALI, A. W., *Appl. Phys. Lett.* **17**, 383 (1970).
122. BAZHULIN, P. A., KNYAZEV, I. N. and PETRASH, G. G., *Sov. Phys. JETP* **21**, 649 (1965).
123. SPINDLER, R. J., *J. Quant. Spectry. Radiative Transfer* **9**, 597 (1969).
124. ALI, A. W. and ANDERSON, A. D., *H_2 Vacuum-uv-laser Rate Coefficients*, Naval Research Laboratory Report 7282, July 28, 1971, Washington, D.C.
125. WAYNANT, R. W., A travelling wave vacuum ultraviolet laser, in *Proceedings of the Technical Program, Electro-Optical Systems Design Conference—1971 West, Anaheim, California, May 18–20* (1971), pp. 1–5.
126. WAYNANT, R. W., SHIPMAN, J. D. Jr., ELTON, R. C. and ALI, A. W., *Proc. IEEE* **59**, 679 (1971).
127. SHIPMAN, J. D. Jr., *Appl. Phys. Letters* **10**, 3 (1967).
128. WAYNANT, R. W. and ELTON, R. C., *Bull. Am. Phys. Soc.* **16**, 593 (1971).
129. CASPERSON, L. and YARIV, A., *Phys. Rev. Lett.* **26**, 293 (1971).
130. PATEL, C. K. N., *Phys. Rev.* **141**, 71 (1966).
131. BARRY, J. D. and BONEY, W. E., *IEEE J. Quant. Electr.* **QE-7**, 101 (1971).
132. HODGSON, R. T., *J. Chem. Phys.* **55**, 5378 (1971).
133. SUART, R. D., KIMBELL, G. H. and ARNOLD, S. J., *Chem. Phys. Lett.* **5**, 519 (1970).
134. WITTIG, C., HASLER, J. C. and COLEMAN, P. D., *Appl. Phys. Lett.* **16**, 117 (1970).
135. PETRASH, G. G., *J. Appl. Spectry.* **4**, 290 (1966).
136. PATEL, C. K. N. and KERL, R. J., *Appl. Phys. Lett.* **5**, 81 (1964).
137. PATEL, C. K. N., *Appl. Phys. Lett.* **7**, 246 (1965).
138. LEGAY-SOMMAIRE, H., HENRY, L. and LEGAY, F., *Compt. Rend.*, **260**, 3339 (1965).
139. EPPERS, W. C., OSGOOD, R. M. Jr. and GREASON, P. R., *IEEE J. Quant. Electr.* **QE-6**, 4 (1970).
140. OSGOOD, R. H. Jr., EPPERS, W. C. Jr. and NICHOLS, E. R., *IEEE J. Quant. Electr.* **QE-6**, 145 (1970).
141. OSGOOD, R. M. Jr., GOLDHAR, J. and McNAIR, R., *IEEE J. Quant. Electr.* **QE-7**, 253 (1971).
142. JEFFERS, W. Q. and WISWALL, C. E., *IEEE J. Quant. Electr.* **QE-7**, 407 (1971).
143. BHAUMIK, M. L., LACINA, W. B. and MANN, M. M., *IEEE J. Quant. Electr.* **QE-8**, 150 (1972).
144. GRAHAM, W. J., KERSHENSTEIN, J., JENSEN, J. T. Jr. and KERSHENSTEIN, K., *Appl. Phys. Lett.* **17**, 194 (1970).
145. DAWSON, P. H. and TAM, W. G., *Can. J. Phys.* **50**, 889 (1972).
146. NIGHAN, W. L., *Appl. Phys. Lett.* **20**, 96 (1972).
147. NIGHAN, W. L., *Phys. Rev.* **2A**, 1989 (1970).
148. JEFFERS, W. Q. and WISWALL, C. E., *J. Appl. Phys.* **42**, 5059 (1971).
149. YARDLEY, J. T., *J. Chem. Phys.* **52**, 3983 (1970).
150. SHARMA, R. D. and BRAU, C. A., *J. Chem. Phys.* **50**, 924 (1969).
151. HERZBERG, G., *Spectra of Diatomic Molecules*, pp. 90–99 and 522, Van Nostrand, New York, 1950.
152. TREANOR, C. E., RICH, J. W. and REHM, R. G., *J. Chem. Phys.* **48**, 1798 (1967).
153. KARL, G., KRUUS, P. and POLANYI, J. C., *J. Chem. Phys.* **46**, 224 (1967).
154. MILLIKAN, R. C., *J. Chem. Phys.* **40**, 2594 (1964).
155. HANCOCK, G. and SMITH, I. M. W. (private communication in ref. 149).
156. JEFFERS, W. Q. and KELLEY, J. D., *Bull. Am. Phys. Soc.* **17**, 394 (1972).
157. RICH, J. W., *J. Appl. Phys.* **42**, 2719 (1971).
158. LEGAY, F., TAIEB, G. and LEGAY-SOMMAIRE, N., Vibrational and electronic processes involved in the mechanisms of the CO–N_2 laser, presented at the 2nd Conference on Chemical and Molecular Lasers, St. Louis, Mo., May 22–24, 1969. Published in conference proceedings in *IEEE J. Quant. Electr.* **QE-6**, 181 (1970).
159. FISHER, E. R. and KUMMLER, R. H., *J. Chem. Phys.* **49**, 1085 (1968).
160. LOTKOVA, E. N., MERCER, G. N. and SOBOLEV, N. N., *Appl. Phys. Lett.* **20**, 309 (1972).

161. CHEN, J. C. Y., *J. Chem. Phys.* **40**, 3513 (1964).
162. SUART, R. D., ARNOLD, S. J. and KIMBELL, G. H., *Chem. Phys. Lett.* **7**, 337 (1970).
163. HARTWICK, T. S. and WALDER, J., *IEEE J. Quant. Electr.* **QE-8**, 455 (1972).
164. FREED, C., *Appl. Phys. Lett.* **18**, 458 (1971).
165. BLAUMIK, M. L., LACINA, W. B. and MANN, M. M., *IEEE J. Quant. Electr.* **QE-6**, 575 (1970).
166. BLAUMIK, M. L., *Appl. Phys. Lett.* **17**, 188 (1970).
167. BLETZINGER, P. and GARSCADDEN, A., *Proc. IEEE* **59**, 675 (1971).
168. HOCKER, G. B., *IEEE J. Quant. Electr.* **QE-7**, 535 (1971).
169. MANN, M. M., BLAUMIK, M. L. and LACINA, W. B., *Appl. Phys. Lett.* **16**, 430 (1970).
170. FISHBURNE, E. S., SEIBERT, G. L. and LAZDINIS, S. S., *J. Chem. Phys.* **48**, 1424 (1968).
171. SCHEER, M. D. and FINE, J., *J. Chem. Phys.* **36**, 1264 (1962).
172. RICH, J. W. and THOMPSON, H. M., *Appl. Phys. Lett.* **19**, 3 (1971).
173. LEGAY, F., LEGAY-SOMMAIRE, N. and TAIEB, G., *Can. J. Phys.* **48**, 1949 (1970).
174. CORCORAN, V. J., CUPP, R. E., SMITH, W. T. and GALLAGHER, J. J., *IEEE J. Quant. Electr.* **QE-7**, 246 (1971).
175. CHEO, P. K. and COOPER, H. G., *Appl. Phys. Lett.* **5**, 42 (1964).
176. COOPER, H. G. and CHEO, P. K., *Appl. Phys. Lett.* **5**, 44 (1964).
177. HERZBERG, G., *Molecular Spectra and Molecular Structure*: I. *Spectra of Diatomic Molecules*, p. 208, Van Nostrand, Princeton, N.J., 1950.
178. McFARLANE, R. A., *IEEE J. Quant. Electr.* **QE-2**, 229 (1966).
179. MASSONE, C. A., CARAVAGLIA, M., GALLARDO, M., CALATRONI, J. A. E. and TAGLIAFERRI, A. A., *Appl. Optics* **11**, 1317 (1972).
180. KASUYA, T. and LIDE, D. R. Jr., *Appl. Optics*, **6**, 69 (1967).
181. KASLIN, V. M. and PETRASH, G. G., *Sov. Phys. JETP-Lett.* **3**, 55 (1966).
182. ALLEN, L., JONES, D. G. C. and SIVARAM, B. M., *Phys. Lett.* **25A**, 280 (1967).
183. LEONARD, D. A., *Appl. Phys. Lett.* **7**, 4 (1965).
184. GERRY, E. T., *Appl. Phys. Lett.* **7**, 6 (1965).
185. GELLER, M., ALTMAN, D. E., DeTEMPLE, T. A., *J. Appl. Phys.* **37**, 3639 (1966).
186. GELLER, M., ALTMAN, D. E., DeTEMPLE, T. A., *Appl. Optics* **7**, 2232 (1968).
187. SHIPMAN, J. D., *Appl. Phys. Lett.* **10**, 1 (1967).
188. WILSON, J., *Appl. Phys. Lett.* **8**, 159 (1966).
189. TARG, R., *IEEE J. Quant. Electr.* **QE-8**, 726 (1972).
190. NILSSON, N. R., STEINVALL, O., SUBRAMANIAN, C. K. and HOGBERG, L., *Physica Scripta* **1**, 153 (1970).
191. DREYFUS, R. W. and HODGSON, R. T., *Appl. Phys. Lett.* **20**, 195 (1972).
192. DREYFUS, R. W. and HODGSON, R. T., *Bull. Am. Phys. Soc.* **17**, 19 (1972).
193. KASLIN, V. M. and PETRASH, G. G., *Sov. Phys.-JETP* **27**, 561 (1968).
194. GARAVAGLIA, M., GALLARDO, M. and MASSONE, C. A., *Phys. Lett.* **28A**, 787 (1969).
195. BENNETT, R. G. and DALBY, F. W., *J. Chem. Phys.* **31**, 434 (1959).
196. SCHULZ, G. J., *Phys. Rev.* **116**, 1141 (1959).
197. HEIDEMAN, H. G. M., KUYATT, C. E. and CHAMBERLAIN, G. E., *J. Chem. Phys.* **44**, 355 (1966).
198. NICHOLLS, R. W., *J. Quant. Spectry, Radiat. Transfer* **2**, 433 (1962).
199. PILLOW, M. E., *Proc. Phys. Soc.* A **67**, 780 (1954).
200. BATES, D. R., *Proc. R. Soc.* A **196**, 239 (1947).
201. STEWART, D. T. and GABATHULER, E., *Proc. Phys. Soc.* A **72**, 287 (1958).
202. HOWORTH, J. R., *J. Phys.* B: *Atom. Molec. Phys.* **5**, 402 (1972).
203. BROADFOOT, A. L. and HUNTEN, D. M., *Can. J. Phys.* **42**, 1212 (1964).
204. ZAPESOCHNYI, I. P. and SKUBENICH, V. V., *Opt. Spectry.* **21**, 83 (1966).
205. NICHOLLS, R. W., *J. Res. Nat. Bur. Stds.* **65A**, 451 (1961).
206. JARMAIN, W. R. and NICHOLLS, R. W., *Can. J. Phys.* **32**, 201 (1954).
207. BENNETT, W. R. Jr., *Appl. Optics, Supplement* **2**, 3 (1965).
208. ALI, A. W., *Appl. Optics* **8**, 993 (1969).
209. VON ENGEL, A., *Ionized Gases*, 2nd ed., p. 182, Clarendon Press, Oxford, 1965.
210. LEONARD, D. A., *Laser Focus* **2**, 26 (1967).
211. KOBAYASHI, T., TAKEMURA, M., SHIMIZU, H. and INABA, H., Paper M9, Quantum Electronics Conference, Montreal (1972), *Conference Proceedings*, p. 61.
212. GLEASON, T. J., WILLETT, C. S., CURNUTT, R. M. and KRUGER, J. S., *Appl. Phys. Lett.* **21**, 276 (1972)..
213. WOOD, O. R., BURKHARDT, E. G., POLLACK, M. A. and BRIDGES, T. J., *Appl. Phys. Lett.* **18**, 261 (1971)
214. BASOV, N. G., DANILYCHEV, V. A., POPOV, YU. M. and KHODEVICH, D. D., *Sov. Phys.-JETP Lett.* **12**, 329 (1970).

215. BASOV, N. G., DANILYCHEV, V. A. and POPOV, YU. M., *Sov. J. Quant. Electr.* **1**, 18 (1971).
216. LORENTS, D. C., "Efficient electron excitation in high pressure rare gas lasers", paper presented at the *3rd Conference on Chemical and Molecular Lasers, May 1–3, St. Louis* (1972).
217. HOPFIELD, J. J., *Astrophysics J.* **72**, 133 (1930).
218. TANAKA, Y., JURSA, A. S. and LEBLANC, F. J., *J. Opt. Soc. Am.* **48**, 304 (1958).
219. HUFFMAN, R. E., TANAKA, Y. and LARRABIE, J. C., *J. Opt. Soc. Am.* **52**, 85 (1962).
220. WILKINSON, P. G., *J. Opt. Soc. Am.* **45**, 1044 (1955).
221. VENKATESWARLU, P., *Proc. Ind. Acad.* **25**, 119 (1947).
222. VENKATESWARLU, P., *Proc. Ind. Acad.* **25**, 138 (1947).
223. MATHIAS, L. E. S. and CROCKER, A., *Phys. Lett.* **13**, 35 (1964).
224. COLEMAN, P. D., *Laser Focus* **9**, 37 (1969).
225. LARGE, L. N. and HILL, H., *Appl. Optics* **4**, 625 (1965).
226. JEFFERS, W. Q. and COLEMAN, P. D., *Appl. Phys. Lett.* **10**, 7 (1967).
227. SARJEANT, W. J. and BRANNEN, E., *IEEE J. Quant. Electr.* **QE-5**, 620 (1969).
228. BENNEDICT, W. S., POLLACK, M. A. and TOMLINSON, W. J., III, *IEEE J. Quant. Electr.* **QE-5**, 108 (1969)
229. PICHAMUTHU, J. P., HASSLER, J. C. and COLEMAN, P. D., *Appl. Phys. Letters* **19**, 510 (1971).
230. JEFFERS, W. Q., *Appl. Phys. Lett.* **13**, 104 (1968).
231. MCFARLANE, R. A. and FRETZ, L. H., *Appl. Phys. Lett.* **14**, 385 (1969).
232. JOHNSON, C. J., *IEEE J. Quant. Electr.* **QE-4**, 701 (1968).
233. JEFFERS, W. Q. and COLEMAN, P. D., *Appl. Phys. Lett.* **13**, 250 (1968).
234. FLESHER, G. T. and MULLER, W. M., *Proc. IEEE* **54**, 543 (1966).
235. MULLER, W. M. and FLESHER, G. T., *Appl. Phys. Lett.* **10**, 93 (1967).
236. FRENKEL, L., SULLIVAN, T., POLLACK, M. A. and BRIDGES, T. J., *Appl. Phys. Lett.* **11**, 344 (1967).
237. DYUBKO, S. F., SVICH, V. A. and VALITOV, R. A., *Sov. Phys.-JETP Lett.* **6**, 80 (1967).
238. BYER, R. L., HERBST, R. L., KILDAL, H. and LEVENSON, M. D., *Appl. Phys. Lett.* **20**, 463 (1972).
239. CHANG, T. Y. and MCGEE, J. D., *Appl. Phys. Lett.* **19**, 103 (1971).
240. WAGNER, R. J. and ZELANO, A. J., New submillimeter laser lines in organic molecules, post-deadline paper W. 4, Q.E.C. Montreal (1972). (Wagner and Zelano are at the U.S. Naval Research Laboratory, Washington, D.C. 20390, U.S.A.)
241. WIEDER, I., *Phys. Lett.* **A24**, 759 (1967).
242. CHANG, T. Y. and WOOD, O. R., *Appl. Phys. Lett.* **21**, 19 (1972).
243. WOOD, O. R. and CHANG, T. Y., *Appl. Phys. Lett.* **20**, 77 (1972).
244. DEUTSCH, T. F., *IEEE J. Quant. Electr.* **QE-3**, 419 (1967).
245. CHANG, T. Y., BRIDGES, T. J. and BURKHARDT, E. G., *Appl. Phys. Lett.* **17**, 357 (1970).
246. MULLIKEN, R. S., *J. Chem. Phys.* **55**, 288 (1971).
247. GEBBIE, H. A., STONE, N. W. B. and FINDLEY, F. D., *Nature* **202**, 685 (1964).
248. STEFFEN, H., STEFFEN, J., MOSER, J. F. and KNEUBUHL, F. K., *Phys. Lett.* **20**, 20 (1966).
249. STEFFEN, H., STEFFEN, J., MOSER, J. F. and KNEUBUHL, F. K., *Phys. Lett.* **21**, 425 (1966).
250. MATHIAS, L. E. S., CROCKER, A. and WILLS, M. S., *Electr. Lett.* **1**, 45 (1965).
251. MATHIAS, L. E. S., CROCKER, A. and WILLS, M. S., *IEEE J. Quant. Electr.* **QE-4**, 205 (1968).
252. STAFSUDD, O. M., HAAK, F. A. and RADISAVLJEVIE, K., *IEEE J. Quant. Electr.* **QE-3**, 618 (1967).
253. FLESHER, G. T. and MULLER, W. M., *Proc. IEEE* **54**, 543 (1966).
254. CHANTRY, G. W., GEBBIE, H. A. and CHAMBERLAIN, J. E., *Nature* **205**, 377 (1965).
255. STEFFEN, H., SCHWALLER, P., MOSER, J. F. and KUEUBUHL, F. K., *Phys. Lett.* **23**, 313 (1966).
256. BROIDA, H. P., EVENSON, K. M. and KIKUCHI, T. T., *J. Appl. Phys.* **36**, 3355 (1965).
257. LIDE, D. R. and MAKI, A. G., *Appl. Phys. Lett.* **11**, 62 (1967).
258. HOCKER, L. O. and JAVAN, A., *Phys. Lett.* **25A**, 489 (1967).
259. MAKI, A. G., *Appl. Phys. Lett.* **12**, 122 (1968).
260. EVENSON, K. M., *Appl. Phys. Lett.* **12**, 253 (1968).
261. GEBBIE, H. A., STONE, N. W. B., SLOUGH, W., CHAMBERLAIN, J. E. and SHERATON, W. A., *Nature* **211**, 62 (1966).
262. MULLER, W. M. and FLESHER, G. T., *Appl. Phys. Lett.* **8**, 217 (1966).
263. HOCKER, L. O., JAVAN, A., RAO, D. R., FRENKEL, L. and SULLIVAN, T., *Appl. Phys. Lett.* **10**, 147 (1967).
264. SHARP, L. E. and WETHERELL, A. T., *Appl. Optics* **11**, 1737 (1972).
265. KON, S., YAMANAKA, M., YAMAMOTO, Y. and YOSHINAGA, H., *Japan J. Appl. Phys.* **6**, 612 (1967).
266. HOCKER, L. O. and JAVAN, A., *Appl. Phys. Lett.* **12**, 124 (1968).
267. TURNER, R. and POEHLER, T. O., *J. Appl. Phys.* **42**, 3819 (1971).
268. STAFSUDD, O. M. and YEH, V. C., *IEEE J. Quant. Electr.* **QE-5**, 377 (1962).

269. TOWNES, C. H. and SCHAWLOW, A. L., *Microwave Spectroscopy*, pp. 29–35, McGraw-Hill, New York, 1955.
270. McCAUL, B. W. and SCHAWLOW, A. L., Plasma refractive effect in HCN lasers, in Digest of Technical Papers presented at the 2nd Conference on Chemical and Molecular Lasers, St. Louis, Missouri, May 22–24 (1969). *IEEE J. Quant. Electr.* **QE-6,** 178 (1970).
271. TURNER, R., HOCHBERG, A. K. and POEHLER, T. O., *Appl. Phys. Lett.* **12,** 104 (1968).
272. TURNER, R. and POEHLER, T. O., *J. Appl. Phys.* **39,** 5726 (1968).
273. HUBNER, G., HASSLER, J. C., COLEMAN, P. D. and STEENBECKELIERS, G., *Appl. Phys. Lett.* **18,** 511 (1971).
274. DYUBKO, S. F., SVICH, V. A. and VALITOV, R. A., *Sov. Phys. JETP Lett.* **1,** 320 (1968).
275. HASSLER, J. C. and COLEMAN, P. D., *Appl. Phys. Lett.* **14,** 135 (1969).
276. HARD, T. M., *Appl. Phys. Lett.* **14,** 130 (1969).
277. Reference 269, pp. 35–40.

Added in Proof

278. RICHARDSON, M. C., ALCOCK, A. J., LEOPOLD, K. and BURTYN, P., *IEEE J. Quant. Elect.* **QE-9,** 236 (1973).
279. See also: CASON, C., DEZENBERG, G. J. and HUFF, R. J., *Appl. Phys. Lett.* **23,** 110 (1973).
280. CHAMPAGNE, *Appl. Phys. Lett.* **3,** 158 (1973).
281. BILLINGSLEY, C. A., HARVEY, J. F. and WILLETT, C. S., unpublished work.
282. WOODWARD, B. W., EHLERS, V. J. and LINEBERGER, W. C., *Rev. Sci. Instr.* **44,** 882 (1973).
283. HANCOCK, G. and SMITH, I. W. M., *Appl. Optics* **10,** 1827 (1973).
284. See also GERALDO, J. B. and JOHNSON, A. W., *IEEE J. Quant. Electr.* **QE-9,** 748 (1973) and references therein.

Appendix

Appendix Figures 1–9: Electron Temperatures in Mixtures of the Noble Gases for Various Values of pD.

Appendix Figures 10–25: Partial Energy-level Diagrams showing Laser Transitions.

Appendix Tables 1–46: Laser Transitions in Atomic Species.

Appendix Tables 47–85: Laser Transitions in Molecular Species.

Electron Temperatures in Mixtures of the Noble Gases for Various Values of pD [Figs. A.1 to A.9]

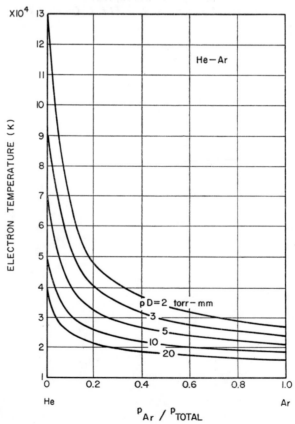

FIG. A.1. Electron temperature as a function of the ratio of Ar pressure to total pressure of He and Ar for various values of pD. (After R. T. Young, Jr., unpublished work.)

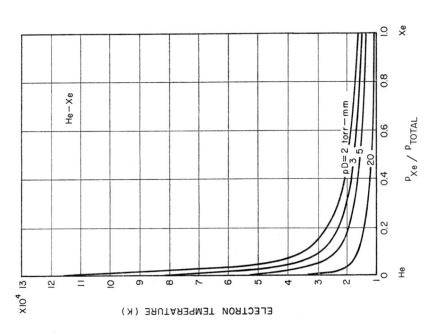

FIG. A.3. Electron temperature as a function of the ratio of Xe pressure to total pressure of He and Xe for various values of pD. (After R. T. Young, Jr., unpublished work.)

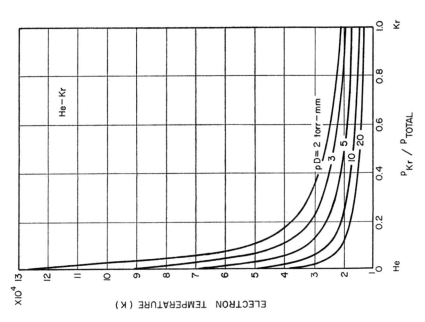

FIG. A.2. Electron temperature as a function of the ratio of Kr pressure to total pressure of He and Kr for various values of pD. (After R. T. Young, Jr., unpublished work.)

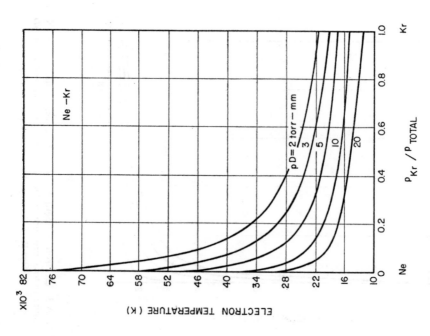

Fig. A.5. Electron temperature as a function of the ratio of Kr pressure to total pressure of Ne and Kr for various values of pD. (After R. T. Young, Jr., unpublished work.)

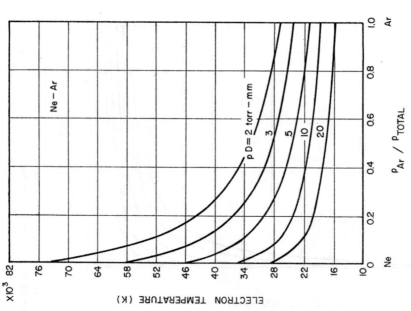

Fig. A.4. Electron temperature as a function of the ratio of Ar pressure to total pressure of Ne and Ar for various values of pD. (After R. T. Young, Jr., unpublished work.)

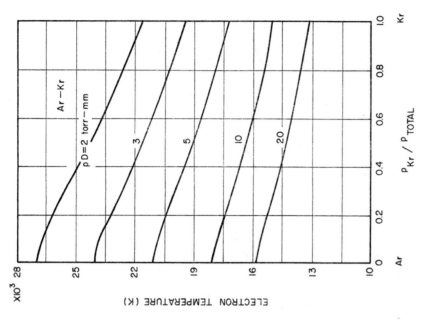

FIG. A.7. Electron temperature as a function of the ratio of Kr pressure to total pressure of Ar and Kr for various values of pD. (After R. T. Young, Jr., unpublished work.)

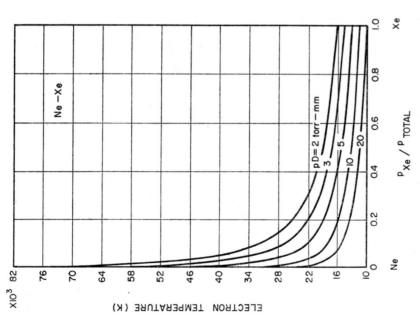

FIG. A.6. Electron temperature as a function of the ratio of Xe pressure to total pressure of Ne and Xe for various values of pD. (After R. T. Young, Jr., unpublished work.)

393

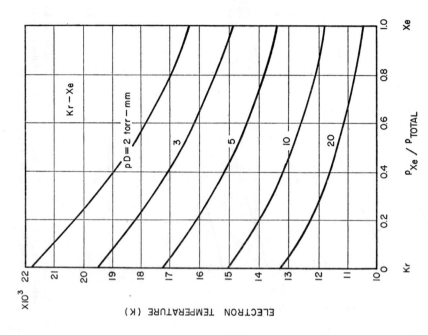

Fig. A.9. Electron temperature as a function of the ratio of Xe pressure to total pressure of Kr and Xe for various values of Dp. (After R. T. Young, Jr., unpublished work.)

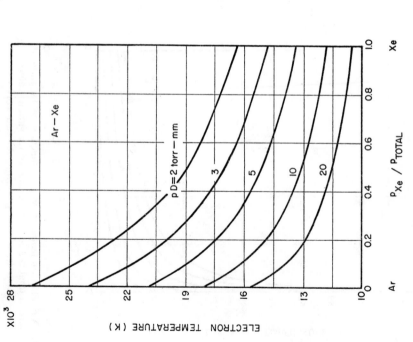

Fig. A.8. Electron temperature as a function of the ratio of Xe pressure to total pressure of Ar and Xe for various values of pD. (After R. T. Young, Jr., unpublished work.)

Partial Energy-level Diagrams showing Laser Transitions [Figs. A.10 to A.25]

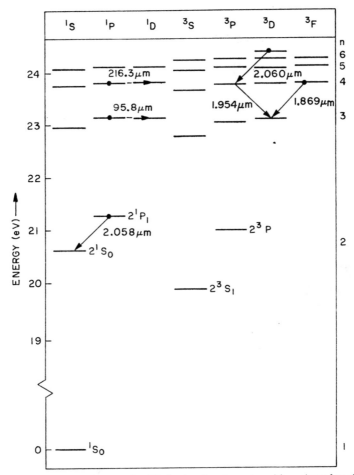

FIG. A.10. Partial energy-level diagram for He-I, showing laser transitions (wavelengths in μm). The 2.058-μm line is a transient laser line.

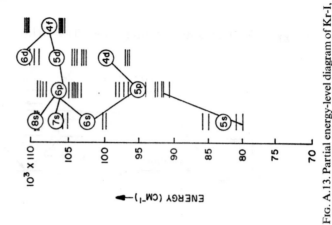

FIG. A.13. Partial energy-level diagram of Kr-I, showing laser transition groups.

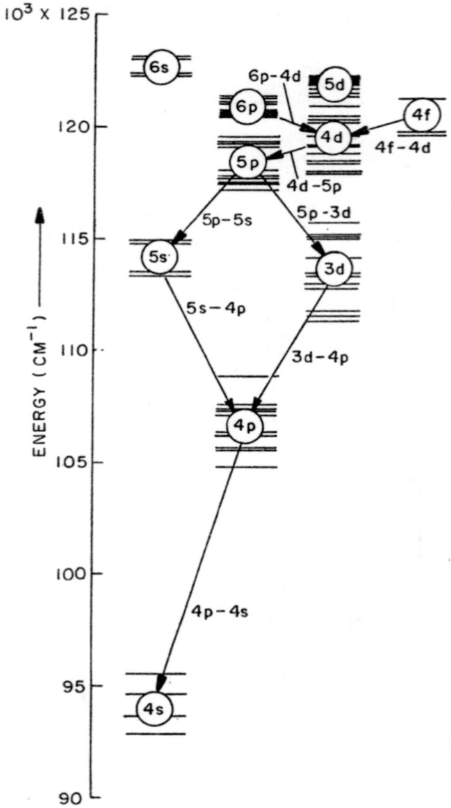

FIG. A.12. Partial energy-level diagram of Ar-I, showing laser transition groups.

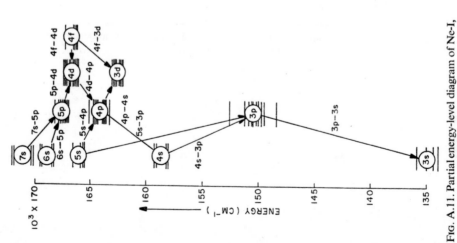

FIG. A.11. Partial energy-level diagram of Ne-I, showing laser transition groups.

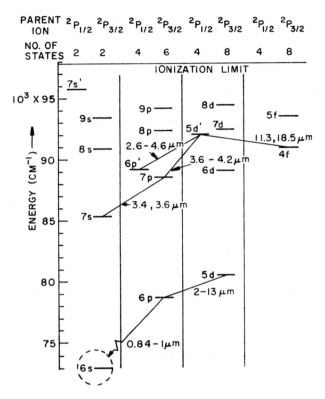

FIG. A.14. Partial energy-level diagram of Xe-I, showing groups of reported laser transitions (wavelengths in μm).

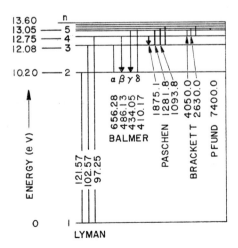

FIG. A.15. Term diagram of H-I, showing the Balmer β and γ laser transitions at 486.13 nm and 434.05 nm, and the 1875.1-nm laser line in the Paschen series. (After G. J. Dezenberg and C. S Willett[172] by courtesy of the IEEE.)

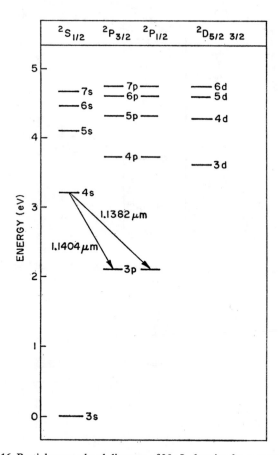

F<small>IG</small>. A.16. Partial energy-level diagram of Na-I, showing laser transitions.

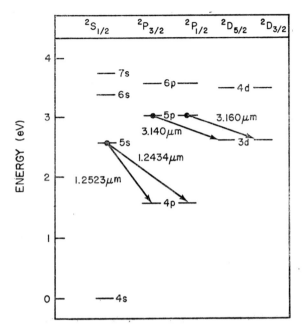

FIG. A.17. Partial energy-level diagram of K-I, showing laser transitions.

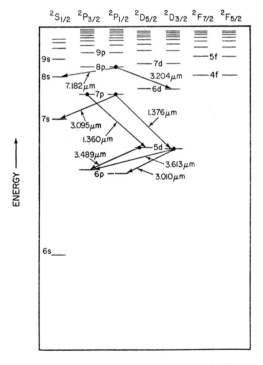

FIG. A.18. Partial energy-level diagram of Cs-I, showing laser transitions.

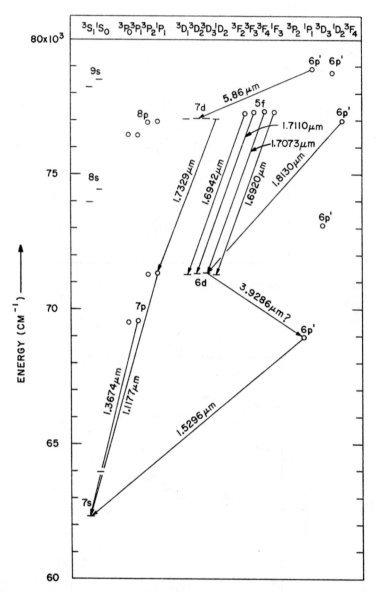

FIG. A.19. Partial term scheme of Hg-I, showing the majority of the reported laser transitions in neutral mercury. Circles represent odd levels; horizontal lines, even levels. (After K. Bockasten, M. Garavaglia, B. A. Lengyel, and T. Lundholm,[225] by courtesy of the Optical Society of America.)

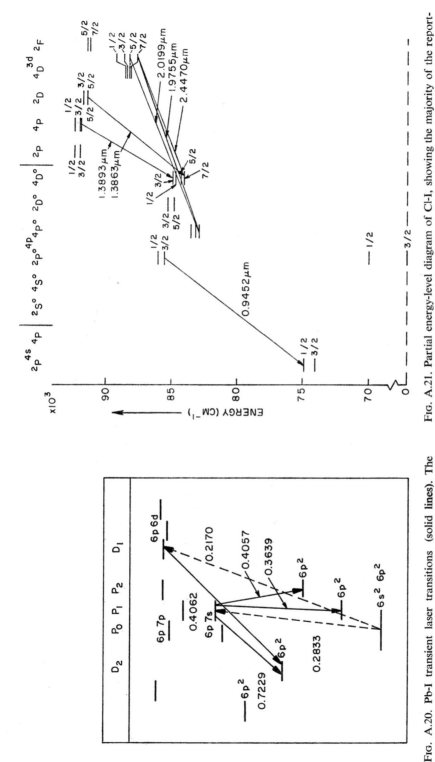

FIG. A.21. Partial energy-level diagram of Cl-I, showing the majority of the reported laser transitions in neutral chlorine (wavelengths in μm). (After S. M. Jarrett, J. Nunez, and G. Gould,[282] by courtesy of the American Institute of Physics.)

FIG. A.20. Pb-I transient laser transitions (solid lines). The broken lines indicate the excitation paths (wavelengths in μm).

401

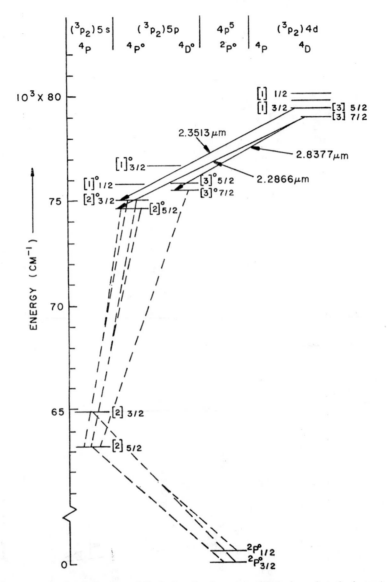

FIG. A.22. Partial energy-level diagram of Br-I, showing laser transitions (wavelengths in μm). The broken lines indicate transitions of strong spectral lines. (After S. M. Jarrett, J. Nunez, and G. Gould,[288] by courtesy of the American Institute of Physics.)

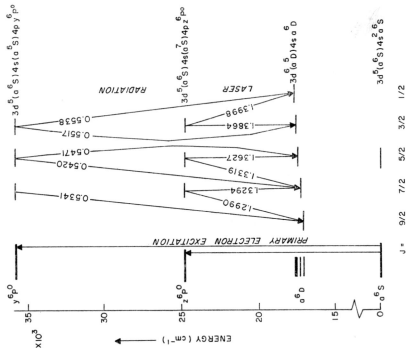

Fig. A.24. Partial energy-level diagram of Mn-I, showing visible and near infrared transient laser transitions and primary electron excitation paths (wavelengths in μm). (After M. Piltch, W. T. Walter, N. Solimene, G. Gould, and W. R. Bennett, Jr.,[306] by courtesy of the American Institute of Physics.)

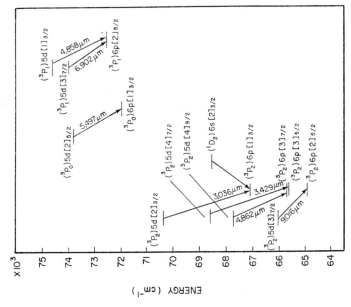

Fig. A.23. Laser transitions in neutral atomic iodine. (After H. Kim, R. A. Paananen, and P. Haust,[305] by courtesy of the IEEE.)

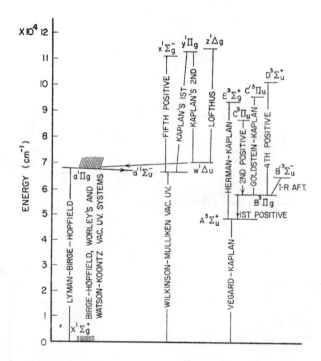

FIG. A.25. **Partial** energy-level diagram of N_2, showing laser transitions (arrow-headed lines). (Adapted from a figure given by R. A. McFarlane.[67])

Appendix Tables 1 to 46

Laser Transitions in Atomic Species

Description of Tables

Laser transitions are arranged according to group in order of increasing atomic number, with Group A elements preceding Group B elements. Because laser transitions in the noble gases constitute the majority of the laser lines reported, they have been placed under the heading of Group O and subdivided into separate tables of neutral and ion-laser transitions.

Where the laser transition has been identified, the calculated wavelength in air is given as reported in the literature. Wavelengths of lines *in vacuo* are given in *italics*. In cases of unidentified laser transitions, and where it is not clear whether or not the lines reported are in vacuum or air wavelength, the wavelength that has been reported is given. Where it is important to identify lines accurately in a small wavelength range, the work that is referenced should be referred to before any reliance is placed on the accuracy of the stated wavelength or on the identification that is listed. In common with what appears to be well-established laser practice, the upper state of a transition is given before the lower state.

The first reference given to a particular laser transition is that reference in which that laser transition was first reported. The other references that follow are in numerical order. The references included are not intended to be exhaustive and do not refer necessarily to investigations on actual laser systems—though most do. The references that are included reflect the intention of the author to provide a source of references to the processes in gas lasers that lead to selective excitation of the upper states involved in population inversion.

Since the emphasis throughout is to give a source of information on the physics of the atomic collision processes involved in gaseous laser systems, limited attention has been paid to provide information on technological characteristics such as high average power, peak power, laser output duration, or gain values.

Because of the difficulty that most people will have in obtaining and translating work published in the Russian literature on gas lasers (with a few exceptions) references to Russian work has been taken only from the Russian-to-English, cover-to-cover translated journals. Up to recently, the majority of Russian gas laser papers have been published in *Optics and Spectroscopy (Optics Spectrosc.)*; *Soviet Physics*; *JETP*; *JETP Letters*; *Technical Physics*; the *Journal of Applied Spectroscopy*; and *Radio Engineering and Electronic Physics*. A recent addition to the cover-to-cover translated Russian laser literature is *Opto-Electronics*.

The tables are arranged as follows:

GROUP O

Table	Element
1	Helium
2	Neon
3	Argon
4	Krypton
5	Xenon

GROUP IA and IB

Table	Element
6	Hydrogen
7	Sodium
8	Potassium
9	Rubidium
10	Cesium
11	Copper
12	Gold

GROUP IIA and IIB

Table	Element
13	Magnesium
14	Calcium
15	Strontium
16	Barium
17	Zinc
18	Cadmium
19	Mercury

GROUP III

Table	Element
20	Boron
21	Indium
22	Thallium

GROUP IVA

Table	Element
23	Carbon
24	Silicon
25	Germanium
26	Tin
27	Lead

GROUP VA

Table	Element
28	Nitrogen
29	Phosphorus
30	Arsenic
31	Antimony
32	Bismuth

GROUP VIA

Table	Element
33	Oxygen
34	Sulfur
35	Selenium
36	Tellurium
37	Thulium

GROUP VIIA and VIIB

Table	Element
(See 6)	Hydrogen
38	Fluorine
39	Chlorine
40	Bromine
41	Iodine
42	Ytterbium
43	Manganese

GROUP VIII

Table	Element
44	Samarium
45	Europium

Miscellaneous (and unidentified) laser transitions, Table 46.

TABLE 1. *Laser transitions in helium (Fig. A.10)*

Wavelength (μm)	Spectrum	Identification	Notes	Excitation	References
1.8685	I	$4f\,{}^3F\!-\!3d\,{}^3D$	0.4 torr He, $D = 6$ mm	CW	1
1.9543	I	$4p\,{}^3P\!-\!3d\,{}^3D$	,,	CW	2, 1, 3
2.0603	I	$7d\,{}^3D\!-\!4p\,{}^3P$	8.0 torr He, with trace of N_2 or Ar, $D = 7$ mm	CW	4, 5, 6
2.05813	I	$2p\,{}^1P_1\!-\!2s\,{}^1S_0$	2.7 torr of He, $D = 1.3$ mm	Short rise-time pulses	7, 8
95.788	I	$3p\,{}^1P_1\!-\!3d\,{}^1D_1$	0.5 torr He, $D = 75$ mm pulsed, 0.1 torr He, $D = 60$ mm (CW)	CW	9, 10
216.3	I	$4p\,{}^1P\!-\!4d\,{}^1D$	0.1 torr He, $D = 60$ mm	CW	10

TABLE 2a. *Neon-neutral laser lines (Fig. A.11)*

Wavelength (μm)	Identification		Notes	References
	Racah	Paschen		
0.540056	$3p'[1/2]_0\!-\!3s[3/2]_1^\circ$	$2p_1\!-\!1s_4$	Transient	11, 12, 13, 14
0.5433	$5s'[1/2]_1^\circ\!-\!3p[1/2]_1$	$3s_2\!-\!2p_{10}$	CW	49
0.58525	$3p'[1/2]_0\!-\!3s'[1/2]_1^\circ$	$2p_1\!-\!1s_2$	Transient	15, 16
0.59393	$5s'[1/2]_1^\circ\!-\!3p[5/2]_2$	$3s_2\!-\!2p_8$	CW[a]	16, 6, 17
0.594483	$3p'[3/2]_2\!-\!3s[3/2]_2^\circ$	$2p_4\!-\!1s_5$	Transient	18, 11, 19, 20, 309
0.60461	$5s'[1/2]_1^\circ\!-\!3p[3/2]_1$	$3s_2\!-\!2p_7$	CW[a]	16, 7
0.61180	$5s'[1/2]_1^\circ\!-\!3p[3/2]_2$	$3s_2\!-\!2p_6$	CW[a]	21, 6, 16, 17
0.614306	$3p[3/2]_2\!-\!3s[3/2]_2^\circ$	$2p_6\!-\!1s_5$	Transient	18, 11, 19, 20, 309
0.6266	$3p'[3/2]_1\!-\!3s'[1/2]_0^\circ$	$2p_5\!-\!1s_3$	Transient	22
0.62937	$5s'[1/2]_1^\circ\!-\!3p'[3/2]_1$	$3s_2\!-\!2p_5$	CW	21, 16
0.63282	$5s'[1/2]_1^\circ\!-\!3p'[3/2]_2$	$3s_2\!-\!2p_4$	CW	23, 6, 16, 24–42, 312
0.63518	$5s'[1/2]_1^\circ\!-\!3p[1/2]_0$	$3s_2\!-\!2p_3$	CW	16
0.64011	$5s'[1/2]_1^\circ\!-\!3p'[1/2]_1$	$3s_2\!-\!2p_2$	CW	21, 16, 43–46
0.73048	$5s'[1/2]_1^\circ\!-\!3p[1/2]_0$	$3s_2\!-\!2p_1$	CW	16
0.88653	$4s'[1/2]_1^\circ\!-\!3p[1/2]_1$	$2s_2\!-\!2p_{10}$	CW	47, 48
0.89886	$4s'[1/2]_0^\circ\!-\!3p[1/2]_1$	$2s_3\!-\!2p_{10}$	CW	47, 48
0.9846	$4s[3/2]_1^\circ\!-\!3p[1/2]_1$	$2s_4\!-\!2p_{10}$	Pulsed	50, 51
0.9665	$4s[3/2]_2^\circ\!-\!3p[1/2]_1$	$2s_5\!-\!2p_{10}$	Pulsed	50, 51
1.0295	$4s'[1/2]_1^\circ\!-\!3p[5/2]_2$	$2s_2\!-\!2p_8$	CW	47, 48
1.0621	$4s'[1/2]_1^\circ\!-\!3p[3/2]_1$	$2s_2\!-\!2p_7$	CW	52, 6, 47, 48, 53, 54
1.0798	$4s'[1/2]_0^\circ\!-\!3p[3/2]_1$	$2s_3\!-\!2p_7$	CW	55, 26, 48, 56
1.0844	$4s'[1/2]_1^\circ\!-\!3p[3/2]_1$	$2s_2\!-\!2p_6$	CW	55, 26, 48, 54, 56
1.1143	$4s[3/2]_1^\circ\!-\!3p[5/2]_2$	$2s_4\!-\!2p_8$	CW	55, 26, 48, 50, 56–61
1.1177	$4s[3/2]_2^\circ\!-\!3p[5/2]_3$	$2s_5\!-\!2p_9$	CW	6, 26, 48–53, 57, 59, 61–63
1.1390	$4s[3/2]_2^\circ\!-\!3p[5/2]_2$	$2s_5\!-\!2p_8$	CW	55, 26, 48, 50, 51, 56
1.1409	$4s'[1/2]_1^\circ\!-\!3p[3/2]_1$	$2s_2\!-\!2p_5$	CW	26, 48, 50, 51, 56

TABLE 2a *(cont.)*

Wavelength (μm)	Identification		Notes	References
	Racah	Paschen		
1.15228	$4s'[1/2]_1^\circ - 3p'[3/2]_2$	$2s_2 - 2p_4$	CW	62, 6, 26, 28, 48, 50, 56–60, 64–68
1.15250	$4s[3/2]_1^\circ - 3p'[3/2]_1$	$2s_4 - 2p_7$	CW	65, 3, 6, 8, 48, 50, 57, 59, 61, 62
1.1602	$4s'[1/2]_1^\circ - 3p[1/2]_0$	$2s_2 - 2p_3$	CW	26, 48, 53, 56, 62
1.1614	$4s[1/2]_0^\circ - 3p'[3/2]_1$	$2s_3 - 2p_5$	CW	62, 6, 26, 48, 56–58, 63
1.1767	$4s'[1/2]_1^\circ - 3p'[1/2]_1$	$2s_2 - 2p_2$	CW	55, 26, 48, 50, 51, 57–60, 68
1.1790	$4s[3/2]_1^\circ - 3p[3/2]_2$	$2s_4 - 2p_6$	CW	47, 48, 50, 51, 61
1.1985	$4s'[1/2]_0^\circ - 3p'[1/2]_1$	$2s_3 - 2p_2$	CW	62, 8, 26, 48, 56–58
1.2066	$4s[3/2]_0^\circ - 3p'[3/2]_2$	$2s_5 - 2p_6$	CW	62, 48, 50, 51, 53, 56–58, 63
1.2460	$4s[3/2]_1^\circ - 3p'[3/2]_1$	$2s_4 - 2p_5$	CW	47, 8, 47, 50
1.2586	$5f[9/2]_{4,\,5} - 3d[7/2]_4^\circ$	$5V - 3d_4'$	CW	69
1.2594	$4s[3/2]_1^\circ - 3p'[3/2]_2$	$2s_4 - 2p_4$	Pulsed	50
1.2689	$4s[3/2]_1^\circ - 3p[1/2]_0$	$2s_4 - 2p_3$	CW	2, 3, 48, 50, 59, 64, 68
1.2767	$4s[3/2]_2^\circ - 3p'[3/2]_1$	$2s_5 - 2p_5$	Pulsed	48
1.2887	$4s[3/2]_1^\circ - 3p'[1/2]_1^\circ$	$2s_4 - 2p_2$	CW	47
1.2912	$4s[3/2]_2^\circ - 3p'[3/2]_2$	$2s_5 - 2p_4$	CW	47, 3, 48, 50, 64, 68
1.3912	$4s[3/2]_2^\circ - 3p'[1/2]_1$	$2s_5 - 2p_2$	Pulsed	50
1.4276	—	—	CW	70
1.4304	—	$5d_3 - 3p_{10}(?)$	CW	70
1.4330	—	—	CW	70
1.4346	—	$5d_5 - 3p_{10}(?)$	CW	70
1.4368	—	$5d_6 - 3p_{10}(?)$	CW	70
1.5231	$4s'[1/2]_1^\circ - 3p'[1/2]_0$	$2s_2 - 2p_1$	CW	55, 8, 26, 48, 56, 71
1.6405	$6s[3/2]_2^\circ - 4p[5/2]_3$	$4s_5 - 3p_9$	CW	69
1.7162	$4s[3/2]_1^\circ - 3p'[1/2]_0$	$2s_4 - 2p_1$	CW	47, 48
1.8210	$4p'[1/2]_1 - 4s[3/2]_1^\circ$	$3p_1 - 2s_4$	CW	48
1.8253	$4f[5/2]_{2,\,3} - 3d[7/2]_3^\circ$	$4Y_{2,\,3} - 3d_4$	CW	6[b]
1.8281	$4f[9/2]_{4,\,5} - 3d[7/2]_4^\circ$	—	CW	72, 69, 73–75
1.8287	$4f[9/2]_4 - 3d[7/2]_3^\circ$	$4V - 3d_4$	CW	72, 69, 74, 75
1.8309	$4f[5/2]_{3,\,3} - 3d[3/2]_2^\circ$	$4Y - 3d_3$	CW	72, 69, 75
1.8408	$4f[5/2]_2 - 3d[3/2]_1^\circ$	$4Y - 3d_2$	CW	72, 69
1.8596	$4f[7/2]_3 - 3d[5/2]_2^\circ$	$4Z - 3d_1''$	CW	72, 69, 75
1.8602	$4f[7/2]_{3,\,4} - 3d[5/2]_3^\circ$	$4Z - 3d_1'$	CW	72, 71, 75
1.9573	$4p'[3/2]_2 - 4s'[3/2]_2^\circ$	$3p_4 - 2s_5$	CW	68, 6, 48, 64, 76
1.9577	$4p'[1/2]_1 - 4s[3/2]_2^\circ$	$4p_2 - 2s_5$	CW	48, 3, 6, 77
2.0351	$4p'[3/2]_2 - 4s[3/2]_1^\circ$	$3p_4 - 2s_4$	CW	73, 2, 3, 6, 48, 59, 64, 76–78
2.0356	$4p'[1/2]_1 - 4s[3/2]_1^\circ$	$3p_2 - 2s_4$	CW	48, 6, 59, 64, 68, 78
2.1019	$4d'[5/2]_2^\circ - 4p[3/2]_2$	$4s'''' - 3p_6(?)$	CW	4, 6, 48
2.1041	$4p'[1/2]_0 - 4s'[1/2]_1^\circ$	$3p_1 - 2s_2$	CW	68, 3, 6, 48, 59, 64, 73
2.1708	$4p[1/2]_0 - 4s[3/2]_1^\circ$	$3p_3 - 2s_4$	CW	3, 6, 48, 59, 68
2.326	$4p[5/2]_2 - 4s[3/2]_2^\circ$	$3p_8 - 3s_5$	CW	48
2.37	—	$4d_{1-4} - 3p_9$	CW	71, 6
2.3951	$4p'[3/2]_2 - 4s[1/2]_1^\circ$	$3p_4 - 2s_2$	CW	71, 3, 48, 64, 76, 79–81
2.3956	$4p'[1/2]_1 - 4s[1/2]_1^\circ$	$3p_2 - 2s_2$	CW	82, 6, 64, 81
2.42	$5s'[1/2]_1^\circ - 4p[1/2]_1$	$3s_2 - 3p_{10}$	CW	83
2.4219	$4d[3/2]_1^\circ - 4p[5/2]_2$	$4d_2 - 3p_8$	CW	84, 6
2.4250	$4p'[3/2]_1 - 4s'[1/2]_1^\circ$	$3p_5 - 2s_2$	CW	48, 3, 64

Table 2a *(cont.)*

Wavelength (μm)	Identification		Notes	References
	Racah	Paschen		
2.5400	$4d[1/2]_1^\circ - 4p[3/2]_2$	$4d_5 - 3p_6$	CW	85, 78
2.5524	$4p[1/2]_1 - 4s[3/2]_2^\circ$	$3p_{10} - 2s_5$	CW	48
2.7582	$4d[3/2]_1^\circ - 4p[1/2]_0$	$4d_2 - 3p_3$	CW	85, 78
2.7826	$5s'[1/2]_0^\circ - 4p[3/2]_1$	$3s_3 - 3p_7$	CW	85, 6, 78
2.864	$4d'[3/2]_1^\circ - 4p[3/2]_2$	$4s_1' - 3p_4$	CW	85
2.9456	$5s[3/2]_1^\circ - 4p[1/2]_1$	$3s_4 - 3p_{10}$	CW	85, 6, 78
2.9676	$4d[3/2]_1^\circ - 4p'[3/2]_1$	$4d_2 - 3p_5$	CW	85, 78
2.9813	$4d[3/2]_2^\circ - 4p'[3/2]_1$	$4d_3 - 3p_5$	CW	85, 78
3.0278	$\cdot\,4d[3/2]_2^\circ - 4p'[1/2]_1$	$4d_3 - 3p_2$	CW	55, 6, 78
3.0720	$5s'[1/2]_2^\circ - 4p[1/2]_0$	$3s_2 - 3p_3$	CW	86
3.3183	$5s[3/2]_1^\circ - 4p[5/2]_2$	$3s_4 - 3p_8$	CW	85, 75, 78, 84
3.3342	$5s'[1/2]_2^\circ - 4p'[3/2]_1$	$3s_2 - 3p_5$	CW	85, 64, 75, 78, 84
3.3362	$5s[3/2]_2^\circ - 4p[5/2]_3$	$3s_5 - 3p_9$	CW	85, 75, 78, 84
3.3813	$7s'[1/2]_0^\circ - 5p[3/2]_1$	$5s_3 - 4p_5$	CW	85, 78
3.3840	$7s'[1/2]_0^\circ - 5p'[1/2]_1$	$5s_3 - 4p_2$	CW	85, 78
3.3903	$5s'[1/2]_1^\circ - 4p'[1/2]_1$	$3s_2 - 3p_2$	CW	78, 75, 81, 84
3.3913	$5s'[1/2]_1^\circ - 4p'[3/2]_2$	$3s_2 - 3p_1$	CW	87, 2, 3, 8, 16, 21, 26, 31, 36, 39, 71, 78, 81, 312
3.4481	$5s[3/2]_1^\circ - 4p[3/2]_1$	$3s_4 - 3p_7$	CW	85, 61, 75, 78
3.5845	$5s[3/2]_2^\circ - 4p[3/2]_2$	$3s_5 - 3p_6$	CW	78, 6, 75
3.7746	$4p'[1/2]_0 - 3d[3/2]_1^\circ$	$3p_1 - 3d_2$	CW	78, 75
3.9817	$5s[3/2]_1^\circ - 4p[1/2]_0$	$3s_4 - 3p_3$	CW	78, 75
4.2183	$5s'[1/2]_1^\circ - 4p'[1/2]^\circ$	$3s_2 - 3p_1$	CW	86
5.4048	$4p'[1/2]_0 - 3d'[3/2]_1^\circ$	$3p_1 - 3s_1$	CW	78, 75, 84
5.6667	$4p[1/2]_0 - 3d[3/2]_1^\circ$	$3p_3 - 3d_2$	CW	78, 84, 85
7.3228	$6s[3/2]_2^\circ - 5p[5/2]_3$	$4s_5 - 4p_9$	CW	78, 84, 85
7.4221	$6s'[1/2]_0^\circ - 5p'[1/2]_1$	$4s_3 - 4p_2(?)$	CW	78, 85
	or	or		
	$6s'[1/2]_0^\circ - 5p'[3/2]_1$	$4s_3 - 4p_5$		
7.4237	$5p'[1/2]_1 - 4d[3/2]_2^\circ$	$4p_2 - 4d_3$	CW	78
7.4699	$4p[3/2]_2 - 3d[5/2]_2^\circ$	$3p_6 - 3d_1$	CW	85
7.4799	$4p[3/2]_2 - 3d[5/2]_3^\circ$	$3p_6 - 3d_1'$	CW	78, 75
7.4994	$6s[3/2]_2^\circ - 5p[5/2]_2$	$4s_5 - 4p_8$	CW	85, 6, 78
7.5292	$6s[3/2]_1^\circ - 5p[3/2]_1$	$4s_4 - 4p_7$	CW	84, 6
7.6163	$4p[3/2]_1 - 3d[5/2]_2^\circ$	$3p_7 - 3d_1$	CW	85, 75, 78, 84
7.6461	$5p'[3/2]_1 - 4d[5/2]_2^\circ$	$4p_5 - 4d_1$	CW	85
7.6510	$4p[5/2]_2 - 3d[7/2]_3^\circ$	$3p_8 - 3d_4$	CW	75, 78, 84
7.6925	$4p'[3/2]_2 - 3d'[5/2]_3^\circ$	$3p_4 - 3s_1'''$	CW	85
7.7015	$4p'[3/2]_2 - 3d[5/2]_3^\circ$	$3p_4 - 3d_1'$	CW	75, 78, 84
7.7407	$4p[5/2]_2 - 3d[3/2]_2^\circ$	$3p_8 - 3d_3$	CW	85, 78
7.7655	$4p'[1/2]_1 - 3d'[5/2]_2^\circ$	$3p_2 - 3s_1''''$	CW	78, 75, 84
7.7815	$6s[3/2]_2^\circ - 5p[3/2]_1$	$4s_5 - 4p_7$	CW	85, 78
7.8368	$6s[3/2]_2^\circ - 5p[3/2]_2$	$4s_5 - 4p_6$	CW	85, 75, 78, 84
8.0088	$4p'[3/2]_1 - 3d'[5/2]_2^\circ$	$3p_5 - 3s_1''''$	CW	85, 75, 78, 84
8.0621	$4p[5/2]_3 - 3d[7/2]_4^\circ$	$3p_9 - 3d_4'$	CW	85, 75, 78, 84
8.3370	$4p'[3/2]_2 - 3d[5/2]_2^\circ$	$3p_5 - 3d_1''$	CW	85, 78
or	or	or		
8.3495	$4p[5/2]_2 - 3d[5/2]_3^\circ$	$3p_8 - 3d_1'$		
8.8413	$4p[5/2]_3 - 3d[5/2]_2^\circ$	$3p_9 - 3d_1''$	CW	85, 78
or	or	or		
8.8553	$4p[5/2]_3 - 3d[5/2]_3^\circ$	$3p_9 - 3d_1'$		

TABLE 2a *(cont.)*

Wavelength (μm)	Identification		Notes	References
	Racah	Paschen		
9.0896	$6s[3/2]_1^\circ - 5p[1/2]_0$	$4s_4 - 4p_3$	CW	85, 75, 78, 84
10.063	$4p[1/2]_1 - 3d[1/2]_1^\circ$	$3p_{10} - 3d_5$	CW	85, 78
10.981	$4p[1/2]_1 - 3d[3/2]_2^\circ$	$3\boldsymbol{p}_{10} - 3d_3$	CW	85, 75, 78, 84
11.861	$5p[1/2]_1 - 5s'[1/2]_0^\circ$	$4p_{10} - 3s_5$	CW	78
11.893	$5p[1/2]_0 - 4d[3/2]_1^\circ$	$4p_3 - 4d_2$	CW	85
12.835	$5p'[1/2]_0 - 4d'[3/2]_1^\circ$	$4p_1 - 4s_1'$	CW	85
12.832	$5p'[1/2]_0 - 4d[3/2]_1^\circ$	$4p_1 - 4d_2$	CW	78
13.759	$7s'[1/2]_1^\circ - 6p'[3/2]_1$	$5s_2 - 5p_4$	CW$^\circ$	75, 78, 88
or	or	or		
13.759	$4d'[5/2]_3^\circ - 4f[5/2]_3$	$4s_1'''' - 4Y$		
14.93	—	—	CW	78
16.638	$5p[3/2]_2 - 4d[5/2]_2^\circ$	$4p_6 - 4d_1''$	CW	85, 78
16.668	$5p[3/2]_2 - 4d[5/2]_3^\circ$	$4p_6 - 4d_1'$	CW	78
16.893	$5p[3/2]_1 - 4d[5/2]_2^\circ$	$4p_7 - 4d_1''$	CW	75, 78
16.947	$5p[5/2]_2 - 4d[7/2]_3^\circ$	$4p_8 - 4d_4$	CW	85, 75, 78
17.158	$5p'[3/2]_2 - 4d'[5/2]_3^\circ$	$4p_4 - 4s_1''''$	CW	75, 78
17.189	$5p'[3/2]_2 - 4d'[3/2]_2^\circ$	$4p_4 - 4s_1''$	CW	78
17.804	$5p'[1/2]_1 - 4d'[3/2]_2^\circ$	$4p_2 - 4s_1''$	CW	75, 78
17.841	$5p'[3/2]_1 - 4d'[5/2]_2^\circ$	$4p_5 - 4s_1''''$	CW	75, 78
17.888	$5p[5/2]_3 - 4d[7/2]_4^\circ$	$4p_9 - 4d_4'$	CW	85, 75, 78
18.396	$5p[5/2]_2 - 4d[5/2]_2^\circ$	$4p_8 - 4d_1''$	CW	78, 75
20.480	$6p[1/2]_0 - 5d[1/2]_1^\circ$	$5p_3 - 5d_5$	CW	75, 78
21.752	$6p[1/2]_0 - 5d[3/2]_1^\circ$	$5p_3 - 5d_2$	CW	75, 78
22.836	$5p[1/2]_1 - 4d[3/2]_2^\circ$	$4p_{10} - 4d_3$	CW	75, 78
25.423	$6p'[1/2]_0 - 5d'[3/2]_1^\circ$	$5p_1 - 5s_5'$	CW	75, 78
28.053	$6p[3/2]_1 - 5d[1/2]_0^\circ$	$5p_7 - 5d_6$	CW	75, 78
31.553	$6p[3/2]_2 - 5d[5/2]_3^\circ$	$5p_5 - 5d_1'$	CW	78
31.928	$6p[3/2]_1 - 5d[5/2]_2^\circ$	$5p_7 - 5d_1''$	CW	89, 90
32.016	$6p[5/2]_2 - 5d[7/2]_3^\circ$	$5p_8 - 5d_4$	CW	87
32.516	$6p'[3/2]_2 - 5d'[5/2]_3^\circ$	$5p_4 - 5s_1'''$	CW	78
33.824	$6p'[3/2]_1 - 5d'[5/2]_2^\circ$	$5p_5 - 5s_1'''$	CW	78
or	or	or		
33.837	$6p[5/2]_3 - 5d[7/2]_4^\circ$	$5p_9 - 5d_4'$		
34.542	$6p'[1/2]_1 - 5d[3/2]_2^\circ$	$5p_2 - 5s_1''$	CW	78
34.679	$6p[5/2]_2 - 5d[5/2]_2^\circ$	$5p_8 - 5d_1''$	CW	89, 90
35.602	$7p[1/2]^\circ - 6d[3/2]_1^\circ$	$6p_3 - 6d_2$	CW	89, 6, 90
37.231	$7p'[1/2]_0 - 6d'[3/2]_1^\circ$	$6p_1 - 6s_1$	CW	89, 90
41.741	$6p[1/2]_1 - 5d[3/2]_2^\circ$	$5p_{10} - 5d_3$	CW	89, 90
50.705	$7p[3/2]_2 - 6d[3/2]_2^\circ$	$6p_6 - 6d_3$	CW	91
52.425	$7p'[1/2]_1 - 6d'[3/2]_2^\circ$	$6p_2 - 6s_1''$	CW	91
53.486	$7p[3/2]_2 - 6d[5/2]_3^\circ$	$6p_6 - 6d_1'$	CW	89, 90
54.019	$7p[3/2]_1 - 6d[5/2]_2^\circ$	$6p_7 - 6d_1''$	CW	89, 90
54.117	$7p[5/2]_2 - 6d[7/2]_3^\circ$	$6p_8 - 6d_4$	CW	89, 90
55.537	$7p'[3/2]_1 - 6d'[5/2]_2^\circ$	$6p_5 - 6s_1''''$	CW	91
57.355	$7p[5/2]_3 - 6d[7/2]_4^\circ$	$6p_9 - 6d_4'$	CW	89, 90
68.329	$7p[1/2]_1 - 6d[3/2]_2^\circ$	$6p_{10} - 6d_3$	CW	92, 6
72.108	$8p'[1/2]_0 - 7d'[3/2]_1^\circ$	$7p_1 - 7s_1'$	CW	91
85.047	$8p[3/2]_2 - 7d[5/2]_3^\circ$	$7p_6 - 7d_1'$	CW	92, 6
86.962	$8p'[3/2]_2 - 7d[5/2]_2^\circ$	$7p_4 - 7s_1''''$	CW	91

TABLE 2a *(cont.)*

Wavelength (μm)	Identification		Notes	References
	Racah	Paschen		
88.471	$8p[3/2]_1-7d[5/2]_2^\circ$	$7p_9-7d_1''$	CW	91, 6
89.859	$8p[5/2]_3-7d[7/2]_3^\circ$	$7p_9-7d_4$	CW	91
93.02	—	—	CW	91
106.07	$10p[1/2]_1-9d[3/2]_1^\circ$	$9p_3-9d_2$	CW	91, 6
124.52	$9p[3/2]_0-8d[5/2]_2^\circ$	—	CW	91
or	or			
124.76	$9p[3/2]_2-8d[5/2]_3^\circ$			
126.1	—	—	CW	91
132.8	—	—	CW	91

ªAlthough oscillation between the Ne $3s_2$ and $2p$ levels is normally observed only in a He–Ne mixture it has been reported to have been observed on the 0.59393 and 0.61180-μm lines under pulsed conditions in pure neon.[17] In view of the excitation used and the inability to obtain oscillation at 0.6328 μm it is likely that the two lines mentioned above should have been identified as 0.5944-μm and 0.6143-μm transient laser lines.

ᵇThis line appears in ref. 6 but no reference to it appears to have been made.

ᶜThis is the most likely transition to give the 13.8-μm line,[78] selective excitation of the Ne-$5s_2$ state occurs via excitation transfer from He $2p\ ^1P_1$ atoms.[88]

TABLE 2b. *Neon-ion laser lines*

Wavelength (μm)	Spectrum	Identification	References
0.235796	IV	$3p\ ^4D_{7/2}^\circ-3s\ ^4P_{5/2}$	93, 94
0.247340	III(?)	—	95, 93, 94
0.267790	III	$3p\ ^3P_{2,0}-3s\ ^3S_1^\circ$	93, 15, 94
0.267864	III	$3p\ ^3P_1-3s\ ^3S_1^\circ$	93, 15, 94, 96, 309
0.277765	III	$3p'\ ^3D_3-3s'\ ^3D_3^\circ$	15, 94
0.286688	IV(?)	—	93, 15, 94
0.307983	III	$3p''\ ^3P_0-3s''\ ^1P_1^\circ$	97
0.331975	II	$3p'\ ^2P_{1/2}^\circ-3s'\ ^2D_{3/2}$	97, 93, 94, 96, 98, 309
0.332377	II	$3p\ ^2P_{3/2}^\circ-3s\ ^2P_{3/2}$	97, 15, 19, 93, 94, 98–101, 309
0.332437	—	—	97, 94, 96, 101
0.332717	II	$3p\ ^4D_{3/2}^\circ-3s\ ^4P_{3/2}$	15, 94, 98
0.332923	II	$3d\ ^4D_{7/2}-3p\ ^4D_{7/2}^\circ$	97, 94
0.333067	II	$3d\ ^4F_{5/2}-3p\ ^2D_{5/2}^\circ$	97
0.333114	III	$3p''\ ^3D_2-3s''\ ^1P_1^\circ$	97, 94
0.334556	II	$3p'\ ^2P_{3/2}^\circ-3s'\ ^2D_{5/2}$	97, 19, 93, 94, 96, 101, 309
0.337830	II	$3p\ ^2P_{1/2}^\circ-3s\ ^2P_{1/2}$	97, 15, 19, 93, 94, 100, 101,
0.337888	II	$3d\ ^4D_{1/2}-3p\ ^4D_{1/2}^\circ$	97
0.339286	II	$3p\ ^2P_{3/2}^\circ-3s\ ^2P_{1/2}$	93, 15, 94
0.339320	II	$3d\ ^2D_{3/2}-3p\ ^2D_{5/2}^\circ$	97, 94, 101
0.364495	II	$3d\ ^2P_{3/2}-3p\ ^2P_{1/2}^\circ$	97
0.371309	II	$3p\ ^2D_{5/2}^\circ-3s\ ^2P_{3/2}$	99, 101

TABLE 3a. *Argon-neutral laser lines (Fig. A.12)*

Wavelength in air (nm)	Identification		Notes	References
	Racah	Paschen		
0.706721	$4p'[3/2]_2-4s[3/2]_2^\circ$	$2p_3-1s_5$	Transient	19
0.7503	$4p'[1/2]_0-4s'[1/2]_1^\circ$	$3p_1-1s_2$	Pulsed	15
0.8780[a]	—	—	Pulsed	102
1.0935	—	—	Pulsed	102
1.21[b]	$3d[5/2]_3^\circ-4p[5/2]_3$	$3d_1'-2p_9(?)$	Pulsed	53
1.213962	$3d'[3/2]_1^\circ-4p'[3/2]_1$	$3s_1'-2p_4$	Transient	103
1.240275	$3d[3/2]_1^\circ-4p[3/2]_1$	$3d_2-2p_7$	Transient	103
1.270221	$3d'[3/2]_1^\circ-4p'[1/2]_1$	$3s_1'-2p_2$	Transient	103, 8, 104, 105, 313
1.28[e]	$3d[5/2]_2^\circ-4p[5/2]_2$	$3d_1''-2p_8$	Pulsed	53, 313
1.3472	$7d[3/2]_2^\circ-5p[3/2]_2$	$7d_3-3p_6$	CW	106
1.4093640	$3d[3/2]_1^\circ-4p[1/2]_0$	$3d_2-2p_5$	Pulsed	60
1.618	$5s[3/2]_2^\circ-4p'[3/2]_2$	$2s_5-2p_3$	CW	4, 56
1.694058	$3d[3/2]_1^\circ-4p[3/2]_2$	$3d_3-2p_6$	CW	4, 56, 60, 313
1.791473	$3d[1/2]_1^\circ-4p[3/2]_2$	$3d_5-2p_6$	CW	4, 6, 8, 56, 104, 105, 106, 107, 313
	or	or		
	$3d[1/2]_0^\circ-4p[3/2]_1$	$3d_6-2p_7$		
1.8167	$3d[7/2]_4^\circ-4p[5/2]_3$	$3d_4'-2p_9$	CW	3
2.0616	$3d[3/2]_2^\circ-4p'[3/2]_2$	$3d_3-2p_3$	CW	56, 4, 107, 313
2.0986	$3d[1/2]_1^\circ-4p[1/2]_0$	$3d_5-2p_5$	CW	79, 3, 6
2.133	$3d[1/2]_1^\circ-4p[3/2]_1$	$3d_5-2p_4$	CW	85
2.1534	$3d[3/2]_2^\circ-4p'[1/2]_1$	$3d_3-2p_2$	CW	79, 3
2.2044	$3d[1/2]_1^\circ-4p'[3/2]_1$	$3d_6-2p_4$	CW	85, 8, 78, 106, 107
or	or	or		
2.2083	$3d[1/2]_1^\circ-4p'[3/2]_2$	$3d_5-2p_3$		
2.313320	$3d[1/2]_1^\circ-4p'[1/2]_1$	$3d_5-2p_2$	CW	78, 60, 85
2.3973	$3d[1/2]_0^\circ-4p'[1/2]_1$	$3d_6-2p_2$	CW	85, 78, 104, 107, 313
2.5014	$6d'[3/2]_2^\circ-6p[1/2]_1$	$6s_1''-4p_{10}$	CW	85, 78
2.5494	$5p[5/2]_3-3d[7/2]_3^\circ$	$3p_9-3d_4$	CW	85, 78
2.5512	$5p[1/2]_0-5s[3/2]_1^\circ$	$3p_5-2s_4$	CW	78, 313
2.5634	$6d[3/2]_2^\circ-6p[5/2]_3$	$6s_1-4p_9$	CW	85
2.5668	$5p[1/2]_0-5s[1/2]_1^\circ$	$3p_1-2s_2$	CW	78, 313
2.6843	$5p[3/2]_1-3d[5/2]_2^\circ$	$3p_7-3d_1$	CW	85, 78
2.7364	$5p'[1/2]_1-3d'[3/2]_2^\circ$	$3p_2-3s_1$	CW	85, 78
2.8202	$5p'[3/2]_1-5s'[1/2]_0^\circ$	$3p_4-2s_3$	CW	78, 107
or	or	or		
2.8245	$5p[3/2]_2-5s[3/2]_1^\circ$	$3p_6-2s_4$		
2.8625	$5p'[3/2]_2-5s'[1/2]_1^\circ$	$3p_4-2s_2$	CW	107
2.8783	$5p[5/2]_3-5s[3/2]_2^\circ$	$3p_9-2s_5$	CW	78
2.8843	$5p[3/2]_2-3d[5/2]_3^\circ$	$3p_6-3d_1$	CW	85, 78
2.9280	$5p[1/2]_0-3d[3/2]_1^\circ$	$3p_5-3d_2$	CW	85, 78
2.9796	$5p[5/2]_2-5s[3/2]_1^\circ$	$3p_8-2s_4$	CW	85, 78
3.0462	$5p[5/2]_2-3d[5/2]_3^\circ$	$3p_8-3d_1$	CW	85, 78
3.0996	$5p[5/2]_3-3d[5/2]_3^\circ$	$3p_9-3d_1$	CW	85, 78
3.1333	$5p[1/2]_1-5s[3/2]_2^\circ$	$3p_{10}-2s_5(?)$	CW	78
3.1346	$6p'[3/2]_1-4d[5/2]_2^\circ$	$4p_3-4d_1$	CW	85, 313
3.708	$4d[3/2]_1^\circ-5p[3/2]_1$	$3d_2-2p_7$	Pulsed	313
4.7330	$5p[1/2]_1-3d'[3/2]_1^\circ$	$3p_{10}-3s_1$	CW	85
4.9160	$6p'[3/2]_2-4d'[3/2]_2^\circ$	$4p_3-4s_1$	CW	85, 78
4.9213	$5d[5/2]_2^\circ-4f[7/2]_3$	$5d_1''-4U_3$	CW	78
5.1216	$6p[5/2]_3-4d[7/2]_3$	$4p_9-4d_4$	CW	85, 78

TABLE 3a (*cont.*)

Wavelength in air (nm)	Identification		Notes	References
	Racah	Paschen		
5.1218	$5d[7/2]_3^\circ - 4f[9/2]_4$	$5d_4 - 4V$	CW	75, 78
5.4680	$5d[7/2]_4^\circ - 4f[9/2]_5$	$5d_4' - 4V$	CW	75, 78
or	or	or		
5.4694	$5d[7/2]_4^\circ - 4f[9/2]_4$	$5d_4' - 4V$		
5.8038	—	—	Pulsed	8
5.8477	$6p[1/2]_0 - 6s[3/2]_1^\circ$	$4p_5 - 3s_4$	CW	75, 78
5.8666	$6p[3/2]_2 - 4d[5/2]_3^\circ$	$4p_6 - 4d_1$	CW	85
6.0531	$4d[1/2]_1^\circ - 5p[5/2]_2$	$4d_5 - 3p_8$	CW	85, 78
6.9429	$4d[3/2]_1^\circ - 5p'[3/2]_1$	$4d_2 - 3p_4$	CW	85, 78
6.9448	$6p'[1/2]_1 - 6s'[1/2]_1^\circ$	$4p_2 - 3s_2$	CW	78
7.2166	$6p[1/2]_1 - 6s[3/2]_2^\circ$	$4p_{10} - 3s_5$	CW	75, 78
7.2932	—	—	Pulsed	8
7.8003	$4f[3/2]_2 - 4d[3/2]_2^\circ$	$4X_2 - 4d_3$	CW	78
or	or	or		
7.8023	$4f[3/2]_1 - 4d[3/2]_2^\circ$	$4X_1 - 4d_3$		
7.8063	$7s'[1/2]_1^\circ - 4p'[1/2]_1$	$4s_2 - 2P_2$	CW	85
12.141	$4d'[3/2]_1^\circ - 4f[3/2]_{1,2}$	$4s_1' - 4X$	CW	85, 75, 78
or				
12.146				
15.037	$5d'[3/2]_2^\circ - 5f[5/2]_3$	$5s_1'' - 5Y$	CW	75, 78
or				
15.042				
26.944	$4d'[3/2]_2^\circ - 4f[5/2]_3$	$4s_1'' - 4Y$	CW	75, 78

[a]This line is probably the Ar-II line at 0.877186 μm.[94]
[b]The Ar-I line at 1.21 μm under pulsed conditions is possibly the Ar-I line reported at 1.21396 μm.
[c]The Ar-I line at 1.28 μm is possibly the same as the Ar-I line reported at 1.27022 μm.
Unidentified laser lines at 1.222, 1.246 and 1.276 μm observed in a pulsed Hg–Ar mixture[109] have been placed under Laser Transitions in Mercury.

TABLE 3b. *Argon-ion laser lines*

Wavelength (μm)	Spectrum	Identification	References
0.262493	IV	$4p'\ ^2D_{5/2}^\circ - 4s'\ ^2D_{5/2}$	95, 93, 94
0.275360	III	—	93, 15, 94
0.288416	III	$4p'\ ^3P_2 - 4s\ ^3D_3^\circ$	97, 93, 94
0.291300	IV	$4p\ ^2D_{5/2}^\circ - 4s\ ^2P_{3/2}$	97, 15, 93, 94, 96, 309
0.292627	IV	$4p\ ^2D_{3/2}^\circ - 4s\ ^2P_{1/2}$	97, 15, 93, 94
0.300296	II(?)	$5f[2]_{5/2}^\circ - 3d'\ ^2D_{3/2}$	97, 15, 93, 94, 96, 309
0.302405	III	$4p''\ ^3D_3 - 4s''\ ^3P_2^\circ$	15, 94
0.304705	II(?)	$4f[3]_{5/2}^\circ - 3d'\ ^2D_{7/2}$	15
0.305484	III	$4p''\ ^3D_2 - 4s''\ ^3P_1^\circ$	15, 94
0.330616	II(?)	—	97
0.332365	III	$4p'\ ^3D_3 - 3d''\ ^3D_3^\circ$	97
0.333613	III	$4p'\ ^3F_4 - 4s'\ ^3D_3^\circ$	97, 15, 93, 94, 101
0.334472	III	$4p'\ ^3F_3 - 4s'\ ^3D_2^\circ$	97, 15, 93, 94, 101

TABLE 3b (cont.)

Wavelength (μm)	Spectrum	Identification	References
0.335849	III	$4p'\,^3F_2\text{–}4s'\,^3D_1^\circ$	97, 15, 93, 94, 101
0.337857	II	$(^1D)6p\,^4P_{1/2}^\circ\text{–}$ $\text{–}(^1D)3d\,^2D_{3/2}$	97
0.35112	III	$4p\,^3P_2\text{–}4s\,^3S_1^\circ$	110, 15, 19, 93, 94, 96, 97, 99, 100, 101, 111, 309
0.351418	III	$4p^3\,P_1\text{–}4s\,^3S_1^\circ$	110, 15, 19, 94, 97
0.351446	II	$4d\,^4D_{5/2}\text{–}4p\,^4P_{3/2}^\circ$	97
0.3545	II	—	112
0.357661	II	$4d\,^4F_{7/2}^\circ\text{–}4p\,^4D_{5/2}$	15, 94
0.363733	II	$4d^4\,^4F_{7/2}\text{–}4p\,^2D_{3/2}^\circ$	97
0.363789	III	$4p'\,^1F_3\text{–}4s'\,^1D_2^\circ$	110, 15, 93, 94, 96, 97, 99, 101, 113, 309
0.37052	III(?)	—	15, 94
0.364548	II	$\{\ 4p'\,^2F_{7/2}^\circ\text{–}3d\,^5F_{5/2}$	97
	II	$\ \ 4d\,^2F_{5/2}\text{–}4p\,^2D_{3/2}^\circ$	97
0.379532	III	$4p''\,^3D_3\text{–}3d''\,^3P_2^\circ$	97, 15, 94
0.385829	III	$4p''\,^3D_2\text{–}3d''\,^3P_1^\circ$	15, 94
0.40648	II(?)	—	97
0.4089[a]	III(?)	—	97
0.414671	III	$4p'\,^3P_2\text{–}4s''\,^3P_2^\circ$	15, 94
0.418292	III(?)	—	15, 94, 101, 132
0.437075	II	$4p'\,^2D_{3/2}^\circ\text{–}3d\,^2D_{3/2}$	15, 94, 98, 114
0.438375	II	$4p\,^4S_{2/2}^\circ\text{–}4s\,^2D_{3/2}$	115
0.448181	II	$4p'\,^2D_{5/2}^\circ\text{–}3d\,^2D_{5/2}$	98, 94, 116
0.454504	II	$4p\,^2P_{3/2}^\circ\text{–}4s\,^2P_{3/2}$	117, 15, 94, 98, 118, 119, 120
0.457934	II	$4p\,^2S_{1/2}^\circ\text{–}4s\,^2P_{1/2}$	117, 15, 94, 98, 111, 112, 118–23
0.460955	II	$4p'\,^2F_{7/2}^\circ\text{–}4s'\,^2D_{5/2}$	15, 94, 118
0.46453	—	—	15, 94
0.465789	II	$4p\,^2P_{1/2}^\circ\text{–}4s\,^2P_{3/2}$	117, 15, 94, 96, 98, 118–21
0.472685	II	$4p\,^2D_{3/2}^\circ\text{–}4s\,^2P_{3/2}$	117, 15, 94, 98, 120
0.476486	II	$4p\,^2P_{3/2}^\circ\text{–}4s\,^2P_{1/2}$	117, 15, 94, 98, 111, 112, 119–25, 309
0.487986	II	$4p\,^2D_{5/2}^\circ\text{–}4s\,^2P_{3/2}$	117, 15, 20, 94, 98, 111, 113, 118–23, 125–31
0.488903	II	$4p\,^2P_{1/2}^\circ\text{–}4s\,^2P_{1/2}$	15, 94, 114
0.496507	II	$4p\,^2D_{3/2}^\circ\text{–}4s\,^2P_{1/2}$	117, 15, 94, 98, 111, 112, 118–23
0.499255	III(?)	—	94
0.501715	II	$4p'\,^2F_{5/2}^\circ\text{–}3d\,^2D_{3/2}$	117, 15, 94, 98, 111, 118–23
0.514178	II	$4p'\,^2F_{7/2}^\circ\text{–}3d\,^2D_{5/2}$	15, 94, 114
0.514530	II	$4p\,^4D_{5/2}^\circ\text{–}4s\,^2P_{3/2}$	117, 15, 94, 98, 111, 112, 118–23, 125, 126, 128, 129
0.528688	II	$4p\,^4D_{3/2}^\circ\text{–}4s\,^2P_{1/2}$	117, 15, 94, 98
0.550220	III	$4p'\,^3D_3\text{–}4s''\,^3P_2^\circ$	15, 94
0.6730	—	—	132
0.7065	—	—	112
0.734805	II	$3d'\,^2D_{5/2}\text{–}4p\,^2D_{5/2}^\circ$	19
0.750514	II	$3d'\,^2P_{3/2}\text{–}4p\,^2S_{1/2}^\circ$	19
0.877186	II	$4p\,^2P_{3/2}\text{–}4s'\,^2D_{5/2}$	102, 94
1.092344	II	$4p\,^2P_{3/2}^\circ\text{–}3d\,^2D_{5/2}$	133, 94

[a]Possibly a Si-IV-line at 0.408886 μm.[15]

TABLE 4a. *Krypton-neutral laser lines (Fig. A.13)*

Wavelength (μm)	Identification		Notes	References
	Racah	Paschen		
0.3050	—	—	Pulsed	15, 6, 93
0.7603	$5p[3/2]_2-5s[3/2]_2^\circ$	$2p_6-1s_5$	Pulsed	313
0.810436	$5p[3/2]_2-5s[3/2]_2^\circ$	$2p_6-1s_5$	Pulsed	20, 19, 60, 313
0.8589	—	—	Pulsed	102
1.1457481	$6s[3/2]_1^\circ-5p[1/2]_1$	$2s_4-2p_{10}$	Pulsed	60
1.3177412	$6s[3/2]_1^\circ-5p[5/2]_2$	$2s_4-2p_8$	Pulsed	60
1.3622416	$4d[3/2]_1^\circ-5p[5/2]_2$	$3d_2-2p_8$	Pulsed	60
1.4426793	$6s[3/2]_1^\circ-5p[3/2]_1$	$2s_4-2p_7$	Pulsed	60, 313
1.4765471	$6s[3/2]_1^\circ-5p[3/2]_2$	$2s_4-2p_8$	Pulsed	60, 313
1.6853498	$4d[7/2]_3^\circ-5p[5/2]_3$	$3d_4-2p_9$	Pulsed	60, 313
1.68968	$4d[1/2]_1^\circ-5p[1/2]_1$	$3d_5-2p_{10}$	CW	56, 60, 313
1.694	$4d[5/2]_2^\circ-5p[3/2]_1$	$3d_1''-2p_7$	CW	4, 56
1.784	$4d[1/2]_0^\circ-5p[1/2]_1$	$3d_6-2p_{10}$	CW	4, 56, 313
1.819	$4d'[5/2]_2^\circ-5p'[3/2]_2$	$3s_1''''-2p_2$	CW	4, 56, 313
1.921	$8s[3/2]_1^\circ-6p[5/2]_2$	$4s_5-3p_8$	CW	4, 56
2.116	$4d[3/2]_2^\circ-5p[3/2]_1$	$3d_3-2p_7$	CW	56, 4, 6, 84
2.1902532	$4d[3/2]_2^\circ-5p[3/2]_2$	$3d_3-2p_6$	CW	4, 6, 56, 60, 84, 134, 313
2.42587	$4d[1/2]_1^\circ-5p[3/2]_1$	$3d_5-2p_7$	CW	79, 3
2.523385	$4d[1/2]_1^\circ-5p[3/2]_2$	$3d_5-2p_6$	CW	75, 8, 56, 60, 84, 313
2.6267	$4d[1/2]_0^\circ-5p[3/2]_1$	$3d_6-2p_7$	CW	85, 78
2.6288	$7p[3/2]_2-4d'[5/2]_2^\circ$	$4p_6-3s_1''''(?)$	CW	78
2.8618	$6p[5/2]_2-6s[3/2]_2^\circ$	$3p_8-2s_5$	CW	85, 60, 78
or	or	or		
2.8663	$6p[5/2]_3-6s[3/2]_2^\circ$	$3p_9-2s_5$		
2.9845	$6p'[1/2]_1-5d[5/2]_2^\circ$	$3p_3-4d_1''$	CW	85, 78
2.9878	$6p'[3/2]_1-6s'[1/2]_0^\circ$	$3p_4-2s_3$	CW	78
3.0536	$6p'[3/2]_1-5d[5/2]_2^\circ$	$3p_4-4d_1''$	CW	78, 85
3.0672	$6p[1/2]_1-6s[3/2]_2^\circ$	$3p_{10}-2s_5$	CW	75, 8, 78, 84
3.1515	$6p'[1/2]_0-5d[3/2]_1^\circ$	$3p_1-4d_2$	CW	85, 78
3.3409	$6p'[1/2]_1-6s[3/2]_{11}^\circ$	$3p_{10}-2s_4$	CW	78
3.3419	$4d[1/2]_1^\circ-5p[1/2]_0$	$3d_5-2p_5$	CW	85, 84
3.4680	$7s[3/2]_1^\circ-6p[1/2]_1$	$3s_4-3p_{10}$	CW	85, 78
3.4883	$6p'[1/2]_1-7s[3/2]_2^\circ$	$3p_3-3s_5$	CW	85, 78
3.4895	$6p'[1/2]_1-5d[3/2]_1^\circ$	$3p_3-4d_2$	CW	78
3.774	$7s[3/2]_1^\circ-6p[5/2]_2$	$3s_4-3p_8$	Pulsed	313
3.956	$5d[3/2]_1^\circ-6p[5/2]_2$	$4d_2-3p_8$	Pulsed	313
4.068	$7s[3/2]_1^\circ-6p[3/2]_1$	$3s_3-3p_7$	Pulsed	313
4.142	$7s[3/2]_1^\circ-6p[3/2]_2$	$3s_4-3p_6$	Pulsed	313
4.3748	$5d[3/2]_1^\circ-6p[3/2]_2$	$4d_2-3p_6$	CW	78, 75, 84, 313
4.3767	$7s[3/2]_2^\circ-6p[3/2]_2$	$3s_5-3p_5$	CW	85
4.8773	$4d[3/2]_1^\circ-5p'[3/2]_1$	$3d_2-2p_4$	CW	85, 78
4.8832	$5d[5/2]_2^\circ-6p[5/2]_3$	$4d_1''-3p_9$	CW	78
5.3000	$5d[3/2]_1^\circ-6p[1/2]_0$	$4d_2-3p_5$	CW	85, 75, 78
or	or	or		
5.3019	$5d[3/2]_2^\circ-6p[5/2]_2$	$4d_3-3p_8$		
5.5700	$5d[7/2]_3^\circ-6p[5/2]_2$	$4d_4-3p_8$	CW	85, 6, 78
5.5863	$6d[7/2]_4^\circ-4f[9/2]_5$	$5d_1'-4U$	CW	78, 75, 84
5.6306	$6d[3/2]_2^\circ-4f[5/2]_3$	$5d_3-4T_3$	CW	78, 75, 84
7.0581	$4f[7/2]_{3,4}-5d[7/2]_4^\circ$	$4W_{3,4}-4d_4$	CW	85, 78

TABLE 4b. *Krypton-ion laser lines*

Wavelength in air (nm)	Spectrum	Identification	References
0.264941	—	—	95, 93, 94
0.266441	II(?)	$5d'\,^2P_{3/2}-5p\,^4D_{1/2}$	93, 94
0.274139	—	—	93, 94
0.304970	—	—	93, 15, 94
0.312438	III	$5p'\,^1D_2-5s'\,^1D_2^\circ$	93, 94
0.323951	III	$5p''\,^1D_2-5s''\,^1P_1^\circ$	93, 15, 94
0.337490	III	$5p''\,^2D_3-5s''\,^3P_2^\circ$	15, 94, 101
0.350742	III	$5p\,^3P_2-5s\,^3S_1^\circ$	93, 15, 94, 99, 100, 101, 113
0.356423	III	$5p\,^3P_1-5s\,^3S_1^\circ$	93, 94, 101
0.377134	II	$5d\,^4P_{1/2}-5p\,^4P_{3/2}^\circ$	19
0.406737	III	$5p'\,^1F_3-5s'\,^1D_2^\circ$	15, 94, 99, 101, 113
0.413133	III	$5p\,^5P_2-5s\,^3S_1^\circ$	15, 94, 101, 113
0.415444	III	$5p'\,^3F_3-5s'\,^1D_2^\circ$	15, 94
0.417179	III	$5p\,^5P_1-5s\,^3S_1^\circ$	15, 94
0.422658	III	$5p'\,^3F_2-4d\,^3D_1^\circ$	15, 94
0.431781[a]	II	$6s\,^4P_{5/2}-5p\,^4P_{5/2}^\circ$	135, 94, 136
0.438654[a]	II	$6s\,^4P_{5/2}-5p\,^4P_{3/2}^\circ$	135, 94, 136
0.444329	III	$5p'\,^3D_2-4d'\,^3D_1^\circ$	15, 94
0.457720	II	$5p'\,^2F_{7/2}^\circ-5s'\,^3D_{5/2}$	137, 15, 94, 114
0.458285[a]	II	$6s\,^4P_{3/2}-5p\,^4D_{5/2}^\circ$	135, 90, 136
0.461528	II	$5p^2\,P_{3/2}-5s\,^2P_{3/2}$	138
0.469151	II	$5p^2\,D_{5/2}^\circ-5s\,^2D_{3/2}$	137, 15, 94, 98
0.463386	II	$5p'\,^2F_{5/2}^\circ-5s'\,^2D_{3/2}$	137, 15, 94, 114
0.465016	II	$5p\,^2P_{1/2}^\circ-5s\,^4P_{1/2}$	15, 94
0.468041	II	$5p\,^2S_{1/2}^\circ-5s\,^2P_{1/2}$	137, 94, 98
0.469444[a]	II	$6s\,^4P_{5/2}-5p\,^4D_{7/2}^\circ$	135, 94, 136
0.471048	III	$5p'\,^3F_4-4d'\,^3D_3^\circ$	115
0.475448	III	$4d'\,^3D_3^\circ-5p'\,^1F_3$	115
0.476243	II	$5p\,^2D_{3/2}^\circ-5s\,^2P_{1/2}$	137, 15, 94, 98, 114
0.476573	II	$5p\,^4D_{5/2}^\circ-5s\,^4P_{3/2}$	137, 15, 94, 98
0.479633	II	$6s\,^2P_{1/2}-5p\,^4S_{3/2}^\circ$	115
0.482517	II	$5p\,^4S_{3/2}^\circ-5s\,^2P_{1/2}$	137, 15, 94, 98, 114
0.484659	II	$5p\,^2P_{1/2}^\circ-5s\,^2P_{3/2}$	15, 94, 98
0.501645	III	$5p'\,^1D_2-4d''\,^1F_3^\circ$	139, 132
0.502240	II	$5p\,^4D_{3/2}^{\prime\circ}-5s\,^2P_{3/2}'$	114 and 116, 94
0.503747	II	—	115
0.512714[a]	II	$6s\,^4P_{3/2}-5p\,^4D_{3/2}^\circ$	135, 94, 136
0.520831	II	$5p\,^4P_{3/2}^\circ-5s\,^4P_{3/2}$	137, 15, 94, 98
0.521793	II	$5p\,^4D_{3/2}^\circ-5s\,^2P_{1/2}$	115
0.530865	II	$5p\,^4P_{5/2}^\circ-5s\,^4P_{3/2}$	137, 15, 94, 98
0.550143	II	$5p'\,^3F_4-4d'\,^3D_2^\circ$	115
0.559732	III	$5p'\,^3P_2-5s''\,^3P_2^\circ$	115
0.568188	II	$5p\,^4D_{5/2}^\circ-5s\,^2P_{3/2}$	137, 15, 94, 98
0.575298	II	$5p\,^4D_{3/3}^\circ-5s\,^2P_{1/2}$	94 and 116, 132
0.593503	III	$5p'\,^3P_2-5s''\,^1P_1^\circ$	115
or	or	or	
0.593503	II	$4d\,^4F_{3/2}-5p'\,^2P_{3/2}^\circ$	
0.603767	III	$5p'\,^3P_1-4d'\,^3D_1^\circ$	141
or	or	or	
0.60381	II	$5f\,^4F_{3/2}-4d'\,^2P_{3/2}$	
0.6072	—	—	142
0.616880	II	$5p'\,^2F_{5/2}^\circ-4d\,^2D_{3/2}$	114

TABLE 4b *(cont.)*

Wavelength in air (nm)	Spectrum	Identification	References
0.631022	III	$5p'\,{}^3P_2\text{--}4d''\,{}^3D_1^\circ$	141
0.631276	—	—	141, 132
0.641661	II	$5p'\,{}^2P_{3/2}^\circ\text{--}4d\,{}^2P_{3/2}$	142
0.647088	II	$5p\,{}^4P_{5/2}^\circ\text{--}5s\,{}^2P_{3/2}$	137, 15, 94, 98, 114
0.657012	II	$5p'\,{}^2D_{5/2}^\circ\text{--}4d\,{}^2F_{5/2}$	137, 15, 94, 98, 114
0.66029	II	$5p'\,{}^2P_{1/2}^\circ\text{--}4d\,{}^3F_{3/2}$	141, 132
0.67442	II	$5p\,{}^4P_{1/2}^\circ\text{--}5s\,{}^2P_{1/2}$	137, 15, 94, 98, 114
0.687084	II	$5p'\,{}^2F_{5/2}^\circ\text{--}4d\,{}^2P_{3/2}$	137, 15, 94, 98, 114
0.752546	II	$5p\,{}^4P_{3/2}^\circ\text{--}5s\,{}^2P_{1/2}$	143
0.79314	II	$5p'\,{}^2F_{7/2}^\circ\text{--}4d\,{}^2F_{5/2}$	98
0.799322	II	$5p\,{}^4P_{3/3}^\circ\text{--}4d\,{}^4D_{1/2}$	137, 15, 94, 98, 102
0.828037	II	$5p'\,{}^2F_{5/2}^\circ\text{--}4d\,{}^2F_{5/2}$	98, 143
0.8589	—	—	102
0.86901	II	$5p\,{}^2P_{1/2}^\circ\text{--}5s'\,{}^2D_{3/2}$	98

[a]The selective excitation of the upper state of this laser line is one involving excitation transfer between an excited atom and an ion:

$$\text{He}^*\,2^3S_1 + \text{Kr}^+ \rightarrow \text{He}\,{}^1S_0 + \text{Kr}^{+\prime}\,(6s\,{}^4P_{3/2,\,5/2}) + \Delta E_\infty.^{(136)}$$

TABLE 5a. *Xenon-neutral laser lines (Fig. A.14)*

Wavelength (μm)	Identification		Notes	References
	Racah	Paschen		
0.840919	$6p[3/2]_1\text{--}6s[3/2]_2^\circ$	$2p_7\text{--}1s_5$	Pulsed	102, 60
0.904545	$6p[5/2]_2\text{--}6s[3/2]_2^\circ$	$2p_9\text{--}1s_5$	Pulsed	20, 60, 313
0.97997	$6p[1/2]_1\text{--}6s[3/2]_2^\circ$	$2p_{10}\text{--}1s_5$	Pulsed	20
1.0634	—	—	Pulsed	102
1.0950	—	—	Pulsed	102
1.36562	$7s[3/2]_1^\circ\text{--}6p[5/2]_2$	$2s_4\text{--}2p_9$	Pulsed	60
1.60519	$7s[3/2]_1^\circ\text{--}6p[3/2]_2$	$2s_4\text{--}2p_6$	Pulsed	60
1.7326	$5d[3/2]_1^\circ\text{--}6p[5/2]_2$	$3d_2\text{--}2p_9$	CW	144, 60, 149, 313
2.026229	$5d[3/2]_1^\circ\text{--}6p[3/2]_1$	$3d_2\text{--}2p_7$	CW	4, 6, 56, 60, 134, 144–9, 313
2.3200	$5d[5/2]_3^\circ\text{--}6p[5/2]_2$	$3d_1'\text{--}2p_9$	CW	147, 75, 134
2.4825	$5d[5/2]_3^\circ\text{--}6p[5/2]_3$	$3d_1'\text{--}2p_8$	CW	144, 149
2.6269	$5d[5/2]_2^\circ\text{--}6p[5/2]_2$	$3d_1''\text{--}2p_9$	CW	147, 75, 134, 150
2.651093	$5d[3/2]_1^\circ\text{--}6p[1/2]_0$	$3d_2\text{--}2p_5$	CW	147, 60, 134, 144, 149, 151, 313
2.6608	$5d[3/2]_1^\circ\text{--}6p'[1/2]_0$	$3s_1'\text{--}2p_1$	CW	75, 134, 149, 313
3.1078	$5d[5/2]_3^\circ\text{--}6p[3/2]_2$	$3d_1'\text{--}2p_6$	CW	147, 75, 134, 149, 150, 151, 313
3.27459	$5d[3/2]_2^\circ\text{--}6p[1/2]_1$	$3d_3\text{--}2p_{10}$	CW	134, 3, 149, 150, 313
3.3676	$5d[5/2]_2^\circ\text{--}6p[3/2]_1$	$3d_1''\text{--}2p_7$	CW	147, 75, 134, 151, 313
3.4345	$7p[5/2]_2\text{--}7s[3/2]_1^\circ$	$3p_9\text{--}2s_4$	CW	75, 6, 78
3.5080	$5d[7/2]_3^\circ\text{--}6p[5/2]_2$	$3d_A\text{--}2p_9$	CW	147, 56, 75, 84, 100, 134, 148, 149, 150, 151, 152, 153, 154, 155, 313
3.6219	$5d'[3/2]_3^\circ\text{--}7p[3/2]_2$	$3s_1''''\text{--}3p_6$	CW	75, 313
3.6158	$7p[1/2]_1\text{--}7s[3/2]_2^\circ$	$3p_{10}\text{--}2s_5$	CW	75, 78
3.651	$7p[1/2]_1\text{--}7s[3/2]_2^\circ$	$3p_{10}\text{--}2s_5$	Pulsed	313
3.6798	$5d[1/2]_1^\circ\text{--}6p[1/2]_1$	$3d_5\text{--}2p_{10}$	CW	147, 75, 150, 313

TABLE 5a *(cont.)*

Wavelength (μm)	Identification		Notes	References
	Racah	Paschen		
3.6859	$5d[5/2]_2^\circ$–$6p[3/2]_2$	$3d_1''$–$2p_6$	CW	147, 75, 151, 313
3.8697	$5d'[5/2]_3^\circ$–$6p'[3/2]_2$	$3s_1''$–$2p_3$	CW	75, 150
3.8950	$5d[7/2]_3^\circ$–$6p[5/2]_3$	$3d_4$–$2p_8$	CW	147, 75, 313
3.9966	$5d[1/2]_0^\circ$–$6p[1/2]_1$	$3d_6$–$2p_{10}$	CW	147, 62, 150, 151, 313
4.1527	$5d'[5/2]_2^\circ$–$7p[3/2]_1$	$3s_1'$–$3p_7$	CW	75, 150, 313
4.5393	$5d[3/2]_2^\circ$–$6p[5/2]_2$	$3d_3$–$2p_9$	CW	86, 150, 313
4.6109	$5d'[3/2]_2^\circ$–$6p'[1/2]_1$	$3s_1''''$–$2p_2$	CW	75, 313
5.3568	$5d[1/2]_1^\circ$–$6p[5/2]_2$	$3d_5$–$2p_9$	CW	86
5.5754	$5d[7/2]_4^\circ$–$6p[5/2]_3$	$3d_4'$–$2p_8$	CW	147, 56, 75, 84, 150, 151, 313
7.3167	$5d[3/2]_2^\circ$–$6p[3/2]_1$	$3d_3$–$2p_7$	CW	147, 75, 150, 151, 313
9.0065	$5d[3/2]_2^\circ$–$6p[3/2]_2$	$2d_3$–$2p_6$	CW	147, 6, 75, 150, 151, 313
9.7029	$5d[1/2]_1^\circ$–$6p[3/2]_1$	$3d_5$–$2p_7$	CW	147, 75
11.289	$8p[3/2]_1^\circ$–$5d'[3/2]_2$	$4p_7$–$3s_1'''$	CW	75, 78
or	or	or		
11.299	$5d'[5/2]_3^\circ$–$4f[9/2]_4$	$3s_1'''$–$4Z$		
12.266	$5d[1/2]_0^\circ$–$6p[3/2]_1$	$3d_6$–$2p_7$	CW	147, 75
12.917	$5d[1/2]_1^\circ$–$6p[3/2]_2$	$3d_5$–$2p_6$	CW	147, 75
18.506	$5d'[3/2]_2^\circ$–$4f[5/2]_3$	$3s_1''''$–$4U$	CW	75, 78
75.5778	$6p'[3/2]_1$–$5d[1/2]_2^\circ$	$2p_5$–$3d_5$	CW	156

TABLE 5b. *Xenon-ion laser lines*

Wavelength (μm)	Spectrum	Identification	References
0.247718	—	—	93, 94, 157
0.269182	—	—	93, 94, 157
0.298385	III	$6p''\,32_1$–$6s''\,^3P_0^\circ$	15, 94
0.307971	II, IV(?)	—	93, 15, 94, 97, 157
0.324684	III	$6p''\,^3D_3$–$5d'\,^3D_3^\circ$	93, 94
0.330604	III, IV(?)	—	93, 94, 97
0.330653	III	nx' $37_{3,\,2}$–$6p'\,^3D_3$	94, 15
0.331963	III(?)	$6p''\,32_1$–$6s''\,^3P_2^\circ$	97
0.333078	IV(?)	—	93, 6, 15, 94, 95, 157, 158
0.335004	—	—	93, 94, 157
0.345434	III	$6p'\,^1D_2$–$6s'\,^1D_2^\circ$	93, 94, 97, 101
0.348296	—	—	93, 94, 157
0.34836	—	—	15
0.354233	III	$6p'\,^3P_2$–$6s'\,^3D_3^\circ$	19
0.3597	III	—	101
0.364546	IV(?)	—	93, 94, 97, 127, 157, 167
0.366920	—	—	93, 94, 157
0.374571	III	$6p'\,^1D_1$–$5d'\,^1D_2^\circ$	93, 94, 97
0.375994	—	—	15, 160
0.378097	III	$6p\,^3P_2$–$6s\,^3S_1^\circ$	97, 15, 93, 94, 101, 157, 159
0.380327	—	—	93, 94, 157
0.397293	—	—	93, 94, 157
0.399285	III	$6p'\,^3P_1$–$6s'\,^1D_2$	115
0.405005	III	$6p\,^3P_1$–$6s\,^3S_1^\circ$	141
0.406041	III	$6p''\,32_1$–$5d''\,25_1^\circ$	97, 15, 94, 101, 127, 159

TABLE 5b *(cont.)*

Wavelength (μm)	Spectrum	Identification	References
0.414573	III	$6p'\ ^3D_2-5d'\ ^2D_1^\circ$	115
0.421401	III	$6p'\ ^3D_2-5d'\ ^3D_3^\circ$	15, 94, 101
0.424024	III	$6p'\ ^1D_2-5d''\ 17_2^\circ$	15, 94, 101
0.427259	III	$6p'\ ^3F_4-5d'\ ^3D_3^\circ$	15, 94, 101, 132
0.428588	III	$6p'\ ^3D_3-6s'\ ^1D_2^\circ$	15, 6, 94, 101
0.429639	II	$7s\ ^4P_{1/2}-6p\ ^4P_{3/2}^\circ$	19
0.430585	III	$6p'\ ^3D_3-5d'\ ^3D_3^\circ$	15, 94, 127, 157, 158, 160
0.44130	II(?)	—	115
0.443415	III	$6p'\ ^3F_2-5d'\ ^3D_1^\circ$	15, 94, 132
0.450346	III	$6p''\ 32_1-5d''\ 27_5$	115
0.455874	—	—	140
0.460302	II	$6p\ ^4D_{3/2}^\circ-6s\ ^4P_{3/2}$	137, 15, 94, 98, 161
0.464740	CIII(?)	—	160
0.465022	CIII(?)	—	160
0.467368	III	$6p'\ ^1F_3-6s'\ ^1D_2^\circ$	15, 94, 98, 114, 158, 159, 162
0.468354	III	$6p\ ^5P_2-6s\ ^3S_1^\circ$	15, 94, 158
0.472357	III	$6p\ ^5P_1-6s\ ^3S_1^\circ$	138
0.47492	III	—	138, 114, 157
0.479448	III	$6p'\ ^3D_1-5d'\ ^3D_1^\circ$	115
0.486388[a]	II	$7s\ ^4P_{5/2}-6p\ ^4P_{5/2}^\circ$	136, 94, 163
0.486946	III	$6p'\ ^3F_3-5d'\ ^3D_2^\circ$	15, 6, 94, 98, 114
0.488730	II	$6p\ ^2P_{3/2}^\circ-6s\ ^2P_{3/2}$	94
0.49196	II	$6p\ ^2P_{1/2}^\circ-6s\ ^2P_{1/2}$	165, 6
0.495416	IV(?)	—	15, 94, 111, 127, 157, 158, 159, 162, 166, 167
0.496508	II	$7s'\ ^2D_{3/2}-6p'\ ^2P_{3/2}^\circ$	15, 94
0.497270	II	$6p'\ ^2P_{2/2}^\circ-5d\ ^2D_{5/2}$	165, 19, 159
0.500778	IV(?)	—	15, 94, 127, 157, 158, 159, 162, 167
0.50128	III	$10_{5/2}-6p\ ^4D_{3/2}^\circ$	165
0.504492	II	$6p'\ ^2P_{1/2}^\circ-6s'\ ^2D_{3/2}$	137, 15, 98, 161, 165
0.515704	—	—	140
0.515904	IV(?)	—	15, 94, 127, 157, 158, 159, 162, 166
0.51881	II	$6p'\ ^2D_{5/2}-6s\ ^2D_{3/2}$	159, 6
0.522366	III	$6p'\ ^1F_3-5d'\ ^1D_2$	115
0.523893	III	$6p'\ ^3P_2-5d''\ 13_1^\circ$	15, 94, 114, 158, 168
0.525650	III(?)	—	115
0.525992	II	$7s\ ^4P_{5/2}-6p\ ^4D_{5/2}^\circ$	94, 111, 157, 160, 167
0.526017	IV(?)	—	15, 157, 159, 160
0.526043	II	$6p\ ^2P_{3/2}^\circ-6s\ ^2P_{1/2}$	165, 94, 114, 140, 157, 162, 166
0.526195	II	$6p'\ ^2D_{3/2}^\circ-6s'\ ^2D_{3/2}$	137, 15, 94, 98, 127, 158, 159, 161
0.52922	II	$6d\ ^4D_{3/2}-6p\ ^4D_{3/2}^\circ$	165
0.531536[a]	II	$7s\ ^4P_{5/2}-6p\ ^4D_{7/2}^\circ$	136, 94, 163
0.53394	II	$6d\ ^4D_{5/2}-6p\ ^4D_{3/2}^\circ$	159, 6
0.5340	—	—	166
0.534334	—	—	140
0.535288	IV(?)	—	15, 94, 111, 127, 140, 157, 162, 166, 158, 159, 167, 169
0.536706	III	$6p'\ ^3F_3-5d\ ^3D_2^\circ$	115
0.539459	IV(?)	—	15, 94, 127, 140, 157, 158, 159, 162, 166, 167

TABLE 5b *(cont.)*

Wavelength (μm)	Spectrum	Identification	References
0.540104	III	$6p'\ ^3P_2 - 5d''\ 15_2^\circ$	168
0.541356	III	$6p'\ ^3P_2 - 5d''\ 17_3^\circ$	115
0.541915	II	$6p\ ^4D_{5/2}^\circ - 6s\ ^4P_{3/2}$	137, 15, 94, 98, 161, 165
0.54543	III	$6d\ ^5D_0^\circ - 6p'\ ^1P_1$	115, 162
0.549931	—	—	140
0.5501	—	—	166
0.552439	III or IV	$6p'\ ^1D_2 - s''\ ^3P_0^\circ$	115, 158
0.5590	—	—	166
0.559235	II	$7s'\ ^2D_{3/2} - 6p'\ ^2D_{3/2}^\circ$	140
0.565938	II	$6p\ ^2P_{1/2}^\circ - 5d\ ^2P_{1/2}$	94
0.572691	II	$6p^2\ D_{5/2}^\circ - 5d\ ^2F_{5/2}$	94
or	or	or	
0.572855[a]	II	$5d'\ ^2S_{1/2} - 6p\ ^4P_{3/2}^\circ$	136
0.575103	II	$6p\ ^2D_{3/2}^\circ - 5d\ ^2P_{1/2}$	94
0.589328	II	$6p'\ ^2P_{3/2}^\circ - 5d'\ ^2D_{5/2}$	19
0.595573	IV(?)	—	15, 94, 140, 157, 159, 162, 167
0.597111	II	$6p'\ ^2P_{3/2}^\circ - 6s'\ ^2D_{3/2}$	15 and 137, 94, 98, 160
0.59765	II	—	165
0.609529[a]	II	$7s\ ^4P_{3/2} - 6p\ ^4D_{3/2}$	136, 94
0.617617	—	—	141, 132, 157
0.623824	III	$6p'\ ^1F_3 - 5d''\ 17_3^\circ$	141, 132
0.627081	II	$6p'\ ^2F_{5/2}^\circ - 6s'\ ^2D_{3/2}$	137, 15, 94, 98, 161
0.62866	II	—	115, 162
0.634318	II	—	162
0.65286	II	$6p'\ ^2F_{7/2} - 5d'\ ^2F_{5/2}$	98, 114, 161
0.66943	II	$6p\ ^4P_{3/2}^\circ - 5d\ ^4D_{1/2}$	98
0.669950	—	—	140
0.6702[b]	II	$6p'\ ^2P_{3/2}^\circ - 5d'\ ^2F_{5/2}$	161
0.70723	II	$37_{5/2}^\circ - 6d\ ^4D_{3/2}$	98
0.71490	II	$6p\ ^4D_{3/2}^\circ - 5d\ ^2P_{3/2}$	114, 98, 138, 161
0.782763	II	$35_{5/2}^\circ - 16_{3/2}$	102, 94
0.798800	II	$6p\ ^4P_{1/2}^\circ - 6s\ ^4P_{1/2}$	102, 94
0.833270	II	$27_{5/2}^\circ - 6d\ ^4D_{5/2}$	102, 94
0.8408	—	—	102, 94
0.8443	—	Oxygen-I Line	102, 94
0.8569	—	—	102, 94, 157
0.858251	II	$31_{3/2}^\circ - 10_{5/2}$	102, 94
0.871617	II	$6p\ ^4D_{3/2}^\circ - 5d\ ^2P_{3/2}$	102, 94, 98
0.905930	II	$27_{5/2}^\circ - 16_{3/2}$	102, 94
0.926539	II	$5d\ ^2D_{3/2} - 6p\ ^4D_{5/2}^\circ$	102, 94
0.928854	II	$13_{2/2}^\circ - 5d'\ ^2S_{1/2}$	102, 94
0.969859	II	$6p\ ^4D_{3/2}^\circ - 5d\ ^4P_{5/2}$	102, 94, 170
1.063385	II	$6p\ ^4D_{3/3}^\circ - 5d\ ^4P_{3/2}$	102, 94
1.0950	—	—	102, 94

[a] The selective excitation of the upper state of this line in a pulsed discharge in a Xe–Ne mixture occurs via the excited atom-ion reaction

$$Ne^* + Xe^+ \rightarrow Ne\ ^1S_0 + Xe^{+\prime}(7s\ ^4P_{3/2,\,5/2}) + \Delta E_\infty .$$

[b] This line has been observed only as a CW-line in a microwave excited discharge in xenon to which helium had been added. Selective excitation of the $6p'^2P_{3/2}^\circ$ level is possibly by the atom–atom excitation transfer reaction

$$He^*2\ ^3S_1 + Xe^*6s[3/2]_2^\circ \rightarrow He\ ^1S_0 + Xe^+6p'\ ^2P_{3/2}^\circ - \Delta E_\infty\ (0.07\ eV). \text{[171]}$$

TABLE 6. *Laser transitions in atomic hydrogen (Fig. A.15)*

Wavelength (μm)	Spectrum	Identification	Notes	Excitation	References
0.4340	I	H-α line	As impurity in pulsed Ne discharge at 1.5 torr, $D = 25$ mm, $E/p = 140$ V/cm-torr	Pulsed	172
0.4861	I	H-β line	,,	Pulsed	172
1.8751	I	First member of the Paschen series	As impurity in 3.5 torr He; or 0.01 torr of H_2, $D = 7$ mm	Pulsed	173

TABLE 7. *Laser transitions in sodium (Fig. A.16)*

Wavelength (μm)	Spectrum	Identification	Notes	Excitation	References
1.1382[a]	I	$4s\ ^2S_{1/2}-3p\ ^2P_{1/2}$	0.001–0.003 torr V. Pressure of sodium, 1–10 torr H_2, $D = 12$ mm	Pulsed	174, 175, 176
1.1404[a]	I	$4s\ ^2S_{1/2}-3p\ ^2P_{3/2}$	,,	Pulsed	174, 175, 176

[a]Selective excitation occurs via the charge neutralization reaction: $Na^+ + H^- \rightarrow Na'(4s\ ^2S_{1/2}) + H$.[174, 175]

TABLE 8. *Laser transitions in potassium (Fig. A.17)*

Wavelength (μm)	Spectrum	Identification	Notes	Excitation	References
1.2434	I	$5s\ ^2S_{1/2}-4p\ ^2P_{1/2}$	0.1 torr of potassium with 3–5 torr H_2 ($D = 12$ mm?)	Pulsed	176
1.2523	I	$5s\ ^2S_{1/2}-4p\ ^2P_{3/2}$	,,	Pulsed	176
3.140	I	$5p-3d$	Potassium vapor excited with λ694.3-nm giant pulse from a ruby laser	Pulsed	177
3.160	I	$5p-3d$	,,	Pulsed	177

TABLE 9. *Laser transitions in rubidium*

Wavelength (μm)	Spectrum	Identification	Notes	Excitation	References
2.254	I	6p–4d	Rubidium vapor excited with a λ694.3-nm giant pulse from a ruby laser	Pulsed	177, 178
2.293	I	6p–4d	,,	Pulsed	177, 178

TABLE 10. *Laser transitions in cesium (Fig. A.18)*

Wavelength (μm)	Spectrum	Identification	Notes	Excitation	References
1.360	I	$7p\,^2P_{3/2}-5d\,^2D_{5/2}$	Cesium vapor excited by λ7658-nm nitrobenzene Raman laser; He absence required	Pulsed	177
1.376	I	$7p\,^2P_{1/2}-5d\,^2D_{3/2}$	,,	Pulsed	177
3.095	I	$7p\,^2P_{1/2}-7s\,^2S_{1/2}$	Cesium vapor excited by λ694.4, 765.8, λ(740.0–900.0)-nm or 1.06-μm laser pulses; He buffer required	Pulsed	177, 178
3.010	I	$5d\,^2D_{3/2}-6p\,^2P_{1/2}$	Cesium vapor excited by 1.06-μm great pulse	Pulsed	177
3.2040[a]	I	$8p\,^2P_{1,2}-6d\,^2D_{3/2}$	Vapor pressure of Cs at 175°C; $D = 10$ mm	CW	179, 6, 180
3.489	I	$5d\,^2D_{5/2}-6p\,^2P_{3/2}$	Cesium vapor excited with 1.06-μm giant pulse	Pulsed	177
3.613	I	$5d\,^2D_{3/2}-6p\,^2P_{3/2}$	Cesium vapor excited with 1.06-μm giant pulse	Pulsed	177
7.1821[a]	I	$8p\,^2P_{1/2}^\circ-8s\,^2S_{1/2}$	Vapor pressure of cesium at 175°C; $D = 10$ mm	CW	181, 6, 179, 180

[a]Selective excitation is by optical pumping through resonance absorption on the Cs $6s\,^2S_{1/2}-8p\,^2P_{1/2}^\circ$ transition by the strong He–I line at 0.3889 μm.

TABLE 11. *Laser transitions in copper (Fig. 4.50)*

Wavelength (μm)	Spectrum	Identification	Notes	Excitation	References
0.510554[a]	I	$4p\ ^2P^{\circ}_{3/2}$–$4s^2\ ^2D_{5/2}$	> 0.1 torr copper at 1420°C, 1–3 torr He; $D = 10$ mm. Also in an electrically accelerated plasma	Short rise-time H.V. pulses	182, 183, 184, 185, 186, 187
0.578213[a]	I	$4p\ ^2P^{\circ}_{1/2}$–$4s^2\ ^2D_{3/2}$	,,	,,	182, 183, 184, 187

[a]Very high gain, greater than 40 dB/m transient laser lines.

TABLE 12. *Laser transition in gold*

Wavelength (μm)	Spectrum	Identification	Notes	Excitation	References
0.627818	I	$6p\ ^2P^{\circ}_{1/2}$–$6s^2\ ^2D_{3/2}$	At a vapor pressure of gold at temp. of approx. 1500°C; $D = 10$ mm	Short rise-time pulses	186, 182, 187[a]

[a]Excitation was in an electrically accelerated plasma.

TABLE 13. *Laser transitions in magnesium (Fig. 5.33 for some of the transitions)*

Wavelength (μm)	Spectrum	Identification	Notes	Excitation	References
0.6630	II	$7p\ ^1P_1$–$4s\ ^1S_0$	In Mg vapor with He or Ne at 6–8 torr; $D = 12$ mm	Pulsed	314
0.8049	II	$9d\ ^3D_{3,2,1}$–$4p\ ^3P_2$	,,	Pulsed	314
0.8736	II	$7f\ ^3F_{2,3,4}$–$3d\ ^3D_{3,2,1}$	,,	Pulsed	314
0.9218	II	$4p\ ^2P^{\circ}_{3/2}$–$4s\ ^2S_{1/2}$	In Mg vapor at 395°C, with He at 0.5 to 5.5 torr; $D = 4$ mm, $I = 260$ mA[a]	CW	188
0.9244	II	$4p\ ^2P^{\circ}_{1/2}$–$4s\ ^2S_{1/2}$	,,	CW	188
0.9820[c]	II	—	In Mg vapor with He or Ne at 6–8 torr; $D = 12$ mm	Pulsed	314
0.9993	II	$7s\ ^3S_1$–$4p\ ^3P_2$	,,	Pulsed	314
1.0731	II	—	,,	Pulsed	314
1.0914	II	$4p\ ^2P^{\circ}_{3/2}$–$3d\ ^2D_{3/2,\,5/2}$	As for 0.9218-μm line	CW	188
1.0951	II	$4p\ ^2P^{\circ}_{1/2}$–$3d\ ^2D_{3/2}$	,,	CW	188

TABLE 13 *(cont.)*

Wavelength (μm)	Spectrum	Identification	Notes	Excitation	References
1.0953	II	$5d\,^3D_{3,\,2,\,1}-4p\,^3P_2$	In Mg vapor with He or Ne at 6–8 torr; $D = 12$ mm	Pulsed	314
1.6800	II	$5s\,^3S_1-4p\,^3P_0$	,,	Pulsed	314
1.9201	II	$7f\,^3F_{2,\,3,\,4}-4d\,^3D_{3,\,2,\,1}$	In Mg vapor with He or Ne at 6–8 torr; $D = 12$ mm	Pulsed	314
2.40415[b]	II	$5p\,^2P_{3/2}-4d\,^2D_{5/2}$	In Mg vapor above 450°C with peak current of 150 A, in He, Ne, or Ar; $D = 10$ mm	Pulsed	310
2.41245[b]	II	$5p\,^2P_{1/2}-4d\,^2D_{3/2}$	,,	Pulsed	310
3.6779[b]	I	$5d\,^3D-5p\,^3P_1$	In Mg vapor above 450°C in He, Ne, or Ar; $D = 10$ mm, $I = 40$ to 100 mA	CW	310
3.68154[b]	I	$5d\,^3D-5p\,^3P_2$	,,	CW	310
3.86573[c]	—	—	,,	CW	310
4.20018[b]	I	$5p\,^3P_2-5s\,^3S_1$	,,	CW	310
4.36269[b]	I	$5d\,^3D-4f\,^3F$	,,	CW	310

[a]Selective excitation of the upper laser level is by the Penning reaction $He^* + Mg \rightarrow He + Mg^{+\prime}\,(4p) + e^-$.
[b]Air wavelength.
[c]Measured wavelength in air $(+0.00001\ \mu m)$.

TABLE 14. *Laser transitions in calcium*

Wavelength (μm)	Spectrum	Identification	Notes	Excitation	References
0.854209[a]	II	$4p\,^2P^\circ_{3/2}-3d\,^2D_{5/2}$	> 0.1 torr vapor pressure of calcium at temp. of 690°C, 1–3 torr of He, $D = 10$ mm	Short risetime pulses	182
0.866214[a]	II	$4p\,^2P^\circ_{1/2}-3d\,^2D_{3/2}$	,,	,,	182
5.5457[b]	I	$4p\,^1P^\circ_1-3d\,^1D_2$	Vapor pressure (min.) of calcium at temp. of 460°C, 3 torr He or Ne (in small bore tubes?)	,,	189

[a]Saturated gain greater than 58 dB/m on these lines.
[b]Single pass gain greater than 300 dB/m.

TABLE 15. *Laser transitions in strontium*

Wavelength (μm)	Spectrum	Identification	Notes	Excitation	References
1.033014[a]	II	$5p\ ^2P^\circ_{3/2}-4d\ ^2D_{5/2}$	Vapor pressure of strontium at a temp. $> 460°C$ with 3 torr He or Ne (in small bore discharge tubes?)	Short risetime pulses	189
1.091797[a]	II	$5p\ ^2P^\circ_{1/2}-4d\ ^2D_{3/2}$	,,	,,	189
6.4567[a]	I	$5p\ ^1P^\circ_1-4d\ ^1D_2$	,,	,,	189

[a]Single-pass gain for these lines typically 300 dB/m.

TABLE 16. *Laser transitions in barium*

Wavelength (μm)	Spectrum	Identification	Notes	Excitation	References
1.1303[a, b]	I	$6p\ ^1P^\circ_1-5d\ ^3D_2$	Barium vapor at 500–850°C, with He, Ne, Ar, or H_2 at 1 to 3 torr; $D = 5$ to 10 mm	Pulsed	190
1.500[a, c]	I	$6p\ ^1P^\circ_1-5d\ ^1D_2$	,,	Pulsed	190
1.9017[a]	I	$6p\ ^1D_2-6p'\ ^3D^\circ_3$	,,	Pulsed	190
2.1568[a]	I	$6p'\ ^1P^\circ_1-5d^2\ ^3P_2$	,,	Pulsed	190
2.3254	I	$6p\ ^3P^\circ_2-5d\ ^3D_3$	,,	Pulsed	190
2.4758[a]	I	$7s\ ^1S_0-6p'\ ^3D^\circ_1$	,,	Pulsed	190
2.5515	I	$6p\ ^3P^\circ_2-5d\ ^3D_3$	,,	Pulsed	190
2.5924	II	$7p\ ^2P^\circ_{3/2}-6d\ ^2D_{5/2}$	,,	Pulsed	190
2.9057	II	$7p\ ^2P^\circ_{1/2}-6d\ ^2D_{3/2}$	,,	Pulsed	190
2.9227	I	$6d\ ^1D_2-6p'\ ^1F^\circ_3$	,,	Pulsed	190
3.9578	I	$7s\ ^1S_0-6p'\ ^3P^\circ_1$	,,	Pulsed	190
4.0069	I	$6p'\ ^3P^\circ_1-5d^2\ ^3P_0$	,,	Pulsed	190
4.6706	I	—	,,	Pulsed	190
4.7156	I	$10d\ ^3D_2-9p\ ^1P^\circ_1$	,,	Pulsed	190
5.0309	I	$8p\ ^1P^\circ_1-8s\ ^3S_1$	,,	Pulsed	190
5.4798	—	—	,,	Pulsed	190
5.5636	—	—	,,	Pulsed	190
5.8899	—	—	,,	Pulsed	190
6.4548	—	—	,,	Pulsed	190

[a]Superradiant line.
[b]Gain of 65 dB/m measured.
[c]The most intense line, gain of 40 dB/m measured.

TABLE 17. *Laser transitions in zinc (Figs. 5.13 and 5.14)*

Wavelength (μm)	Spectrum	Identification	Notes	Excitation	References
0.49116[a]	II	$4f\ ^2F^\circ_{5/2}\!-\!4d\ ^2D_{5/2}$	0.001–0.01 torr vapor pressure of zinc, 1–2 torr of He ($D = 4$ mm?)	CW	101, 192, 193, 197
0.492404[a]	II	$4f\ ^2F^\circ_{7/2}\!-\!4d\ ^2D_{5/2}$	,,	CW	194, 192, 193, 195, 196, 197
0.5894	II	$4s^2\ ^2D_{3/2}\!-\!4p\ ^2P^\circ_{1/2}$	Zinc temp. 325°C, 4 torr of He ($D = 4$ mm?)	CW	192, 197, 212
0.6021[a]	II	$5d\ ^2D_{3/2}\!-\!5p\ ^2P^\circ_{1/2}$	In HCD He–Zn laser	CW	197, 198
0.610253[a]	II	$5d\ ^2D_{5/2}\!-\!5p\ ^2P^\circ_{3/2}$	Zinc temp. 290–360°C, 6 torr of He ($D = 4$ mm?)	Pulsed	199, 192, 193, 197
0.747879	II	$4s^2\ ^2D_{5/2}\!-\!4p\ ^2P^\circ_{3/2}$	0.004 torr vapor pressure of zinc, 4 torr of He; $D = 4$ mm	CW	199, 192, 197, 200, 201, 212
0.758848	II	$5p^2\ ^2P^\circ_{3/2}\!-\!5s\ ^2S_{1/2}$	0.0002–0.1 torr of zinc, 0.5–4 torr He; $D = 5$ mm	CW	199, 192, 197, 200, 201
0.761290	II	$6s\ ^2S_{1/2}\!-\!5p\ ^2P^\circ_{1/2}$	0.001–0.1 torr of zinc, 1–2 torr He; $D = 6$ mm	Pulsed	199
0.775786	II	$6s\ ^2S_{1/2}\!-\!5p\ ^2P^\circ_{3/2}$	0.0002–0.1 torr of zinc,[b] 0.5–4 torr He; $D = 5$ mm	Pulsed	194
1.0269	II	$8p\ ^3P_2\!-\!4d\ ^3D_1$	In zinc vapor with He or Ne at 6–8 torr; $D = 12$ mm	Pulsed	314
1.1054	II	$5s\ ^1S_0\!-\!4p\ ^1P_1$	,,	Pulsed	314
1.1880	II	$7s\ ^3S_1\!-\!5p\ ^3P_2$	,,	Pulsed	314
1.2169	II	$7p\ ^3P_2\!-\!4d\ ^3D_3$	,,	Pulsed	314
1.2840[c]	II	—	In zinc vapor with He or Ne at 6–8 torr, $D = 12$ mm	Pulsed	314
1.3055	II	$5p\ ^3P_2\!-\!5s\ ^3S_1$	,,	Pulsed	314
1.3152	II	$5p\ ^3P_1\!-\!5s\ ^3S_1$	,,	Pulsed	314
1.3637	II	$5d\ ^3D_1\!-\!5p\ ^3P_0$	,,	Pulsed	314
1.5430[c]	II	—	,,	Pulsed	314
1.5680	II	$4f\ ^1F_3\!-\!4d\ ^1D_2$	,,	Pulsed	314
1.6260	II	$6p\ ^1P_1\!-\!4d\ ^1D_2$	,,	Pulsed	314
1.6499	II	$4f\ ^3F_{2,\,3,\,4}\!-\!4d\ ^3D_3$	,,	Pulsed	314

[a]Selective excitation is by charge transfer from the $He^{+2}\ S_{1/2}$ ion.
[b]It is assumed here that the excitation conditions will be similar for those for the 0.7588-μm line.
[c]Measured wavelength.

TABLE 18. *Laser transitions in cadmium* (*Figs. 5.11 and 5.12*)

Wavelength (μm)	Spectrum	Identification	Notes	Excitation	References
0.3250	II	$5s^2\,^2D_{3/2}-5p\,^2P^\circ_{1/2}$	Approx. 0.003 torr vapor pressure of cadmium at a temp. of 240°C, with 3 torr of He; $D = 3.6$ mm[a]	CW	202, 200, 203–6
0.441563[b]	II	$5s^2\,^2D_{3/2}-5p\,^2P^\circ_{3/2}$	,,	CW	199, 193, 203–10
0.4882	II	$6p\,^2P_{3/2}-5s^2\,^2D_{3/2}$	—	CW	195
0.5025	II	$4f\,^2F-5d\,^2D$	—	CW	195
0.533749	II	$4f\,^2F^\circ_{5/2}-5d\,^2D_{3/2}$	Approx. 0.01 to 0.3 torr Cd with He in a HCD; cathode diam. 5.6 mm	CW	104, 193, 196, 211
0.537804	II	$4f\,^2F^\circ_{7/2}-5d\,^2D_{5/2}$	,,	CW	194, 193, 196, 211
0.63548	II	$6g\,^2G_{7/2}-4f\,^2F^\circ_{5/2}$	,,	CW	211, 213
0.63600	II	$6g\,^2G_{9/2}-4f\,^2F^\circ_{7/2}$	,,	CW	211, 213
0.72369	II	$6f\,^2F^\circ_{5/2}-6d\,^2D_{3/2}$	,,	CW	211, 213
0.72842	II	$6f\,^2F^\circ_{7/2}-6d\,^2D_{5/2}$	,,	CW	211, 213
0.80669	II	$6p\,^2P^\circ_{3/2}-6s\,^2S_{1/2}$	In a HCD	CW	196
0.8390	II	$7s\,^2S_{1/2}-6p\,^4P^\circ_{3/2}$	0.03–0.05 torr of cadmium, optimum 0.01 torr; 2–20 torr of He, or Ne; $D = 12$ mm	Pulsed	214, 314
0.9840	I	$9s\,^1S_0-6p\,^3P_2$	As for 0.8390-μm line	Pulsed	314
1.0865	I	$7d\,^1D_2-6p\,^3P_2$	,,	Pulsed	314
1.1485[b]	I	—	,,	Pulsed	314
1.1554[c]	I	—	,,	Pulsed	314
1.1665	I	$8p\,^1P_1-5d\,^3D_1$	,,	Pulsed	314
1.1742	I	$8s\,^1S_0-6p\,^3P_0$	,,	Pulsed	314
1.1869	I	$6p\,^1P_1-6s\,^3S_1$	0.03–0.05 torr of cadmium, 0.5–1.0 torr of He or Ne; $D = 12$ mm	Pulsed	214, 314
1.3979	II	$6p\,^3P_2-6s\,^3S_1$	,,	Pulsed	314
1.4328	II	$6p\,^3P_1-6s\,^3S_1$	0.03–0.05 torr of cadmium, 2–20 torr of He; $D = 12$ mm	Pulsed	214, 314
1.6402	II	$4f\,^3F_{2,\,3,\,4}-5d\,^3D_1$	0.03–0.05 torr of cadmium, 2–20 torr of Ne or He; $D = 12$ mm	Pulsed	314
1.6434	II	$4f\,^3F_{2,\,3,\,4}-5d\,^3D_2$	,,	Pulsed	314
1.6482	II	$4f\,^3F_{4,\,3,\,2}-5d\,^3D_3$	,,	Pulsed	214, 314
1.9123	II	$6d\,^1D_2-6p\,^1P_1$	,,	Pulsed	314

[a]The gain varies inversely as the tube diameter as in He–Ne lasers operating at 1.15 μm and 0.6328 μm. The selective excitation is by a Penning reaction.

[b]The first CW metallic ion laser.

[c]Measured wavelength.

TABLE 19. *Laser transitions in mercury* (*Figs. A.19 and 5.4*)

Wavelength (μm)	Spectrum	Identification	Notes	Excitation	References
0.3545	Hg[a] or Ar II	—	In Ar–Hg mixture	Pulsed	112
0.479701	III[b]	$5d^8\,6s^2\ (J = 4)$ $5d^9\,6p_{1/2}\ (J = 3)$	0.001 torr of mercury, with a trace of He (> 0.5 torr); $D = 6$ mm	Pulsed	215, 6, 94
0.5225*	(?)	—	In pure mercury with < 0.01 torr of Ar; $D = 7$ mm	Pulsed	109
0.56772	II	$5f^2\,F^\circ_{7/2}$–$6d^2\,D_{5/2}$	0.001 torr of mercury, 0.5 torr of He, also in Hg–Ne or Hg–Ar mixtures; $D = 3$–15 mm	Pulsed	216, 6, 94, 112, 124, 215, 217, 218
0.61495[c]	II	$7p\,^2P^\circ_{3/2}$–$7s\,^2S_{1/2}$	0.001 torr of mercury, 0.5–3.0 torr of He, $D = 3$–15 mm	CW	216, 6, 94, 109, 112, 124, 171, 215, 216, 217, 218, 219, 220, 221, 222, 223 224[d]
0.73466	II	$7d\,^2D_{5/2}$–$7p\,^2P^\circ_{3/2}$	0.001 torr of mercury, 0.5 torr of He, $D = 3$–15 mm	CW	216, 217
0.7065	Ar I (or Hg)	—	—	—	112
0.74181	II	$5d^9\,6s_{1/2}\,6p_{3/2}\,^2P^\circ_{3/2}$ $-7s\,^2S_{1/2}$	0.001 torr of mercury, 1–30 torr of He, $D = 24$ mm	Pulsed[d]	223, 171
0.79447	II	$7p\,^2P^\circ_{1/2}$–$7s\,^2S_{1/2}$	0.001 torr of mercury, 1–30 torr of He, $D = 24$ mm	CW[e]	221, 171, 223, 224
0.8548	II	$5g\,^2G_{7/2}$–$'C'\,^2F^\circ$	0.001 torr of mercury, 0.8–1.2 torr of He, $D = 15$ mm	Pulsed	217
0.8622	II	$8p\,^2P^\circ_{3/2}$–$^4D_{5/2}$	,,	Pulsed	217
0.8677	II(?)	—	,,	Pulsed	217
0.9396	II	$10s\,^2S_{1/2}$–$8p\,^2P^\circ_{3,2}$	,,	Pulsed	217
1.0583	II	$8s\,^2S_{1/2}$–$7p\,^2P^\circ_{3/2}$	,,	Pulsed	216, 217
1.117686	I	$7p\,^1P^\circ_1$–$7s\,^3S_1$	0.09–0.12 torr of mercury, 0.005–0.05 torr of He, $D = 6$ mm	Pulsed	217, 225
1.1181	II[f]	$7g\,^2G$–$6f\,^2F_{5/2}$	0.001 torr of mercury, 0.8–1.2 torr of He, $D = 15$ mm	Pulsed	217

TABLE 19 (*cont.*)

Wavelength (μm)	Spectrum	Identification	Notes	Excitation	References
1.222	I(?)	—	0.001 torr of mercury, in argon, or in He–Ar mixtures (contaminated with mercury (?)), $D = 5$ mm	Pulsed	109, 6, 112
1.2246	I(?)	—	0.001 torr of mercury in Hg–Ar mixture, $D = 5$ mm	Pulsed	109, 6, 112
1.2545	(?)	—	0.001 torr of mercury, 0.8–1.2 torr of He or Ar, $D = 15$ mm	Pulsed	217
1.2760	I(?)	—	0.001 torr of mercury, 0.2 torr Ar, $D = 5$ mm	Pulsed	109, 112
1.2981	(?)	—	0.001 torr of mercury, 0.8 torr of Ar or 1.2 torr of He, $D = 15$ mm	Pulsed	217
1.3655	(?)	(Probably the following transition)	0.001 torr of mercury, 0.8–1.2 torr of He, $D = 15$ mm	Pulsed	217
1.367351	I	$7p\ ^3P_1^\circ - 7s\ ^3S_1$	0.09–0.12 torr of mercury, 0.005–0.05 torr He, $D = 6$ mm	Pulsed	225
1.529582	I	$6p'\ ^3P_2^\circ - 7s\ ^3S_1$	0.09–0.12 torr of mercury, 0.1–1.0 torr of He, Ne, Kr or Ar, $D = 6$ mm	CW	226, 6, 217, 225–30
1.5554	II	$7p\ ^2P_{3/2}^\circ - 6d\ ^2D_{5/2}$	0.001 torr of mercury, 0.8–1.2 torr of He, $D = 15$ mm	Pulsed	217
1.692016	I	$5f\ ^1F_3^\circ - 6d\ ^1D_2$	0.09–0.12 torr of mercury, 0.005–0.05 torr of He, $D = 6$ mm	Pulsed	217, 6, 225
1.694200	I	$5f\ ^3F_2^\circ - 6d\ ^3D_1$	,,	Pulsed	217, 6, 225
1.707279	I	$5f\ ^3F_4^\circ - 6d\ ^3D_3$	,,	Pulsed	217, 6, 225
1.710993	I	$5f\ ^3F_3^\circ - 6d\ ^3D_2$	,,	Pulsed	217, 6, 225
1.732941	I	$7d\ ^1D_2 - 7p\ ^1P_1^\circ$	0.09–0.3 torr of mercury, 0.005–0.01 torr of He, Ne, Kr or air, $D = 6$ mm	Pulsed	229, 225
1.813038	I	$6p'\ ^3F_4^\circ - 6d\ ^3D_3$	,,	CW	226, 6, 109, 217 225, 227, 229, 230

TABLE 19 (*cont.*)

Wavelength (μm)	Spectrum	Identification	Notes	Excitation	References
3.34[g]	I	$6p'\ ^3F_4-8s\ ^3S_1$	0.3 torr of mercury, 0.1–1.0 torr of He, Ne, Kr or air, $D = 8$ mm	Pulsed	229
3.9286	I	$6d\ ^3D_3-6p'\ ^3P_2^\circ$ or $5g\ G-5f\ F^\circ$	,,	Pulsed	229, 225
5.86	I	$6p'\ ^1P_1^\circ-7d\ ^3D_2$	,,	Pulsed	229, 124, 225
6.4 or 6.4916	I	$9s\ ^1S_0-8p\ ^1P_1^\circ$ or $11p\ ^3P_1^\circ-10s\ ^3S_1$	,,	Pulsed	124, 225

[a] Possibly line at 0.354506 given as Hg in *M.I.T. Wavelength Tables* (John Wiley, Inc., New York, 1939), p. 247.

[b] The first doubly ionized laser line to be reported.

[c] The first ionized laser line to be reported; observed in a pulsed Hg–He mixture.

[d] CW oscillation.

[e] Observed only in a pulsed hollow cathode discharge laser in a Hg–He mixture.

[f] A new classification has been given for this line.[225]

[g] This line is possibly not a Hg–I line but a Kr–I line.[225]

*This line at 0.5225 μm has not been reported by any other researcher, or by the original reporters in subsequent more extensive work. It appears likely that this line was actually the 0.5677 μm Hg–II laser line that was observed, particularly as it appeared in conjunction with the 0.6150-μm Hg–II laser line.

TABLE 20. *Laser transition in boron*

Wavelength (μm)	Spectrum	Identification	Notes	Excitation	References
0.345134	II	$2p^2\ ^1D_2-2p\ ^1P_1^\circ$	0.01–0.05 torr of BCl_3; $D = 4$ mm	Pulsed	231

TABLE 21. *Laser transition in indium*

Wavelength (μm)	Spectrum	Identification	Notes	Excitation	References
0.468082	II	$4f\ ^3F_4^\circ-5d\ ^3D_3$	0.003–0.03 torr of indium, 1–2 torr of He or Ne; $D = 8$ mm	Pulsed	199

APPENDIX

TABLE 22. *Laser transition in thallium*

Wavelength (μm)	Spectrum	Identification	Notes	Excitation	References
0.5350	I	$7s\ ^2S_{1/2}-6p\ ^2P^\circ_{3/2}$	More than 0.01 torr of thallium with Ne or He at several torr; $D = 1.3, 2.0$ or 3 mm	Short risetime pulses	7

TABLE 23. *Laser transitions in carbon*

Wavelength (μm)	Spectrum	Identification	Notes	Excitation	References
0.464745	III	$3p\ ^3P^2_\circ-3s\ ^3S_1$	In 0.02–0.05 torr carbon dioxide, $D = 7$ mm, or as impurity in noble gases	Pulsed	232, 6, 15, 94 160, 233
0.465016	III	$3p\ ^3P^\circ_1-3s\ ^3S_1$	,,	Pulsed	232, 160, 233
0.49541[a]	II(?)	$3d\ ^2P^\circ_{1/2}-3p\ ^2P_{1/2}$	As carbon impurity in discharge in Xe	Pulsed	94
0.833515	I	$3p\ ^1S_0-3s\ ^1P^\circ_1$	In mixture He (Ne), N_2 (CO_2) at 4 torr, $D = 15$ mm	Pulsed	234
0.940575	I	$3p\ ^1D_2-3s\ ^1P^\circ_1$	As for 0.8335-μm line, also in 0.05 torr CO with He at 2–16 torr	Pulsed	234, 235
1.068318	I	$3p\ ^3D_2-3s\ ^3P^\circ_1$	,,	Pulsed	234, 235
1.06853	I	$3p\ ^3D_1-3s\ ^3P_0$	0.05 torr CO with 2–16 torr of He	Pulsed	235
1.06936	I	$3p\ ^3D_3-3s\ ^3P^\circ_2$	0.01 torr of CO_2 or CO, with 2 torr of He, $D = 5$ mm	CW	66, 6, 85, 234, 235, 236
1.45425	I	$3p\ ^1P_1-3s\ ^1P^\circ_1$	,,	CW	66, 6, 8, 85, 234 235, 236
2.0645	I	$5d\ ^1D^\circ_2-4p\ ^3P_2$	0.02 torr of carbon monoxide, 1 torr of He, $D=10$ mm	CW	85
			,,		
3.4046	I	$4d\ ^1D^\circ_2-4p\ ^1P_1$	,,	CW	85
3.5155	I	$6d\ ^3P^\circ_2-5p\ ^3D_3$	,,	CW	85
5.5956	I	$4p\ ^3S_1-3d\ ^3P^\circ_1$	,,	CW	85, 6

[a]Possibly a xenon ion-laser line.[94]

APPENDIX

TABLE 24. *Laser transitions in silicon*

Wavelength (μm)	Spectrum	Identification	Notes	Excitation	References
0.40889	IV	$4p\ ^2P^\circ_{3/2}-4s\ ^2S_{1/2}$	Observed in discharge in argon	Pulsed	15
0.455259	III	$4s\ ^3S_1-4p\ ^3P_2$	Observed when using SF_6 or PF_5	Pulsed	237, 238
0.456784	III	$4s\ ^3S_1-4p\ ^3P_1$	"	Pulsed	237, 238
0.634724	II	$4s\ ^2S_{1/2}-4p\ ^2P_{3/2}$	"	Pulsed	237, 238
0.637148	II	$4s\ ^2S_{1/2}-4p\ ^2P_{1/2}$	"	Pulsed	237, 238
0.667193	II	$4s\ ^4P_{5/2}-4p\ ^4D_{7/2}$	"	Pulsed	237, 238
1.198418	I	$4p\ ^3D_2-4s\ ^3P^\circ_1$	0.03 torr of $SiCl_4$, 0.5 torr of Ne, $D = 6$ mm	CW	239
1.203148	I	$4p\ ^3D_3-4s\ ^3P^\circ_2$	0.04 torr of $SiCl_4$, 0.5 torr of Ne, $D = 6$ mm	CW	239
1.588402	I	$5s\ ^3P^\circ_1-4p\ ^3D_1$	0.03–0.05 torr of $SiCl_4$, 1–5 torr of Ne, $D = 6$ mm	CW	239

TABLE 25. *Laser transitions in germanium*

Wavelength (μm)	Spectrum	Identification	Notes	Excitation	References
0.513175	II	$4f\ ^2F^\circ_{5/2}-4d\ ^2D_{3/2}$	$3\times10^{-7}-5\times10^{-6}$ torr of germanium,[a] 1–2 torr of He or Ne, $D = 5$ mm	Pulsed	199
0.517865	II	$4f\ ^2F^\circ_{7/2}-4d\ ^2D_{5/2}$	"	Pulsed	199

[a] The vapor pressure of germanium is deduced from vapor pressure data for the more common elements; R. E. Honig, *R.C.A. Review* **18**, 195–204 (June 1957).

TABLE 26. *Laser transitions in tin*

Wavelength (μm)	Spectrum	Identification	Notes	Excitation	References
0.579918	II	$4f\ ^2F^\circ_{7/2}-5d\ ^2D_{5/2}$	0.00001–0.0002 torr vapor pressure of tin,[a] 1–2 torr of He or Ne, $D = 6$ mm	Pulsed	199
0.645350	II	$6p\ ^2P^\circ_{3/2}-6s\ ^2S_{1/2}$	0.0015 torr vapor pressure of tin,[a] 8 torr of He, $D = 5$ mm	CW	199

432

TABLE 26 *(cont.)*

Wavelength (μm)	Spectrum	Identification	Notes	Excitation	References
0.657926	I	$10d\ ^3D_{0,\,2}-6p\ ^3P_1$	In SnCl$_4$ vapor at room temp., $D = 5.6$ mm	Pulsed	231, 240
0.684405	II	$6p\ ^2P^{\circ}_{1/2}-6s\ ^2S_{1/2}$	0.0015 torr vapor pressure of tin,[a] 8 torr of He; $D = 5$ mm	CW	199

[a] The vapor pressure of tin is deduced from vapor pressure data for the more common elements, R. E. Honig, *R.C.A. Review* **18**, 195–204 (June 1957).

TABLE 27. *Laser transitions in lead (Fig. A.20)*

Wavelength (μm)	Spectrum	Identification	Notes	Excitation	References
0.363954	I	$6p7s\ ^3P^{\circ}_1-6p^2\ ^3P_1$	Vapor pressure of Pb[208] at a temp. of 800–900°C, with He, Ne or Ar, $D = 2$ mm	Pulsed	241
0.405779	I	$6p7s\ ^3P^{\circ}_1-6p^2\ ^3P_2$	"	Pulsed	241
0.406213	I	$6p6d\ ^3D^{\circ}_1-6p^2\ ^1D_2$	"	Pulsed	241
0.56088	II	$7p\ ^2P_{3/2}-7s\ ^2S_{1/2}$	A few torr of lead, 6 torr of He; $D = 3.5$ mm	CW	—[c]
0.53721	II	$5f\ ^2F^{\circ}_{7/2}-6s6p^2\ ^4P_{5/2}$	0.002–0.05 torr vapor pressure of lead,[a] 1–2 torr of He, $D = 5$ mm	Pulsed	199
0.66600	II	$7p\ ^2P_{1/2}-7s\ ^2S_{1/2}$	A few m torr of lead with 6 torr of He, $D = 3.5$ mm	CW	—[c]
0.72290[b]	I	$7s\ ^3P^{\circ}_1-6p^2\ ^1D_2$	0.2–2.0 torr vapor pressure of lead,[a] 3 torr of He; $D = 10$ mm	Pulsed	242, 182, 241, 243

[a] The vapor pressure of lead is deduced from vapor pressure data for the more common elements, R. E. Honig, *R.C.A. Review*, **18**, 195–204 (June 1957).

[b] A single pass gain of 6 dB/cm has been reported on this line.[243]

[c] Selective excitation is via a Penning reaction with He* 2^3S_1 metastables. (W. T. Silfvast, private communication.)

TABLE 28. *Laser transitions in atomic nitrogen*

Wavelength (μm)	Spectrum	Identification	Notes	Excitation	References
0.336734	III	$3p\ ^4P_{5/2}-3s\ ^4P^{\circ}_{5/2}$	In pulsed low pressure noble gases (impurity), $D = 4$ mm. Also in ammonia at 0.02–0.05 torr, $D = 7$ mm	Pulsed	95, 93, 94
0.347867	IV	$3p\ ^3P^{\circ}_2-3s\ ^3S_1$	"	Pulsed	232, 93, 94, 309
0.348296	IV	$3p\ ^3P^{\circ}_1-3s\ ^3S_1$	"	Pulsed	93, 94
0.3500	II(?)	—	In 0.001–0.02 torr mixture of nitrogen and mercury, $D = 3$ mm	Pulsed	244
0.399501	II	$3p\ ^1D_2-3s\ ^1P^{\circ}_1$	0.01 torr of nitrogen, $D = 6$ mm	Pulsed	233
0.409732	III	$3p\ ^2P^{\circ}_{3/2}-3s\ ^2S_{1/2}$	In 0.02–0.05 torr of ammonia, $D = 7$ mm	Pulsed	232, 94, 245, 309
0.410338	III	$3p\ ^2P^{\circ}_{1/2}-3s\ ^2S_{1/2}$	"	Pulsed	232, 94, 233
0.4120	(?)	—	In mercury–nitrogen mixture at 0.001–0.02 torr, $D = 3$ mm	Pulsed	244
0.4321	(?)	—	—	Pulsed	246
0.4329	(?)	—	—	Pulsed	246
0.451088	III	$3p\ ^4D_{5/2}-3s\ ^4P^{\circ}_{3/2}$	In 0.02–0.05 torr of ammonia, $D = 7$ mm. In 1 torr of nitrogen, $D = 10$ mm	Pulsed	232, 94
0.451488	III	$3p\ ^4D_{7/2}-3s\ ^4P^{\circ}_{5/2}$	"	Pulsed	232, 94, 233
0.4525	(?)	—	In mercury–nitrogen mixture at 0.001–0.02 torr, $D = 3$ mm	Pulsed	244
0.4621	II(?)	—	> 0.03 torr of nitrogen, $D = 1.5$ mm	Pulsed	115
0.463055	II	$3p\ ^3P_2-3s\ ^3P^{\circ}_2$	0.02–0.05 torr of air or nitrogen, $D = 7$ mm	Pulsed	232, 94, 233, 245
0.46439	II(?)	—	> 0.03 torr of nitrogen, $D = 1.5$ mm, or in Hg–N_2 mixture	Pulsed	115
0.4750	(?)	—	In mercury–nitrogen mixture at 0.001–0.02 torr, $D = 3$ mm	Pulsed	244
0.5440	(?)	—	"	Pulsed	244
0.5500	(?)	—	"	Pulsed	244
0.5540	(?)	—	"	Pulsed	244

TABLE 28 *(cont.)*

Wavelength (μm)	Spectrum	Identification	Notes	Excitation	References
0.5610	III or IV	—	In air at low pressure, $D = 5$–10 mm	Pulsed	245
0.5620	III or IV	—	"	Pulsed	245
0.566663	II	$3p\ ^3D_2$–$3s\ ^3P_1^\circ$	—	Pulsed	94ᵃ
0.567601	II	$3p\ ^3D_1$–$3s\ ^3P_0^\circ$	—	Pulsed	94ᵃ
0.567956	II	$3p\ ^3D_3$–$3s\ ^3P_2^\circ$	In air, and in nitrogen–mercury mixture at 0.001–0.02 torr, $D = 3$ mm	Pulsed	244, 15, 94ᵃ
0.56869	II(?)	—	> 0.03 torr of nitrogen (or CO), $D = 1.5$ mm	Pulsed	115
0.648207	II	$3p\ ^1P_1$–$3s\ ^1P_1$	"	Pulsed	115, 247
0.859400	I	$3p\ ^2P_{1/2}^\circ$–$3s\ ^2P_{1/2}$	He(Ne) + N_2(CO_2) at 4 torr, $D = 15$ mm	Pulsed	234
0.862924	I	$3p\ ^2P_{3/2}^\circ$–$3s\ ^2P_{3/2}$	0.2–0.7 torr of nitrogen, 3.0 torr of He or 0.15 torr of nitrogen, $D = 15$ mm	Pulsed	248, 234, 235
0.91876	I	$3p'\ ^2D_{5/2}^\circ$–$3s'\ ^2D_{3/2}$	He(Ne) + N_2(CO_2) at 4 torr, $D = 15$ mm	Pulsed	234
0.938681	I	$3p\ ^2D_{3/2}^\circ$–$3s\ ^2P_{1/2}$	0.02–0.2 torr of nitrogen or nitrous oxide, 0.01–0.1 torr of O_2, H_2, He or Ne, $D = 3$ mm	CW	248, 234, 235, 249
0.939279	I	$3p\ ^2D_{5/2}^\circ$–$3s\ ^2P_{3/2}$	"	CW	248, 234, 235, 249
1.34295	I	$3p\ ^2S_{1/2}^\circ$–$3s\ ^2P_{1/2}$	0.15 torr of nitrogen with 3.0 torr of He, or 0.2–0.7 torr of nitrogen, $D = 15$ mm (or 75 mm ?)	Pulsed	248
1.35813ᵇ	I	$3p\ ^2S_{1/2}^\circ$–$3s\ ^2P_{3/2}$	0.03 torr NO or N_2O with 2 torr of He, or 1 torr of Ne, $D = 5$ mm	CW	248, 6, 85, 235, 236
1.45423	I	—	0.2–0.7 torr of nitrogen, or 0.15 torr of nitrogen with 3 torr of He, $D = 16$ mm (or 75 mm ?)	Pulsed	248
1.4577	I	$4s\ ^4P_{5/2}$–$3p\ ^2D_{5/2}^\circ$	0.03 torr NO or N_2O with 2 torr of He or 1 torr of Ne, $D = 5$ mm	CW	85, 236

TABLE 28 *(cont.)*

Wavelength (μm)	Spectrum	Identification	Notes	Excitation	References
3.7942	I(?)	—	0.2–0.7 torr of nitrogen, or 0.15 torr of nitrogen with 3 torr of He, $D = 15$ mm (or 75 mm ?)	Pulsed	248
3.8154	I(?)	—	0.15 torr of nitrogen, 3.0 torr of He or 0.2–0.7 torr nitrogen, $D = 15$ mm (or 75 mm?)	Pulsed	248

[a]Reference 27 in [94] R. A. McFarlane (private communication), W. B. Bridges and A. N. Chester.
[b]Very strong oscillation obtainable at this wavelength.[248]

TABLE 29. *Laser transitions in phosphorus*

Wavelength (μm)	Spectrum	Identification[a]	Notes	Excitation	References
0.334769	IV	$4p\ ^3P_1^\circ - 4s\ ^3S_1$	0.04 torr vapor pressure of phosphorus, $D = 4$ mm	Pulsed	237
0.422208	III	$4p\ ^2P_{3/2}^\circ - 4s\ ^2S_{1/2}$	0.4 torr vapor pressure of phosphorus, $D = 3$ mm	Pulsed	237
0.602421	II	$4p\ ^3D_2 - 4s\ ^3P_1^0$	0.04 torr of phosphorus, also with 0.01–1.0 torr He or Ne, $D = 3$ mm	Pulsed	250, 237
0.603421	II	$4p\ ^3D_1 - 4s\ ^3P_0^\circ$	0.04 torr of phosphorus, $D = 3$ mm	Pulsed	237
0.604325	II	$4p\ ^3D_1 - 4s\ ^3P_2^\circ$	0.04 torr of phosphorus, also with 0.01–1.0 torr He or Ne	Pulsed	250, 237
0.608786	II	$4p\ ^3D_1 - 4s\ ^3P_0^\circ$	0.04 torr of phosphorus, $D = 3$ mm	Pulsed	237
0.616577	II	$4p\ ^3D_2 - 3s\ ^3P_2^\circ$	,,	Pulsed	237
0.667193	(?)	—	,,	Pulsed	237
0.784563	II	$4p\ ^1P_1 - 4s\ ^1P_1^\circ$	Small amount of phosphorus with 0.01–1.0 torr of He or Ne, $D = 5$ mm	Pulsed	250

[a]Classifications are taken from W. C. Martin, Atomic energy levels and spectra of neutral and singly ionized phosphorus (PI and PII), *J. Opt. Soc. Am.* **49**, 1071–85 (Nov. 1959).

TABLE 30. *Laser transitions in arsenic*

Wavelength (μm)	Spectrum	Identification	Notes	Excitation	References
0.54980	II	$5p\ ^3D_1 - 5s\ ^3P_0^\circ$	Small amount of arsenic with various noble gases, Ne best at 0.1 torr, $D = 5$ mm	Pulsed	251
0.55583	II	$5p\ ^3D_2 - 5s\ ^3P_1^\circ$	"	Pulsed	251
0.56516	II	$5p\ ^3D_3 - 5s\ ^3P_2^\circ$	"	Pulsed	251
0.61703	II	$5p\ ^1P_1 - 5s\ ^3P_1^\circ$	"	Pulsed	251

TABLE 31. *Laser transition in antimony*

Wavelength (μm)	Spectrum	Identification	Notes	Excitation	References
0.61299	II	$6p\ ^3D_3 - 6s\ ^3P_2^\circ$	0.001–0.002 torr of of antimony, 0.2 torr of He, $D = 3$ mm	Pulsed	252

TABLE 32. *Laser transitions in bismuth*

Wavelength (μm)	Spectrum	Identification	Notes	Excitation	References
0.456084	III	$6s^2(^1S)7p\ ^2P_{1/2}^\circ - 6s^2(^1S)7s\ ^2S_{1/2}$	Bismuth vapor at 0.05–0.5 torr with He or Ne at 0.01–0.1 torr, $D = 7$ mm[a]	Pulsed	253
0.571921	II	$6p_{1/2}\ 7p_{1/2}\ ^3P_0 - 6p_{1/2}\ 7s\ ^3P_1^\circ$	—	Pulsed	253
0.75990	III	$6s^2(^1S)6f\ ^2F_{5/2}^\circ - 6s^2(^1S)7d\ ^2D_{3/2}$	—	Pulsed	253
0.80689	III	$6s^2(^1S)6f\ ^2F_{7/2}^\circ - 6s^2(^1S)7d\ ^2D_{5/2}$	—	Pulsed	253

[a]Generation does not depend on the nature of the buffer gas.

TABLE 33. *Laser transitions in atomic oxygen*

Wavelength (μm)	Spectrum	Identification	Notes	Excitation	References
0.298378	III	$3p\ ^1D_2\text{–}3s\ ^1P_1^\circ$	As an impurity in noble gases at low partial pressures in the range 0.001–0.1 torr, $D = 4$ mm	Pulsed	95, 15, 93, 94
0.298461	III	$4p\ ^5P_2^\circ\text{–}3d\ ^3D_2$	"	Pulsed	15
0.304713	III	$3p\ ^3P_2^\circ\text{–}3s\ ^3P_2^\circ$	"	Pulsed	93, 94
0.306346	IV	$3p\ ^2P_{3/2}^\circ\text{–}3s\ ^2S_{1/2}$	"	Pulsed	93, 94
0.338134	IV	$3p\ ^4D_{3/2}\text{–}3s\ ^4P_{1/2}^\circ$	"	Pulsed	93, 94
0.338554	IV	$3p\ ^4D_{7/2}\text{–}3s\ ^4P_{5/2}^\circ$	"	Pulsed	93, 94
0.374949	II	$3p\ ^4S_{3/2}^\circ\text{–}3s\ ^4P_{5/2}$	As impurity in noble gases, or in 0.02–0.05 torr of CO, $D = 7$ mm	Pulsed	232, 15, 93, 94
0.375467	III	$3p\ ^3D_2\text{–}3s\ ^3P_1^\circ$	"	Pulsed	232, 15, 93, 94
0.375988	III	$3p\ ^3D_3\text{–}3s\ ^3P_2^\circ$	"	Pulsed	232, 15, 93, 94
0.4120	II(?)	—	In mercury–nitrogen mixture at 0.001–0.02 torr, $D = 3$ mm	Pulsed	244
0.434738	II	$3p'\ ^2D_{3/2}^\circ\text{–}3s'\ ^2D_{3/2}$	In 0.02–0.05 torr of ammonia $D = 7$ mm; or as an impurity in noble gases	Pulsed	246, 15, 94
0.435128	II	$3p'\ ^2D_{5/2}^\circ\text{–}3s'\ ^2D_{5/2}$	"	Pulsed	232, 15, 94
0.441488	II	$3p\ ^2D_{5/2}^\circ\text{–}3s\ ^2P_{3/2}$	"	Pulsed	232, 15, 94
0.441697	II	$3p\ ^2D_{3/2}^\circ\text{–}3s\ ^2P_{1/2}$	"	Pulsed	232, 15, 94
0.4525	II(?)	—	As an impurity in mercury–nitrogen mixture	Pulsed	244
0.460552	II(?)	—	In 0.02–0.05 torr of ammonia $D = 7$ mm; or as impurity in noble gases	Pulsed	232, 94
0.464914	II	$3p\ ^4D_\circ^{7/2}\text{–}3s\ ^4P_{5/2}$	As a CO impurity in a mercury–nitrogen mixture, at 0.001–0.02 torr, $D = 3$ mm	Pulsed	244, 15, 94
0.4750	II(?)	—	"	Pulsed	244
0.477291	I	—	In flowing oxygen, at 0.6–0.015 torr; $D = 5, 8,$ or 12 mm	Pulsed	316
0.5016387	II	—	"	Pulsed	316
0.514606	I	—	"	Pulsed	316
0.5440	II(?)	—	As a CO inpurity in mercury–nitrogen mixture, at 0.001–0.02 torr, $D = 3$ mm		
0.5500	II(?)	—		Pulsed	244
0.5540	II(?)	—			

TABLE 33 *(cont.)*

Wavelength (μm)	Spectrum	Identification	Notes	Excitation	References
0.559257	III	$3p\ {}^1P_1 - 3s^1P_1^\circ$	0.02–0.05 torr of oxygen, $D = 7$ mm	Pulsed	232, 15, 94, 247, 316
0.566662	II	—	In flowing oxygen at 0.6–0.015 torr, $D = 5$, 8, or 12 mm	Pulsed	316
0.5679[a]	II (or N II)	—	In mercury–nitrogen at 0.001–0.02 torr, $D = 3$ mm	Pulsed	244, 316
0.5690	II (or O II)	—	,,	Pulsed	244
0.657745	III(?)	$3d\ {}^5F_2 - 4d\ {}^3D_3^\circ$	As impurity in noble gases in range 0.001–0.1 torr, $D = 3$ mm	Pulsed	15
0.66490	II	$3p\ {}^2S_{1/2}^\circ - 3s\ {}^2P_{1/2}$	In 0.016–0.035 torr O_2, $D = 6$ mm	Pulsed	247
0.666694	II	—	In flowing oxygen at 0.015–0.6 torr, $D = 5$, 8, or 12 mm	Pulsed	316
0.672136	II	$3p\ {}^2S_{1/2}^\circ - 3s\ {}^2P_{3/2}$	In 0.02–0.05 torr of ammonia, $D = 7$ mm	Pulsed	232, 94, 316
0.844628 ⎫ 0.844638 ⎪ 0.844672 ⎬ 0.844686 ⎭	I[b]	$3p\ {}^3P_{0,\,2,\,1} - 3s\ {}^3S_1^\circ$	Approximately 0.01–0.04 torr of oxygen with 0.35 torr of Ne, 1.4 torr Ar, CO or CO_2, or He; $D = 7$ mm. As an impurity in bromine with He or Ne, also in nitrous oxide, and in pure oxygen at 0.1–2 torr, $D = 10$ mm	CW	254, 6, 8, 56, 85, 234, 236, 248, 255–62
0.882045	I(?)	$3p'\ {}^1F_3 - 3s'\ {}^1D_2^\circ$	In He(Ne)–$N_2(CO_2)$ at 4 torr, $D = 15$ mm	Pulsed	234
2.890	I	$4p\ {}^3P - 4s\ {}^3S^0$	0.08 torr of oxygen with 0.5–1.0 torr of He or Ne (probably $D = 5$ or 7 mm)	CW	263
4.563	I	$4p\ {}^3P - 3d\ {}^3D^\circ$	,,	CW	263, 264, 265
5.981	I	$7d\ {}^3D^\circ - 6p\ {}^3P$	,,	CW	263
6.8161	I	$3s'\ {}^3D_2^\circ - 4p\ {}^3P_{2,\,1,\,0}$	In pure oxygen, $D = 15$ mm (or 75 mm?)	Pulsed	248

TABLE 33 *(cont.)*

Wavelength (μm)	Spectrum	Identification	Notes	Excitation	References
6.858	I	$5p\ ^3P-5s\ ^3S°$	0.08 torr of oxygen with 0.5–1.0 torr of He or Ne (probably $D = 5$ or 7 mm)	CW	263
6.8731	I	$3s'\ ^3D_3°-4p\ ^3P_2$	In pure oxygen, $D = 15$ mm (or 75 mm ?)	Pulsed	248
10.400	I	$5p\ ^3P-4d\ ^3D°$	0.08 torr of oxygen with 0.5–1.0 torr of He or Ne (probably $D = 5$ or 7 mm)	CW	263

[a]This line is possibly a Hg-II line at 0.5677 μm, or a N-II line at 0.5679 μm.

[b]This is the O-I *triplet*, as shown in ref. 255. The *quartet* oscillation is due to the large Doppler width of the line (caused by excitation and dissociation of oxygen molecules) and radiation trapping at the line center causing gain to occur only in the wings of the line.

TABLE 34. *Laser transitions in sulfur*

Wavelength (μm)	Spectrum	Identification	Notes	Excitation	References
0.263898	V[a]	—	0.015–0.06 torr of SO_2, $D = 1.5$–6 mm	Pulsed	266
0.332486	III[a]	$4p\ ^3P_2-3d\ ^3P_2°$	"	Pulsed	266
0.349737	III[a]	—	0.015–0.06 torr of SO_2, or in H_2S, $D = 1.5$–6 mm	Pulsed	266
0.370937	III[a]	$4p\ ^3D_2-3d\ ^3P_1°$	In SO_2, SF_6 or H_2S at 0.015–0.06 torr, $D = 1.5$–6 mm	Pulsed	266
0.492532	II	$4p\ ^4P_{3/2}°-4s\ ^4P_{1/2}$	In SO_2 at 0.015–0.06 torr, $D = 1.5$–6 mm	Pulsed	266
0.501401	II	$4p\ ^2P_{3/2}°-4s\ ^2P_{3/2}$	"	Pulsed	266
0.503239	II	$4p\ ^4P_{5/2}°-4s\ ^4P_{5/2}$	In SO_2 or SF_6 at 0.015–0.06 torr, $D = 1.5$–6 mm	Pulsed	266
0.516032	—	—	"	Pulsed	266
0.521962	—	—	"	Pulsed	266
0.532070	II	$4p'\ ^2F_{7/2}°-4s'\ ^2D_{5/2}$	In S, SO_2, SF_6 or H_2S at 0.015–0.06 torr, also with He, Ne, Ar, O_2, H_2, N_2 or air, at 0.01–1 torr, $D = 5$ mm	Pulsed	250, 266
0.534567	II	$4p'\ ^2F_{5/2}°-4s'\ ^2D_{3/2}$	"	CW	250, 266

TABLE 34. (*cont.*)

Wavelength (μm)	Spectrum	Identification	Notes	Excitation	References
0.542864	II	$4p\ ^4D^\circ_{3/2}\text{–}4s\ ^4P_{1/2}$	In S, SO$_2$, SF$_6$ or H$_2$S at 0.015–0.06 torr, also with He, Ne, Ar, O$_2$, H$_2$, N$_2$ or air, at 0.01–1 torr, $D = 5$ mm	Pulsed	250, 218
0.543277	II	$4p\ ^4D^\circ_{5/2}\text{–}4s\ ^4P_{3/2}$	,,	CW	250, 218
0.545380	II	$4p\ ^4D^\circ_{7/2}\text{–}4s\ ^4P_{5/2}$	,,	CW	250, 218
0.547374	II	$4p\ ^4D^\circ_{1/2}\text{–}4s\ ^4P_{1/2}$	,,	Pulsed	250, 218
0.550990	II	$4p\ ^4D^\circ_{3/2}\text{–}4s\ ^4P_{3/2}$	,,	Pulsed	218, 266
0.556495	II	$4p\ ^4D^\circ_{5/2}\text{–}4s\ ^4P_{5/2}$	In SO$_2$ at 0.015–0.06 torr, $D = 1.5$–6 mm	Pulsed	266
0.563999	II	$4p\ ^2D^\circ_{5/2}\text{–}4s\ ^2P_{3/2}$	In S, SO$_2$, SF$_6$, H$_2$S, also with He, Ne, Ar, O$_2$, N$_2$, H$_2$ or air as above, $D = 5$ mm	CW	250, 266
0.564698	II	$4p\ ^2D^\circ_{3/2}\text{–}4s\ ^2P_{1/2}$	,,	Pulsed	250, 266
0.581919	II	$4p\ ^2D^\circ_{3/2}\text{–}4s\ ^2P_{3/2}$	In SO$_2$ at 0.015–0.06 torr $D = 1.5$–6 mm	Pulsed	266
1.0455	I	$4p\ ^3P_2\text{–}4s\ ^3S^\circ_1$	0.03 torr of SF$_6$ or 0.03 torr of SF$_6$ with 2 torr of helium. Also with H$_2$S, with He, Ne and Ar, $D = 5$ mm	CW	236, 85, 267
1.0636	I(?)	$4p'\ ^1F_3\text{–}4s'\ ^1D^\circ_2$	,,	CW	236, 85, 267
2.2799	I	$4p\ ^5P_2\text{–}4s'\ ^3D^\circ_3$	—	CW	264
3.3892	I	$4s'\ ^3D^\circ_3\text{–}3d\ ^5D^\circ_2$	—	CW	264

[a]Oscillation from multiple-charged sulfur lines is found to be strongest in 6-mm bore discharge tubes in the range 0.15–6.0 mm.[266]

TABLE 35. *Laser transitions in selenium*

Wavelength (μm)	Spectrum	Identification	Notes	Excitation	References
0.446760	II	$5p\ ^2P^\circ_{1/2}\text{–}5s\ ^2P_{1/2}$	In selenium vapor at about 270°C, with He at 6 to 8 torr; $D = 3$ mm (see Figs. 5.17, 5.18)	CW	268
0.460434	II	$5p\ ^2D^\circ_{5/2}\text{–}5s\ ^4P_{5/2}$	,,	CW	269
0.461877	II	$5p\ ^4P^\circ_{5/2}\text{–}5s\ ^4P_{3/2}$	,,	CW	268
0.464844	II	$5p\ ^4P^\circ_{3/2}\text{–}5s\ ^4P_{1/2}$	,,	CW	269
0.471823	II	$5p\ ^4S^\circ_{3/2}\text{–}3^{a}$	,,	CW	268

TABLE 35 *(cont.)*

Wavelength (μm)	Spectrum	Identification	Notes	Excitation	References
0.474097	II	$5p\ ^2P^\circ_{3/2}-4s\ 4p^4\ ^2P_{3/2}$	In selenium vapor at about 270°C, with He at 6 to 8 torr; $D = 3$ mm (see Figs. 5.17, 5.18)	CW	268
0.476365	II	$5p\ ^2D^\circ_{3/2}-5s\ ^4P_{3/2}$	,,	CW	269
0.476552	II	$5p\ ^2P^\circ_{1/2}-5s\ ^2P_{3/2}$	,,	CW	268
0.484063	II	$5p\ ^2S^\circ_{1/2}-5s\ ^4P_{3/2}$	,,	CW	269
0.484496	II	$5p\ ^4S^\circ_{1/2}-5s\ ^4P_{5/2}$	,,	CW	269
0.497566	II	$5p\ ^2D^\circ_{5/2}-4s\ 4p^4\ ^2P_{3/2}$	,,	CW	269
0.499275	II	$5p\ ^4P^\circ_{3/2}-5s\ ^4P_{3/2}$	,,	CW	269
0.506865	II	$5p\ ^4P^\circ_{5/2}-5s\ ^4P_{5/2}$	,,	CW	269
0.509650	II	$5p\ ^4D^\circ_{7/2}-4d\ ^4F_{9/2}{}^b$	,,	CW	251, 269
0.514214	II	$5p\ ^4D^\circ_{3/2}-5s\ ^4P_{1/2}$	,,	CW	269
0.517598	II	$5p\ ^4D^\circ_{3/2}-5s\ ^4P_{3/2}$	,,	CW	269
0.522751	II	$5p\ ^4D^\circ_{7/2}-5s\ ^4P_{5/2}$	,,	CW	251, 269
0.525307	II	$5p\ ^2D^\circ_{5/2}-5s\ ^2P_{3/2}$	,,	CW	269
0.525363	II	$5p\ ^4D^\circ_{1/2}-5s\ ^4P_{1/2}$	,,	CW	269
0.527111	II	$5p\ ^4D^\circ_{5/2}-4d\ ^4F_{7/2}{}^b$	,,	CW	269
0.530535	II	$5p\ ^2D^\circ_{3/2}-5s\ ^2P_{1/2}$	,,	CW	269
0.552242	II	$\begin{cases} 5p\ ^4P^\circ_{3/2}-5s\ ^4P_{5/2}(?) \\ 5p\ ^4P^\circ_{5/2}-4s\ 4p^4\ ^2P_{3/2}(?) \end{cases}$	,,	CW	269
0.556693	II	$5p\ ^4D^\circ_{3/2}-5s\ ^4P_{3/2}$	,,	CW	268
0.559116	II	$5p\ ^4P^\circ_{3/2}-5s\ ^2P_{1/2}$	,,	CW	269
0.562313	II	$5p\ ^4P^\circ_{1/2}-5s\ ^4P_{1/2}$	,,	CW	268
0.569788	II	$5p\ ^4D^\circ_{1/2}-5s\ ^4P_{3/2}$	,,	CW	269
0.574762	II	$5p\ ^4D^\circ_{5/2}-5s\ ^4P_{5/2}$	,,	CW	269
0.584268	II	$5p\ ^2S^\circ_{1/2}-5s\ ^2P_{3/2}$	,,	CW	268
0.586627	II	$5p\ ^4P^\circ_{5/2}-5s\ ^2P_{3/2}$	,,	CW	268
0.605596	II	$5p\ ^2P^\circ_{3/2}-7$	,,	CW	269
0.606583	II	$5p\ ^4P^\circ_{3/2}-5s\ ^2P_{3/2}$	,,	CW	268
0.610196	II	$5p\ ^2D^\circ_{3/2}-5s\ ^2P_{3/2}$	,,	CW	268
0.644425	II	$5p\ ^2D^\circ_{5/2}-7$	,,	CW	269
0.649048	II	$5p\ ^4D^\circ_{1/2}-5s\ ^2P_{1/2}$	,,	CW	269
0.653495	II	$5p\ ^2P^\circ_{1/2}-7$	,,	CW	269
0.706389[c]	II	$5p\ ^4P^\circ_{1/2}-5s\ ^2P_{1/2}$	,,	CW	268
0.739199	II	$5p\ ^4P^\circ_{5/2}-7$	,,	CW	268
0.767482	II	$5p\ ^2P^\circ_{3/2}-5s'\ ^2D_{5/2}$	,,	CW	268
0.772404	II	$5p\ ^4D^\circ_{1/2}-5s\ ^2P_{3/2}$	,,	CW	268
0.779615	II	$5p\ ^2P^\circ_{1/2}-5s'\ ^2D_{3/2}$	,,	CW	268
0.783881	II	$5p\ ^4P^\circ_{1/2}-4s4p^4\ ^2P_{3/2}$	,,	CW	268
0.830952	II	$5p\ ^2D^\circ_{3/2}-5s'\ ^2D_{5/2}$	,,	CW	268
0.92493[d]	—	Unknown	,,	CW	268
0.995515[c]	II	$5p\ ^4P^\circ_{5/2}-5s'\ ^2D_{5/2}$	,,	CW	268
1.040881[c]	II	$5p\ ^4D^\circ_{3/2}-6$	,,	CW	268
1.258678[c]	II	$5p\ ^4P^\circ_{3/2}-10$	,,	CW	268

[a]Incorrect identification in D. C. Martin, *Phys. Rev.* **48**, 938 (1935).
[b]Identification: S. G. Krishnamurty and K. R. Rao, *Proc. Roy. Soc.* (London) A **149**, 56 (1935).
[c]Calculated from energy-level differences.
[d]Measured wavelength (± 0.0001 μm).

TABLE 36. *Laser transitions in tellurium*

Wavelength (μm)	Spectrum	Identification	Notes	Excitation	References
0.48429	II	$122'_{5/2}$–$102_{3/2}$	In tellurium vapor at 420°C, with 5 torr He; $D = 3$ mm	CW	270
0.50204	II	Unknown	,,	CW	270
0.52564	II	Unknown	,,	CW	270
0.54498	II	$103_{3/2}$–$85'_{5/2}$	In tellurium vapor at 420°C, with 2 torr He; $D = 3$ mm	CW	270
0.54540	II(?)	—	In tellurium at 125 to 250°C, with 0.1–0.25 torr Ne; $D = 6$ mm	Pulsed	271
0.54791	II	$105_{3/2}$–$86_{3/2}$	In tellurium vapor at 420°C, with 2 torr He; $D = 3$ mm	CW	270
0.55763	II	$112_{7/2}$–$94_{5/2}$	In tellurium vapor at 420°C, with 5 torr He; $D = 3$ mm	CW	252, 270, 271
0.56404	II(?)	—	In tellurium at 125 to 250°C, with 0.2 torr Ne; $D = 3$ mm	Pulsed	271
0.56662	II	$101_{3/2}$–$83_{1/2}$	As for 0.48429-μm line	CW	270
0.57081	II	$103_{7/2}$–$85'_{5/2}$	,,	CW	272, 270
0.57416	II	$112_{3/2}$–$94_{5/2}$	As for 0.54498-μm line	CW	270
0.57559	II	$100_{5/2}$–$82_{3/2}$	As for 0.48429-μm and 0.54498-μm lines	CW	270
0.57653	II	$112_{5/2}$–$95_{3/2}$	As for 0.48429-μm line	CW	270
0.58511	II	$111_{5/2}$–$95_{5/2}$	As for 0.48429-μm and 0.54498-μm lines	CW	270
0.59361	II	$99_{3/2}$–$82_{3/2}$	As for 0.54498-μm line	CW	271, 270
0.59726	II	$111_{5/2}$–$95_{3/2}$	As for 0.48429-μm and 0.54498-μm lines	CW	270
0.59747	II	$102_{5/2}$–$85'_{5/2}$	As for 0.48429-μm line	CW	270
0.60145	II	$105_{5/2}$–$88_{3/2}$	As for 0.54498-μm line	CW	270
0.60823	II	Unknown	As for 0.48429-μm line	CW	270
0.62307	II	$105_{3/2}$–$88_{3/2}$	As for 0.54498-μm line	CW	270

443

TABLE 36 *(cont.)*

Wavelength (μm)	Spectrum	Identification	Notes	Excitation	References
0.62454	II	$99_{3/2}$–$83_{1/2}$	As for 0.48429-μm and 0.54498-μm lines	CW	271, 270
0.6350	I(?)	—	0.001 to 0.002 torr of tellurium with 0.2 torr of Ne; $D = 3$ mm	Pulsed	252
0.65851	—	Unknown	As for 0.48429-μm line	CW	270
0.66486	II	$97_{1/2}$–$82_{3/2}$	As for 0.54498-μm line	CW	270
0.66761	II	$103_{3/2}$–$88_{3/2}$	"	CW	270
0.68851	II	$100_{5/2}$–$85'_{5/2}$	"	CW	270
0.70391	II	$97_{1/2}$–$83_{1/2}$	"	CW	271, 270
0.78017	II	$105_{3/2}$–$92_{3/2}$	"	CW	270
0.79217	II	$97_{1/2}$–$85_{3/2}$	"	CW	270
0.86046	—	Unknown	As for 0.48429-μm line	CW	270
0.87338	—	Unknown	"	CW	270
0.89721	II	$103_{3/2}$–$92_{5/2}$	As for 0.54498-μm line	CW	270
0.89982[a]	—	Unknown	"	CW	270
0.93785[b]	II	$99_{3/2}$–$88_{5/2}$	"	CW	270

[a]Measured wavelength ($\pm 0.00006\ \mu$m).
[b]Obtained from energy-level differences.

TABLE 37. *Laser transitions in thulium*

Wavelength (μm)	Spectrum	Identification[a]	Notes	Excitation	References
1.3040	—	—	In thulium vapor above 800°C, with 2.5 torr He; 1.5 torr Ne; or 0.8 torr Ar	Pulsed	273
1.3101	I	—	"	Pulsed	273
1.3380	I	—	"	Pulsed	273
1.4340	I	—	"	Pulsed	273
1.4485	I	—	"	Pulsed	273
1.5000	—	—	"	Pulsed	273
1.6379	I	—	"	Pulsed	273
1.6754	I	—	"	Pulsed	273
1.7320[b]	I	—	"	Pulsed	274
1.9584	I	—	"	Pulsed	273
1.9730	I	—	"	Pulsed	273
1.9942	I	—	"	Pulsed	273
2.1070	—	—	"	Pulsed	273
2.3845	I	—	"	Pulsed	273

[a]To be published: J. L. Verges and P. Camus, *Revue Spectrochimica Acta.*
[b]$\pm 0.0003\ \mu$m.

TABLE 38. *Laser transitions in atomic fluorine*

Wavelength (μm)	Spectrum	Identification	Notes	Excitation	References
0.275959	III	$3p\ ^2D^\circ_{5/2}-3s\ ^2D_{5/2}$	0.02 torr of fluorine; $D = 4$ mm	Pulsed	237
0.282608	IV	$3p\ ^3D_3-3s\ ^3P^\circ_2$	”	Pulsed	237
0.312156	III	$3p\ ^4D^\circ_{7/2}-3s\ ^4P_{5/2}$	”	Pulsed	237
0.317418	III	$3p\ ^2D^\circ_{5/2}-3s\ ^2P_{3/2}$	”	Pulsed	237
0.320274	II	$3p'\ ^1D_2-3s'\ ^1D^\circ_2$	”	Pulsed	237
0.402478	II	$3p\ ^3P_2-3s\ ^3S^\circ_1$	”	Pulsed	237
0.7039[a, b]	I	$3p\ ^2P^\circ_{3/2}-3s\ ^2P_{3/2}$	In flowing CF_4, SF_6 or C_2F_6 at 0.03–0.1 torr with He at 2–10 torr; $D = 25$ mm. Also in 0.3 torr He with 0.05 torr HF; $D = 10$ cm; He essential with HF	Pulsed	275, 276
0.7129[a, b]	I	$3p\ ^2P^\circ_{1/2}-3s\ ^2P_{1/2}$	”	Pulsed	275, 276
0.7204[a, b]	I	$3p\ ^2P_{3/2}-3s\ ^2P_{1/2}$	”	Pulsed	275, 276
0.7311	I	—	In flowing fluorine and He at 5 torr	Pulsed	277
0.7802[a, b]	I	$3p\ ^2D^\circ_{3/2}-3s\ ^2P_{1/2}$		Pulsed	276
1.5900 to 9.3462		Twenty unidentified lines		Pulsed	276

[a]Selective excitation of the upper laser level is by the dissociative excitation reaction:

$$\text{He*}\ 2^1S_0 + \text{HF}(^1\Sigma^+, v = 0) \rightarrow \text{He}\ ^1S_0 + \text{H} + \text{F}'(3P) \pm \Delta E_\infty.$$

[b]See Fig. 4.43. Laser lines at 0.455259, 0.456784, 0.634724, 0.637148 and 0.667193 μm ascribed to fluorine in ref. 237 have been identified as laser lines in silicon.[238]

TABLE 39. *Laser transitions in atomic chlorine (Fig. A.21)*

Wavelength (μm)	Spectrum	Identification	Notes	Excitation	References
0.263270	III	$4p'\ ^2F_{7/2}-4d'\ ^2D_{5/2}$	Less than 0.02 torr of chlorine; $D = 4$ mm	Pulsed	237, 238
0.319143	III	$4p\ ^4S^\circ_1-4s\ ^4P_2$	”	Pulsed	237
0.339287	III	$4p'\ ^2D^\circ_1-4s'\ ^2D_1$	”	Pulsed	237
0.339345	III	$4p'\ ^2D^\circ_2-4s'\ ^2D_2$	”	Pulsed	237
0.353003	III	$4p'\ ^2F^\circ_3-4s'\ ^2D_2$	”	Pulsed	237
0.356069	III	$4p'\ ^2F^\circ_2-4s'\ ^2D_1$	”	Pulsed	237
0.360210	III	$4p\ ^4D^\circ_2-4s\ ^4P_2$	”	Pulsed	237
0.361210	III	$4p\ ^1D^\circ_2-4s\ ^4P_1$	”	Pulsed	237
0.362268	III	$4p\ ^2D^\circ_1-4s\ ^4P_0$	”	Pulsed	237
0.372046	III	$4p\ ^2D^\circ_2-4s\ ^2P_1$	”	Pulsed	237
0.374878	III	$4p\ ^2D^\circ_1-4s\ ^2P_0$	”	Pulsed	237

TABLE 39 (*cont.*)

Wavelength (μm)	Spectrum	Identification	Notes	Excitation	References
0.413250	II	$4p'\ ^1D_2 - 4s'\ ^1D_2^\circ$	Approx. 0.05 torr of chlorine, $D = 3$ mm	CW	278
0.474042	II	$4p''\ ^1P_1 - 3d''\ ^1D_2^\circ$	"	CW	278
0 476874	II	$4p''\ ^3D_2 - 4s''\ ^3P_1^\circ$	"	CW	237, 278
0.478134	II	$4p''\ ^3D_3 - 4s''\ ^3P_2^\circ$	Approx. 0.05 torr of chlorine, $D = 3$ mm[251]	CW	110, 94, 251
0.489685	II	$4p'\ ^3F_4 - 4s'\ ^3D_3^\circ$	"	CW	110, 94, 251
0.490483	II	$4p'\ ^3F_3 - 4s'\ ^3D_2^\circ$	"	CW	110, 94, 251
0.491781	II	$4p'\ ^3F_2 - 4s'\ ^3D_1^\circ$	"	CW	110, 94, 251
0.507829	II	$4p'\ ^3D_3 - 4s'\ ^3D_3^\circ$	"	CW	110, 94, 251
0.510310	II	$4p'\ ^3D_2 - 4s'\ ^3D_2^\circ$	"	CW	278
0.521792	II	$4p\ ^3P_2 - 4s\ ^3S_1^\circ$	Approx. 0.05 torr of chlorine, or $SnCl_4$ with $D = 5.6$ mm;[251] or 0.5 torr of chlorine, $D = 2$ mm	CW	110, 94, 240, 251, 279
0.522135	II	$4p\ ^3P_1 - 4s\ ^3S_1^\circ$	Approx. 0.05 torr of chlorine, $D = 3$ mm[251]	CW	110, 94, 251
0.539216	II	$4p'\ ^1F_3 - 4s'\ ^1D_2^\circ$	"	CW	110, 94, 251, 278, 279
0.609473	II	$4p'\ ^1P_1 - 4s'\ ^1D_2^\circ$	Approx. 0.05 torr of chlorine, or $SnCl_4$ with $D = 5.6$ mm[240]	CW	110, 94, 240, 251, 278
0.945206	I	$4p\ ^2P_{3/2}^\circ - 4s\ ^2P_{1/2}$	In chlorine at 0.010–0.080 torr with 0.3–3 torr He or Ne, $D = 6$ mm. Also in freon at 0.001 torr with 0.8 torr of Ne, $D = 7$ mm[281]	CW	280, 281, 282
1.3863[a]	I	$3d\ ^2D_{3/2} - 4p\ ^4D_{5/2}^\circ$	In 0.3 torr of HCl with 0.1 torr He or Ne, $D = 14$ mm	CW	282
1.3893[a]	I	$3d\ ^4P_{3/2} - 4p\ ^4D_{3/2}^\circ$	"	Pulsed	282
1.5874	I	$3d\ ^4P_{9/2} - 4p\ ^4D_{7/2}^\circ$	In 250 torr of He plus NOCl at 1%	Pulsed	283, 8, 284
1.975528[a]	I	$3d\ ^4D_{7/2} - 4p\ ^4P_{5/2}$	In HCl, or 0.3 torr of $SiCl_4$ with 0.1 of He or Ne, or 0.1 torr of chlorine, $D = 6$ mm	CW	285, 6, 106, 107 239, 285, 315
2.019936[a]	I	$3d\ ^4D_{5/2} - 4p\ ^4P_{3/2}$	"	CW	285, 6, 239, 285, 315

TABLE 39 *(cont.)*

Wavelength (μm)	Spectrum	Identification	Notes	Excitation	References
2.4470[a]	I	$3d\ ^4D_{7/2}-4p\ ^4D^\circ_{7/2}$	In 0.3 torr of HCl with 0.1 torr of He or Ne, $D = 14$ mm. Also in chlorine at 0.09 torr with He at 7.2 torr, $D = 10$ mm(?)	CW	282, 106, 107
3.0672	I	$5p\ ^2D^\circ_{5/2}-5s\ ^2P_{3/2}$	In 0.09 torr of chlorine with 2.1 torr Ar; $D = 25$ mm	CW	106, 107

[a]Classification of these lines is as in C. J. Humphrey and E. Paul, Jr., First spectrum of chlorine; an extension based on observations in the 7000 to 25,000 Å region, *J. Opt. Soc. Am.* **49**, 1180–87 (Dec. 1959).
Note: Lines at 1.5891, 2.499, 2.535, 2.602, 2.784, 3.801 and 10.604 μm observed CW in discharges in He–CCl_2F_2 and He-$CBrF_3$ mixtures are possibly also Cl-I lines.[285]

TABLE 40. *Laser transitions in atomic bromine (Fig. A.22)*

Wavelength (μm)	Spectrum	Identification	Notes	Excitation	References
0.474266	II	$5p''\ ^3D_3-5s''\ ^3P^\circ_2$	0.04 torr bromine, $D = 7$ mm	Pulsed	286, 287
0.505463	II	$5p'\ ^3F_3-4d\ ^3D_2$	”	Pulsed	286, 287
0.518228	II	$5p\ ^3P_2-5s\ ^3S_1$	Small amount of bromine with various noble gases, Ne (best) at about 0.1 torr, $D = 5$ mm	CW	251, 286, 287
0.523826	II	$5p\ ^3P_1-5s\ ^3S_1$	0.04 torr bromine, $D = 7$ mm	Pulsed	286, 287
0.533223	II	$5p'\ F_3-5s'\ ^1D_2$	Small amount of bromine with Ne (best) at about 0.1 torr, $D = 5$ mm	CW	251, 286, 287
0.611786	II	$5p\ ^5P_2-5s\ ^3S^\circ_1$	Approx. 0.04 torr bromine, $D = 7$ mm	Pulsed	231, 287
0.616878	II	$5p\ ^5P_1-5s\ ^3S^\circ_1$	0.04 torr bromine, $D = 7$ mm	Pulsed	287
0.8446[a]	OI	See oxygen lines	—	CW	6
2.2866[b]	I	$4d[3]_{7/2}-5p[2]^\circ_{5/2}$	0.3 torr HBr, $D = 12$ mm	CW	288
2.3513[b]	I	$4d[3]_{5/2}-5p[2]^\circ_{3/2}$	”	CW	288
2.8377[b]	I	$4d[3]_{7/2}-5p[3]^\circ_{7/2}$	”	CW	288

[a]This line originally thought to be a Br-I laser line has been shown to be an O-I line.
[b]Identification based on term-values given by J. L. Tech, Analysis of the spectrum of neutral bromine, *J. Res. Natl. Bur. Stand.* A**67**, 505–54 (Nov.–Dec. 1963).

TABLE 41. *Laser transitions in atomic iodine* (*Figs. A.23 and 5.15*)

Wavelength (μm)	Spectrum	Identification	Notes	Excitation	References
0.453379	III or IV(?)	—	0.001 torr iodine, noble gases suppress oscillation; $D = 10$–12 mm	Pulsed	289, 293, 298
0.467440	III or IV(?)	—	"	Pulsed	289
0.493467	III or IV(?)	—	"	Pulsed	289
0.498692[a]	II	$6p'\ ^3D_2$–$5d\ ^3D_1^\circ$	0.1–0.4 torr of iodine, with a few torr of He, $D = 5$ mm	Pulsed	290, 94
0.521627[a]	II	$6p'\ ^3F_2$–$5d\ ^3D_1^\circ$	"	Pulsed	290, 94, 291
0.540736[a]	II	$6p'\ ^3D_2$–$6s'\ ^3D_2^\circ$	"	CW[f]	292, 94, 290, 291, 293
0.5419[b]	Xe II(?)	—	—	Pulsed	290
0.562569	II	$6p\ ^3P_2$–$6s\ ^3S_1^\circ$	0.1 torr of iodine, with a few torr of He or Ne, $D = 5$ mm	Pulsed	290, 94
0.567808[a]	II	$6p'\ ^3F_2$–$6s'\ ^3D_2^\circ$	0.1–0.4 torr of iodine, with a few torr of He, $D = 5$ mm	CW[f]	292, 290, 291, 293
0.567072[a]	II	$6p'\ ^3D_2$–$6s'\ ^3D_1^\circ$	0.1–0.4 torr of iodine, with a few torr of He 0.2–0.5	CW[f]	294, 94, 290, 291 292, 293
0.606893[a]	II	$6p'\ ^3F_2$–$6s'\ ^3D_1^\circ$	0.2–0.5 torr of iodine, with a few torr of He, $D = 2.0$ cm, need to suppress 0.5760 μm line	Pulsed	291, 295, 296
0.612749[a]	II	$6p'\ ^3D_1$–$6s'\ ^3D_2^\circ$	0.1–0.4 torr of iodine with a few torr of He	CW[f]	294, 94, 290, 291, 292, 293,
0.651618[a]	II	$6p'\ ^3F_2$–$5p^5\ ^1P_1^\circ$	0.2–0.5 torr of iodine, with a few torr of He, $D = 2.0$ cm	CW[f]	291, 293, 298
0.658521[a]	II	$6p'\ ^3D_1$–$6s'\ ^3D_1^\circ$	0.1–0.4 torr of iodine, with a few torr of He	CW[f]	290, 94, 291, 292, 293, 298
0.682523[a]	II	$6p'\ ^3F_2$–$6s'\ ^3D_3^\circ$	0.2–0.5 torr of iodine, with a few torr of He, need to suppress 0.5678 μm line	Pulsed	295

TABLE 41. *(cont.)*

Wavelength (μm)	Spectrum	Identification	Notes	Excitation	References
0.690477[a]	II	$6p'\,^3D_2-6s'\,^1D_2^\circ$	0.1 torr of iodine, with a few torr of He or 0.1 torr of Kr, $D = 5$ mm	Pulsed combined with RF	290, 84
0.703299[a]	II	$6p'\,^3D_2-5d'\,^3D_3^\circ$	0.1–0.5 torr of iodine with a few torr of He, $D = 5$ mm	CW[f]	292, 94, 290, 292, 293, 295, 296
0.713897[a]	II	$6p'\,^3D_2-5d'\,^3D_3^\circ$	—	CW[f]	295, 293
0.825384	II	$6p'\,^3D_1-5d'\,^3P_0^\circ$	0.1 torr of iodine, with a few torr of He, $D = 5$ mm	CW[f]	290, 94, 293
0.880423[a]	II	$6p'\,^3F_2-5d'\,^3F_3^\circ$	"	CW[f]	290, 94, 293
0.980	I(?)	—	"	Pulsed	290
1.010	I(?)	—	"	Pulsed	290
1.030	I(?)	—	"	Pulsed	290
1.04172	III or IV	—	0.001 torr of iodine, inert gases suppress oscillation, $D = 10$–12 mm	Pulsed	289
1.060	I	—	0.1 torr of iodine with a few torr of He, $D = 5$ mm	Pulsed	290
1.3152	I	$5p^5\,^2P_{1/2}^\circ-5p^5\,^2P_{3/2}$	Tens of torr of CF_3I or CH_3I, excitation is by flash, photolysis, $D = 7$ mm[(299)]	Pulsed	299, 300, 301, 302, 303
1.4542[c]	I	$7p[1]_{3/2}^\circ-6d[2]_{3/2}$	In HI at 0.3 torr, $D = 14$ mm	Pulsed	282
1.553	I	$7s[2]_{3/2}-6p[1/2]_{1/2}$	In CH_2I_2 at 0.05 torr, $D = 5$ cm	Pulsed	304
2.5986[c]	I	$5d[2]_{3/2}-6p[1]_{3/2}^\circ$	0.3 torr HI with 0.3 torr Ne, $D = 14$ mm	CW	282
2.7572	I	$(^1D_2)5d[2]_{5/2}-(^1D_2)6p[1]_{3/2}^\circ$	In HI, $D = 12$ mm	CW	288
3.036[d]	I	$5d[2]_{3/2}-6p[1]_{3/2}^\circ$	In CH_3I and I_2 or CF_3I with He, Ar Ar or Xe, $D = 50$ mm	CW	305
3.2363	I	$(^3P_2)5d[2]_{5/2}-(^3P_2)6p[1]_{3/2}^\circ$	In HI, or I_2, $D = 12$ mm	CW	226, 288, 304, 315
3.4296[d]	I	$(^3P_2)5d[4]_{7/2}-(^3P_2)6p[3]_{3/2}^\circ$	In CH_3I and I_2 or CF_3I and I_2, $D = 12$ mm	CW	226, 288, 304, 305, 315

TABLE 41 *(cont.)*

Wavelength (μm)	Spectrum	Identification	Notes	Excitation	References
4.331	I	$5d[1/2]_{1/2}-6p[2]_{3/2}$	In CH_2I_2 and Ar at 0.05 torr, $D = 5$ cm	Pulsed	304
4.858	I	$5d[1]_{3/2}-6p[2]_{5/2}$	In CH_3I, CF_3I or I_2 with He, Ar or Xe, $D = 50$ mm	CW	305
4.862	I	$5d[4]_{9/2}-6p[3]_{7/2}$	,,	CW	305, 304
5.497	I	$5d[2]_{5/2}-6p[1]_{3/2}$	,,	CW	305, 304
6.720	I	$6s[2]_{3/2}-6p[1]_{3/2}$	,,	CW	305
6.902[e]	I	$5d[3]_{7/2}-6p[2]_{5/2}$	,,	CW	305
9.016[e]	I	$5d[3]_{7/2}-6p[2]_{5/2}$	,,	CW	305

Note: The classification of singly ionized laser lines is taken from W. C. Martin and C. H. Corliss, The spectrum of singly ionized atomic iodine (I-II). *J. Res. Natl. Bur. Stand.* **64**A, 443–79 (Nov.–Dec. 1960).

The classification of neutral laser lines, except where otherwise stated, is from C. C. Kiess and C. H. Corliss, Description and analysis of the first spectrum of iodine, *J. Res. Natl. Bur. Stand.* **63**A, 1–18 (July–Aug. 1959).

[a]Selective excitation largely by asymmetric charge transfer with $He^{+2}S_{1/2}$ ground-state ions: large-bore discharge tubes and short, high-voltage pulses favor the selective excitation.

[b]This 5419-Å line is possibly a Xe-II laser line as xenon has been used in the discharge tube used in the work reported in ref. 290; there are no likely iodine-I or II lines at this measured wavelength.

[c]Identification from L. Minnhagen, *Arkiv. Physik.* **21**, 415 (1962).

[d]Identification as in ref. 315.

[e]Reported at 1968 Q.E.C. Miami in paper 20T-5.[(305)]

[f]In a HCD, I_2 vapor pressure 0.7 torr with He at 10 torr, cathode diameter 6.3 mm (flowing mixture).

TABLE 42. *Laser transitions in ytterbium*

Wavelength (μm)	Spectrum	Identification[a]	Notes	Excitation	References
1.0322	I	—	In Yb vapor above 500°C, with 2.5 torr He; 1.5 torr Ne; or 0.8 torr Ar	Pulsed	274
1.2548	I	—	,,	Pulsed	274
1.4280	I	—	,,	Pulsed	274
1.6499[b]	II	—	,,	Pulsed	273, 274
1.7454	I	—	,,	Pulsed	274
1.9832[b]	I	—	,,	Pulsed	273, 274
2.0038[b]	I	—	,,	Pulsed	273, 274
2.1181[b]	I	—	,,	Pulsed	273, 274
2.4377[b]	II	—	,,	Pulsed	274
2.7082[b]	I	—	,,	Pulsed	273
4.8009	I	—	,,	Pulsed	274

[a]Energy of levels given in refs. 273 and 274 as per P. Camus, G. Guelachvili and J. Verges, *Spectrochim. Acta* **24**B, 373 (1969).

[b]Measured to ± 0.0002 μm.

TABLE 43. *Laser transitions in manganese* (*Fig. A.24*)

Wavelength (μm)	Spectrum	Identification[a]	Notes	Excitation	References
0.5341065[b]	I	$y\ ^6P^\circ_{7/2}-a\ ^6D_{7/2}$	0.1–2.0 torr of manganese (at temp. of. 1100–1300°C), with 1.0–2.0 torr of He or Ne, $D = 10$ mm	Short rise-time pulses	306, 182, 307
0.5420368	I	$y\ ^6P^\circ_{5/2}-a\ ^6D_{7/2}$	,,	,,	306, 182, 307
0.5470640	I	$y\ ^6P^\circ_{5/2}-a\ ^6D_{5/2}$	,,	,,	306, 182, 307
0.5481345	I	$y\ ^6P^\circ_{3/2}-a\ ^6D_{5/2}$	,,	,,	307
0.5516777	I	$y\ ^6P^\circ_{3/2}-a\ ^6D_{3/2}$	,,	,,	306, 182, 307
0.5537149	I	$y\ ^6P^\circ_{3/2}-a\ ^6D_{1/2}$	,,	,,	306, 182, 307
1.28997	I	$z\ ^6P^\circ_{7/2}-a\ ^6D_{7/2}$	,,	,,	306, 182, 307
1.32938	I	$z\ ^6P^\circ_{7/2}-a\ ^6D_{9/2}$	,,	,,	306, 182, 307
1.33179	I	$z\ ^6P^\circ_{5/2}-a\ ^6D_{7/2}$	,,	,,	306, 182, 307
1.36257	I	$z\ ^6P^\circ_{5/2}-a\ ^6D_{1/2}$	,,	,,	306, 182, 307
1.38638	I	$z\ ^6P^\circ_{3/2}-a\ ^6D_{3/2}$	,,	,,	306, 182
1.39975	I	$z\ ^6P^\circ_{3/2}-a\ ^6D_{1/2}$	,,	,,	306, 182

[a]Identification is based on measurements of lines in spontaneous emission in Mn-I (M.A. Catalan, W. F. Meggers and O. Garcia-Riquelme). The first spectrum of manganese, Mn-I, *J. Res. Natl. Bur. Stand.* vol. **68A**, 9–56 (Jan.–Feb. 1964).

[b]Oscillation has been observed on six hyperfine components of this line.[307]

TABLE 44. *Laser transitions in samarium*

Wavelength (μm)	Spectrum	Identification	Notes	Excitation	References
1.9120±0.0050	—	—	In samarium vapor at about 0.1 torr, with He, Ne, or Ar	Pulsed	274
2.0482±0.0001	—	—	,,	Pulsed	274
2.6998[a]	I	—	,,	Pulsed	274
2.9663±0.0002	—	—	,,	Pulsed	274
3.4654±0.0002	—	—	,,	Pulsed	274
3.5361±0.0002	—	—	,,	Pulsed	274
4.1368±0.0002	—	—	,,	Pulsed	274
4.8658±0.0004	—	—	,,	Pulsed	274

[a]Calculated wavelength; from J. Blaise, C. Morillon, M. G. Schweighofer, and J. Verges, *Spectrochim. Acta* **24B**, 405 (1969).

TABLE 45. *Laser transitions in europium*

Wavelength (μm)	Spectrum	Identification[a]	Notes	Excitation	References
1.7596	I	—	In europium vapor at about 0.1 torr, with He, Ne, or Ar	Pulsed	274
2.5811	I	—	"	Pulsed	274
2.7174	I	—	"	Pulsed	274
4.3202	I	—	"	Pulsed	274
4.6935	I	—	"	Pulsed	274
5.0647	I	—	"	Pulsed	274
5.2811	I	—	"	Pulsed	274
5.4292	I	—	"	Pulsed	274
5.7706	I	—	"	Pulsed	274
5.9479	I	—	"	Pulsed	274
6.0576	I	—	"	Pulsed	274

[a]Values of energy levels involved are given in ref. 274, as per H. N. Russell and A. King, *Astrophys. J.* **90**, 155 (1939).

TABLE 46. *Miscellaneous (and unidentified) laser transitions*

Wavelength (μm)	Likely spectrum	Occurrence	Excitation	References
0.247350	Ne-III		Pulsed	93, 94
0.247718	Xe		Pulsed	93, 94
0.263898	S-V	In SO_2	Pulsed	266
0.264941	Kr-II		Pulsed	93, 94
0.269182	Xe		Pulsed	93, 94
0.274139	Kr		Pulsed	93
0.275391	Ar-IV	In Ar at low pressures	Pulsed	15, 93, 94
0.286688	—	In Ne at low pressures, impurity likely	Pulsed	15, 93
0.304970	—	In Kr at low pressures, impurity likely	Pulsed	15, 93
0.307971	Xe-II, III, IV	In Xe at low pressures	Pulsed	15, 93, 95
0.330604	Xe-III, Ar-II	In Xe at low pressures	Pulsed	15, 95
0.332437	—		Pulsed	95
0.3300	Ne	In Ne, using fast-pulse excitation	Pulsed	311
0.333078	Xe-III, IV	In Xe at low pressures	CW	15, 93, 95, 156
0.335004	Xe		Pulsed	95, 94
0.34836	Xe	In Xe at low pressure	Pulsed	15
0.349737	S-III	In SO_2 or H_2S	Pulsed	266
0.3500	N-II	In N_2 at 1–2000 mtorr, $D = 3$ mm	Pulsed	244
0.351039	Ar-II, Xe-II		Pulsed	95
0.3545	Ar, Hg	In Hg–Ar mixture at low pressure	Pulsed	112
0.363789	Ar	In Ar	Pulsed	110, 15, 95
0.364548	Ar, Xe		Pulsed	95, 93, 94
0.366920	Xe		Pulsed	93, 94
0.37052	Ar-II, III	In Ar at low pressures	Pulsed	15
0.3760	—	In Xe at low pressures, impurity likely	Pulsed	15, 94
0.380327	Xe		Pulsed	93, 94
0.397293	Xe		Pulsed	93, 94

TABLE 46 *(cont.)*

Wavelength (μm)	Likely spectrum	Occurrence	Excitation	References
0.406048	Xe-III, Ar-II		CW	95, 15, 156
0.4089	Ar-III, Si-IV	In Ar at low pressure, Si-IV $(0.408886$-μm $(4p^2 P^\circ_{3/2} - 4s\ ^2 S_{1/2}))$ line	Pulsed	15, 94
0.4120	O-II	In Hg–N_2 mixture at 1–20 mtorr, $D = 3$ mm	Pulsed	244
0.418292	Ar-III	In Ar	Pulsed	15
0.44130	Xe-II	In Xe at 7 m torr, $D = 2,4$ mm	Pulsed	115
0.4525	O-II	In Hg–N_2 mixture at 1–20 m torr, $D = 3$ mm	Pulsed	244
0.453379	I-III, IV	In pure iodine vapor	Pulsed	289
0.46055	O-II	At impurity in noble gases	Pulsed	233
0.4621	N-II	In N_2 at 40 mtorr, $D = 1.5$ mm	Pulsed	115
0.46439	N-II	In N_2 at 40 m torr, $D = 1.5$ mm	Pulsed	115
0.46453	—	In argon at low pressure	Pulsed	15, 94
0.46504	C-III	In Xe at low pressure, impurity likely	Pulsed	15, 94, 244
0.467440	I-III, IV	In pure iodine vapor	Pulsed	289
0.47487	Xe-III		CW	138, 114
0.4750	N, Hg, O	In Hg–N_2 mixture at 1–20 mtorr, $D = 3$ mm	Pulsed	244
0.4764	Hg-II, Ar-I		Pulsed	124
0.493467	I-III, IV	In pure iodine vapor	Pulsed	289
0.495410	Xe, C-II	In Xe at low pressure	CW ·	15, 94, 156
0.499255	Ar-III		?	94ª
0.50072	—	In Xe at low pressure, impurity likely	CW	15, 94, 156
0.503747	Kr-II	In Kr at 11 mtorr, $D = 2.4$ mm	Pulsed	115
0.51590	Xe-III	In Xe	CW	15, 94, 156
0.516032	S-III	In SO_2 or SF_6	Pulsed	266
0.521962	S	In SO_2 or SF_6	Pulsed	266
0.5225	Hg-II	In Hg with Ar at low pressure, possibly 0.5677-μm Hg-II line}	Pulsed	109
0.526305	Xe-III, IV	In Xe at 8 mtorr, $D = 2.4$ mm	Pulsed	115
0.526017	Xe-II	In Xe	Pulsed	15
0.535289	Xe-II	In Xe	CW	15, 94, 95, 156, 159, 169
0.539459	Xe-II	In Xe	CW	15, 94, 156
0.5440	—	In Hg–N_2 mixture	Pulsed	244
0.5500	O	In Hg–N_2 mixture	Pulsed	244
0.5540	O	In Hg–N_2 mixture	Pulsed	244
0.5610	N-III, IV	In Mg–N_2 mixture or in air	Pulsed	245
0.5620	N-III, IV	In Hg–N_2 mixture or in air	Pulsed	245
0.56796	O-II, Hg, N-II	In O_2, Hg or N_2, or air is likely N-II $3p\ ^3 D_3 - 3s\ ^3 P^\circ_2$-line at 0.567956 μm	Pulsed	15, 244, 316
0.56869	N-II	In N_2 at 40 mtorr, $D = 1.5$ mm	Pulsed	115
0.5690	O	In CO at 11–20 mtorr, $D = 3$ mm	Pulsed	244
0.58525	Ne-I	In Ne–Ar mixture, requires Ar, is likely Ne-I, $2p_1 - 1s_2$-line at 0.58525 μm	Pulsed	15
0.595573	Xe, O	In air or Xe	Pulsed	15, 94
0.6072	Kr	In Kr, $D = 5$ mm	Pulsed	142
0.61762	Xe	In Xe at 8 mtorr, $D = 1.5$ mm	Pulsed	141
0.62866	Xe-II	In Xe at 14 mtorr, $D = 1.5$ mm	Pulsed	115
0.631276	Kr-II	In Kr at 12 mtorr, $D = 1.5$ mm	Pulsed	141

TABLE 46 (*cont.*)

Wavelength (μm)	Likely spectrum	Occurrence	Excitation	References
0.6349	Te I	In Te–Ne mixture	CW	252
0.657745	O-III	In He, Ne or O_2; possibly O-III, $3d\ ^5F_2$–$4d\ ^3D_3^\circ$-line at 0.657750 μm	Pulsed	15, 94
0.667193	P	In PF_5 vapor, $D = 3$ mm	Pulsed	237
0.7065	Ar, Hg	In Hg–Ar mixture at low pressure	Pulsed	112
0.75035	Ar-I	In Ar–Ne, –He, possibly Ar-I, $2p_1$–$2s_2$-line at 0.7503867 μm	Pulsed	15
0.8350	H_2	As impurity in N_2 and/or noble gases	Pulsed	234
0.8390	Cd-II	In He–Cd mixture	Pulsed	214
0.8408	Xe	In Xe	Pulsed	102
0.8569	Xe	In Xe	Pulsed	102
0.8589	Kr-III	In Kr	Pulsed	102
0.8677	Hg-II	In Hg–He mixture	Pulsed	217
0.8780	Ar-II	In Ar, likely Ar-II, 0.877186-μm line	Pulsed	102
0.882045		In He (Ne) + $N_2(CO_2)$ at 4 torr, $D = 15$ mm	Pulsed	234
0.980	II	In He–I_2 mixture	Pulsed	290
1.01	I	In He–I_2 mixtures	Pulsed	290
1.03	I	,,	Pulsed	290
1.04172	I–III, IV	In pure I_2 vapor	Pulsed	289
1.060	I	In He–I_2 mixtures	Pulsed	290
1.065	S–I	In SF°, or SF° and He	CW	85, 267
1.0935	Ar	In Ar	Pulsed	102
1.0950	Xe	In Ar	Pulsed	102
1.1869	Cd-II	In He, Ne and Cd mixtures, $D = 12$ mm	Pulsed	214
1.2222	Hg-I	In Hg–Ar mixture	Pulsed	109, 112
1.2246	Hg-I	In Hg–Ar mixture	Pulsed	109, 112
1.2545	Hg	In Hg–He mixture	Pulsed	217
1.2760	Hg	In Hg–Ar mixture	Pulsed	109, 112
1.2981	Hg	In Hg–He mixture	Pulsed	217
1.3583	He	In 150 torr He with TEA laser excitation	Pulsed	8
1.3655	Hg	In Hg–He mixture	Pulsed	217
1.45423	N-I	In He–N_2 mixture	Pulsed	248
1.589	—	In He–CCl_2F_2 mixture	CW	285
1.975	—	In He–CCl_2F_2 mixture	CW	285
1.977	Cl	In He–CCl_2F_2 mixture	CW	285
2.021	Cl	In He–CCl_2F_2 mixture	CW	285
2.499	—	In He–CCl_2F_2 mixture	CW	285
2.535	—	In He–CCl_2F_2 mixture	CW	285
2.602	—	In He–CCl_2F_2 mixture	CW	285
2.784	—	In He–CCl_2F_2 mixture	CW	285
3.7942	N-I	In N_2–He mixture	Pulsed	248
3.801	—	In He–CCl_2F_2 mixture	CW	285
3.8154	N-1	In N_2–He mixture	Pulsed	248
10.604	—	In He–$CBrF_3$ mixture	CW	285
14.78	—	In NH_3 at 0.5 to 1.0 torr, $D = 10$ cm	Pulsed	308
15.04	—	In NH_3 at 0.5 to 1.0 torr, $D = 10$ cm	Pulsed	308
15.08	—	In NH_3 at 0.5 to 1.0 torr, $D = 10$ cm	Pulsed	308
15.41	—	In NH_3 at 0.5 to 1.0 torr, $D = 10$ cm	Pulsed	308
15.47	—	In NH_3 at 0.5 to 1.0 torr, $D = 10$ cm	Pulsed	308

TABLE 46 *(cont.)*

Wavelength (μm)	Likely spectrum	Occurrence	Excitation	References
18.21	—	In NH_3 at 0.5 to 1.0 torr, $D = 10$ cm	Pulsed	308
21.46	—	In NH_3 at 0.5 to 1.0 torr, $D = 10$ cm	Pulsed	308
22.54	—	In NH_3 at 0.5 to 1.0 torr, $D = 10$ cm	Pulsed	308
22.71	—	In NH_3 at 0.5 to 1.0 torr, $D = 10$ cm	Pulsed	308
23.68	—	In NH_3 at 0.5 to 1.0 torr, $D = 10$ cm	Pulsed	308
23.86	—	In NH_3 at 0.5 to 1.0 torr, $D = 10$ cm	Pulsed	308
24.92	—	In NH_3 at 0.5 to 1.0 torr, $D = 10$ cm	Pulsed	308
25.12	—	In NH_3 at 0.5 to 1.0 torr, $D = 10$ cm	Pulsed	308
26.27	—	In NH_3 at 0.5 to 1.0 torr, $D = 10$ cm	Pulsed	308
30.69	—	In NH_3 at 0.5 to 1.0 torr, $D = 10$ cm	Pulsed	308
31.47	—	In NH_3 at 0.5 to 1.0 torr, $D = 10$ cm	Pulsed	308
31.92	—	In NH_3 at 0.5 to 1.0 torr, $D = 10$ cm	Pulsed	308
32.13	—	In NH_3 at 0.5 to 1.0 torr, $D = 10$ cm	Pulsed	308
126.100	Ne-I	In pure Ne	CW	91
132.800	Ne-I	In pure Ne	CW	91

[a] M. Birnbaum and T. L. Stocker, private communication.[94]

References to Figs. A.1 to A.25 and Tables 1 to 46

1. ABRAHAM, R. L. and WOLGA, G. L., *IEEE J. Quant. Electr.* **QE-5**, 368 (1967).
2. CAGNARD, R., AGOBIAN, R. DER, OTTO, J. L. and ECHARD, R., *Compt. Rend.* **257**, 1044 (1963).
3. AGOBIAN, R. DER, OTTO, J. L., CAGNARD, R. and ECHARD, R., *J. Phys. Rad.* **25**, 887 (1964).
4. PATEL, C. K. N., BENNETT, W. R. Jr., FAUST, W. L. and MCFARLANE, R. A., *Phys. Rev. Lett.* **9**, 102 (1962).
5. BENNETT, W. R. Jr. and KINDLMANN, P. J., *Bull. Am. Phys. Soc.* **8**, 87 (1963).
6. BENNETT, W. R. Jr., Inversion mechanisms in gas lasers, in *Appl. Optics, Supplement on Chemical Lasers* (1965), pp. 3–33.
7. ISAEV, A. A., ISCHENKO, P. I. and PETRASH, G. G., *Sov. Phys. JETP Lett.* **6**, 118 (1967).
8. WOOD, O. R., BURKHARDT, E. G., POLLACK, M. A. and BRIDGES, T. J., *Appl. Phys. Lett.* **18**, 261 (1971).
9. MATHIAS, L. E. S., CROCKER, A. and WILLS, M. S., *IEEE J. Quant. Electr.* **QE-3**, 170 (1967).
10. LEVINE, J. S. and JAVAN, A., *Appl. Phys. Lett.* **14**, 348 (1969).
11. CLUNIE, D. M., THORN, R. S. A. and TREZISE, K. E., *Phys. Lett.* **14**, 28 (1965).
12. LEONARD, D. A., NEAL, R. A. and GERRY, E. T., *Appl. Phys. Lett.* **7**, 175 (1965).
13. LEONARD, D. A., *IEEE J. Quant. Electr.* **QE-3**, 133 (1967).
14. KNYAZEV, I. N. and PETRASH, G. G., *J. Appl. Spectrosc.* **4**, 401 (1966).
15. BRIDGES, W. B. and CHESTER, A. N., *Appl. Optics* **4**, 573 (1965).
16. WHITE, A. D. and RIGDEN, J. D., *Appl. Phys. Lett.* **2**, 211 (1963).
17. HEARD, H. G. and PETERSON, J., *Proc. IEEE* **52**, 1258 (1964).
18. ROSENBERGER, D., *Phys. Lett.* **13**, 228 (1964).
19. ERICSSON, K. G. and LIDHOLT, L. R., *IEEE J. Quant. Electr.* **QE-3**, 94 (1967).
20. ROSENBERGER, D., *Phys. Lett.* **14**, 32 (1965).
21. BLOOM, A. L., *Appl. Phys. Lett.* **2**, 101 (1963).
22. ISAEV, A. A. and PETRASH, G. G., *Zh. éksp. teor. Fiz.* **56**, 1132 (1969).
23. WHITE, A. D. and RIGDEN, J. D., *Proc. IRE* **50**, 1697 (1962).
24. ALLEN, L. and JONES, D. G. C., *Principles of Gas Lasers*, pp. 73–103, Plenum Press, New York, and Butterworths, London, 1967.
25. BLOOM, A. L., *Gas Lasers*, pp. 52–59, John Wiley, Inc., New York, 1968.
26. RIGDEN and WHITE, A. D., *Proc. IEEE* **51**, 943 (1963).
27. WHITE, A. D. and GORDON, E. I., *Appl. Phys. Lett.* **3**, 197 (1963).
28. YOUNG, R. T. Jr., WILLETT, C. S. and MAUPIN, R. T., *J. Appl. Phys.* **41**, 2936 (1970).

29. LABUDA, E. F. and GORDON, E. I., *J. Appl. Phys.* **35**, 1647 (1964).
30. YOUNG, R. T. Jr., *J. Appl. Phys.* **36**, 2324 (1965).
31. GORDON, E. I. and WHITE, A. D., *Appl. Phys. Lett.* **3**, 199 (1963).
32. SUZUKI, N., *Japan. J. Appl. Phys.* **4**, 285 (1965).
33. SUZUKI, N., *Japan. J. Appl. Phys.* **4**, Supplement 1 (*Proceedings of the Conference on Photographic and Spectroscopic Optics, 1964*), pp. 642-7 (1965).
34. FIELD, R. L. Jr., *Rev. Sci. Instr.* **38**, 1720 (1967). .
35. GONCHUKOV, G. A., ERMAKOV, G. A., MIKHENKO, G. A. and PROTSENKO, E. D., *Optics Spectrosc.* **20**, 601 (1966).
36. ALEKSEEVA, A. N. and GORDEEV, D. V., *Optics Spectrosc.* **23**, 520 (1967).
37. SMITH, P. W., *IEEE J. Quant. Electr.* **QE-2**, 62 (1966).
38. SMITH, P. W., *IEEE J. Quant. Electr.* **QE-2**, 77 (1966).
39. BELL, W. E. and BLOOM, A. L., *Appl. Optics* **3**, 413 (1964).
40. BELOUSOVA, I. M., DANILOV, O. B., ELKINA, I. A. and KISELYOV, K. M., *Optics Spectrosc.* **16**, 44 (1969).
41. HERZIGER, G., HOLZAPFEL, W. and SIELIG, W., *Z. für Physik* **189**, 385 (1966).
42. JONES, C. R. and ROBERTSON, W. W., *Bull. Am. Phys. Soc.* **13**, 198 (1968). (At the 20th Gaseous Electronics Conference, San Francisco, Oct. 1967, data was presented on the potential-barrier behavior of the reaction $He^{*1}S_0 + Ne \rightarrow He + Ne\ 3s_2 - \Delta E_\infty$ (386 cm^{-1}), as well as the reaction involving $He^*2^3S_1$ metastables stated in the abstract in the above reference.)
43. WHITE, A. D., *Proc. IEEE* **52**, 721 (1964).
44. BLOOM, A. L. and HARDWICK, D. L., *Phys. Lett.* **20**, 373 (1966).
45. TOBIAS, I. and STROUSE, W. M., *Appl. Phys. Lett.* **10**, 342 (1967).
46. SCHLIE, L. A. and VERDEYN, J. T., *IEEE J. Quant. Electr.* **QE-5**, 21 (1969).
47. ZITTER, R. N., *Bull. Am. Phys. Soc.* **9**, 500 (1964).
48. ZITTER, R. N., *J. Appl. Phys.* **35**, 3070 (1964).
49. PERRY, D. L., *IEEE J. Quant. Electr.* **QE-7**, 102 (1971).
50. CHEBOTAYEV, V. P. and VASILENKO, L. S., *Sov. Phys. JETP* **21**, 515 (1965).
51. AFANASEVA, V. L., LUKIN, A. V. and MUSTAFIN, K. S., *Sov. Phys. Tech. Phys.* **12**, 233 (1967).
52. MCCLURE, R. M., PIZZO, R., SCHIFF, M. and ZAROWIN, C. B., *Proc. IEEE* **52**, 851 (1964).
53. SHTYRKOV, E. I. and SUBBES, E. V., *Optics Spectrosc.* **21**, 143 (1966).
54. ITZKAN, I. and PINCUS, G., *Appl. Optics* **5**, 349 (1966).
55. MCFARLANE, R. A., PATEL, C. K. N., BENNETT, W. R. Jr. and FAUST, W. L., *Proc. IRE* **50**, 2111 (1962).
56. BENNETT, W. R. Jr., Gaseous optical masers, in *Appl. Optics*, Supplement I: *Optical Masers* (O. S. HEAVENS, ed.), pp. 24–61 (1962).
57. CHEBOTAYEV, V. P. and POKOSOV, V. V., *Radio Engng Electr. Phys.* **10**, 817 (1965).
58. PETRASH, G. G. and KNYAZEV, I. N., *Sov. Phys. JETP* **18**, 571 (1964).
59. AGOBIAN, R. DER, OTTO, J. L., CAGNARD, R. and ECHARD, R., *Compt. Rend.* **259**, 323 (1964).
60. ANDRADE, O., GALLARDO, M. and BOCKASTEN, K., *Appl. Phys. Letters*, **11**, 99 (1967).
61. CHEBOTAYEV, V. P., *Radio Engng Electr., Phys.* **10**, 316 (1965).
62. JAVAN, A., BENNETT, W. R. Jr. and HERRIOTT, D. R., *Phys. Rev. Lett.* **6**, 106 (1961).
63. BOOT, H. A. H., CLUNIE, D. M. and THORN, R. S. A., *Nature* **4882**, 773 (1963).
64. AGOBIAN, R. DER, CAGNARD, R., ECHARD, R. and OTTO, J. L., *Compt. Rend.* **258**, 3661 (1964).
65. BENNETT, W. R. Jr. and KNUTSON, J. W., *Proc. IEEE* **52**, 861 (1964).
66. BOOT, H. A. H. and CLUNIE, D. M., *Nature* **197**, 173 (1963).
67. SMITH, J., *J. Appl. Phys.* **35**, 723 (1964).
68. AGOBIAN, R. DER, OTTO, J. L., ECHARD, R. and CAGNARD, R., *Compt. Rend.* **257**, 3344 (1963).
69. LISITSYN, V. N., FEDCHENKO, A. I. and CHEBOTAYEV, V. P., *Optics Spectrosc.* **27**, 157 (1969).
70. BLAU, E. J., HOCHHEIMER, B. F., MASSEY, J. T. and SCHULZ, A. G., *J. Appl. Phys.* **34**, 703 (1963).
71. GIRES, F., MAYER, H. and PAILETTE, M., *Compt. Rend.* **256**, 3428 (1963).
72. MCFARLANE, R. A., FAUST, W. L. and PATEL, C. K. N., *Proc. IEEE* **51**, 468 (1963).
73. SMILEY, V. N., *Appl. Phys. Lett.* **4**, 123 (1964).
74. LISITSYN, V. N. and CHEBOTAYEV, V. P., *Optics Spectrosc.* **20**, 603 (1966).
75. MCFARLANE, R. A., FAUST, W. L., PATEL, C. K. N. and GARRETT, C. G. B., Gas maser operation at wavelengths out to 28 microns, in *Quantum Electronics*, vol. III (P. GRIVET and N. BLOEMBERGEN, eds.), pp. 573–86, Columbia University Press, New York, 1964.
76. ROSENBERGER, D., *Phys. Lett.* **9**, 29 (1964).
77. BENNETT, W. R. Jr., PAWLIKOWSKI, A. T. and KNUTSON, J. W., *Bull. Am. Phys. Soc.* **9**, 500 (1964).
78. FAUST, W. L., MCFARLANE, R. A., PATEL, C. K. N. and GARRETT, C. G. B., *Phys. Rev.* **133**, A1476 (1964).

APPENDIX

79. OTTO, J. L., CAGNARD, R., ECHARD, R. and AGOBIAN, R. DER, *Compt. Rend.* **258**, 2779 (1964).
80. GRUDZINSKI, R., PAILETTE, M. R. and BECRELLE, J., *Compt. Rend.* **258**, 1452 (1964).
81. BERGMAN, K. and DEMTRODER, W., *Phys. Lett.* **29A**, 94 (1969).
82. GERRITSEN, J. J. and GOEDERTIER, P. V., *Appl. Phys. Lett.* **4**, 20 (1964).
83. SMILEY, V. N., A long gas phase optical maser cell, in *Quantum Electronics*, vol. III (P. GRIVET and N BLOEMBERGEN, eds.), pp. 587–91, Columbia University Press, New York, 1964.
84. MCMULLIN, P. G., *Appl. Optics* **3**, 641 (1964).
85. PATEL, C. K. N., MCFARLANE, R. A. and FAUST, W. L., Further infrared spectroscopy using stimulated emission techniques, in *Quantum Electronics*, vol. III (P. GRIVET and N. BLOEMBERGEN, eds.), pp. 561–72, Columbia University Press, New York, 1964.
86. BRUNET, H. and LAURES, P., *Phys. Lett.* **12**, 106 (1964).
87. BLOOM, A. L., BELL, W. E. and REMPEL, R. C., *Appl. Optics* **2**, 317 (1963).
88. WILLETT, C. S. and YOUNG, R. T. Jr., *J. Appl. Phys.* **73**, 725 (1972).
89. PATEL, C. K. N., MCFARLANE, R. A. and GARRETT, C. G. B., *Appl. Phys. Lett.* **4**, 18 (1964).
90. PATEL, C. K. N., MCFARLANE, R. A. and GARRETT, C. G. B., *Bull. Am. Phys. Soc.* **9**, 65 (1964).
91. PATEL, C. K. N., FAUST, W. L., MCFARLANE, R. A. and GARRETT, C. G. V., *Proc. IEEE* **52**, 713 (1964).
92. MCFARLANE, R. A., FAUST, W. L., PATEL, C. K. N. and GARRETT, C. G. B., *Proc. IEEE* **52**, 318 (1964).
93. CHEO, P. K. and COOPER, H. G., *J. Appl. Phys.* **36**, 1862 (1965).
94. BRIDGES, W. B. and CHESTER, A. N., *IEEE J. Quant. Electr.* **QE-1**, 66 (1965).
95. CHEO, P. K. and COOPER, H. G., *Bull. Am. Phys. Soc.* **9**, 626 (1964).
96. HOSHINO, Y., KATSUYAMA, Y., FUKUDA, K., *Japan. J. Appl. Phys.* **11**, 907 (1972).
97. DANA, L., LAURES, P. and ROCHEROLLES, R., *Compt. Rend.* **260**, 481 (1965).
98. LABUDA, E. F. and JOHNSON, A. M., *IEEE J. Quant. Electr.* **QE-2**, 700 (1966).
99. BRIDGES, W. B., FREIBURG, R. J. and HALSTEAD, A. S., *IEEE J. Quant. Electr.* **QE-3**, 339 (1967).
100. PAANANEN, R. A., *Appl. Phys. Lett.* **9**, 34 (1966).
101. FENDLEY, J. R. Jr., *IEEE J. Quant. Electr.* **4**, 627 (1968).
102. SINCLAIR, D. C., *J. Opt. Soc. Am.* **55**, 571 (1965).
103. BOCKASTEN, K., LUNDHOLM, T. and ANDRADE, O., *Phys. Lett.* **22**, 145 (1966).
104. BRISBANE, A. D., *Nature*, **214**, 75 (1967).
105. BOCKASTEN, K. and ANDRADE, O., *Nature* **215**, 382 (1967).
106. DAUGER, A. B. and STAFSUDD, O. M., *IEEE. J. Quant. Electr.* **QE-6**, 572 (1970).
107. DAUGER, A. B. and STAFSUDD, O. M., *Appl. Optics* **10**, 2690 (1971).
108. WILLETT, C. S., *Appl. Optics* **11**, 1429 (1972).
109. HEARD, H. G. and PETERSON, J., *Proc. IEEE* **52**, 414 (1964).
110. MCFARLANE, R. A., *Appl. Optics* **3**, 1196 (1964).
111. CHEO, P. K. and COOPER, H. G., *Appl. Phys. Lett.* **6**, 177 (1965).
112. HEARD, H. G. and PETERSON, J., *Proc. IEEE* **52**, 1049 (1964).
113. LATIMER, I. D., *Appl. Phys. Lett.* **13**, 333 (1968).
114. BRIDGES, W. B. and HALSTEAD, A. S., *IEEE J. Quant. Electr.* **QE-2**, 84 (1966).
115. NEUSEL, R. H., *IEEE J. Quant. Electr.* **QE-3**, 207 (1967).
116. BELL, W. E. and BLOOM, A. L., (see ref. 9) Spectroscopy of ion lasers, by W. B. Bridges and A. N. Chester, ref. 94.
117. BRIDGES, W. B. and CHESTER, A. N., *Appl. Phys. Lett.* **4**, 128 (1964).
118. TRACH, YU. V., FAINBERG, RA. B., BOLATIN, L. I., BESSAREB, YA., GADETSKII, N. P., CHERNEN'KIL, YA. N. and BEREZIN, A. K., *JETP Lett.* **6**, 371 (1967).
119. CONVERT, G., ARMAND, M. and MARTINOT LAGARDE, P., *Compt. Rend.* **258**, 4467 (1964).
120. GORDON, E. I., LABUDA, E. F. and BRIDGES, W. B., *Appl. Phys. Lett.* **4**, 178 (1964).
121. BENNETT, W. R. Jr., KNUTSON, J. W. Jr., MERCER, G. N. and DETCH, J. L., *Appl. Phys. Lett.* **4**, 180 (1964).
122. KOBAYASHI, S., IZAWA, I., KAWAMURA, K. and KAMIYAMA, M., *IEEE J. Quant. Electr.* **QE-2**, 699 (1966).
123. GLAZUNOV, V. K., KITAEVA, V. P., OSTROVSKAYA, L. YA. and SOBOLEV, N. N., *JETP Lett.* **5**, 215 (1967).
124. CONVERT, G., ARMAND, M. and MARTINOT-LAGARDE, P., *Compt. Rend.* **258**, 3259 (1964).
125. PREOBRAZHENSKII, N. G. and SHAPAREV, N. YA., *Optics Spectrosc.* **25**, 172 (1968).
126. MERKELO, H., WRIGHT, R. H., KAPLOFKA, J. P. and BIALECKE, E. P., *Appl. Phys. Lett.* **13**, 401 (1968).
127. PAPAYOANOU, A. and GUMEINER, I., *Appl. Phys. Lett.* **16**, 5 (1970).
128. LABUDA, E. F. and GORDON, E. I., *IEEE J. Quant. Electr.* **QE-1**, 273 (1965).

457

129. GORDON, E. I., LABUDA, E. F., MILLER, R. C. and WEBB, C. E., Excitation mechanisms of the argon-ion laser, in *Physics of Quantum Electronics* (P. L. KELLEY, B. LAX and P. E. TANNENWALD, eds.), pp. 664–73, McGraw-Hill, New York, 1966.
130. KOOZEKANANI, S. A., *Appl. Phys. Lett.* **11**, 107 (1967).
131. HATTORI, S. and GOTO, T., *IEEE J. Quant. Electr.* **QE-5**, 531 (1969).
132. HODGES, D. T. and TANG, C. L., *IEEE J. Quant. Electr.* **QE-6**, 757 (1970).
133. HORRIGAN, F. A., KOOZEKANANI, S. H. and PAANANEN, R. A., *Appl. Phys. Lett.* **6**, 41 (1965).
134. WALTER, W. T. and JARRETT, J. M., *Appl. Optics* **3**, 789 (1964).
135. LAURES, P., DANA, L. and FRAPARD, C., *Compt. Rend.* **258**, 6363 (1964).
136. DANA, L. and LAURES, P., *Proc. IEEE* **53**, 78 (1965).
137. BRIDGES, W. B., *Proc. IEEE* **52**, 843 (1964).
138. NEUSEL, R. H., *IEEE J. Quant. Electr.* **QE-2**, 334 (1966).
139. NEUSEL, R. H., *IEEE J. Quant. Electr.* **QE-2**, 106 (1966).
140. HOFFMAN, V. and TOSCHEK, P., *IEEE J. Quant. Electr.* **QE-6**, 757 (1970).
141. NEUSEL, R. H., *IEEE J. Quant. Electr.* **QE-2**, 758 (1966).
142. COTTRELL, T. H. E., SINCLAIR, D. C. and FORSYTH, J. M., *IEEE J. Quant. Electr.* **QE-2**, 703 (1966).
143. JOHNSON, A. M. and WEBB, C. E., *IEEE J. Quant. Electr.* **QE-3**, 369 (1967).
144. COURVILLE, G. E., WALSH, P. D. and WASKO, J. H., *J. Appl. Phys.* **35**, 2547 (1964).
145. PATEL, C. K. N., FAUST, W. L. and McFARLANE, R. A., *Appl. Phys. Lett.* **1**, 84 (1962).
146. PATEL, C. K. N., McFARLANE, R. A. and FAUST, W. L., High gain medium for gaseous optical masers, in *Quantum Electronics*, vol. III (P. GRIVET and N. BLOEMBERGEN, eds.), pp. 507–14, Columbia University Press, New York, 1964.
147. FAUST, W. L., McFARLANE, R. A., PATEL, C. K. N. and GARRETT, C. G. B., *Appl. Phys. Lett.* **1**, 85 (1962).
148. MOSKALENKO, V. F., OSTAPCHENKO, E. P. and PUGNIN, V. I., *Optics Spectrosc.* **23**, 94 (1967).
149. DAVIES, C. C. and KING, T. A., *Phys. Lett.* **39A**, 186 (1972).
150. LIBERMAN, S., *Compt. Rend.* **266**, 236 (1968).
151. KUIZNETSOV, A. A. and MASH, D. I., *Radio Engng Electr. Phys.* **10**, 319 (1965).
152. BRIDGES, W. B., *Appl. Phys. Lett.* **3**, 45 (1963).
153. PAANANEN, R. A. and BOBROFF, D. L., *Appl. Phys. Lett.* **2**, 99 (1963).
154. CLARK, P. O., *IEEE J. Quantum Electronics* **QE-1**, 109 (1965).
155. ARMSTRONG, D. R., *IEEE J. Quant. Electr.* **QE-4**, 968 (1968).
156. PETROV, YU. N. and PROKHOROV, A. M., *Sov. Phys. JETP Lett.* **1**, 24 (1965).
157. GALLARDO, M., GARAVAGLIA, M., TAGLIAFERRI, A. A. and LLEUSMA, E. G., *IEEE J. Quant. Electr.* **QE-6**, 745 (1970).
158. BRIDGES, W. B. and MERCER, G. N., *IEEE J. Quant. Electr.* **QE-5**, 476 (1969).
159. DAHLQUIST, J. A., *Appl. Phys. Lett.* **6**, 193 (1965).
160. BRIDGES, W. B. and CHESTER, A. N., *IEEE J. Quant. Electr.* **QE-7**, 471 (1971).
161. GOLDSBOROUGH, J. P. and BLOOM, A. L., *IEEE J. Quant. Electr.* **QE-3**, 96 (1967).
162. WHEELER, J. P., *IEEE J. Quant. Electr.* **QE-7**, 429 (1971).
163. LAURES, P., DANA, L. and FRAPARD, C., *Compt. Rend.* **259**, 745 (1964).
164. SMIRNOV, YU. M. and SHATONOV, YA. D., *Optics Spectrosc.* **32**, 333 (1972).
165. HEARD, H. G. and PETERSON, J., *Proc. IEEE* **52**, 1050 (1964).
166. SIMMONS, W. W. and WITTE, R. S., *IEEE J. Quant. Electr.* **QE-6**, 466 (1970).
167. DAVIS, C. C. and KING, T. A., *IEEE J. Quant. Electr.* **QE-8**, 755 (1972).
168. NEUSEL, R. H., *IEEE J. Quant. Electr.* **QE-2**, 70 (1966).
169. JARRETT, S. M. and BARKER, G. C., *IEEE J. Quant. Electr.* **QE-5**, 166 (1969).
170. TELL, B., MARTIN, R. J. and McNAIR, D., *IEEE J. Quant. Electr.* **QE-3**, 96 (1967).
171. WILLETT, C. S., *IEEE J. Quant. Electr.* **QE-6**, 469 (1970).
172. DEZENBERG, G. J. and WILLETT, C. S., *IEEE J. Quant. Electr.* **QE-7**, 491 (1971).
173. BOCKASTEN, K., LUNDHOLM, T. and ANDRADE, O. T., *J. Opt. Soc. Am.* **56**, 1260 (1966).
174. TIBILOV, A. S. and SHUKHTIN, A. M., *Optics Spectrosc.* **21**, 69 (1966).
175. POGORELYI, P. A. and TIBILOV, A. S., *Optics Spectrosc.* **25**, 301 (1968).
176. TIBILOV, A. S. and SHUKHTIN, A. M., *Optics Spectrosc.* **25**, 221 (1968).
177. SOROKIN, P. P. and LANKARD, J. R., *J. Chem. Phys.* **54**, 2184 (1971).
178. SOROKIN, P. P. and LANKARD, J. R., *J. Chem. Phys.* **51**, 2929 (1969).
179. JACOBS, S., RABINOWITZ, P. and GOULD, G., *Phys. Rev. Lett.* **7**, 415 (1961).
180. RABINOWITZ, P. and JACOBS, S., The optically pumped cesium laser, in *Quantum Electronics*, vol III (P. GRIVET and N. BLOEMBERGEN, eds.), pp. 489–98, Columbia University Press, New York, 1964.

458

181. RABINOWITZ, P., JACOBS, S. and GOULD, G., *Appl. Optics* **1**, 513 (1962).
182. WALTER, W. T., SOLIMENE, N., PILTCH, M. and GOULD, G., *IEEE J. Quant. Electr.* **QE-2**, 474 (1966)
183. WALTER, W. T., SOLIMENE, N., PILTCH, M. and GOULD, G., *Bull. Am. Phys. Soc.* **11**, 113 (1966).
184. WALTER, W. T., *Bull. Am. Phys. Soc.* **12**, 90 (1967).
185. LEONARD, D. A., *IEEE J. Quant. Electr.* **QE-3**, 380 (1967).
186. WALTER, W. T., *IEEE J. Quant. Electr.* **QE-4**, 355 (1968).
187. ASMUS, J. F. and MONCUR, N. K., *Appl. Phys. Lett.* **13**, 384 (1968).
188. HODGES, D. T., *Appl. Phys. Letters* **18**, 454 (1971).
189. DEECH, J. S. and SANDERS, J. H., *IEEE J. Quant. Electr.* **QE-4**, 474 (1968).
190. CAHUZAC, P., *Phys. Letters* **32A**, 150 (1970).
191. JENSEN, R. C. and BENNETT, W. R. Jr., *IEEE J. Quant. Electr.* **QE-4**, 356 (1968).
192. JENSEN, R. C., COLLINS, G. J. and BENNETT, W. R. Jr., *Phys. Rev. Lett.* **23**, 363 (1969).
193. SUGAWARA, Y. and TOKIWA, Y., *Japan. J. Appl. Phys.* **9**, 588 (1970).
194. FOWLES, G. R. and SILFVAST, W. T., *IEEE J. Quant. Electr.* **QE-1**, 131 (1965).
195. BLOOM, A. L. and GOLDSBOROUGH, J. P., *IEEE. J. Quant. Electr.* **QE-6**, 164 (1970).
196. KARABUT, E. K., MIKHALEVSKII, V. S., PAPAKIN, V. F. and SEM, M. F., *Sov. Phys. Tech. Phys.* **14**, 1447 (1970).
197. COLLINS, G. J., Ph.D. Dissertation, Yale University (1971).
198. JENSEN, R. C., COLLINS, G. J. and BENNETT, W. R. Jr., *Appl. Phys. Lett.* **18**, 50 (1971).
199. SILFVAST, W. T., FOWLES, G. R. and HOPKINS, R. D., *Appl. Phys. Lett.* **8**, 318 (1966).
200. SILFVAST, W. T., *Appl. Phys. Lett.* **15**, 23 (1969).
201. BENNETT, W. R. Jr., Laser sources, in *Atomic Physics* (B. BEDERSON, V. W. COHEN, and F. M. J. PICHANICK, eds.), pp. 435–73, Plenum Press, New York, 1969.
202. GOLDSBOROUGH, J. P., *IEEE J. Quant Electr.* **QE-5**, 133 (1969).
203. FENDLEY, J. R., GOROG, I., HERNQUIST, K. G. and SUN, C., *R.C.A. Rev.* **30**, 361 (1969).
204. GOLDSBOROUGH, J. P. and HODGES, E. B., *IEEE J. Quant. Electr.* **QE-5**, 361 (1969).
205. TOMPKINS, J. D., The commercial helium–cadmium laser, in *Laser Focus*, 4 Aug. 1969, pp. 32–35.
206. GOLDSBOROUGH, J. P., *Appl. Phys. Lett.* **5**, 159 (1969).
207. FOWLES, G. R. and HOPKINS, B. D., *IEEE J. Quant. Electr.* **QE-3**, 419 (1967).
208. SILFVAST, W. T., *Appl. Phys. Lett.* **13**, 169 (1968).
209. SCHEARER, L. D., *Phys. Rev. Lett.* **22**, 629 (1969).
210. SILFVAST, W. T., *Phys. Rev. Lett.* **27**, 1489 (1971).
211. SCHUEBEL, W. K., *Appl. Phys. Lett.* **16**, 470 (1970).
212. JENSEN, R. C., COLLINS, G. J. and BENNETT, W. R. Jr., *Appl. Phys. Lett.* **18**, 282 (1971); see also erratum **19**, 512 (1971).
213. SUGAWARA, Y. and TOKIWA, Y., CW hollow cathode laser oscillation in Zn^+ and Cd^+, *Technology Reports of the Seikei University* **9**, Mar. 1970, pp. 759–60.
214. TIBILOV, A. S., *Optics Spectrosc.* **19**, 463 (1965).
215. GERRITSEN, H. J. and GOLDERTIER, P. V., *J. Appl. Phys.* **35**, 3060 (1964).
216. BELL, W. E., *Appl. Phys. Lett.* **4**, 34 (1964).
217. BLOOM, A. L., BELL, W. E. and LOPEZ, F. O., *Phys. Rev.* **135**, A578 (1964).
218. BELL, W. E., *Appl. Phys. Lett.* **3**, 100 (1965).
219. SUZUKI, N., *Japan. J. Appl. Phys.* **4**, 452 (1965).
220. WIEDER, H., MYERS, R. A., FISCHER, C. L., POWELL, C. G. and COLOMBO, J., *Rev. Sci. Instr.* **38**, 1538 (1967).
221. BYER, R. L., BELL, W. E., HODGES, E. and BLOOM, A. L., *J. Opt. Soc. Am.* **55**, 1598 (1965).
222. DYSON, D. J., *Nature* **207**, 361 (1965).
223. GOLDSBOROUGH, J. P. and BLOOM, A. L., *IEEE J. Quant. Electr.* **QE-5**, 450 (1969).
224. SCHUEBEL, W. K., *IEEE J. Quant. Electr.* **QE-7**, 39 (1971).
225. BOCKASTEN, K., GARAVAGLIA, M., LENGYEL, B. A. and LUNDHOLM, T., *J. Opt. Soc. Am.* **55**, 1051 (1965).
226. RIGDEN, J. D. and WHITE, A. D., *Nature* **198**, 774 (1963).
227. ARMAND, M. and LAGARDE-MARTINOT, P., *Compt. Rend.* **258**, 867 (1964).
228. PAANANEN, R. A., TANG, C. L., HORRIGAN, P. A. and STATZ, H., *J. Appl. Phys.* **34**, 3148 (1963).
229. DOYLE, W. M., *J. Appl. Phys.* **35**, 1348 (1964).
230. CHEBOTAYEV, V. P., *Optics Spectrosc.* **25**, 267 (1968).
231. COOPER, H. G. and CHEO, P. K., *IEEE J. Quant. Electr.* **QE-2**, 785 (1966).
232. McFARLANE, R. A., *Appl. Phys. Lett.* **5**, 91 (1964).
233. ALLEN, R. B., STARNES, R. B. and DOUGALL, A. A., *IEEE J. Quant. Electr.* **QE-2**, 334 (1966).

234. TUNITSKII, L. N. and CHERKASOV, E. M., *Sov. Phys.* **13**, 1696 (1969).
235. ATKINSON, J. P. and SANDERS, J. H., *J. Phys.* B (Proc. Phys. Soc.) **1**, 1171 (1968).
236. PATEL, C. K. N., McFARLANE, R. A. and FAUST, W. L., *Phys. Rev.* **133**, A1244 (1964).
237. CHEO, P. K. and COOPER, H. G., *Appl. Phys. Lett.* **7**, 202 (1965).
238. PALENIUS, H., *Appl. Phys. Lett.* **8**, 82 (1966).
239. SHIMAZU, M. and SUZAKI, Y., *Japan. J. Appl. Phys.* **4**, 819 (1965).
240. CARR, W. C. and GROW, R. W., *Proc. IEEE* **55**, 1198 (1967).
241. ISAEV, A. A. and PETRASH, G. G., *Sov. Phys. JETP Lett.* **10**, 119 (1969).
242. FOWLES, G. R. and SILFVAST, W. T., *Appl. Phys. Lett.* **6**, 236 (1965).
243. SILFVAST, W. T. and DEECH, J. S., *Appl. Phys. Lett.* **11**, 97 (1967).
244. HEARD, H. G. and PETERSON, J., *Proc. IEEE* **52**, 1258 (1964).
245. LEONOV, R. K., PROTSENKO, E. D. and SAPUNOV, YU. M., *Optics Spectrosc.* **21**, 141 (1966).
246. HITT, J. S. and HASWELL, W. T., *IEEE J. Quant. Electr.* **QE-2**, p. xlii, April 1966.
247. BIRNBAUM, M., TUCKER, A. W., GELBWACHS, J. A. and FINCHER, C. L., *IEEE J. Quant. Electr.* **QE-7**, 208 (1971).
248. McFARLANE, R. A., Stimulated emission spectroscopy of some diatomic molecules, in *Physics of Quantum Electronics* (P. L. KELLEY, B. LAX, and P. E. TANNENWALD, eds.), pp. 655–63, McGraw-Hill, New York, 1966.
249. JANNEY, G. M., *IEEE J. Quant. Electr.* **QE-3**, 133 (1967); see also **QE-3**, 339 (1967).
250. FOWLES, W. R., SILFVAST, W. T. and JENSEN, R. C., *IEEE J. Quant. Electr.* **QE-1**, 183 (1965).
251. BELL, W. E., BLOOM, A. L. and GOLDSBOROUGH, J. P., *IEEE J. Quant. Electr.* **QE-2**, 400 (1965).
252. BELL, W. E., BLOOM, A. L. and GOLDSBOROUGH, J. P., *IEEE J. Quant. Electr.* **QE-2**, 154 (1966).
253. KEIDAN, V. F. and MIKHALEVSKII, V. S., Pulsed generation in bismuth vapor, *Zh. Prikl. Spectrosk.* **9**, 713 (1968).
254. BENNETT, W. R. Jr., FAUST, W. L., McFARLANE, R. A. and PATEL, C. K. N., *Phys. Rev. Lett.* **8**, 470 (1962).
255. TUNITSKII, L. N. and CHERKASOV, E. M., *J. Opt. Soc. Am.* **56**, 1783 (1966).
256. RAUTIN, S. G. and RUBIN, P. L., *Optics Spectrosc.* **18**, 180 (1965).
257. TUNITSKII, L. N. and CHERKASOV, E. M., *Sov. Phys. Tech. Phys.* **13**, 993 (1969).
258. FELD, M. S., FELDMAN, B. J. and JAVAN, A., *Bull. Am. Phys. Soc.* **12**, 15 (1967).
259. TUNITSKII, L. N. and CHERKASOV, E. M., *Optics Spectrosc.* **26**, 344 (1969).
260. TUNITSKII, L. N. and CHERKASOV, E. M., *Optics Spectrosc.* **23**, 154 (1967).
261. TUNITSKII, L. N. and CHERKASOV, E. M., *Sov. Phys. Tech. Phys.* **12**, 1500 (1968).
262. KOLPAKOVA, I. V. and REDKO, T. P., *Optics Spectrosc.* **23**, 351 (1967).
263. FLYNN, G. W., FELD, M. S. and FELDMAN, B. J., *Bull. Am. Phys. Soc.* **12**, 15 (1967).
264. HUBNER, G. and WITTIG, C., *J. Opt. Soc. Am.* **61**, 415 (1971).
265. POWELL, F. X. and DJEU, N., *IEEE J. Quant. Electr.* **QE-7**, 176 (1971).
266. COOPER, H. G. and CHEO, P. K., Ion laser oscillation in sulphur, in *Physics of Quantum Electronics* (P. L. KELLEY, B. LAX, and P. E. TANNENWALD, eds.), pp. 690–7, McGraw-Hill, New York, 1966.
267. MARTINELLI, R. U. and GERRITSEN, H. J., *J. Appl. Phys.* **37**, 444 (1966).
268. KLEIN, M. B. and SILFVAST, W. T., *Appl. Phys. Lett.* **18**, 482 (1971).
269. SILFVAST, W. T. and KLEIN, M. B., *Appl. Phys. Lett.* **17**, 400 (1970).
270. SILFVAST, W. T. and KLEIN, M. B., *Appl. Phys. Lett.* **20**, 501 (1972).
271. WEBB, C. E., *IEEE J. Quant. Electr.* **QE-4**, 426 (1968).
272. WATANABE, S., CHIRARA, M. and OGURA, I., *Japan. J. Appl. Phys.* **11**, 60 (1972).
273. CAHUZAC, PH., *Phys. Lett.* **27A**, 473 (1968).
274. CAHUZAC, PH., *Phys. Lett.* **31A**, 541 (1970).
275. KOVACS, M. A. and ULTEE, C. J., *Appl. Phys. Lett.* **17**, 39 (1970).
276. JEFFERS, W. Q. and WISWALL, C. E., *Appl. Phys. Lett.* **17**, 144 (1970).
277. FLORIN, A. E. and JENSEN, R. J., *IEEE J. Quant. Electr.* **QE-7**, 472 (1971).
278. ZAROWIN, C. B., *Appl. Phys. Lett.* **9**, 241 (1966).
279. GOLDSBOROUGH, J. P., HODGES, E. B. and BELL, W. E., *Appl. Phys. Lett.* **8**, 137 (1966).
280. PAANANEN, R. A. and HORRIGAN, F. A., *Proc. IEEE* **52**, 1261 (1964).
281. SHIMAZU, M. and SUZAKI, Y., *Japan. J. Appl. Phys.* **4**, 381 (1965).
282. JARRETT, S. M., NUNEZ, J. and GOULD, G., *Appl. Phys. Lett.* **8**, 150 (1966).
283. TRUSTY, G. L., YIN, P. K. and KOOZEKANANI, S. K., *IEEE J. Quant. Electr.* **QE-3**, 368 (1967).
284. Identification by M. A. Pollack (private communication, August 1971).
285. PAANANEN, R. A., TANG, C. L. and HORRIGAN, F. A., *Appl. Phys. Lett.* **3**, 154 (1963).
286. KEEFE, W. M. and GRAHAM, W. J., *Appl. Phys. Lett.* **7**, 263 (1965).

287. KEEFE, W. M. and GRAHAM, W. J., *Phys. Lett.* **20**, 643 (1966).
288. JARRETT, S. M., NUNEZ, J. and GOULD, G., *Appl. Phys. Lett.* **7**, 294 (1965).
289. KOVAL'CHUK, V. M. and PETRASH, G. G., *Sov. Phys. JETP Letters* **4**, 144 (1966).
290. JENSEN, R. C. and FOWLES, G. R., *Proc. IEEE* **52**, 1350 (1964).
291. WILLETT, C. S. and HEAVENS, O. S., *Optica Acta* **14**, 195 (1967).
292. FOWLES, G. R. and JENSEN, R. C., *Appl. Optics* **3**, 1191 (1964).
293. PIPER, J. A., COLLINS, G. J. and WEBB, C. E., *Appl. Phys. Lett.* **21**, 203 (1972).
294. FOWLES, G. R. and JENSEN, R. C., *Proc. IEEE* **52**, 851 (1964).
295. WILLETT, C. S., *IEEE J. Quant. Electr.* **QE-3**, 33 (1967).
296. WILLETT, C. S., Population inversion in hollow cathode, pulsed, and positive column discharges, Ph.D. thesis, University of London, Feb. (1967).
297. COLLINS, G. J., KUNO, H., HATTORI, S., TOKUTOME, K., ISHIKAWA, M. and KAMIIDE, N., *IEEE J. Quant. Electr.* **QE-8**, 679 (1972).
298. WILLETT, C. S. and HEAVENS, O. S., *Optica Acta* **13**, 271 (1966).
299. KASPER, J. V. and PIMENTAL, G. C., *Appl. Phys. Lett.* **5**, 231 (1964).
300. POLLACK, M. A., *Appl. Phys. Lett.* **8**, 36 (1966).
301. O'BRIEN, D. E. and BOWEN, J. E., *J. Appl. Phys.* **40**, 4767 (1969).
302. ZALESSKI, V. YU. and VENEDIKTOV, A. A., *Zh. Prikl. Spectrosk.* **9**, 713 (1968).
303. LANZEROTTI, M. Y., *IEEE J. Quant. Electr.* **QE-7**, 207 (1971).
304. POWELL, F. X. and DJEU, N., *IEEE J. Quant. Electr.* **QE-7**, 537 (1971).
305. KIM, H., PAANANEN, R. A. and HAUST, P., *IEEE J. Quant. Electr.* **QE-4**, 385 (1968).
306. PILTCH, M., WALTER, W. T., SOLIMENE, N., GOULD, G. and BENNETT, W. R. Jr., *Appl. Phys. Lett.* **7**, 309 (1965).
307. SILFVAST, W. T. and FOWLES, G. R., *J. Opt. Soc. Am.* **56**, 832 (1966).
308. AKITT, D. P. and WITTIG, C. F., *J. Appl. Phys.* **40**, 902 (1969).
309. HASKINO, Y., KATSUYAMA, Y. and FUKUDA, K., *Japan. J. Appl. Phys.* **11**, 907 (1972).
310. CAHUZAC, P., *IEEE J. Quant. Electr.* **QE-8**, 500 (1972).
311. DREYFUS, R. W. and HODGSON, R. T., *Bull. Am. Phys. Soc.* **17**, 19 (1972).
312. SMITH, P. W., *Appl. Phys. Lett.* **19**, 132 (1971).
313. LINFORD, G. J., *IEEE J. Quant. Electr.* **QE-8**, 477 (1972).
314. DUBROVIN, A. N., TIBILOV, A. S. and SHEVTSOV, M. K., *Optics Spectrosc.* **32**, 685 (1972).
315. BOCKASTEN, K., *Appl. Phys. Lett.* **4**, 118 (1964).
316. GADETSKII, N. P., TKACH, YU. V., BESSAREB, YA. YA. and MAGDA, I. I., *Sov. Phys. JETP Lett.* **14**, 101 (1971).

Appendix Tables 47 to 85

Laser Transitions in Molecular Species

Description of Tables

Laser transitions in molecular species are subdivided into the following: Diatomic, Tri-atomic, Polyatomic, and Unidentified species. Each species is further subdivided in alpha-betical order of the chemical formula of the molecule to which the laser transitions belong, with, for example, H_2 preceding HD.

The wavelengths given are specified as being observed or calculated in air or *in vacuo* as taken from the literature.

Because of the uncertainty in the actual pressure and nature of the chemical constituents in the numerous excitation techniques used to produce the tabulated laser transitions, except in certain cases, only the wavelengths, transitions, excitation (pulsed or CW), and references are tabulated.

The tables are arranged as follows:

DIATOMIC SPECIES:

Table	Molecule	Table	Molecule
47	CN	56	HD
48	CO	57	HF
49	D_2	58	I_2
50	DBr	59	N_2
51	DCl	60	NO
52	DF	61	OD
53	H_2	62	OH
54	HBr	63	Xe_2
55	HCl		

TRIATOMIC SPECIES:

Table	Molecule	Table	Molecule
64	CO_2	70	HCN
65	CS_2	71	ICN
66	D_2O	72	N_2O
67	DCN	73	OCS
68	H_2O	74	SO_2
69	H_2S		

463

POLYATOMIC SPECIES:

Table	Molecule
75	BCl_3
76	$CBrF_3$
77	CCl_2F_2
78	CH_3CCH
79	CH_3CN
80	CH_3F
81	CH_3OH
82	$H_2C : CHCl$
83	H_2CO
84	NH_3

Unidentified Species: Table 85.

Diatomic Species: TABLE 47a. *CN laser*

Transitions in the $X^2\Sigma$ (4–3) band

Wavelength λ_{vac} (μm)	Transition	Excitation	References
5.1838	$P(9)$	Flash photolysis	1
5.1946	$P(10)$	Flash photolysis	1
5.2055	$P(11)$	Flash photolysis	1

Diatomic Species: TABLE 47b. *CN laser*

Transitions between hyperfine components of the $K' = 4$ rotational level of the $A^2\pi_{3/2}$ electronic state of CN

Wavelength λ_{vac} (μm)	Transition	Excitation	References
Approx. 20	$\pi(u)$–$\pi(P)$, six $F \rightarrow F'$ transitions from 5/2, 7/2 and 9/2 levels	CW; microwave excited	2

Diatomic Species: TABLE 48a. *CO laser*

Transitions between the $A^1\pi$–$X^1\Sigma^+$ electronic states—the 4th positive system (Fig. 6.42)

Wavelength λ_{vac} (μm)	Transition	Excitation	References
0.181085	2–6 band $Q(5)$–(13) $R(2)$–(9)	Short rise-time pulsed excitation	3[a]
0.187831	2–7 band $Q(5)$–(13)	Short rise-time pulsed excitation	3[a]
0.1955006	2–8 band $Q(5)$–(11) $R(2)$–(9)	Short rise-time pulsed excitation	3[a]
0.189784	3–8 band $Q(5)$–(12) $R(2)$–(9)	Short rise-time pulsed excitation	3[a]
0.197013	3–9 band $Q(5)$–(11)	Short rise-time pulsed excitation	3[a]

[a] Also observed by R. Waynant (private communication, December 1972).

APPENDIX

Diatomic Species: TABLE 48b. *CO laser*

Transitions between the $B^1\Sigma$–$A^1\pi$ electronic states—the angström band system
(Fig. 6.42)

Wavelength λ_{vac} (μm)	Transition	Excitation	References
	0–2 band(?)		
Four lines between 0.5186 and 0.5198 μm	—	Pulsed	4
	0–3 band		
0.55906	$Q(11)$	Pulsed	4
0.55934	$Q(10)$	Pulsed	4
0.55960	$Q(9)$	Pulsed	4
0.55983	$Q(8)$ or $R(13)$	Pulsed	4
0.56004	$Q(7)$	Pulsed	4
0.56025	$Q(6)$	Pulsed	4
0.56038	$Q(5)$	Pulsed	4
	0–4 band		
0.60629	$Q(9)$	Pulsed	4, 5
0.60657	$Q(8)$	Pulsed	4, 5
0.60682	$Q(7)$	Pulsed	4, 5
0.60705	$Q(6)$	Pulsed	4, 5
0.60725	$Q(5)$	Pulsed	4, 5
0.60742	$Q(4)$	Pulsed	4, 5
	0–5 band		
0.65955	$Q(10)$	Pulsed	4
0.65995	$Q(9)$	Pulsed	4
0.66031	$P(13)$ or $Q(8)$	Pulsed	4
0.66064	$Q(7)$	Pulsed	4
0.66091	$Q(6)$	Pulsed	4
0.66115	$Q(5)$	Pulsed	4
0.66135	$Q(4)$	Pulsed	4

Diatomic Species: TABLE 48c. *CO laser*

Transitions in the $X^1\Sigma^+$ electronic ground state (Fig. 6.42)

Wavelength λ_{vac} (μm)	Transition	Excitation	References
	3–2 band		
4.8836[a]	$P(11)$	CW	6, 7
4.8935[a]	$P(12)$	CW	6, 7
4.914[a]	$P(14)$	CW	8, 9
4.924[a]	$P(15)$	CW	9
	4–3 band		
4.8974[a]	$P(6)$	CW	6, 7

TABLE 48c *(cont.)*

Wavelength λ_{vac} (μm)	Transition	Excitation	References
4.9072[a]	P(7)	CW	6, 7
4.9366[a]	P(10)	CW	6, 7
4.9466[a]	P(11)	CW	6, 7
4.9670[a]	P(13)	CW	6, 7, 9
4.9778[a]	P(14)	CW	6, 7, 9
4.988[a]	P(15)	CW	9
4.998[a]	P(16)	CW	9
	5–4 band		
5.08691	P(16)	CW	10
5.09806	P(19)	CW	10
5.10937	P(20)	CW	10
5.12079	P(21)	CW	10
5.13237	P(22)	CW	10
5.14407	P(23)	CW	10
5.15597	P(24)	CW	10
5.16794	P(25)	CW	10
5.18009	P(26)	CW	10
5.19236	P(27)	CW	10
	6–5 band		
5.03755	P(7)	CW	10–16
5.04750	P(8)	CW	10–16
5.05755	P(9)	CW	10–16
5.06773	P(10)	CW	10–16
5.07807	P(11)	CW	10–16
5.08845	P(12)	CW	10–16
5.09905	P(13)	CW	10–16
5.10985	P(14)	CW	10–16
5.12030	P(15)	CW	10–16
5.13157	P(16)	CW	10–16
5.14268	P(17)	CW	10–16
5.15390	P(18)	CW	10–16
5.16527	P(19)	CW	10–16
5.17681	P(20)	CW	10–16
5.18848	P(21)	CW	10–16
5.20026	P(22)	CW	10–16
5.21218	P(23)	CW	10–16
5.22422	P(24)	CW	10–16
5.23649	P(25)	CW	10–16
5.24882	P(26)	CW	10–16
5.26137	P(27)	CW	10–16
5.27396	P(28)	CW	10–16
	7–6 band		
5.10410	P(7)	CW	10–16
5.11418	P(8)	CW	10–16
5.12445	P(9)	CW	10–16
5.13485	P(10)	CW	10–16
5.14530	P(11)	CW	10–16
5.15595	P(12)	CW	10–17
5.16666	P(13)	CW	10–17

T<small>ABLE</small> 48c *(cont.)*

Wavelength λ_{vac} (μm)	Transition	Excitation	References
5.17765	P(14)	CW	10–17
5.18865	P(15)	CW	10–16
5.19980	P(16)	CW	10–16
5.21110	P(17)	CW	10–16, 63
5.22256	P(18)	CW	10–16, 63
5.23420	P(19)	CW	10–16, 63
5.24590	P(20)	CW	10–16, 63
5.25776	P(21)	CW	10–16, 63
5.26981	P(22)	CW	10–16, 63
5.28189	P(23)	CW	10–16, 63
5.29423	P(24)	CW	10–16, 63
5.30674	P(25)	CW	10–16, 63
5.31924	P(26)	CW	10–16, 63
5.33204	P(27)	CW	10–16, 63
5.34494	P(28)	CW	10–16, 63
	8–7 band		
5.17220	P(7)	CW	10–16, 63
5.18250	P(8)	CW	10–16, 63
5.19290	P(9)	CW	10–16, 63
5.20345	P(10)	CW	10–16, 63
5.21410	P(11)	CW	10–16, 63
5.22498	P(12)	CW	10–17
5.23600	P(13)	CW	10–17
5.24710	P(14)	CW	10–16
5.25835	P(15)	CW	10–16
5.26966	P(16)	CW	10–16
5.28118	P(17)	CW	10–16
5.29284	P(18)	CW	10–16
5.30467	P(19)	CW	10–16
5.31663	P(20)	CW	10–16
5.32871	P(21)	CW	10–16
5.34095	P(22)	CW	10–16
5.35334	P(23)	CW	10–16
5.36585	P(24)	CW	10–16
5.37860	P(25)	CW	10–16
5.39141	P(26)	CW	10–16
5.40442	P(27)	CW	10–16
5.41751	P(28)	CW	10–16
	9–8 band		
5.24195	P(7)	CW	10–16, 63
5.25250	P(8)	CW	10–16, 63
5.26310	P(9)	CW	10–16, 63
5.27380	P(10)	CW	10–16, 63
5.28465	P(11)	CW	10–17
5.29570	P(12)	CW	10–16
5.30695	P(13)	CW	10–17
5.31820	P(14)	CW	10–17
5.32964	P(15)	CW	10–17
5.34127	P(16)	CW	10–17
5.35298	P(17)	CW	10–16

TABLE 48c *(cont.)*

Wavelength λ_{vac} (μm)	Transition	Excitation	References
5.36485	*P*(18)	CW	10–16
5.37692	*P*(19)	CW	10–16
5.38906	*P*(20)	CW	10–16
5.40138	*P*(21)	CW	10–16
5.41385	*P*(22)	CW	10–16
5.42648	*P*(23)	CW	10–16
5.43926	*P*(24)	CW	10–16
5.45225	*P*(25)	CW	10–16
5.46533	*P*(26)	CW	10–16
5.47852	*P*(27)	CW	10–16
5.49191	*P*(28)	CW	10–16
	10–9 band		
5.32415	*P*(8)	Pulsed	10, 18, 63
5.33490	*P*(9)	Pulsed	10, 18, 63
5.34590	*P*(10)	Pulsed	10, 18, 63
5.35695	*P*(11)	Pulsed	10, 18, 63
5.36820	*P*(12)	CW	10–17
5.37950	*P*(13)	CW	10–16
5.39110	*P*(14)	CW	10–17
5.40274	*P*(15)	CW	10–17
5.41457	*P*(16)	CW	10–16
5.42651	*P*(17)	CW	10–16
5.45087	*P*(19)	CW	10–16
5.46328	*P*(20)	CW	10–16
5.47582	*P*(21)	CW	10–16
5.48850	*P*(22)	CW	10–16
5.50138	*P*(23)	CW	10–16
5.51442	*P*(24)	CW	10–16
5.52762	*P*(25)	CW	10–16
5.54091	*P*(26)	CW	10–16
5.55438	*P*(27)	CW	10–16
	11–10 band		
5.4080	*P*(9)	Pulsed	19, 63
5.4196	*P*(10)	Pulsed	19, 63
5.4299	*P*(11)	CW	18
5.4425	*P*(12)	Pulsed	19
5.45402	*P*(13)	CW	10–17
5.46571	*P*(14)	CW	10–16
5.47763	*P*(15)	CW	10–16
5.48968	*P*(16)	CW	10–16
5.50189	*P*(17)	CW	10–16
5.51421	*P*(18)	CW	10–16
5.52667	*P*(19)	CW	10–16, 19
5.53927	*P*(20)	CW	10–16, 19
5.55207	*P*(21)	CW	10–16
5.56503	*P*(22)	CW	10–16
5.59147	*P*(24)	CW	10–16
5.60494	*P*(25)	CW	10–16

TABLE 48c *(cont.)*

Wavelength λ_{vac} (µm)	Transition	Excitation	References
	12–11 band		
5.4842	*P*(9)	Pulsed	19, 63
5.4946	*P*(10)	CW	10–17, 19
5.5072	*P*(11)	CW	19
5.5187	*P*(12)	Pulsed	19
5.5299	*P*(13)	CW	11, 19
5.5424	*P*(14)	CW	11, 19
—	*P*(15)	CW	11
5.57904	*P*(17)	CW	10, 18
5.59158	*P*(18)	CW	10, 11
5.60436	*P*(19)	CW	10, 11
5.61725	*P*(20)	CW	10
5.64350	*P*(22)	CW	10
5.65687	*P*(23)	CW	10
5.67044	*P*(24)	CW	10
5.68414	*P*(25)	CW	10
	13–12 band		
5.5971	*P*(12)	CW	17, 19, 63
5.6087	*P*(13)	CW	17, 19, 63
5.63304	*P*(15)	CW	10, 11
5.65816	*P*(17)	CW	10, 11
5.67098	*P*(18)	CW	10, 11
5.68396	*P*(19)	CW	10
5.69712	*P*(20)	CW	10
5.71042	*P*(21)	CW	10
5.73754	*P*(23)	CW	10
5.75142	*P*(24)	CW	10
	14–13 band		
5.6546	*P*(10)	Pulsed	19, 63
5.6654	*P*(11)	CW	17, 19
5.6780	*P*(12)	Pulsed	19
–	*P*(13)	CW	11
–	*P*(14)	CW	11
5.71561	*P*(15)	CW	10
5.72642	*P*(16)	CW	10
5.73931	*P*(17)	CW	10, 11
5.75243	*P*(18)	CW	10
5.77911	*P*(20)	CW	10
5.79264	*P*(21)	CW	10
5.80636	*P*(22)	CW	10
5.82031	*P*(23)	CW	10
5.83441	*P*(24)	CW	10
5.84874	*P*(25)	CW	10
	15–14 band		
5.78346	*P*(14)	CW	10
5.79633	*P*(15)	CW	10
5.80927	*P*(16)	CW	10
5.83581	*P*(18)	CW	10
5.84935	*P*(19)	CW	10

TABLE 48c *(cont.)*

Wavelength λ_{vac} (µm)	Transition	Excitation	References
5.86300	P(20)	CW	10
5.87689	P(21)	CW	10
5.89088	P(22)	CW	10
5.90507	P(23)	CW	10
5.91951	P(24)	CW	10
	16–15 band		
5.89450	P(16)	CW	10
5.90689	P(17)	CW	10
5.92156	P(18)	CW	10
5.94923	P(20)	CW	10
5.96338	P(21)	CW	10
5.97768	P(22)	CW	10
	17–16 band		
5.98177	P(16)	CW	10
5.99553	P(17)	CW	10
6.00939	P(18)	CW	10
6.02348	P(19)	CW	10
6.03774	P(20)	CW	10
	18–17 band		
6.05755	P(15)	CW	10
6.07145	P(16)	CW	10
6.0856	P(17)	CW	10
6.09961	P(18)	CW	10
6.12845	P(20)	CW	10
	19–18 band		
6.14904	P(15)	CW	10
6.16288	P(16)	CW	10
6.17712	P(17)	CW	10
6.1924	P(18)	CW	10
6.2068	P(19)	CW	10
	20–19 band		
6.24320	P(15)	CW	10
6.25712	P(16)	CW	10
6.27228	P(17)	CW	10
6.2870	P(18)	CW	10
	21–20 band		
6.3260	P(14)	CW	10
6.33936	P(15)	CW	10
6.3552	P(16)	CW	10
6.38476	P(17)	CW	10
	22–21 band		
6.4252	P(14)	CW	10
6.43968	P(15)	CW	10
6.45488	P(16)	CW	10
6.4704	P(17)	CW	10

TABLE 48c *(cont.)*

Wavelength λ_{vac} (µm)	Transition	Excitation	References
	23–22 band		
6.5120	P(13)	CW	10
6.5268	P(14)	CW	10
6.5424	P(15)	CW	10
6.5584	P(16)	CW	10
	24–23 band		
6.6476	P(15)	CW	10
6.6632	P(16)	CW	10

[a] Air wavelength.

Diatomic Species: TABLE 49a. D_2 *laser*

Transitions in the $B^1\Sigma_u^+ - X^1\Sigma_g^+$ system

Wavelength λ_{vac} (µm)[a]	Transition	Excitation	References
0.158675	(10–19) P(2)	Short risetime pulses	20–22
0.158694	(10–19) P(3)	Short risetime pulses	20–22
0.158714	(10–19) P(4)	Short risetime pulses	20–22
0.159130	(9–18) P(2)	Short risetime pulses	20–22
0.159257	(9–18) P(4)	Short risetime pulses	20–22
0.160086	(5–15) P(2)	Short risetime pulses	20–22
0.160354	(5–15) P(4)	Short risetime pulses	20–22
0.160848	(6–16) P(2)	Short risetime pulses	20–22
0.161080	(6–16) P(4)	Short risetime pulses	20–22
0.161198	(8–18) P(2)	Short risetime pulses	20–22
0.161236	(7–17) P(2)	Short risetime pulses	20–22
0.161251	(8–18) P(3)	Short risetime pulses	20–22
0.161320	(8–18) P(4)	Short risetime pulses	20–22
0.161412	(7–17) P(4)	Short risetime pulses	20–22

[a] Wavelength as in ref. 23.

Diatomic Species: TABLE 49b. D_2 *Laser*

Transitions in the $2s\sigma^1\Sigma_g^+ - 2p\sigma^1\Sigma_u^+$ system

Wavelength λ_{air} (µm)	Transition	Excitation	References
0.827752	(2–0) P(3)	Pulsed	24–27
0.944156		Pulsed	25
0.952367		Pulsed	25
0.953005	(1–0) P(3)	Pulsed	24–26, 28
1.477548		Pulsed	25

Diatomic Species: TABLE 50. *DBr laser*

Wavelength λ_{vac} (μm)	Transition DBr^{79}	DBr^{81}	Excitation	References
	2–1 band			
5.8049		P(8)	Pulsed	29
	3–2 band			
5.8620	P(5)		Pulsed	29
5.8626		P(5)	Pulsed	29
5.8928	P(6)		Pulsed	29
5.8944		P(6)	Pulsed	29
5.9246	P(7)		Pulsed	29
5.9261		P(7)	Pulsed	29
5.9573	P(8)		Pulsed	29
5.9590		P(8)	Pulsed	29
	4–3 band			
6.0209	P(5)		Pulsed	29
6.0225		P(5)	Pulsed	29
6.0529	P(6)		Pulsed	29
6.0544		P(6)	Pulsed	29
6.0858	P(7)		Pulsed	29
6.0873		P(7)	Pulsed	29
6.1200	P(8)		Pulsed	29
6.1216		P(8)	Pulsed	29
6.1546	P(9)		Pulsed	29
6.1562		P(9)	Pulsed	29
6.1903	P(10)		Pulsed	29
6.1918		P(10)	Pulsed	29
6.2272	P(11)		Pulsed	29
6.2289		P(11)	Pulsed	29
	5–4 band			
6.2237		P(6)	Pulsed	29
6.2566	P(7)		Pulsed	29
6.2581		P(7)	Pulsed	29
6.2916	P(8)		Pulsed	29
6.2932		P(8)	Pulsed	29
6.3279	P(9)		Pulsed	29
6.3294		P(9)	Pulsed	29

Diatomic Species: TABLE 51. *DCl laser*

Wavelength λ_{vac} (μm)	Transition DCl[35] DCl[37]		Excitation	References
	2–1 band			
5.0455	$P(5)$		Pulsed	29
5.0514		$P(5)$	Pulsed	29
5.0743	$P(6)$		Pulsed	29
5.0811		$P(6)$	Pulsed	29
5.1049	$P(7)$		Pulsed	29, 30
5.1118		$P(7)$	Pulsed	29
5.1363	$P(8)$		Pulsed	29, 30
5.1431		$P(8)$	Pulsed	29
5.1688	$P(9)$		Pulsed	29, 30
	3–2 band			
5.1511	$P(4)$		Pulsed	29, 30
5.1811	$P(5)$		Pulsed	29, 30
5.1879		$P(5)$	Pulsed	29
5.2118	$P(6)$		Pulsed	29
5.2186		$P(6)$	Pulsed	29
5.2435	$P(7)$		Pulsed	29
5.2503		$P(7)$	Pulsed	29
5.2760	$P(8)$		Pulsed	29
5.2829		$P(8)$	Pulsed	29
5.3097	$P(9)$		Pulsed	29
5.3443	$P(10)$		Pulsed	29
5.3799	$P(11)$		Pulsed	29
	4–3 band			
5.3244	$P(5)$		Pulsed	29
5.3562	$P(6)$		Pulsed	29
5.3629		$P(6)$	Pulsed	29
5.3889	$P(7)$		Pulsed	29
5.3956		$P(7)$	Pulsed	29
5.4295		$P(8)$	Pulsed	29
5.4577	$P(9)$		Pulsed	29
5.4935	$P(10)$		Pulsed	29
5.5304	$P(11)$		Pulsed	29
	5–4 band			
5.5084	$P(6)$		Pulsed	29
5.5423	$P(7)$		Pulsed	29
5.5776	$P(8)$		Pulsed	29
5.6137	$P(9)$		Pulsed	29

APPENDIX

Diatomic Species: TABLE 52. *DF laser*

Wavelength λ_{vac} (µm)	Transition	Excitation	References
	1–0 band		
3.8298	P(12)	Pulsed	31
3.9572	P(15)	Pulsed	31
4.0032	P(16)	Pulsed	31
	2–1 band		
3.6363	P(3)	Pulsed	31
3.6665	P(4)	Pulsed	31, 32
3.6983	P(5)	Pulsed	31, 32
3.7310	P(6)	Pulsed	31, 32
3.7651	P(7)	Pulsed	31, 32
3.8007	P(8)	Pulsed	31
3.8375	P(9)	Pulsed	31
3.8757	P(10)	Pulsed	31
3.9155	P(11)	Pulsed	31
3.9565	P(12)	Pulsed	31
3.9995	P(13)	Pulsed	31
4.1369	P(16)	Pulsed	31
4.1862	P(17)	Pulsed	31
	3–2 band		
3.7563	P(3)	Pulsed	31
3.7878	P(4)	Pulsed	31
3.8206	P(5)	Pulsed	31, 32
3.8547	P(6)	Pulsed	31, 32
3.8903	P(7)	Pulsed	31, 32
3.9272	P(8)	Pulsed	31
3.9654	P(9)	Pulsed	31
4.0054	P(10)	Pulsed	31
4.0464	P(11)	Pulsed	31
4.0895	P(12)	Pulsed	31
4.1798	P(14)	Pulsed	31
	4–3 band		
3.9487	P(5)	Pulsed	31
3.9843	P(6)	Pulsed	31
4.0212	P(7)	Pulsed	31

Diatomic Species: TABLE 53a. H_2 *laser*

Transitions in the Werner band $(C^1\pi_u - X^1\Sigma_g^+)$ system

Wavelength λ_{vac} (μm)	Transition	Excitation	References
0.116132*	$v' = 1$ to $v'' = 4, Q(1)$	Pulsed; transient laser system[a, b, d]	33, 34
0.117587	$v' = 2$ to $v'' = 5, P(3)$	,, [b, d]	34
0.121894[c]	$v' = 2$ to $v'' = 6, P(3)$	,, [b]	34
0.122998†	$v' = 3$ to $v'' = 7, Q(1)$	,, [a, b, d]	33
0.12395[d]	$v' = 4$ to $v'' = 8, Q(1)$	,,	—[d]

* Strongest line as predicted.

† Abnormally strong.

[a] Travelling-wave excitation using Blumlein circuitry.

[b] Excitation with 400-keV electron beam.

[c] Doubtful laser line (R. Waynant, private communication, December 1972).

[d] R. Waynant (private communication, December 1972).

Diatomic Species: TABLE 53b. H_2 *laser*[a]

Transitions in the Lyman $(B^1\Sigma_u^+ - X^1\Sigma_g^+)$ system

Wavelength λ_{vac} (μm)[b]	Transition	Excitation	References
0.152325	(2–8) $P(3)$	Pulsed; requires very short, fast risetime pulses	20, 22, 35
0.156725	(8–14) $P(1)$	,,	21, 22, 35
0.15674	(8–14) $P(3)$	,,	21, 22, 35
0.157199	(2–9) $P(1)$	,,	21, 22, 35
0.15771	(7–13) $R(1)$	,,	21
0.157739	(2–9) $P(3)$	,,	22, 35
0.157919	(7–13) $P(1)$	,,	21, 22
0.157998[c]	(7–13) $P(2)$	,,	35
0.158074	(7–13) $P(3)$	,,	21, 22
0.159131	(3–10) $P(1)$	,,	21, 22
0.159340	(3–10) $P(2)$	,,	35
0.159606	(3–10) $P(3)$	,,	20–22
0.160448[d]	(4–11) $P(1)$[c]	,,	22
0.16052	(6–13) $R(1)$	,,	20–22
0.160623	(4–11) $P(2)$	,,	22, 35
0.160750	(6–13) $P(1)$	,,	20–22
0.160839	(4–11) $P(3)$	,,	20–22
0.160902	(6–13) $P(3)$	,,	20, 32
0.161033	(5–12) $P(1)$	,,	20–22
0.161165	(5–12) $P(2)$	,,	20–22
0.161318	(5–12) $P(3)$	,,	20, 22

[a] See Tables 49a and 56a for similar transitions in D_2 and HD.

[b] Wavelengths as in ref. 23.

[c] Questionable laser line (R. Waynant, private communication, December 1972).

[d] Doubtful laser line (R. Waynant, private communication, December, 1972).

Diatomic Species: TABLE 53c. *Para H$_2$ laser*

Transitions in the Lyman ($B^1\Sigma_u^+ - X^1\Sigma_u^+$) system

Wavelength λ_{vac} (μm)[a]	Transition[b]	Excitation	References
0.151994	(2–8) P(2)	Pulsed; requires short, fast risetime pulses	35
0.156753	(8–14) P(2)	”	35
0.157434	(2–9) P(2)	”	35
0.157771	(7–13) R(0)	”	35
0.157998	(7–13) P(2)	”	35
0.158110	(2–9) P(4)	”	35
0.158140	(7–13) P(4)	”	35
0.158899	(3–10) R(0)	”	35
0.159340	(3–10) P(2)	”	35
0.159925	(3–10) P(4)	”	35
0.160236	(4–11) R(0)	”	35
0.160594	(6–13) R(0)	”	35
0.160623	(4–11) P(2)	”	35
0.160829	(6–13) P(2)	”	35
0.160961	(6–13) P(4)	”	35
0.161033	(5–12) P(1)	”	35
0.161091	(4–11) P(4)	”	35
0.161165	(5–12) P(2)	”	35
0.161318	(5–12) P(3)	”	35
0.161485	(5–12) P(4)	”	35

[a] Wavelength as in ref. 23.
[b] As in ref. 23.

Diatomic Species: TABLE 53d. *H$_2$ laser*

Transitions in the $2s\sigma^1\Sigma_g^+ - 2p\sigma^1\Sigma_u^+$ system (Fig. 6.36)

Wavelength λ_{vac} (μm)	Transition	Excitation	References
0.75250[a]	(2–0) P(1)	Pulsed	36
0.835190[b]	(2–1) P(2)	Pulsed	37, 27, 42
0.887868[b]	(1–0) P(4)	Pulsed	37, 37
0.890128[b]	(1–0) P(2)	Pulsed	37, 27, 28
1.116520[b]	(0–0) P(4)	Pulsed	37, 27
1.122507[b]	(0–0) P(2)	Pulsed	37, 27, 28
1.30613[b]	(0–1) P(4)	Pulsed	37, 27, 28
1.31658[b]	(0–1) P(2)	Pulsed	26

[a] λ_{air}, 0.75250$\pm$0.00005 μm.
[b] Wavelength as in ref. 23.

Diatomic Species: TABLE 54a. *HBr laser*

Wavelength λ_{vac} (μm)	Transition HBr[79]	HBr[81]	Excitation	References
	1–0 band			
4.0170	P(4)		Pulsed	29
4.0176		P(4)	Pulsed	29
4.0470	P(5)		Pulsed	29
4.0475		P(5)	Pulsed	29
4.0783	P(6)		Pulsed	29
4.0788		P(6)	Pulsed	29
4.1107	P(7)		Pulsed	29
4.1112		P(7)	Pulsed	29
4.1442	P(8)		Pulsed	29
4.1448		P(8)	Pulsed	29
4.1796		P(9)	Pulsed	29
	2–1 band			
4.1653	P(4)		Pulsed	29
4.1658		P(4)	Pulsed	29
4.1970	P(5)		Pulsed	29
4.1975		P(5)	Pulsed	29
4.2295	P(6)		Pulsed	29
4.2633	P(7)		Pulsed	29
4.2639		P(7)	Pulsed	29
4.2988	P(8)		Pulsed	29
4.2994		P(8)	Pulsed	29
4.3354	P(9)		Pulsed	29
4.3359		P(9)	Pulsed	29
	3–2 band			
4.3250	P(4)		Pulsed	29
4.3255		P(4)	Pulsed	29
4.3579	P(5)		Pulsed	29
4.3585		P(5)	Pulsed	29
4.3925	P(6)		Pulsed	29
4.3931		P(6)	Pulsed	29
4.4281	P(7)		Pulsed	29
4.4307		P(7)	Pulsed	29
4.4652	P(8)		Pulsed	29
4.4658		P(8)	Pulsed	29
4.5041	P(9)		Pulsed	29
4.5047		P(9)	Pulsed	29
	4–3 band			
4.5330	P(5)		Pulsed	29
4.5335		P(5)	Pulsed	29
4.5691	P(6)		Pulsed	29
4.5696		P(6)	Pulsed	29
4.6070	P(7)		Pulsed	29
4.6076		P(7)	Pulsed	29
4.6463	P(8)		Pulsed	29
4.6467		P(8)	Pulsed	29

Diatomic Species: TABLE 54b. *HBr laser*

Pure rotational transitions

Wavelength λ_{vac} (μm)	Transition HBr[79] HBr[81]		Excitation	References
	$v = 0$			
19.399		$R(33)$	Pulsed	38
20.360	$R(31)$ or	$R(31)$	Pulsed	38
20.896	$R(30)$		Pulsed	38
20.949		$R(30)$	Pulsed	38
20.501	$R(29)$ or	$R(29)$	Pulsed	38
22.136	or	$R(28)$	Pulsed	38
30.948	$R(19)$ or	$R(19)$	Pulsed	38
32.469	$R(18)$ or	$R(18)$	Pulsed	38
	$v = 1$			
19.988	$R(33)$ or	$R(33)$	Pulsed	38
21.546	$R(30)$		Pulsed	38
30.445	$R(20)$ or	$R(20)$	Pulsed	38
31.849	$R(19)$ or	$R(19)$	Pulsed	38
33.409	$R(18)$ or	$R(18)$	Pulsed	38
	$v = 2$			
22.226	$R(30)$		Pulsed	38
22.855	$R(29)$ or	$R(29)$	Pulsed	38
31.368	$R(20)$ or	$R(20)$	Pulsed	38
32.799	$R(19)$		Pulsed	38
40.526	$R(15)$ or	$R(15)$	Pulsed	38
	$v = 3$			
23.436	—		Pulsed	38
29.786		$R(22)$	Pulsed	38

Diatomic Species: TABLE 55a. *HCl laser*

Wavelength λ_{vac} (μm)	Transition HCl[35] HCl[37]		Excitation	References
	2–1 band			
3.7071	$P(4)$		Pulsed	29, 30, 39
3.7383	$P(5)$		Pulsed	29, 30, 39–41
3.7408		$P(5)$	Pulsed	29, 41
3.7710	$P(6)$		Pulsed	29, 30, 39–41
3.7735		$P(6)$	Pulsed	29
3.8050	$P(7)$		Pulsed	29, 30, 39–41
3.8074		$P(7)$	Pulsed	29, 41
3.8401	$P(8)$		Pulsed	29, 30, 39, 40
3.8425		$P(8)$	Pulsed	29, 41
3.8768	$P(9)$		Pulsed	29, 40
3.9149	$P(10)$		Pulsed	29, 40

TABLE 55a *(cont.)*

Wavelength λ_{vac} (μm)	Transition HCl³⁵ HCl³⁷		Excitation	References
	3–2 band			
3.8509	P(4)		Pulsed	29, 30, 39, 41
3.8840	P(5)		Pulsed	29, 30, 39, 41
3.9181	P(6)		Pulsed	29, 30, 39
3.9205		P(6)	Pulsed	30, 39
3.9536	P(7)		Pulsed	29, 30, 39
3.9560		P(7)	Pulsed	30, 39
3.9909	P(8)		Pulsed	29, 30, 39
4.0295	P(9)		Pulsed	29, 30, 39

Diatomic Species: TABLE 55b. *HCl laser*

Pure rotational transitions

Wavelength λ_{vac} (μm)	Transition HCl³⁵ HCl³⁷		Excitation	References
	$v = 0$			
13.8720	R(40)		Pulsed	29
14.0994	R(39)		Pulsed	29
14.3434	R(38)		Pulsed	29
16.2125	R(32)		Pulsed	29, 9
16.6085	R(31)		Pulsed	29
16.664		R(31)	Pulsed	38
17.0340	R(30)		Pulsed	29, 38, 9
17.4923	R(29)		Pulsed	29, 38, 9
17.9874	R(28)		Pulsed	29, 38
17.997		R(28)	Pulsed	38
18.522	R(27)		Pulsed	29
19.122		R(26)	Pulsed	38
20.4106	R(24)		Pulsed	29
21.1556	R(23)		Pulsed	29
21.9706	R(22)		Pulsed	29
22.8637	R(21)		Pulsed	29
23.8485	R(20)		Pulsed	29, 38
24.9367	R(19)		Pulsed	29, 9
26.1462	R(18)		Pulsed	29
27.508	R(17)		Pulsed	38
	$v = 1$			
16.765		R(32)	Pulsed	38
17.125	R(31)		Pulsed	38
17.575	R(30)		Pulsed	38
18.035	R(29)		Pulsed	38
18.555	R(28)		Pulsed	38
18.593		R(28)	Pulsed	38
19.145		R(27)	Pulsed	38
19.7002	R(26)		Pulsed	29

TABLE 55b *(cont.)*

Wavelength λ_{vac} (µm)	Transition HCl³⁵	HCl³⁷	Excitation	References
20.3455	R(25)		Pulsed	29
21.0470	R(24)		Pulsed	29
21.8127	R(23)		Pulsed	29
22.6514	R(22)		Pulsed	29
23.5705	R(21)		Pulsed	29, 9
24.6177		R(20)	Pulsed	29
24.5833	R(20)		Pulsed	9
25.7040	R(19)		Pulsed	29
	v = 2			
19.183	R(28)		Pulsed	38
20.9991	R(26)		Pulsed	29
24.3178	R(21)		Pulsed	29, 9
25.3	R(20)?		Pulsed	29, 9
	v = 3			
19.783	R(28)?		Pulsed	38
19.821			Pulsed	38

Diatomic Species: TABLE 56a. *HD laser*

Transitions in the $B^1\Sigma_u^+ - X^1\Sigma_g^+$ system

Wavelength λ_{vac} (µm)[a]	Transition	Excitation	References
0.157242	(9–16) P(2), P(3)	Pulsed; requires short, fast risetime pulses	35
0.159524	(4–12) P(2)	"	35
0.159713	(4–12) P(3)	"	35
0.160365	(5–13) P(1)	"	35
0.160496	(5–13) P(2)	"	35
0.160569	(7–15) P(1)	"	35
0.160631	(7–15) P(2)	"	35
0.160646	(5–13) P(3)	"	35
0.160692	(7–15) P(3)	"	35
0.160747	(7–15) P(4)	"	35
0.160794	(6–14) P(1)	"	35
0.160827	(5–13) P(4)	"	35
0.160893	(6–14) P(2)	"	35
0.161005	(6–14) P(3)	"	35
0.161131	(6–14) P(4)	"	35

[a] Wavelength as in ref. 23.

Diatomic Species: TABLE 56b. *HD laser*

Transitions in the $2s\sigma^1\Sigma_g^+-2p\alpha^1\Sigma_u^+$ system

Wavelength λ_{vac} (μm)[a]	Transition	Excitation	References
0.9163	(1–0)	Pulsed	28, 27

[a] Wavelength as in ref. 23.

Diatomic Species: TABLE 57a. *HF laser*

Wavelength λ_{vac} (μm)[a]	Transition	Excitation	References
	1–0 band		
2.640	$P(4)$	CW	43
2.673	$P(5)$	CW	43
2.7075	$P(6)$	CW	43, 31
2.7441	$P(7)$	CW	43, 31
2.7826	$P(8)$	Pulsed	31, 63
2.8231	$P(9)$	Pulsed	31, 63
2.8657	$P(10)$	Pulsed	31, 63
2.9103	$P(11)$	Pulsed	31, 63
2.9573	$P(12)$	Pulsed	31, 63
3.0064	$P(13)$	Pulsed	31, 63
3.0582	$P(14)$	Pulsed	31
3.1125	$P(15)$	Pulsed	31
	2–1 band		
2.6962	$P(2)$	Pulsed	31
2.7275	$P(3)$	Pulsed	31
2.7604	$P(4)$	CW	43, 31–33, 63
2.7953	$P(5)$	CW	43, 41–33, 63
2.8318	$P(6)$	CW	43, 31–33, 63
2.8706	$P(7)$	CW	43, 31–33, 63
2.9111	$P(8)$	Pulsed	31, 44, 63
2.9539	$P(9)$	Pulsed	31, 44, 63
2.9989	$P(10)$	Pulsed	31, 44, 63
3.0461	$P(11)$	Pulsed	31, 44, 63
3.0958	$P(12)$	Pulsed	31, 63
3.1480	$P(13)$	Pulsed	31, 63
3.2029	$P(14)$	Pulsed	31, 63
3.2603	$P(15)$	Pulsed	31, 63
	3–2 band		
2.8213	$P(2)$	Pulsed	31, 63
2.8540	$P(3)$	Pulsed	31, 63
2.8889	$P(4)$	Pulsed	31, 63
2.9256	$P(5)$	Pulsed	31, 63
2.9643	$P(6)$	Pulsed	31, 63
3.0051	$P(7)$	Pulsed	31, 63
3.0482	$P(8)$	Pulsed	31, 63

Diatomic Species: TABLE 57b. *HF laser*

Pure rotational transitions

Wavelength λ_{vac} (μm)[a]	Transition	Excitation	References
	$v = 0$		
10.1978	$R(27)$	Pulsed	45, 38
10.4578	$R(26)$	Pulsed	45
10.7439	$R(25)$	Pulsed	45
11.0573	$R(24)$	Pulsed	45, 38
11.4033	$R(23)$	Pulsed	45, 38
11.7854	$R(22)$	Pulsed	45, 38
12.2082	$R(21)$	Pulsed	45, 38
12.6781	$R(20)$	Pulsed	45, 38, 63
13.2009	$R(19)$	Pulsed	45, 38
13.7841	$R(18)$	Pulsed	45, 38
14.4406	$R(17)$	Pulsed	45, 38, 63
15.1744	$R(16)$	Pulsed	45, 38, 63
16.0215	$R(15)$	Pulsed	45, 38
16.975	$R(14)$	Pulsed	45, 38
18.085	$R(13)$	Pulsed	45, 38
	$v = 1$		
12.2619	$R(22)$	Pulsed	45, 38
12.7006	$R(21)$	Pulsed	45, 38
13.1877	$R(20)$	Pulsed	45
13.7277	$R(19)$	Pulsed	45
15.0163	$R(17)$	Pulsed	45
16.655	$R(16)$	Pulsed	38
17.645	$R(15)$	Pulsed	38
18.8010	$R(13)$	Pulsed	38
20.1337	$R(12)$	Pulsed	45
21.6986	$R(11)$	Pulsed	45
	$v = 2$		
10.5819	$R(29)$	Pulsed	45
10.8117	$R(28)$	Pulsed	45
13.2211	$R(21)$	Pulsed	45
14.2881	$R(19)$	Pulsed	45
16.444	$R(16)$	Pulsed	38
17.327	$R(15)$	Pulsed	38
20.939	$R(12)$	Pulsed	45
	$v = 3$		
11.5408	$R(27)$	Pulsed	45
17.095	$R(16)$	Pulsed	38
19.1129	$R(14)$	Pulsed	45
20.3513	$R(13)$	Pulsed	45
21.7885	$R(12)$	Pulsed	45

[a] Wavelength as in ref. 23.

Diatomic Species: TABLE 58. I_2 *laser*

Partial listing of transitions in the $B^3\pi_{o+u}-X^1\Sigma^+$ band[a]

Wavelength λ_{vac} (μm)	Transition	Excitation	References
0.5443	$B^3\pi_{o+u}-X^1\Sigma_g^+$	Optically pumped with 0.5306-μm frequency-doubled Nd : YAG laser pulse	46
0.5550	$B^3\pi_{o+u}-X^1\Sigma_g^+$	"	46
0.5567	$B^3\pi_{o+u}-X^1\Sigma_g^+$	"	46
0.5680	$B^3\pi_{o+u}-X^1\Sigma_g^+$	"	46
0.5697	$B^3\pi_{o+u}-X^1\Sigma_g^+$	"	46
0.5745	$B^3\pi_{o+u}-X^1\Sigma_g^+$	"	46
0.5815	$B^3\pi_{o+u}-X^1\Sigma_g^+$	"	46
0.5880	$B^3\pi_{o+u}-X^1\Sigma_g^+$	"	46
0.6025	$B^3\pi_{o+u}-X^1\Sigma_g^+$	"	46
0.6175	$B^3\pi_{o+u}-X^1\Sigma_g^+$	"	46
0.6330	$B^3\pi_{o+u}-X^1\Sigma_g^+$	"	46
0.6490	$B^3\pi_{o+u}-X^1\Sigma_g^+$	"	46
0.6511	$B^3\pi_{o+u}-X^1\Sigma_g^+$	"	46
0.6592	$B^3\pi_{o+u}-X^1\Sigma_g^+$	"	46
0.6645	$B^3\pi_{o+u}-X^1\Sigma_g^+$	"	46
0.6763	$B^3\pi_{o+u}-X^1\Sigma_g^+$	"	46
0.9060	$B^3\pi_{o+u}-X^1\Sigma_g^+$	"	46
0.9288	$B^3\pi_{o+u}-X^1\Sigma_g^+$	"	46
0.9295	$B^3\pi_{o+u}-X^1\Sigma_g^+$	"	46
1.0053	$B^3\pi_{o+u}-X^1\Sigma_g^+$	"	46
1.0225	$B^3\pi_{o+u}-X^1\Sigma_g^+$	"	46
1.0245	$B^3\pi_{o+u}-X^1\Sigma_g^+$	"	46
1.1073	$B^3\pi_{o+u}-X^1\Sigma_g^+$	"	46
1.1255	$B^3\pi_{o+u}-X^1\Sigma_g^+$	"	46
1.1350	$B^3\pi_{o+u}-X^1\Sigma_g^+$	"	46
1.3299	$B^3\pi_{o+u}-X^1\Sigma_g^+$	"	46
1.3310	$B^3\pi_{o+u}-X^1\Sigma_g^+$	"	46
1.3324	$B^3\pi_{o+u}-X^1\Sigma_g^+$	"	46
0.5746	$B^3\pi_{o+u}-X^1\Sigma_g^+$	Optically pumped with 0.5319-μm frequency-doubled Nd : YAG laser pulse	46
0.5905	$B^3\pi_{o+u}-X^1\Sigma_g^+$	"	46
0.6049	$B^3\pi_{o+u}-X^1\Sigma_g^+$	"	46
0.6198	$B^3\pi_{o+u}-X^1\Sigma_g^+$	"	46
0.6305	$B^3\pi_{o+u}-X^1\Sigma_g^+$	"	46
0.6352	$B^3\pi_{o+u}-X^1\Sigma_g^+$	"	46
0.6936	$B^3\pi_{o+u}-X^1\Sigma_g^+$	"	46
0.7114	$B^3\pi_{o+u}-X^1\Sigma_g^+$	"	46
0.8813	$B^3\pi_{o+u}-X^1\Sigma_g^+$	"	46
0.9047	$B^3\pi_{o+u}-X^1\Sigma_g^+$	"	46

[a] Laser oscillation occurs to every second vibrational level of the $X^1\Sigma_g^+$ ground state from $v'' = 2$ to $v'' = 73$. The wavelength range was limited by the cavity mirror reflectance.

TABLE 58 *(cont.)*

Wavelength λ_{vac} (μm)	Transition	Excitation	References
0.9545	$B^3\pi_{0+u}-X^1\Sigma_g^+$	Optically pumped with 0-5319-μm frequency-doubled Nd : YAG laser pulse	46
0.9555	$B^3\pi_{0+u}-X^1\Sigma_g^+$	"	46
0.9963	$B^3\pi_{0+u}-X^1\Sigma_g^+$	"	46
0.9973	$B^3\pi_{0+u}-X^1\Sigma_g^+$	"	46
1.0534	$B^3\pi_{0+u}-X^1\Sigma_g^+$	"	46
1.0775	$B^3\pi_{0+u}-X^1\Sigma_g^+$	"	46
1.0788	$B^3\pi_{0+n}-X^1\Sigma_g^+$	"	46
1.1066	$B^3\pi_{0+u}-X^1\Sigma_g^+$	"	46
1.3153	$B^3\pi_{0+u}-X^1\Sigma_g^+$	"	46
1.3192	$B^3\pi_{0+u}-X^1\Sigma_g^+$	"	46
1.3282	$B^3\pi_{0+u}-X^1\Sigma_g^+$	"	46

Diatomic Species: TABLE 59a. N_2 *laser (Fig. A.25)*

Transitions in the 2nd positive ($C^3\pi_u-B^3\pi_g$) system[a]

Wavelength λ_{vac} (μm)	Transition	Excitation	References[b]
	1–0 band		
0.3159	—	Pulsed; transient laser system	47
	0–0 band		
0.3364909	$R(7)$	Pulsed; transient laser system	48, 49
0.3365425[c]	—	"	48, 49
0.3365478	$R(6)$	"	48
0.3366913	$R(4)$	"	48, 49, 50
0.336789	—	"	50
0.3368428	$P(21)$	"	48, 49
0.3369257	$P(1), Q(1)$	"	48, 49
0.3369769	—	"	48, 49
0.3369844	$P(3), P(17)$	"	48, 49

[a] First reported in ref. 53.
[b] See additional general references to the N_2 UV laser, refs. 47 and 54 to 60 and 137, and those given in Section 6.2.3.
[c] Measured value.

TABLE 59a *(cont.)*

Wavelength λ_{vac} (μm)	Transition	Excitation	References
0.3369907	$P(17)$	Pulsed; transient laser system	48, 50
0.3370088	$P(4)$	"	48
0.3370137	$P(16)$	"	48
0.3370174	$P(16)$	"	48
0.3370302	$P(5)$	"	48
0.3370381	$P(15)$	"	48
0.3370438	$P(15)$	"	48
0.3370480	$P(6)$	"	48
0.3370537	$P(14)$	"	48, 49
0.3370567	$P(14)$, $P(4)$	"	48, 49
0.3370623	$P(7)$	"	48, 49
0.3370665	$P(14)$	"	48, 49
0.3370728	$P(8)$	"	48, 49
0.3370767	$P(5)$, $P(3)$	"	48, 49
0.3370807	$P(9)$	"	48, 49
0.3370824	$P(11)$, $P(10)$	"	48, 49
0.3370841	$P(11)$, $P(13)$	"	48, 49
0.3370932	$P(7)$, $P(13)$	"	48, 49
0.3370943	—	"	49
0.3370997	$P(12)$, $P(4)$	"	48, 49
0.3371048	$P(7)$	"	48, 49
0.3371088	$P(11)$	"	48, 49
0.3371138	—	"	49
0.3371151	$P(9)$	"	48, 49
0.3371185	$P(5)$	"	48, 49
0.3371278	$P(11)$	"	48, 49
0.3371320	$P(6)$	"	48, 49
0.3371377	$P(10)$	"	48, 49
0.3371403	$P(7)$	"	48, 49
0.3371417	—	"	49
0.3371437	$P(8)$, $P(9)$	"	48, 49
0.337162	P_1–7, 13	"	51
0.337168	P_1'–8	"	51
0.337172	P_1'–12	"	51
0.33718	P_1'–10, P_1–11, 9	"	51
0.337189	P_2–6	"	51
0.337195	P_2–7	"	51
0.337204	P_2–11	"	51
0.33721	P_2–8, 10	"	51
0.337212	P_2–9	"	51
0.337223	P_3–11	"	51
0.337227	P_3–6	"	51
0.337235	P_3–10	"	51
0.337237	P_3–7	"	51
0.33724	P_3–8, 9	"	51
	0–1 band		
0.3575460	—	Pulsed, transient laser system	48, 52
0.3575790	$P(5)$	"	48, 52

TABLE 59a *(cont.)*

Wavelength λ_{vac} (μm)	Transition	Excitation	References
0.3576112	$P(7)$ or $P(4)$	Pulsed; transient laser system	48, 52
0.3576613	$P(8)$, $P(9)$ or $P(11)$	"	48, 52
0.3576892	$P(9)$, $P(6)$	"	48, 52
0.3576950	$P(8)$, $P(7)$	"	48, 52

Diatomic Species: TABLE 59b. N_2 *laser (Fig. A.25)*

Transitions in the 1st positive ($B^3\pi_g$–$A^3\Sigma_u^+$) system

Wavelength (λ_{vac} μm)[b]	Transition	Excitation	References[a]
	4–2 band		
0.748480	Q_1–11	Pulsed; transient laser system	51, 52
0.748948	Q_1–9	"	51, 52
0.749189	Q_{23}–7	"	51, 52
0.749377	Q_1–7	"	51, 52
0.749775	Q_1–5	"	51, 52
0.7750363	P_1–11	"	51, 52
	3–1 band		
0.757434	Q_3–17	Pulsed, transient laser system	61, 51
0.758313	Q_3–13	"	61, 51
0.758632	Q_3–11	"	61, 51
0.758855	Q_3–9	"	61, 51
0.758987	Q_2–13	"	61, 51
0.759199	Q_2–12	"	61, 51
0.75939	Q_2–11	"	61, 51
0.75949	Q_1–15 or R_1–8	"	61, 51
0.759524	Q_1–15	"	61, 51
0.75973	Q_2–9	"	61, 51
0.760079	Q_1–15	"	61, 51
0.760608	Q_1–11	"	61, 51
0.761091	Q_1–9	"	61, 51
0.761195	Q_{23}–5	"	61, 51
0.761361	$R_1 Q_{23}$–7	"	61, 51
0.761535	Q_1–7	"	61, 51
0.761946	Q_1–5	"	61, 51
0.762506	Q_{12}–11	"	51
0.762631	Q_{12}–9	"	51
0.762721	Q_{12}–7	"	51
0.76281	P, Q_{23}–5	"	51
	2–0 band		
0.771418	Q_3–9	Pulsed, transient laser system	61, 52
0.772455	Q_1–13	"	61, 62

TABLE 59b *(cont.)*

Wavelength λ_{vac} (μm)[b]	Transition	Excitation	References[a]
0.773003	Q_1–11	Pulsed; transient laser system	61, 62
0.773504	Q_1–9	"	61, 62
0.773963	Q_1–7	"	61, 62
0.774602	Q_1–5	"	61, 51
0.775483	R_{13}–1	"	51
	2–1 band		
0.865569	R_3–7	Pulsed; transient laser system	51, 52
0.865730	Q_3–15	"	51, 52
0.866327	Q_3–13	"	51, 52
0.866494	Q_2–15	"	51, 52
0.866583	Q_3–12	"	51, 52
0.866810	Q_3–11	"	51, 52
0.86700	Q_3–10	"	51, 52
0.867161	Q_3–9	"	51, 52
0.867197	Q_2–13	"	51, 52
0.867371	Q_3–7	"	51, 52
0.867793	Q_2–11	"	51, 52
0.868294	Q_2–9	"	51, 52
0.868520	R_1–7	"	51, 52
0.868613	Q_1–13	"	51, 61
0.859001	Q_1–12	"	51, 61
0.869375	Q_1–11	"	51, 61
0.869729	Q_1–10	"	51, 52
0.870067	Q_1–9	"	61
0.870307	Q_{23}–5	"	61
0.870388	Q_1–8	"	61
0.870494	Q_{23}–7	"	61
0.870570	Q_{23}–11	"	61
0.870696	Q_1–7	"	61
0.870986	Q_1–6	"	61
0.871267	Q_1–5	"	61
0.871533	Q_1–4	"	61
0.871690	Q_1–13	"	61
0.871792	Q_1–3	"	61
0.871884	Q_{12}–11	"	61
0.871977	P_1–11	"	61
0.872106	Q_{12}–9	"	61
0.872193	P_1–9	"	61
0.872266	Q_{12}–7	"	61
0.872374	Q_{12}–5	"	61
0.872460	Q_{12}–3	"	61
	1–0 band		
0.883620	—	Pulsed; transient laser system	51
0.884372	Q_3–15	"	51
0.884662	—	"	61
0.88472	R, Q_{21}–9 or R, Q_{22}–5	"	51
0.885001	Q_3–13	"	61

TABLE 59b *(cont.)*

Wavelength λ_{vac} (μm)[b]	Transition	Excitation	References[a]
0.885163	Q_2–15	Pulsed; transient laser system	51
0.885269	Q_3–12	"	61
0.885504	Q_3–11	"	61
0.885704	Q_3–10	"	51
0.885872	Q_3–9	"	61
0.885893	Q_2–13	"	51
0.885998	Q_3–8	"	51
0.886090	Q_3–7	"	51
0.886093	Q_3–5, R_1–9	"	51
0.88612	R_1–9 or Q_1–4	"	51
0.886397	Q_{21}–5	"	51
0.886500	Q_1–15	"	51
0.886522	Q_2–11	"	61, 51
0.886941	Q_1–14	"	51
0.887043	Q_2–2	"	51
0.887365	Q_1–13	"	61, 51
0.887775	Q_1–12	"	51
0.888162	Q_1–11	"	61, 51
0.888532	Q_1–10	"	51
0.888884	Q_1–9	"	61, 51
0.889150	Q_{23}–5	"	51
0.889219	Q_1–8	"	51
0.889358	Q_{23}–7	"	51
0.889460	Q_{23}–9	"	51
0.889539	Q_1–7	"	61, 51
0.889846	Q_1–6	"	51
0.890137	Q_1–5	"	61, 51
0.89031	P_1–15	"	51
0.890420	Q_1–4	"	51
0.890617	P_1–13	"	51
0.890688	Q_1–3	"	51
0.890811	Q_{12}–11	"	51
0.890911	P_1–11	"	51
0.891038	Q_{12}–9	"	51
0.891135	P_1–9	"	51
0.891199	Q_{12}–7	"	61, 51
0.891308	Q_{12}–5 or P_1–7	"	51
0.891373	Q_{12}–3	"	51
	3–3 band		
0.9599		Pulsed; transient laser system	52
	0–0 band[c, d]		
1.04387	Q_1–15	Pulsed; transient laser system	51
1.04454	Q_1–14	"	51
1.04518	Q_1–13	"	61, 51
1.04580	Q_1–12	"	51
1.04640	Q_1–11	"	61, 51

TABLE 59b *(cont.)*

Wavelength λ_{vac} (μm)[b]	Transition	Excitation	References[a]
1.04695	Q_1–10	Pulsed; transient laser system	51
1.04748	Q_1–9	”	61, 51
1.04797	Q_1–8	”	51
1.04824	Q_{23}–7	”	61, 51
1.04846	Q_1–7	”	51
1.04892	Q_1–6	”	51
1.04934	Q_1–5	”	61, 51
1.04976	Q_1–4	”	61, 51
1.05016	Q_1–3	”	51
1.05051	Q_{12}–9	”	51
1.05080	Q_{12}–7	”	61, 51
1.05100	Q_{12}–5	”	51
1.05111	Q_{12}–3	”	51
1.05254	Q_{13}–3	”	51
1.05291	P_{12}–5	”	51
1.05338	P_{12}–7	”	51
1.05376	P_{12}–9	”	51
	1–2 band		
1.1933	—	Pulsed; transient laser system	52
	0–1 band[c]		
1.230598	Q_1–11	Pulsed; transient laser system	61, 51
1.231430	Q_1–10	”	61, 51
1.232219	Q_1–9	”	61, 51
1.232962	Q_1–8	”	51
1.233671	Q_1–7	”	61, 51
1.234332	Q_1–6	”	51
1.234969	Q_1–5	”	61, 51
	2–4 band		
1.3646	—	Pulsed; transient laser system	52
	0–2 band		
1.4983	—	Pulsed; transient laser system	52

[a] See additional references given in subsection 6.2.3.2.
[b] Wavelengths generally as given in ref. 23.
[c] Observed also in TEA laser system, ref. 63.
[d] Observed in TEA laser, ref. 64.

Diatomic Species: TABLE 59c. N_2 *laser (Fig. A.25)*

Transitions in the $a^1\pi_g - a^1\Sigma_u^-$ system

Wavelength λ_{vac} (μm)[a]	Transition	Excitation	References
	2–1 band		
3.29463	$Q(14)$	Pulsed	65, 66
3.30149	$Q(12)$	Pulsed	65, 66
3.30734	$Q(10)$	Pulsed	65, 66
3.30989	$Q(9)$	Pulsed	66
3.31221	$Q(8)$	Pulsed	65, 66
3.31426	$Q(7)$	Pulsed	66
3.31607	$Q(6)$	Pulsed	65, 66
3.31760	$Q(5)$	Pulsed	66
3.31889	$Q(4)$	Pulsed	65, 66
3.32069	$Q(2)$	Pulsed	66
	1–0 band		
3.45184	$Q(12)$	Pulsed	65, 66
3.45832	$Q(10)$	Pulsed	65, 66
3.46114	$Q(9)$	Pulsed	66
3.46368	$Q(8)$	Pulsed	65, 66
3.46596	$Q(7)$	Pulsed	66
3.46795	$Q(6)$	Pulsed	65, 66
3.46967	$Q(5)$	Pulsed	66
3.47109	$Q(4)$	Pulsed	65, 66
	0–0 band		
8.1483	$Q(10)$	Pulsed	65, 66
8.1827	$Q(8)$	Pulsed	65, 66
8.2102	$Q(6)$	Pulsed	65, 66

[a] Wavelength as in ref. 23.

Diatomic Species: TABLE 59d. N_2 *laser* (*Fig. A.25*)

Transitions in the $w^1\Delta_u - a^1\pi_g$ system

Wavelength λ_{vac} (μm)[a]	Transition	Excitation	References
	0–0 band		
3.62349	$R(4)$	Pulsed	67
3.62614	$R(3)$	Pulsed	67
3.62910	$R(2)$	Pulsed	67
3.64313	$Q(4)$	Pulsed	67
3.64472	$Q(5)$	Pulsed	67
3.64662	$Q(6)$	Pulsed	67, 63
3.64883	$Q(7)$	Pulsed	67, 63
3.65138	$Q(8)$	Pulsed	67, 63
3.65424	$Q(9)$	Pulsed	67
3.65745	$Q(10)$	Pulsed	67
3.66095	$Q(11)$	Pulsed	67

Wavelength λ_{vac} (μm)[a]	Transition	Excitation	References
3.66483	$Q(12)$	Pulsed	67
3.66899	$Q(13)$	Pulsed	67
3.67352	$Q(14)$	Pulsed	67
3.67834	$Q(15)$	Pulsed	67

[a] Wavelength as in ref. 23.

Diatomic Species: TABLE 59e. N_2 *laser*

Unidentified transitions (in N_2)[†]

Wavelength λ_{vac} (μm)	Notes	Excitation	References
5.35650 to 5.54628	24 lines	Pulsed	67
5.90723 to 6.24828	24 lines	Pulsed	67
6.47744 to 6.7441	16 lines	Pulsed	67
7.60821 to 8.06517	20 lines	Pulsed	67

[†] See ref. 142 for identification of these transitions.

Diatomic Species: TABLE 60. *NO laser*

Transitions in the $^2\pi_{1/2}$ and $^2\pi_{3/2}$ electronic ground states

Wavelength λ_{vac} (μm)	Transition	Excitation	References
	6–5 band		
5.8462	$^2\pi_{3/2}\ P(7)$	Pulsed	68
5.8549	$^2\pi_{1/2}\ P(8)$	Pulsed	68
5.8584	$^2\pi_{3/2}\ P(8)$	Pulsed	68
5.8706	$^2\pi_{3/2}\ P(9)$	Pulsed	68
5.8789	$^2\pi_{1/2}\ P(10)$	Pulsed	68
5.9036	$^2\pi_{1/2}\ P(12)$	Pulsed	68
5.9083	$^2\pi_{3/2}\ P(12)$	Pulsed	68
5.9550	$^2\pi_{1/2}\ P(16)$	Pulsed	69
	7–6 band		
5.9423	$^2\pi_{3/2}\ P(7)$	Pulsed	68
5.9546	$^2\pi_{3/2}\ P(8)$	Pulsed	68
5.9632	$^2\pi_{1/2}\ P(9)$	Pulsed	68, 69
5.9673	$^2\pi_{3/2}\ P(9)$	Pulsed	68
5.9756	$^2\pi_{1/2}\ P(10)$	Pulsed	68, 69

TABLE 60 *(cont.)*

Wavelength λ_{vac} (μm)	Transition	Excitation	References
5.9799	$^2\pi_{3/2}\ P(10)$	Pulsed	68
5.9882	$^2\pi_{1/2}\ P(11)$	Pulsed	68
5.9931	$^2\pi_{3/2}\ P(11)$	Pulsed	68
6.0010	$^2\pi_{1/2}\ P(12)$	Pulsed	68, 69
6.0054	$^2\pi_{3/2}\ P(12)$	Pulsed	68
6.0192	$^2\pi_{3/2}\ P(13)$	Pulsed	68
6.0267	$^2\pi_{1/2}\ P(14)$	Pulsed	68, 69
6.0324	$^2\pi_{3/2}\ P(14)$	Pulsed	68, 69
6.0402	$^2\pi_{1/2}\ P(15)$	Pulsed	68, 69
	8–7 band		
6.0386	$^2\pi_{1/2}\ P(7)$	Pulsed	69
6.0419	$^2\pi_{3/2}\ P(7)$	Pulsed	68
6.0543	$^2\pi_{3/2}\ P(8)$	Pulsed	68
6.0628	$^2\pi_{1/2}\ P(9)$	Pulsed	68, 69
6.0673	$^2\pi_{3/2}\ P(9)$	Pulsed	68
6.0801	$^2\pi_{3/2}\ P(10)$	Pulsed	68
6.0884	$^2\pi_{1/2}\ P(11)$	Pulsed	68, 69
6.0934	$^2\pi_{3/2}\ P(11)$	Pulsed	68, 69
6.1015	$^2\pi_{1/2}\ P(12)$	Pulsed	68
6.1204	$^2\pi_{3/2}\ P(13)$	Pulsed	68
6.1417	$^2\pi_{1/2}\ P(15)$	Pulsed	68, 69
6.1546	$^2\pi_{1/2}\ P(16)$	Pulsed	69
6.1973	$^2\pi_{1/2}\ P(19)$	Pulsed	69
	9–8 band		
6.1538	$^2\pi_{1/2}\ P(8)$	Pulsed	68
6.1576	$^2\pi_{3/2}\ P(8)$	Pulsed	68
6.1663	$^2\pi_{1/2}\ P(9)$	Pulsed	68
6.1792	$^2\pi_{1/2}\ P(10)$	Pulsed	68
6.1838	$^2\pi_{3/2}\ P(10)$	Pulsed	68
6.1921	$^2\pi_{1/2}\ P(11)$	Pulsed	68
6.1972	$^2\pi_{3/2}\ P(11)$	Pulsed	68
6.2055	$^2\pi_{1/2}\ P(12)$	Pulsed	68
6.2110	$^2\pi_{3/2}\ P(12)$	Pulsed	68
6.2191	$^2\pi_{1/2}\ P(13)$	Pulsed	68
6.2249	$^2\pi_{3/2}\ P(13)$	Pulsed	68
	10–9 band		
6.2381	$^2\pi_{3/2}\ P(6)$	Pulsed	68
6.2511	$^2\pi_{3/2}\ P(7)$	Pulsed	68
6.2602	$^2\pi_{1/2}\ P(8)$	Pulsed	68
6.2645	$^2\pi_{3/2}\ P(8)$	Pulsed	68
6.2778	$^2\pi_{3/2}\ P(9)$	Pulsed	68
6.2865	$^2\pi_{1/2}\ P(10)$	Pulsed	68
6.2913	$^2\pi_{3/2}\ P(10)$	Pulsed	68
6.2998	$^2\pi_{1/2}\ P(11)$	Pulsed	68
6.3051	$^2\pi_{3/2}\ P(11)$	Pulsed	68
6.3136	$^2\pi_{1/2}\ P(12)$	Pulsed	68
6.3191	$^2\pi_{3/2}\ P(12)$	Pulsed	68
6.3274	$^2\pi_{1/2}\ P(13)$	Pulsed	68
6.3334	$^2\pi_{3/2}\ P(13)$	Pulsed	68

TABLE 60 *(cont.)*

Wavelength λ_{vac} (μm)	Transition	Excitation	References
	11–10 band		
6.3764	$^2\pi_{3/2}\ P(8)$	Pulsed	68
6.3894	$^2\pi_{3/2}\ P(9)$	Pulsed	68
6.3980	$^2\pi_{1/2}\ P(10)$	Pulsed	68
6.4031	$^2\pi_{3/2}\ P(10)$	Pulsed	68
6.4262	$^2\pi_{1/2}\ P(12)$	Pulsed	68
6.4321	$^2\pi_{3/2}\ P(12)$	Pulsed	68

Diatomic Species: TABLE 61. *OD lines*
Pure rotational transitions

Wavelength λ_{vac} (μm)	Transition	Excitation	References
	$v = 0$		
18.121	$R_1\ (20)$	Pulsed	70
18.138	$R_2\ (30)$	Pulsed	70
18.590	$R_1'\ (29)$	Pulsed	70
18.603	$R_1\ (29)$	Pulsed	70
18.624	$R_2\ (29)$	Pulsed	70
19.102	$R_1'\ (28)$	Pulsed	70
19.121	$R_1\ (28)$	Pulsed	70
19.141	$R_2'\ (28)$	Pulsed	70
19.161	$R_2\ (28)$	Pulsed	70
19.662	$R_1'\ (27)$	Pulsed	70
19.681	$R_1\ (27)$	Pulsed	70
19.696	$R_2'\ (27)$	Pulsed	70
19.704	$R_2\ (27)$	Pulsed	70
20.271	$R_1'\ (26)$	Pulsed	70
20.288	$R_1\ (26)$	Pulsed	70
20.296	$R_2'\ (26)$	Pulsed	70
20.313	$R_2\ (26)$	Pulsed	70

Diatomic Species: TABLE 62. *OH laser*
Pure rotational transitions

Wavelength λ_{vac} (μm)	Transition	Excitation	References
	$v = 0$		
12.273	$R_1\ (24)$	Pulsed	70
12.279	$R_2\ (24)$	Pulsed	70
12.656	$R_1\ (23)$	Pulsed	70
12.663	$R_2\ (23)$	Pulsed	70
13.073	$R_2'\ (22)$	Pulsed	70
13.079	$R_1\ (22)$	Pulsed	70
13.088	$R_2\ (22)$	Pulsed	70
13.525	$R_1'\ (21)$	Pulsed	70
13.538	$R_2'\ (21)$	Pulsed	70
13.547	$R_1\ (21)$	Pulsed	70
13.557	$R_2\ (21)$	Pulsed	70

TABLE 62 (*cont.*)

Wavelength λ_{vac} (μm)	Transition	Excitation	References
14.043	R_1' (20)	Pulsed	70
14.059	R_2' (20)	Pulsed	70
14.067	R_1 (20)	Pulsed	70
14.081	R_2 (20)	Pulsed	70
14.620	R_1' (19)	Pulsed	70
14.640	R_2' (19)	Pulsed	70
14.646	R_1 (19)	Pulsed	70
14.662	R_2 (19)	Pulsed	70
15.289	R_2' (18)	Pulsed	70
15.294	R_1 (18)	Pulsed	70
15.313	R_2 (18)	Pulsed	70
18.788	R_1' (14)	Pulsed	70
18.828	R_1 (14)	Pulsed	70
18.849	R_2' (14)	Pulsed	70
18.878	R_2 (14)	Pulsed	70
	$v = 1$		
13.632	R_1 (22)	Pulsed	70
13.642	R_2 (22)	Pulsed	70
14.118	R_1 (21)	Pulsed	70
14.129	R_2 (21)	Pulsed	70
14.655	R_1 (20)	Pulsed	70
14.669	R_2 (20)	Pulsed	70
15.256	R_1 (19)	Pulsed	70
15.274	R_2 (19)	Pulsed	70
18.455	R_1' (15)	Pulsed	70
18.492	R_1 (15)	Pulsed	70
18.502	R_2' (15)	Pulsed	70
18.532	R_2 (15)	Pulsed	70
19.557	R_1' (14)	Pulsed	70
19.594	R_1 (14)	Pulsed	70
19.619	R_2' (14)	Pulsed	70
19.650	R_2 (14)	Pulsed	70
	$v = 2$		
19.273	R_1 (15)	Pulsed	70
19.321	R_2 (15)	Pulsed	70

Diatomic Species: TABLE 63. *Xe$_2$ laser*[a]

Wavelength λ_{vac} (μm)	Transition	Excitation	References
Band emission approximately 0.0017 μm wide at approximately 0.1716 μm	$Xe_2(^{1,3}\Sigma_{u,g}^+)$ to repulsive Xe_2 ground state	Pulsed; by short rise-time pulsed electron beam[b]	71–76, 143

[a] In liquid xenon, and high pressure xenon.

[b] Current densities up to 300 A/cm² at up to 1 MeV and current-pulse durations of 10 nsec.

Triatomic Species: TABLE 64a. CO_2 *laser*
Transitions in the (10^02)–(10^01) band[a]

Wavelength λ_{vac} (μm)	Transition	Excitation	References
4.3203	$R(17)$	Pulsed	77
4.3249	$R(13)$	Pulsed	77
4.3276	$R(11)$	Pulsed	77
4.3549	$P(7)$	Pulsed	77
4.3580	$P(9)$	Pulsed	77
4.3612	$P(11)$	Pulsed	77
4.3644	$P(13)$	Pulsed	77
4.3677	$P(15)$	Pulsed	77
4.3711	$P(17)$	Pulsed	77
4.3745	$P(19)$	Pulsed	77
4.3779	$P(21)$	Pulsed	77
4.3814	$P(23)$	Pulsed	77
4.3849	$P(25)$	Pulsed	77

[a] Observed in Q-switched CO_2 laser, oscillation at 10.6 μm necessary.

Triatomic Species: TABLE 64b. CO_2 *laser*
Transitions in the (00^01)–(02^00) band (R branch)

Wavelength λ_{vac} (μm)[b]	Transition	Excitation	References[a]
9.126866	$R(52)$	CW	
9.134184	$R(50)$	CW	
9.141719	$R(48)$	CW	
9.149471	$R(46)$	CW	
9.157446	$R(44)$	CW	
9.165645	$R(42)$	CW	
9.174070	$R(40)$	CW	
9.182725	$R(31)$	CW	
9.191612	$R(36)$	CW	
9.200733	$R(34)$	CW	
9.210092	$R(32)$	CW	
9.219690	$R(30)$	CW	
9.229530	$R(28)$	CW	
9.239615	$R(26)$	CW	
9.249946	$R(24)$	CW	
9.260526	$R(22)$	CW	
9.271358	$R(20)$	CW	
9.282444	$R(18)$	CW	
9.293786	$R(16)$	CW	
9.305386	$R(14)$	CW	
9.317246	$R(12)$	CW	
9.329370	$R(10)$	CW	
9.341758	$R(8)$	CW	
9.354414	$R(6)$	CW	
9.367339	$R(4)$	CW	

[a] See refs. 78 to 80 for reviews of the CO_2 laser, 81 to 90 for transversely excited high-pressure and e-beam excited CO_2 lasers, and 138 to 140 for CO_2 waveguide lasers.
[b] Wavelengths as in ref. 23.

Triatomic Species: TABLE 64c. *CO$_2$ laser*

Transitions in the (00^01)–(02^00) band (P branch)

Wavelength λ_{vac} $(\mu m)^b$	Transition	Excitation	References[a]
9.428885	$P(4)$	CW	
9.443328	$P(6)$	CW	
9.458052	$P(8)$	CW	
9.473060	$P(10)$	CW	
9.488355	$P(12)$	CW	
9.503937	$P(14)$	CW	
9.519808	$P(16)$	CW	
9.535972	$P(18)$	CW	
9.552428	$P(20)$	CW	
9.569179	$P(22)$	CW	
9.586227	$P(24)$	CW	
9.603573	$P(26)$	CW	
9.621219	$P(28)$	CW	
9.639166	$P(30)$	CW	
9.657416	$P(32)$	CW	
9.675971	$P(34)$	CW	
9.694831	$P(36)$	CW	
9.713998	$P(38)$	CW	
9.733474	$P(40)$	CW	
9.753259	$P(42)$	CW	
9.773356	$P(44)$	CW	
9.793764	$P(46)$	CW	
9.814487	$P(48)$	CW	
9.835523	$P(50)$	CW	
9.856876	$P(52)$	CW	
9.878544	$P(54)$	CW	
9.900531	$P(56)$	CW	
9.922835	$P(58)$	CW	
9.945458	$P(60)$	CW	

[a] See ref. 78 to 80 for reviews of the CO_2 laser, 81 to 90 for transversely excited high pressure CO_2 lasers, and 138 to 140 for waveguide CO_2 lasers.

[b] Wavelengths as in ref. 23.

Triatomic Species: TABLE 64d. *CO$_2$ laser*

Transitions in the (00^01)–(10^00) band (R branch)

Wavelength λ_{vac} $(\mu m)^b$	Transition	Excitation	References[a]
10.057895	$R(54)$	CW	
10.066650	$R(52)$	CW	
10.075698	$R(50)$	CW	
10.085041	$R(48)$	CW	
10.094676	$R(46)$	CW	
10.104605	$R(44)$	CW	
10.114826	$R(42)$	CW	

TABLE 64d *(cont.)*

Wavelength λ_{vac} (μm)[b]	Transition	Excitation	References[a]
10.125340	$R(40)$	CW	
10.136146	$R(38)$	CW	
10.147246	$R(36)$	CW	
10.158637	$R(34)$	CW	
10.170323	$R(32)$	CW	
10.182301	$R(30)$	CW	
10.194574	$R(28)$	CW	
10.207142	$R(26)$	CW	
10.220006	$R(24)$	CW	
10.233167	$R(22)$	CW	
10.246625	$R(20)$	CW	
10.260381	$R(18)$	CW	
10.274438	$R(16)$	CW	
10.288797	$R(14)$	CW	
10.303458	$R(12)$	CW	
10.318424	$R(10)$	CW	
10.336696	$R(8)$	CW	
10.349277	$R(6)$	CW	
10.365168	$R(4)$	CW	

[a] See refs. 78 to 80 for reviews of the CO_2 laser, 81 to 90 for transversely excited high pressure, and e-beam excited CO_2 lasers, and 138 to 140 for waveguide CO_2 lasers.
[b] Wavelengths as in ref. 23.

Triatomic Species: TABLE 64e. CO_2 *laser*

Transitions in the (00^01)–(10^00) band (P branch)

Wavelength λ_{vac} (μm)[b]	Transition	Excitation	References[a]
10.440579	$P(4)$	CW	
10.458220	$P(6)$	CW	
10.476187	$P(8)$	CW	
10.494484	$P(10)$	CW	
10.513114	$P(12)$	CW	
10.532080	$P(14)$	CW	
10.551387	$P(16)$	CW	
10.571037	$P(18)$	CW	
10.591035	$P(20)$	CW	
10.611385	$P(22)$	CW	
10.632090	$P(24)$	CW	
10.653156	$P(26)$	CW	
10.674586	$P(28)$	CW	
10.696386	$P(30)$	CW	
10.718560	$P(32)$	CW	
10.741113	$P(34)$	CW	
10.764052	$P(36)$	CW	

TABLE 64e *(cont.)*

Wavelength λ_{vac} (μm)[b]	Transition	Excitation	References[a]
10.787380	*P*(38)	CW	
10.811105	*P*(40)	CW	
10.835231	*P*(42)	CW	
10.859765	*P*(44)	CW	
10.884713	*P*(46)	CW	
10.910087	*P*(48)	CW	
10.935879	*P*(50)	CW	
10.962110	*P*(52)	CW	
10.988783	*P*(54)	CW	
11.015906	*P*(56)	CW	

[a] See refs. 78 to 80 for reviews of the CO_2 laser, 81 to 90 for transversely excited, high-pressure, and e-beam excited CO_2 lasers, and 138 to 140 for waveguide CO_2 lasers.

[b] Wavelengths as in ref. 23.

Triatomic Species: TABLE 64f. *CO_2 laser*

Transitions in the (01^11)–(11^10) band

Wavelength λ_{vac} (μm)[a]	Transition	Excitation	References
10.9730	*P*(19)	Pulsed	91, 92
10.9856	*P*(20)	Pulsed	91, 92
10.9944	*P*(21)	Pulsed	91, 92
11.0078	*P*(22)	Pulsed	91, 92
11.0164	*P*(23)	Pulsed	91, 92
11.0300	*P*(24)	Pulsed	91, 92
11.0385	*P*(25)	Pulsed	91, 92
11.0529	*P*(26)	Pulsed	91, 92
11.0610	*P*(27)	Pulsed	91, 92
11.0762	*P*(28)	Pulsed	91, 92
11.0840	*P*(29)	Pulsed	91, 92
11.0999	*P*(30)	Pulsed	91, 92
11.1073	*P*(31)	Pulsed	91, 92
11.1238	*P*(32)	Pulsed	91
11.1309	*P*(33)	Pulsed	91
11.1483	*P*(34)	Pulsed	91

[a] Wavelengths as in ref. 23.

Triatomic Species: TABLE 64g. CO_2 *laser*

Transitions in the $(01^01)–(03^10)$ band

Wavelength λ_{vac} (µm)[b]	Transition	Excitation	References[a]
10.9735	P(19)	CW	
10.9951	P(21)	CW	
11.0165	P(23)	CW	
11.0300	P(24)	CW	
11.0385	P(25)	CW	
11.0535	P(26)	CW	
11.0610	P(27)	CW	
11.0760	P(28)	CW	
11.0850	P(29)	CW	
11.1000	P(30)	CW	
11.1070	P(31)	CW	
11.1235	P(32)	CW	
11.1315	P(33)	CW	
11.1485	P(34)	CW	
11.1555	P(35)	CW	
11.1736	P(36)	CW	
11.1791	P(37)	CW	
11.1980	P(38)	CW	
11.2035	P(39)	CW	
11.2235	P(40)	CW	
11.2295	P(41)	CW	
11.2495	P(42)	CW	
11.2545	P(43)	CW	
11.2770	P(44)	CW	
11.2804	P(45)	CW	

[a] Reference 93.
[b] Wavelengths as in ref. 23.

Triatomic Species: TABLE 64h. CO_2^{18} *laser*

Transitions in the $(00^01)–(10^00)$ band

Wavelength λ_{vac} (µm)[a]	Transition	Excitation	References
9.341	P(18)	CW	94
9.355	P(20)	CW	94
9.369	P(22)	CW	94
9.383	P(24)	CW	94
9.397	P(26)	CW	94

[a] As in ref. 23.

Triatomic Species: TABLE 64i. CO_2 *laser*

Miscellaneous laser lines

Wavelength λ_{vac} (μm)	Transition[b]	Excitation	References[a]
13.144	Q branch (14^00)–(05^10)?	Pulsed	
13.154	—	Pulsed	
13.159	—	Pulsed	
13.541	Q branch (21^10)–(12^20)?	Pulsed	
16.585	Q branch (14^00)–(13^10)?	Pulsed	
16.597	—	Pulsed	
17.023	—	Pulsed	
17.029	Q branch (03^11)–(02^21)?	Pulsed	
17.036	—	Pulsed	
17.048	—	Pulsed	
17.370	—	Pulsed	
17.376	Q branch (24^00)–(23^10)?	Pulsed	
17.390	—	Pulsed	

[a] See ref. 92.
[b] As in ref. 23.

Triatomic Species: TABLE 65. CS_2 *laser*

Transitions in the (02^01)–(12^00) band

Wavelength λ_{vac} (μm)	Transition[a]	Excitation	References
11.4283	$P(28)$	CW	95
11.4893[b]	$P(30)$	CW	95
11.5962	$P(32)$	CW	95
11.5031	$P(34)$	CW	95
11.5099	$P(36)$	CW	95
11.5166	$P(38)$	CW	95
11.5237	$P(40)$	CW	95
11.5307	$P(42)$	CW	95
11.5376	$P(44)$	CW	95
11.5446	$P(46)$	CW	95

[a] Identification as per N. Legay-Sommaire, *Appl. Phys. Letters* **12,** 34 (1968), previously made by A. G. Maki, *Appl. Phys. Letters* **11,** 204 (1967).
[b] Strongest line, $P \geqslant 10$ mW.

Triatomic Species: TABLE 66. D_2O laser

Wavelength λ_{vac} (µm)[a]	Transition	Excitation	References
26.36	—	Pulsed	96
33.896	$(100)-(020)\ 13_{59}-12_{66}(?)$	Pulsed	97, 98
35.081	$(100)-(020)\ 12_{58}-11_{65}(?)$	Pulsed	97, 98
36.096	$(100)-(020)\ 11_{56}-10_{65}$	Pulsed	
36.324	$(100)-(020)\ 11_{57}-10_{64}$	Pulsed	97, 98
36.526	$(020)-(020)\ 11_{75}-10_{64}$	Pulsed	97, 98
37.788	$(100)-(020)\ 10_{55}-9_{64}$	Pulsed	97, 98
37.860	$(100)-(020)\ 10_{56}-9_{63}$	Pulsed	98, 99
39.53	$(100)-(020)\ 9_{55}-8_{64}$	Pulsed	96
40.994	$(100)-(020)\ 10_{64}-9_{55}$	Pulsed	97
41.79	$(100)-(020)\ 12_{66}-11_{75}$	Pulsed	99, 100
48.80	—	Pulsed	96
50.71	—	Pulsed	96
54.73	—	Pulsed	96
56.830	$(100)-(020)\ 16_{0,16}-15_{3,13}$	Pulsed	97, 98
61.182	—	Pulsed	99, 100
71.944	$(100)-(020)\ 11_{57}-11_{66}$	CW	97, 98
72.427	$(020)-(020)\ 10_{74}-10_{65}$	Pulsed	97, 98
	or $10_{73}-10_{44}$	Pulsed	97, 98
72.757	$(020)-(020)\ 11_{75}-11_{66}$	Pulsed	97, 98
73.341	$(100)-(020)\ 12_{58}-12_{67}$	Pulsed	97, 98
74.526	$(100)-(020)\ 13_{59}-13_{68}$	Pulsed	97, 98
76.305	—	Pulsed	97
78.16	—	Pulsed	96
83.730	$(020)-(020)\ 10_{65}-10_{56}$	Pulsed	99, 100
84.111	$(020)-(020)\ 11_{66}-11_{57}$	Pulsed	97
84.284	$(100)-(020)\ 12_{1,12}-11_{47}$	CW	97, 98
99.00	—	Pulsed	96
103.33	—	Pulsed	96
107.731	$(100)-(020)\ 11_{66}-11_{75}$	CW	97, 98
107.91	$(100)-(100)\ 13_{68}-13_{59}$	Pulsed	99, 100
108.88	$(100)-(020)\ 11_{65}-11_{74}$	Pulsed	96
110.49	$(100)-(020)\ 12_{66}-12_{75}$	Pulsed	99, 100
111.74	$(100)-(020)\ 13_{68}-13_{77}$	Pulsed	99, 100
170.08	$(020)-(020)\ 11_{47}-11_{38}$	Pulsed	98
171.67	$(100)-(020)\ 11_{0,11}-11_{38}$	CW	98
218.5	—	Pulsed	102

[a] Wavelength as in ref. 23.

APPENDIX

Triatomic Species: TABLE 67. *DCN laser*

Wavelength λ_{vac} (μm)	Transition	Excitation	References
181.789	(22^00)–(22^00) $R(20)$	Pulsed	103
189.9490	(22^00)–$(09^{1c}0)$ $R(21)$	CW	103
190.0080	(22^00)–(22^00) $R(21)$	CW	104
194.7027	(22^00)–$(09^{1c}0)$ $R(20)$	CW	103
194.7644	$(09^{1c}0)$–$(09^{1c}0)$ $R(20)$	CW	104
204.3872	$(09^{1c}0)$–$(09^{1c}0)$ $R(19)$	CW	103

Triatomic Species: TABLE 68a. *H_2O laser*

Wavelength λ_{vac} (μm)[a]	Transition	Excitation	References[b]
2.279	—	Pulsed(?)	66
4.77	—	Pulsed	106
7.458	(020)–(010) $4_{41}-5_{50}$	Pulsed	107
7.596	(020)–(010) $6_{50}-7_{61}$	Pulsed	107, 63
7.7097	(020)–(010) $6_{61}-7_{40}$	Pulsed	107, 66, 106,
9.3938	—	Pulsed	106, 63
9.4747	—	Pulsed	106, 63
9.5674	—	Pulsed	106, 63
11.83	—	Pulsed	106
11.96	—	Pulsed	106
16.932	(010)–(010) $13_{11,2}-12_{10,3}$	Pulsed	97, 98
23.13	(020)–(020) $9_{63}-8_{54}$	Pulsed	96, 66
23.365	(100)–(020) $9_{45}-8_{54}$	Pulsed	97
24.966	(100)–(020) $8_{40}-7_{53}$	Pulsed	66
26.660	(100)–(020) $7_{43}-6_{52}$	Pulsed	97, 98, 105
27.9707	(001)–(020) $6_{33}-5_{50}$	CW	97, 105, 108, 109, 63
28.054	(020)–(020) $6_{61}-5_{50}$	Pulsed	97
28.270	(100)–(020) $8_{08}-7_{35}$	Pulsed	97
28.356	(020)–(020) $8_{44}-7_{35}$	Pulsed	97
28.451	(100)–(020) $6_{42}-5_{51}$	Pulsed	66
32.924	(020)–(020) $5_{50}-4_{41}$	Pulsed	97
33.029	(100)–(020) $5_{14}-4_{41}$	CW	97, 98, 105
34.60	—	Pulsed	96
35.017	(100)–(100) $7_{34}-6_{25}$	Pulsed	97, 98
35.833	(100)–(202) $12_{1,12}-11_{29}$	Pulsed	97, 98
36.606	(010)–(010) $13_{11,2}-13_{10,3}$	Pulsed	97, 98
37.848	(100)–(020) $12_{0,12}-11_{39}$	Pulsed	97, 98
38.086	—	Pulsed	97, 98
39.695	(001)–(020) $7_{44}-6_{61}$	Pulsed	97, 98
40.45	(100)–(020) $13_{1,13}-12_{2,10}$	Pulsed	66
40.638	(020)–(020) $4_{41}-3_{30}$	Pulsed	97, 98
42.51	—	Pulsed	96
45.517	—	Pulsed	97, 98
45.91	—	Pulsed	96
47.244	(020)–(020) $9_{63}-9_{54}$	CW	97, 98
47.39	—	Pulsed	66

TABLE 68a *(cont.)*

Wavelength λ_{vac} (μm)[a]	Transition	Excitation	References[b]
47.468	(001)–(020) 6_{33}–6_{52}	CW	97, 98
47.687	(020)–(020) 6_{61}–6_{52}	CW	97, 98
48.19	(100)–(020) 9_{45}–9_{54}	Pulsed	96
48.676	—	Pulsed	97, 98
49.06	(100)–(020) 7_{43}–7_{52}	Pulsed	96, 66
53.910	—	Pulsed	97, 98
54.853	—	Pulsed	66
55.000	(020)–(020) 6_{52}–6_{43}	Pulsed	66
55.088	(020)–(020) 5_{50}–5_{41}	CW	97, 98
57.659	(100)–(020) 9_{19}–8_{44}	Pulsed	97, 98
57.799	(020)–(020) 9_{54}–9_{45}	Pulsed	66
66.880	—	Pulsed	66
66.903	—	Pulsed	66
67.169	(020)–(020) 4_{41}–4_{32} or (100)–(020) 6_{25}–5_{50}	Pulsed	97, 98
68.344	(020)–(020) 5_{41}–5_{32}	Pulsed	66
72.856	(100)–(020) 8_{17}–8_{44}	Pulsed	66, 98
73.401	(100)–(100) 8_{17}–8_{08}	Pulsed	97, 98
78.443	(100)–(020) 8_{08}–8_{35}	CW	97
79.087	(020)–(020) 8_{44}–8_{35}	CW	97
85.564	(100)–(020) 7_{52}–7_{61}	Pulsed	66, 98
86.478	(100)–(100) 9_{54}–9_{45}	Pulsed	66
87.323	—	Pulsed	66
87.469	(100)–(020) 8_{53}–8_{62}	Pulsed	66
89.772	(100)–(020) 9_{54}–9_{63}	Pulsed	98
89.947	—	Pulsed	66
90.565	—	Pulsed	66
115.32	(020)–(020) 8_{35}–8_{26}	CW	98
118.591	(001)–(020) 6_{42}–6_{61}	CW	98
120.08	(001)–(001) 6_{42}–6_{33}	Pulsed	98
220.230	(100)–(020) 5_{23}–5_{50}	CW	98, 101
791.0	—	Pulsed	101

[a] As in ref. 23.
[b] See references given in Section 6.2.5, "The Far Infrared H_2O Laser".

Triatomic Species: TABLE 68b. H_2O^{18} *laser*

Wavelength λ_{vac} (μm)[a]	Transition	Excitation	References
25.162	(100)–(020) 8_{45}–7_{52}	Pulsed	98
25.595	(100)–(020) 7_{44}–6_{51}	Pulsed	98
28.295	(100)–(020) 6_{43}–5_{50}	Pulsed	98
33.308	(020)–(020) 5_{50}–4_{41}	Pulsed	98
35.383	(100)–(020) $12_{1,12}$–11_{29}(?)	Pulsed	98

TABLE 68b *(cont.)*

Wavelength λ_{vac} (μm)	Transition	Excitation	References
48.366	(100)–(020) 6_{43}–6_{52}	Pulsed	98
48.604	(020)–(020) 6_{61}–6_{52}	Pulsed	98
48.765	(100)–(020) 7_{44}–7_{53}	Pulsed	98
49.430	(100)–(020) 8_{45}–8_{54}	Pulsed	98
56.129	(020)–(020) 5_{50}–5_{41}	Pulsed	98

[a] As in ref. 23.

Triatomic Species: TABLE 69. H_2S *laser*

Wavelength λ_{vac} (μm)	Transition	Excitation	References[a]
33.47	—	Pulsed	
33.64	—	Pulsed	
49.62	—	Pulsed	
52.40	—	Pulsed	
56.84	—	Pulsed	
60.29	—	Pulsed	
61.50[b]	—	Pulsed	
73.52	—	Pulsed	
80.50	—	Pulsed	
83.43	—	Pulsed	
87.47[b]	—	Pulsed	
92.00	—	Pulsed	
96.38	—	Pulsed	
103.3	—	Pulsed	
108.8	—	Pulsed	
116.8	—	Pulsed	
126.2	—	Pulsed	
129.1	—	Pulsed	
130.8	—	Pulsed	
135.5	—	Pulsed	
140.6	—	Pulsed	
162.4	—	Pulsed	
192.9	—	Pulsed	
225.3[a]	—	Pulsed	

[a] Ref. 110.
[b] Lines with equal and largest relative strength.

Triatomic Species: TABLE 70a. *HCN laser*

Wavelength λ_{vac} (μm)[a]	Transition	Excitation	References[b]
12.85	—	Pulsed	106
71.899	—	Pulsed	103, 111
73.101	—	Pulsed	103, 111
76.093	—	Pulsed	103, 111
77.001	—	Pulsed	103, 111
81.554	—	Pulsed	103, 111
96.401	—	Pulsed	103, 111
98.693	—	Pulsed	103, 111
101.257	—	Pulsed	103, 111
112.066	—	Pulsed	103, 111
116.132	—	Pulsed	103
126.164	$(12^{2d}0)-(95^{1d}0)$ $R(26)$	Pulsed	103, 112, 113, 104
128.629	$(12^{2d}0)-(05^{1d}0)$ $R(25)$	CW	,,
130.838	$(12^00)-(05^{1c}0)$ $R(25)$	Pulsed	,,
134.932	$(12^00)-(05^{1c}0)$ $R(24)$	Pulsed	,,
201.059	—	Pulsed	103, 112
211.001	—	CW	103, 112
222.949	—	Pulsed	103, 112
284	$(11^{1c}0)-(11^{1c}0)$ $R(11)$	CW	115
291	—	Pulsed	114
309.7140	$(11^{1c}0)-(11^{1c}0)$ $R(10)$	Pulsed	103, 104, 112, 113, 115
310.8870	$(11^{1c}0)-(04^00)$ $R(10)$	CW	103, 104, 112–16
335.1831	$(04^00)-(04^00)$ $R(9)$	CW	104, 113, 115
336.5578	$(11^{1c}0)-(04^00)$ $R(9)$	Pulsed	103, 104, 112–16
372.5283	$(04^00)-(04^00)$ $R(8)$	CW	103, 104, 112–15
469	—	Pulsed	114
538.2	—	Pulsed	103, 111, 116
545.4	—	Pulsed	103, 111
676	—	Pulsed	103, 111
773.5	—	Pulsed	103, 111, 116

[a] Wavelengths as in ref. 23.
[b] See also references given in Section 6.4.1, "The HCN (and DCN) submillimeter lasers".

Triatomic Species: TABLE 70b. *HCN[15] laser*

Wavelength λ_{vac} (μm)[a]	Transition	Excitation	References
110.240	—	Pulsed	103
113.311	—	Pulsed	103
138.768	—	Pulsed	103
165.150	—	Pulsed	103

[a] As given in ref. 23.

Triatomic Species: TABLE 71. *ICN laser*

Wavelength λ_{air} (μm)[a]	Transition	Excitation	References
538	—	Pulsed	116, 117
676	—	Pulsed	117
773.5 ± 1.0	—	Pulsed	116, 117

[a] It is possible that these lines are HCN lines (see Table 6.18, HCN laser).

Triatomic Species: TABLE 72a. N_2O *laser*

Transitions in the (00^01)–(10^00) band (*R* branch)

Wavelength λ_{vac} (μm)[a]	Transition	Excitation	References
10.3456	*R*(35)	CW	118
10.3532	*R*(34)	CW	118
10.3609	*R*(33)	CW	118
10.3687	*R*(32)	CW	118
10.3765	*R*(31)	CW	118
10.3843	*R*(30)	CW	118
10.3922	*R*(29)	CW	118
10.4001	*R*(28)	CW	118
10.4081	*R*(27)	CW	118
10.4161	*R*(26)	CW	118
10.4242	*R*(25)	CW	118
10.4323	*R*(24)	CW	118
10.4405	*R*(23)	CW	118
10.4487	*R*(22)	CW	118
10.4570	*R*(21)	CW	118
10.4653	*R*(20)	CW	118, 119
10.4737	*R*(19)	CW	118, 119
10.4821	*R*(18)	CW	118, 119, 63
10.4906	*R*(17)	CW	118, 119, 63
10.4991	*R*(16)	CW	118, 119, 63
10.5077	*R*(15)	CW	118, 119, 63
10.5163	*R*(14)	CW	118, 119, 63
10.5250	*R*(13)	CW	118, 119, 63
10.5337	*R*(12)	CW	118, 119, 63
10.5425	*R*(11)	CW	118, 119, 63
10.5513	*R*(10)	CW	118, 63
10.5602	*R*(9)	CW	118
10.5692	*R*(8)	CW	118
10.5781	*R*(7)	CW	118
10.5872	*R*(6)	CW	118
10.5963	*R*(5)	CW	118
10.6054	*R*(4)	CW	118
10.6146	*R*(3)	CW	118
10.6239	*R*(2)	CW	118
10.6332	*R*(1)	CW	118
10.6426	*R*(0)	CW	120

[a] As in ref. 23.

Triatomic Species: TABLE 72b. *N_2O laser*

Transitions in the (00^01)–(10^00) band (P branch)

Wavelength λ_{vac} (μm)[a]	Transition	Excitation	References
10.6614	$P(1)$	CW	120
10.6710	$P(2)$	CW	118
10.6806	$P(3)$	CW	118
10.6903	$P(4)$	CW	118
10.6999	$P(5)$	CW	118
10.7097	$P(6)$	CW	118
10.7195	$P(7)$	CW	118
10.7294	$P(8)$	CW	118
10.7393	$P(9)$	CW	118
10.7493	$P(10)$	CW	118
10.7593	$P(11)$	CW	118
10.7694	$P(12)$	CW	118, 121, 63
10.7796	$P(13)$	CW	118, 121, 63
10.7898	$P(14)$	CW	118, 121, 63
10.8000	$P(15)$	CW	118, 121, 63
10.8104	$P(16)$	CW	118, 121, 63
10.8208	$P(17)$	CW	118, 121, 63
10.8312	$P(18)$	CW	118, 121, 63
10.8418	$P(19)$	CW	118, 121, 63
10.8523	$P(20)$	CW	118, 63, 121, 122
10.8629	$P(21)$	CW	”
10.8736	$P(22)$	CW	118, 121, 122
10.8844	$P(23)$	CW	118, 121, 122
10.8952	$P(24)$	CW	118, 121, 122
10.9061	$P(25)$	CW	118, 121, 122
10.9170	$P(26)$	CW	118, 121, 122
10.9280	$P(27)$	CW	118, 121, 122
10.9390	$P(28)$	CW	118, 121, 122
10.9501	$P(29)$	CW	118, 121
10.9613	$P(30)$	CW	118, 121
10.9726	$P(31)$	CW	118, 121
10.9839	$P(32)$	CW	118, 121
10.9953	$P(33)$	CW	118, 121
11.0067	$P(34)$	CW	118, 121
11.0182	$P(35)$	CW	118, 121
11.0298	$P(36)$	CW	118, 121
11.0415	$P(37)$	CW	118, 121

[a] As in ref. 23.

APPENDIX

Triatomic Species: TABLE 73a. *OCS laser*

Transitions in the (00^01)–(10^00) band

Wavelength λ_{vac} (μm)[a]	Transition	Excitation	References[b]
8.2388	$R(26)$	Pulsed	
8.2416	$R(25)$	Pulsed	
8.2439	$R(24)$	Pulsed	
8.2518	$R(21)$	Pulsed	
8.2543	$R(20)$	Pulsed	
8.2571	$R(19)$	Pulsed	
8.2595	$R(18)$	Pulsed	
8.2623	$R(17)$	Pulsed	
8.2645	$R(16)$	Pulsed	
8.2673	$R(15)$	Pulsed	
8.3625	$P(18)$	Pulsed	
8.3654	$P(19)$	Pulsed	
8.3685	$P(20)$	Pulsed	
8.3715	$P(21)$	Pulsed	
8.3746	$P(22)$	Pulsed	
8.3779	$P(23)$	Pulsed	
8.3809	$P(24)$	Pulsed	
8.3839	$P(25)$	Pulsed	
8.3870	$P(26)$	Pulsed	
8.3900	$P(27)$	Pulsed	
8.3930	$P(28)$	Pulsed	
8.3962	$P(29)$	Pulsed	
8.3999	$P(30)$	Pulsed	
8.4024	$P(31)$	Pulsed	
8.4055	$P(32)$	Pulsed	
8.4085	$P(33)$	Pulsed	
8.4117	$P(34)$	Pulsed	
8.4146	$P(35)$	Pulsed	
8.4178	$P(36)$	Pulsed	
8.4213	$P(37)$	Pulsed	
8.4243	$P(38)$	Pulsed	

[a] As given in ref. 23.
[b] Ref. 123.

Triatomic Species: TABLE 73b. *OCS laser*

Unidentified laser transitions

Wavelength λ_{vac} (μm)	Transition	Excitation	References
123	—	Pulsed	110
132	—	Pulsed	110

Triatomic Species: TABLE 74. *SO$_2$ laser*

Wavelength λ_{vac} (μm)	Transition	Excitation	References
140	$2v_2 27_{16} - 2v_2 26_{15}$	Pulsed	124
141	$v_1 27_{14} - 2v_2 26_{15}$	CW	124, 125
141′	$v_1 26_{14} - 2v_2 25_{15}$	Pulsed	124
142	$2v_2 26_{16} - 2v_2 25_{15}$	Pulsed	124
150	$v_1 28_{15} - v_1 27_{14}$	Pulsed	124
151	$v_1 28_{15} - 2v_2 27_{16}$	Pulsed	110, 124, 125
151′	$v_1 27_{15} - 2v_2 26_{16}$	Pulsed	110, 124, 125
152	$v_1 27_{15} - 2v_2 25_{15}$	Pulsed	124
193	$v_1 28_{16} - v_1 28_{15}$	CW	124, 125, 126
206	$v_1 26_{15} - v_1 26_{14}$	Pulsed	124
215	—	Pulsed	110, 124, 125

Polyatomic Species: TABLE 75. *BCl$_3$ laser*[a]

Wavelength (μm)	Transition $B^{10}Cl_3$	$B^{11}Cl_3$	Excitation	References
18.8	In $B^{10}Cl_3$	—	CW; optically pumped by 10.6-μm radiation in a discharge in CO_2 and BCl_3 mixture	127
19.1	In $B^{10}Cl_3$		”	127
19.4	In $B^{10}Cl_3$		”	127
22.0	—	In $B^{11}Cl_3$	”	127
22.4	In $B^{10}Cl_3$	—	”	127
23.0	—	In $B^{11}Cl_3$	”	127

[a] 100 mW at 20 μm obtained.

Polyatomic Species: TABLE 76. *CBrF$_3$ laser*

Wavelength λ_{air} (μm)	Transition	Excitation	References
1.975	—	CW	128
10.604	—	CW	128

Polyatomic Species: TABLE 77. *CCl$_2$F$_2$ laser*

Wavelength λ_{air} (μm)	Transition	Excitation	References
1.589	—	CW	128
1.977	—	CW	128
2.021	—	CW	128
2.499	—	CW	128
2.535	—	CW	128
2.602	—	CW	128
2.784	—	CW	128
3.801	—	CW	128

Polyatomic Species: TABLE 78. CH_3CCH

Wavelength λ_{vac} (μm)	Transition	Excitation	References
427.89	—	CW; optically pumped by a 10.49448-μm CO_2 laser	129
488.88	—	CW; optically pumped by a 10.51311-μm CO_2 laser	129
563.13	—	CW; optically pumped by a 10.63209-μm CO_2 laser	129
647.89	—	CW; optically pumped by a 10.53208-μm CO_2 laser	129
649.59	—	CW; optically pumped by a 10.74111-μm CO_2 laser	129
757.41	—	CW; optically pumped by a 10.49448-μm CO_2 laser	129
798.55	—	CW; optically pumped by a 10.59104-μm CO_2 laser	129
1174.87	—	CW; optically pumped by a 10.85977-μm CO_2 laser	129

Polyatomic Species: TABLE 79. CH_3CN

Wavelength λ_{vac} (μm)	Transition	Excitation	References
303.54	—	CW; optically pumped by 10.49448-μm CO_2 laser	129
372.87	—	CW; optically pumped by 10.59104-μm CO_2 laser	129
380.71	—	CW; optically pumped by 10.55139-μm CO_2 laser	129
422.14	—	CW; optically pumped by 10.63209-μm CO_2 laser	129
430.55	—	CW; optically pumped by 10.57104-μm CO_2 laser	129
713.72	—	CW; optically pumped by 10.71856-μm CO_2 laser	129
1814.37	—	CW; optically pumped by 10.88471-μm CO_2 laser	129

APPENDIX

Polyatomic Species: Table 80a. CH_3F laser

Pure rotational transitions

Wavelength λ_{vac} (μm)	Transition	Excitation	References
	$\nu = 0$		
451.903	$R(12), K = 1$	Pulsed; optically pumped by a Q-switched 9.5524-μm CO_2 laser	130
451.924	$R(12), K = 2$	Pulsed	130
	$\nu = 3$		
496.072[a]	$R(10), K = 1$	CW	130, 131, 141
496.105[a]	$R(10), K = 2$	CW	130, 131, 141
541.113	$R(11), K = 1$	Pulsed	130
541.147	$R(11), K = 2$	Pulsed	130

[a] CW operation achieved using optical pumping by a 10-W, 9.5524-μm CW CO_2 laser.[129]

Polyatomic Species: Table 80b. $C^{12}H_3F$ laser

Wavelength λ_{vac} (μm)	Transition	Excitation	References
192.78	—	CW; optically pumped by 10.17032-μm CO_2 laser	129
199.14	—	CW; cascade pumped by the 192.78-μm line	129
251.91	—	CW; optically pumped by 10.15864-μm CO_2 laser	129
372.68	—	CW; optically pumped by 9.83552-μm CO_2 laser	129
397.51	—	CW; cascade pumped by the 372.68-μm line	129

Polyatomic Species: Table 80c. $C^{13}H_3F$ laser

Wavelength λ_{vac} (μm)	Transition	Excitation	References
1221.79	—	CW; optically pumped by 9.65742-μm CO_2 laser	129

512

Polyatomic Species: TABLE 81. *CH₃OH laser*

Pure rotational transitions (in C–O(ν_5) stretching mode)

Wavelength λ_{vac} (μm)	Transition	Excitation	References
70.6[a]	—	CW	131
118.8[b]	—	CW	131
164.3[c]	—	CW	131
170.6[b]	—	CW	131
185.5[a]	—	CW	131
190.8[a]	—	CW	131
193.2[d]	—	CW	131
198.8[d]	—	CW	131
202.4[b]	—	CW	131
223.5[c]	—	CW	131
237.6[a]	—	CW	131
253.6[a]	—	CW	131
254.1[a]	—	CW	131
263.7[a]	—	CW	131
264.6[a]	—	CW	131
278.8[d]	—	CW	131
292.2[d]	—	CW	131
292.5[a]	—	CW	131
369.1[c]	—	CW	131
392.3[b]	—	CW	131
417.8[b]	—	CW	131
570.5[c]	—	CW	131
699.5[a]	—	CW	131

[a] Optically pumped by a 10-W, CW, 9.6760-μm CO_2 laser.
[b] Optically pumped by a 10-W, CW, 9.6948-μm CO_2 laser.
[c] Optically pumped by a 10-W, CW, 9.518-μm CO_2 laser.
[d] Optically pumped by a 10-W, CW, 9.7140-μm CO_2 laser.

Polyatomic Species: TABLE 82. *H₂C : CHCl laser*

Wavelength λ_{vac} (μm)	Transition	Excitation	References
386.0	CH wagging	CW; optically pumped by a 10-W, 10.6114-μm CO_2, CW laser	131
507.7	CH wagging	"	131
634.4	CH_2 rocking	CW; optically pumped by a 10-W, 9.5524-μm CO_2, CW laser	131

Polyatomic Species: TABLE 83. H_2CO laser[a]

Wavelength λ_{vac} (μm)	Transition	Excitation	References
102[b]	—	Pulsed	132
119	—	Pulsed	132
122	—	Pulsed	132
159[c]	—	Pulsed	132

[a] Observed in pulsed discharge through formaldehyde vapor (H_2CO) at 0.05 to 0.4 torr; $D = 27$ cm.

[b] Weakest line.

[c] The output power (highest of the four lines) was nearly equal to that of the 118-μm H_2O laser line also observed.

Polyatomic Species: TABLE 84a. NH_3 laser

Transitions in the $(03^a0)–(02^a0)$ band

Wavelength λ_{vac} (μm)[a]	Transition	Excitation	References
14.78	—	Pulsed	133
15.04	—	Pulsed	133
15.08	—	Pulsed	133
15.41	—	Pulsed	133
15.47	—	Pulsed	133
18.21	—	Pulsed	133, 134
21.471	$P(2)1_0$	Pulsed	135, 133
22.542	$P(3)2_2$	Pulsed	135, 133
22.563	$P(3)2_1$	Pulsed	135
22.71	—	Pulsed	133
23.675	$P(4)3_0$ or 3_1	Pulsed	131, 133
23.86	—	Pulsed	133
24.918	$P(5)4_1$	Pulsed	133
25.12	—	Pulsed	133
26.282	$P(6)5_0$ or 5_1	Pulsed	133
30.69	—	Pulsed	133
31.47	—	Pulsed	133
31.951	—	Pulsed	133
32.13	—	Pulsed	133

[a] As given in ref. 23.

Polyatomic Species: TABLE 84b. NH_3 laser

Transition in the $(0, 1, 0, 0)$ band

Wavelength λ_{vac} (μm)	Transition	Excitation	References
81.48	$(8, 7)^s–(7, 7)^a$	CW; optically pumped by a CW H_2O laser	134

Polyatomic Species: TABLE 84c. *NH₃ laser*

Pure rotational transition

Wavelength λ_{vac} (μm)	Transition	Excitation	References
263.3	$(7, 7)^a–(7, 7)^s$	CW; optically pumped by a CW N_2O laser	134

Polyatomic Species: TABLE 85. *Unidentified laser lines*

Wavelength λ (μm)	Transition	Excitation	References
50 lines between 200–750 μm	—	Pulsed[a]	136
A number of lines between 50–100 μm	—	Pulsed[b]	136

[a] CO_2-laser pumped formic acid, acetic acid, methyl chloride, methyl bromide, and ethyl fluoride.

[b] CO_2-laser pumped methyl alcohol.

References to Tables 47 to 85

1. POLLACK, M. A., *Appl. Phys. Lett.* **9**, 230 (1966).
2. EVENSON, K. M., *Appl. Phys. Lett.* **12**, 253 (1968).
3. HODGSON, R. T., *J. Chem. Phys.* **55**, 5378 (1971).
4. MATHIAS, L. E. S. and PARKER, J. T., *Phys. Rev. Lett.* **7**, 194 (1963).
5. CHEO, P. K. and COOPER, H. G., *Appl. Phys. Lett.* **5**, 42 (1964).
6. BARRY, J. D. and BONEY, W. E., *IEEE J. Quant. Electr.* **QE-7**, 648 (1971).
7. BARRY, J. D. and BONEY, W. E., *IEEE J. Quant. Electr.* **QE-7**, 101 (1971).
8. BARRY, J. D., BONEY, W. E. and BRANDELIK, J. E., *IEEE J. Quant. Electr.* **QE-7**, 464 (1971).
9. WEISBACH, M. F. and CHACKERIAN, C. Jr., *IEEE J. Quant. Electr.* **QE-8**, 679 (1972).
10. PATEL, C. K. N., *Appl. Phys. Lett.* **7**, 246 (1965).
11. BHAUMIK, M. L., LACINA, W. B. and MANN, M. M., *IEEE J. Quant. Electr.* **QE-8**, 150 (1972).
12. EPPERS, W. C., OSGOOD, R. M. Jr. and GREASON, P. R., *IEEE J. Quant. Electr.* **QE-6**, 4 (1970).
13. OSGOOD, R. N. Jr., EPPERS, W. C. Jr. and NICHOLS, E. R., *IEEE J. Quant. Electr.* **QE-6**, 145 (1970).
14. OSGOOD, R. N. Jr., GOLDHAR, J. and McNAIR, R., *IEEE J. Quant. Electr.* **QE-7**, 253 (1971).
15. JEFFERS, W. Q. and WISWALL, C. E., *IEEE J. Quant. Electr.* **QE-7**, 407 (1971).
16. JEFFERS, W. Q. and WISWALL, C. E., *J. Appl. Phys.* **42**, 5059 (1971).
17. OSGOOD, R. M., NICHOLS, E. R., EPPERS, W. C. and PETTY, R. D., *Appl. Phys. Lett.* **15**, 69 (1969).
18. PATEL, C. K. N. and KERL, R. J., *Appl. Phys. Lett.* **5**, 81 (1964).
19. POLLACK, M. A., *Appl. Phys. Lett.* **8**, 237 (1966).
20. HODGSON, R. T., *Phys. Rev. Lett.* **25**, 494 (1970).
21. WAYNANT, R. W., SHIPMAN, J. D., ELTON, R. C. and ALI, A. W., *Appl. Phys. Lett.* **17**, 383 (1970).
22. WAYNANT, R. W., *Proceedings of the Technical Program*, Electro-Optical Systems Design Conference—1971 West, Anaheim, California, May 18–20 (1971), pp. 1–5.
23. POLLACK, M. A., Molecular gas lasers, in *Handbook of Lasers* (ed. R. J. PRESSLEY), pp. 298–349, C. R. C. Press, Cleveland, Ohio, 1971.
24. BAZHULIN, P. A., KNYAZEV, I. N. and PETRASH, G. G., *Sov. Phys. JETP* **22**, 11 (1966).

25. McFarlane, R. A., *Physics of Quantum Electronics*, pp. 655–63, McGraw-Hill, 1966.
26. Bockasten, K., Lundholm, T. and Andrade, D., *J. Opt. Soc. Am.* **56,** 1260 (1966).
27. Bazhulin, P. A., Knyazev, I. N. and Petrash, G. G., *Sov. Phys. JETP* **22,** 11 (1966).
28. Fromm, D., *Molecular Hydrogen and Deuterium Laser*, Report No. 3, Night Vision Laboratory, U.S.A.E.C., Fort Belvoir, Va., April (1967).
29. Deutsch, T. F., *IEEE J. Quant. Electr.* **QE-3,** 419 (1967).
30. Corneil, P. H. and Pimental, G. C., *J. Chem. Phys.* **49,** 1379 (1968).
31. Deutsch, T. F., *Appl. Phys. Lett.* **10,** 234 (1967).
32. Kompa, K. L. and Pimental, G. C., *J. Chem. Phys.* **47,** 857 (1967).
33. Waynant, R. W., *Phys. Rev. Lett.* **28,** 533 (1972).
34. Hodgson, R. T. and Dreyfus, R. W., *Phys. Rev. Lett.* **28,** 536 (1972).
35. Hodgson, R. T. (to be published), ref. 40 in ref. 23.
36. Pixton, R. M. and Fowles, G. R., *IEEE J. Quant. Electr.* **QE-5,** 478 (1969).
37. Bazhulin, P. A., Knyazev, I. N. and Petrash, G. G., *Sov. Phys. JETP Lett.* **20,** 1069 (1965).
38. Akitt, D. P. and Yardley, J. J., *IEEE J. Quant. Electr.* **QE-6,** 113 (1970).
39. Kasper, J. V. V. and Pimental, G. C., *Phys. Rev. Lett.* **14,** 352 (1965).
40. Moore, C. B., *IEEE J. Quant. Electr.* **QE-4,** 52 (1968).
41. Airly, J. R., *IEEE J. Quant. Electr.* **QE-3,** 208 (1967).
42. Tunitskii, L. N. and Cherkasov, E. M., *Sov. Phys. Tech. Phys.* **13,** 1696 (1969).
43. Spencer, D. J., Mirels, H., Jacobs, T. A. and Gross, R. W. E., *Appl. Phys. Lett.* **16,** 235 (1970).
44. Basov, N. G., Kulakov, L. V., Markin, E. P., Nikitin, A. I. and Oraevskii, A. N., *Sov. Phys. JETP Lett.* **9,** 375 (1969).
45. Deutsch, T. F., *Appl. Phys. Lett.* **11,** 18 (1967).
46. Byer, R. L., Herbst, R. L., Kildal, H. and Levenson, M. D., *Appl. Phys. Lett.* **20,** 463 (1972).
47. Kaslin, V. M. and Petrash, G. G., *Sov. Phys. JETP* **27,** 561 (1968).
48. Kaslin, V. M. and Petrash, G. G., *Sov. Phys. JETP Lett.* **3,** 55 (1966).
49. Gallardo, M., Massone, C. A. and Garravaglia, M., *Appl. Optics* **7,** 2418 (1968).
50. Janney, G. M. (private communication, 1969).
51. Kasuya, T. and Lide, D. R. Jr., *Appl. Optics* **6,** 69 (1967).
52. Massone, C. A., Garavaglia, M., Gallardo, M., Calatroni, J. A. E. and Tagliaferri, A. A., *Appl. Optics* **11,** 1317 (1972).
53. Heard, H. G., *Nature* **200,** 667 (1963).
54. Leonard, D. A., *Appl. Phys. Lett.* **7,** 4 (1965).
55. Gerry, E. T., *Appl. Phys. Lett.* **7,** 6 (1965).
56. Shipman, J. D. Jr., *Appl. Phys. Lett.* **10,** 3 (1967).
57. Geller, M., Altman, D. E. and DeTemple, T. A., *Appl. Optics* **7,** 2232 (1968).
58. Ali, A. W., *Appl. Optics* **8,** 993 (1969).
59. Dreyfus, R. W. and Hodgson, R. T., *Appl. Phys. Lett.* **20,** 195 (1972), also *Bull. Am. Phys. Soc.* **17,** 19 (1972).
60. Patterson, E. L., Gerardo, J. B. and Johnson, A. W., *Appl. Phys. Lett.* **21,** 293 (1972).
61. Mathias, L. E. S. and Parker, J. T., *Appl. Phys. Lett.* **3,** 16 (1963).
62. Andrade, O., Gallardo, M. and Bockasten, K., *Appl. Optics* **6,** 2006 (1967).
63. Wood, O. R., Burkardt, E. G., Pollack, M. A. and Bridges, T. J., *Appl. Phys. Lett.* **18,** 261 (1971).
64. Gleason, T. J., Willett, C. S., Curnutt, R. M. and Kruger, J. S., *Appl. Phys. Lett.* **21,** 276 (1972).
65. McFarlane, R. A., *Phys. Rev.* **140,** 1070 (1965).
66. McFarlane, R. A. (unpublished work mentioned in ref. 23).
67. McFarlane, R. A., *IEEE J. Quant. Electr.* **QE-2,** 229 (1966).
68. Deutsch, T. F., *Appl. Phys. Lett.* **9,** 295 (1966).
69. Pollack, M. A., *Appl. Phys. Lett.* **9,** 94 (1966).
70. Ducas, T. W., Geoffrion, L. D., Osgood, R. M. Jr. and Javan, A., *Appl. Phys. Lett.* **21,** 42 (1972).
71. Basov, N. G., Danilychev, V. A., Popov, Yu. M. and Khodkevich, D. D., *Sov. Phys. JETP Lett.* **12,** 329 (1970).
72. Koehler, H. A., Ferderber, L. J., Redhead, D. L. and Ebert, P. J., *Appl. Phys. Lett.* **21,** 198 (1972).
73. Basov, N. G., *IEEE J. Quant. Electr.* **QE-2,** 354 (1966).
74. Molchanov, A. G., Poluetov, I. A. and Popov, Yu. M., *Sov. Phys. Solid State* **9,** 2655 (1968).
75. Basov, N. G., Balashov, B. M., Bogdankevitch, O. V., Danilychev, V. A., Kashnikov, G. N., Lantzov, N. P. and Khokevich, D. D., *J. Luminescence* **12,** 834 (1970).
76. Basov, N. G., Bogdankevich, O. V., Danilychev, V. A., Devyatkov, A. G., Kashnikov, G. N. and Lantsov, N. P., *Sov. Phys. JETP Letters* **7,** 317 (1968).

77. Rao, D. R., Hocker, L. O. and Javan, A., *J. Molec. Spectrosc.* **25**, 410 (1968).
78. Cheo, P. K., CO_2 lasers, in *Lasers*, vol. 3 (eds. A. K. Levine and A. J. DeMaria), pp. 111–267, Dekker, N.Y., 1971.
79. Sobolev, N. N. and Sokovikov, V. V., *Sov. Phys. Uspekhi* **10**, 153 (1967).
80. Tychinskii, V. P., *Sov. Phys. Uspekhi* **10**, 131 (1967).
81. Beaulieu, J., *Appl. Phys. Lett.* **16**, 504 (1970).
82. Laurie, K. A. and Hale, M. M., *IEEE J. Quant. Electr.* **QE-6**, 530 (1970).
83. Laurie, K. A. and Hale, M. M., *IEEE J. Quant. Electr.* **QE-7**, 530 (1971).
84. LaFlamme, A. K., *Rev. Sci. Instr.* **41**, 1578 (1970).
85. Dumanchin, R., Michon, M., Farcy, J. C., Boudinet, G. and Rocca-Serra, J., *IEEE J. Quant. Electr.* **QE-8**, 163 (1972) and *Laser Focus*, Aug. 1971, p. 32.
86. Pearson, P. R. and Lamberton, M. M., *IEEE. J. Quant. Electr.* **QE-8**, 145 (1972).
87. Chang, T. Y. and Wood, O. R., *IEEE J. Quant. Electr.* **QE-8**, 721 (1972).
88. Fenstermacher, C. A., Nutter, M. J., Leland, W. T. and Boyer, K., *Appl. Phys. Lett.* **20**, 56 (1972).
89. Marcus, S., *Appl. Phys. Lett.* **21**, 18 (1972).
90. Basov, N. G., *Laser Focus* **9**, 45 (1972).
91. Howe, J. A. and McFarlane, R. A., *J. Molec. Spectrosc.* **19**, 224 (1966).
92. Hartman, B. and Kleman, B., *Can. J. Phys.* **44**, 1609 (1966).
93. Frapard, C., Laures, P., Roulot, M., Ziegler, X. and Legay-Sommaire, N., *Compt. Rend.* B **262**, 1340 (1966).
94. Wieder, I. and McCurdy, G. B., *Phys. Rev. Lett.* **16**, 565 (1966).
95. Patel, C. K. N., *Appl. Phys. Lett.* **7**, 273 (1965).
96. Jeffers, W. Q. and Coleman, P. D., *Proc. IEEE* **55**, 1222 (1967).
97. Mathias, L. E. S. and Crocker, A., *Phys. Lett.* **13**, 35 (1964).
98. Bennedict, W. S., Pollack, M. A. and Tomlinson, W. J., *IEEE J. Quant. Electr.* **QE-5**, 108 (1969).
99. Faust, W. L. and McFarlane (private communication), in M. A. Pollack (ref. 23).
100. According to M. A. Pollack[23], a line at this wavelength has not been verified.
101. Muller, W. M. and Flesher, G. T., *Appl. Phys. Lett.* **8**, 217 (1966).
102. Kasuya, T., Minoh, A. and Shimoda, K., *J. Phys. Soc. Jap.* **25**, 1201 (1968).
103. Mathias, L. E. S., Crocker, A. and Wills, M. S., *IEEE J. Quant. Electr.* **QE-4**, 205 (1968).
104. Maki, A. G., *Appl. Phys. Lett.* **12**, 122 (1968).
105. McFarlane, R. A. and Fretz, L. H., *Appl. Phys. Lett.* **14**, 385 (1969).
106. Turner, R. and Poehler, T. O., *Phys. Lett.* **27A**, 479 (1968).
107. Hartmann, B., Kleman, B. and Spaongstedt, G., *IEEE J. Quant. Electr.* **QE-4**, 296 (1968).
108. Coleman, P. D., *Laser Focus* **9**, 37 (1969).
109. Johnson, C. J., *IEEE J. Quant. Electr.* **QE-4**, 701 (1968).
110. Hassler, J. C. and Coleman, P. D., *Appl. Phys. Lett.* **14**, 135 (1969).
111. Steffen, H. and Kneubuhl, F. K., *IEEE J. Quant. Electr.* **QE-4**, 992 (1968).
112. Mathias, L. E. S., Crocker, A. and Wills, M. S., *Electr. Lett.* **1**, 45 (1965).
113. Lide, D. R. Jr. and Maki, A. G., *Appl. Phys. Lett.* **11**, 62 (1967).
114. Stafsudd, O. M., Haak, F. A. and Radisavljevie, K., *IEEE J. Quant. Electr.* **QE-3**, 618 (1967).
115. Hocker, L. O. and Javan, A., *Phys. Lett.* **25A**, 489 (1967).
116. Steffen, H., Steffen, J., Moser, J. F. and Kneubuhl, F. K., *Phys. Lett.* **21**, 425 (1966).
117. Kneubuhl, F. K., *IEEE. J. Quant. Electr.* **QE-4**, 10 (1968).
118. Moeller, G. and Rigden, J. D., *Appl. Phys. Lett.* **8**, 69 (1966).
119. Howe, J. A., *Phys. Lett.* **17**, 252 (1965).
120. Djeu, N. and Wolga, G. J., *IEEE J. Quant. Electr.* **QE-5**, 50 (1969).
121. Patel, C. K. N., *Appl. Phys. Lett.* **6**, 12 (1965).
122. Mathias, L. E. S., Crocker, A. and Wills, M. S., *Phys. Lett.* **13**, 303 (1964).
123. Deutsch, T. F., *Appl. Phys. Lett.* **8**, 334 (1966).
124. Hubner, G., Hassler, J. C., Coleman, P. D. and Steenbeckeliers, G., *Appl. Phys. Lett.* **18**, 511 (1971).
125. Hard, T. M., *Appl. Phys. Lett.* **14**, 130 (1969).
126. Dyubko, S. F., Svitch, V. A. and Valitov, R. A., *Sov. Phys. JETP Lett.* **7**, 320 (1968).
127. Karlov, N. N., Konev, Y. B., Petrov, Y. N., Prokhorov, A. M. and Stel'makh, G. M., Post-deadline paper 76-7, Q.E.C., Miami (1968), mentioned in *IEEE J. Quant. Electr.* **QE-4**, 23 (1968).
128. Trusty, G. L. and Yin, P. K., *IEEE J. Quant. Electr.* **QE-3**, 368 (1967).
129. Chang, T. Y. and McGee, J. D., *Appl. Phys. Lett.* **19**, 103 (1971).
130. Chang, T. Y. and Bridges, T. J., *Optics Commun.* **1**, 423 (1970).
131. Chang, T. Y., Bridges, T. J. and Burkhardt, E. G., *Appl. Phys. Lett.* **17**, 249 (1970).

132. OKAJIMA, S. and MURAI, A., *IEEE J. Quant. Electr.* **QE-8,** 677 (1972).
133. AKITT, D. P. and WITTIG, C. F., *J. Appl. Phys.* **40,** 902 (1969).
134. CHANG, T. Y., BRIDGES, T. J. and BURKHARDT, E. G., *Appl. Phys. Lett.* **17,** 357 (1970).
135. MATHIAS, L. E. S., CROCKER, A. and WILLS, M. S., *Phys. Lett.* **14,** 33 (1965).
136. WAGNER, R. J. and ZELANO, A. J., New submillimeter laser lines in organic molecules, Post-deadline paper, Q.E.C., Montreal, Canada (1972).
137. CLERC, M. and SCHMIDT, M., *Compt. Rend.* **B272,** 668 (1971).
138. BURKHARDT, E. G., BRIDGES, T. J. and SMITH, P. W., *Optics Commun.* **6,** 193 (1972).
139. BURKHARDT, E. G., BRIDGES, T. J. and SMITH, P. W., *Appl. Phys. Lett.* **20,** 403 (1972).
140. JENSEN, R. E. and TOBIN, M. S., *Appl. Phys. Lett.* **20,** 508 (1972).
141. BROWN, F., SILVER, E., CHASE, C. E., BUTTON, K. J. and LAX, B., *IEEE J. Quant. Electr.* **QE-8,** 499 (1972).
142. WU, H. L. and BENESCH, W. M., *Phys. Rev.* **172,** 31 (1968), and BENESCH, W. M. and SAUM, K. A., *J. Phys. B,* **4,** 732 (1971).
143. GERALDO, J. B. and JOHNSON, A. W., *IEEE J. Quant. Electr.* **QE-9,** 748 (1973) and references therein.

Index

Other Titles in the Series in Natural Philosophy

528